PTOLEMY'S TETRABIBLOS
IN THE TRANSLATION OF WILLIAM OF MOERBEKE

CLAUDII PTOLEMAEI LIBER IUDICIALIUM

ANCIENT AND MEDIEVAL PHILOSOPHY

DE WULF-MANSION CENTRE
Series I

XIX

The "De Wulf-Mansion Centre" is a research centre for Ancient, Medieval, and Renaissance philosophy at the Institute of Philosophy of the KU Leuven,
Kardinaal Mercierplein, 2, B-3000 Leuven (Belgium).
It hosts the international project "Aristoteles latinus" and publishes the "Opera omnia" of Henry of Ghent and the "Opera Philosophica et Theologica" of Francis of Marchia.

PTOLEMY'S TETRABIBLOS
IN THE TRANSLATION OF WILLIAM OF MOERBEKE

CLAUDII PTOLEMAEI LIBER IUDICIALIUM

Edited by
Gudrun Vuillemin-Diem and Carlos Steel
with the assistance of
Pieter De Leemans

LEUVEN UNIVERSITY PRESS

Leuven University Press / Presses Universitaires de Louvain /
Universitaire Pers Leuven
Minderbroedersstraat 4, B-3000 Leuven (Belgium)

ISBN 978 90 5867 962 8
D/2015/1869/9
NUR: 732

PREFACE

... *et facit cum pace et tranquillitate singula euenientium recipere* (1.203-4)

This is the first edition ever of William of Moerbeke's Latin translation of Ptolemy's celebrated astrological handbook, known under the title *Tetrabiblos* or *Quadripartitum* (*opus*). Ptolemy's treatise (composed after 141 A.D.) offers a systematic overview of astrological science and had, together with his *Almagest*, an enormous influence into the 17th century. In the Middle Ages, the work was mostly known through translations from the Arabic. William of Moerbeke's translation was made directly from the Greek: it is a major scholarly achievement that displays not only his genius as a translator, but also as a scientist. This edition is accompanied by extensive Greek-Latin indices, which give evidence of Moerbeke's astonishing enrichment of the Latin vocabulary, which he needed both to translate the work's technical scientific vocabulary and to cope with the many new terms Ptolemy created.

The edition has been announced for many years as 'in preparation' by Fernand Bossier and Luc Anthonis. In fact, it was Bossier and his student Anthonis who discovered this translation, following a comment by A. Björnbo (1909), and attributed it to Moerbeke. They planned a critical edition of the translation. Unfortunately, the premature death of our colleague and friend, Fernand, halted the preparation of this editorial project. In the intervening years, Luc had prepared a provisional edition of the four books; but as he was active in other fields than philology, he too had to abandon the project after Fernand's death. Inspired by our common friendship of Fernand and our admiration for Ptolemy's translator, we decided to continue this editorial project, starting from the provisional text prepared by Anthonis and the handwritten annotations left by Bossier, which Anthonis generously communicated. We greatly profited from their numerous conjectures particularly in the part where only the Oxford manuscript is available.

We at first thought that to accomplish a project so prosperously started would be easy-going, a good pastime for retired academics. In fact it proved immensely difficult, time-devouring, strained. A new collation of the manuscripts was needed, while the comparison of Moerbeke's translation with the Greek text in the Greek-Latin apparatus, and the establishment of the indexes, demanded a considerable effort. We were not experts in astrology and had never read the Ptolemaic text before. Still, reading Ptolemy was undoubtedly worth all trouble: we had never suspected that the *Tetrabiblos* was such a fascinating text, and by the end of the project had nearly become adepts of astrology. Nevertheless, we suffered from Ptolemy's dreadful 'Teutonic style' (as the Loeb editor Robbins called it), and the problematic state of the transmission of the Latin text (with only one mediocre copy for the major part).

We profited from comments and suggestions made by Pieter Beullens, Charles Burnett, David Juste, Dag Hasse, Wolfgang Hübner, Stephan Heilen, Raúl Caballero-Sánchez, Fabio Acerbi, Guy Guldentops, Marwan Rashed and Bojidar Dimitrov. Adriano Oliva and Iacopo Costa helped us obtain photos of needed manuscripts. We are greatly indebted to Pieter De Leemans, who devoted much time to a thorough revision of the edition, indexes and introduction. In expressing our deepest gratitude to all our learned friends, we must include the two research institutions to which we belong: the Thomas Institute in Cologne – and in particular, its director Andreas Speer and librarian Wolfram Klatt – and the De Wulf-Mansion Centre in Louvain. We thank Leuven University Press for their support and patience in the delivery of the final manuscript, and appreciate the wonderful work done at Typeface Leuven by Michiel and Anne Van den Bosch in formatting this complex edition. Finally, we are grateful to David L. Dusenbury for his close reading of our introduction.

The intensive preparation of this edition was in many ways a complex personal experience for both editors. Collaboration at a distance, even with the benefit of modern digital connections, is complicated and troublesome. On certain points the editors had their own, strongly held, divergent views (for instance, regarding the place of the Latin in the complicated Greek tradition), and it took some time to find a common outlook on difficult passages. It was, however, a great pleasure to work together, learn from one another, and modify first versions, all the while sharing our admiration for Moerbeke as translator and Ptolemy as philosopher-astronomer.

Plato is right to praise the intellectual life, which offers great pleasures even in difficult circumstances and makes us forget troubles caused by the body. And both Plato and Ptolemy teach us well to be prepared by foreknowledge, 'to greet with calm and steadiness whatever comes'.

Gudrun Vuillemin-Diem — Les Granges Berrard

Carlos Steel — Leuven

CONTENTS

INTRODUCTION

The astrological treatise of Claudius Ptolemaeus, composed sometime after 141, is generally known as *Tetrabiblos* or *Quadripartitum* (*opus*), so called because it consists of four books. Its original title was Ἀποτελεσματικά (βιβλία), announcing Ptolemy's studies on the influences and effects of the celestial bodies on the sublunary world, and in particular on human life, in so far as they can be 'scientifically' predicted. This treatise is the practical counterpart of Ptolemy's major work, the *Syntaxis mathematica* ('Almagest'), which offers a theoretical study and mathematical explanation of the apparent movements and configurations of planets and stars. The *Apotelesmatica* had already been recognized as a canonical work on astrology in late antiquity: it was studied, commented upon and excerpted. It exerted a great influence in the Byzantine period, notwithstanding Christian invectives against astrology, and had a considerable impact in the Arabic world, where a first translation was made in the ninth century.

In the Latin Middle Ages, the text was almost exclusively known through Latin translations of the Arabic translation. Plato of Tivoli (Plato Tiburtinus) made the first Latin translation in 1138, in Toledo – an excellent work which enjoyed much success. Egidio de' Tebaldi of Parma (Aegidius de Thebaldis) made a new translation around 1271-75, under the patronage of King Alfonso X of Castile. Egidio's translation is less reliable, but it includes the important commentary of Ibn Ridwān (known in Latin as Haly). Tiburtinus' version became the standard translation used by scholars for several centuries, often complemented with Aegidius' translation of Haly's commentary. Besides these two successful translations there exist at least three other Latin translations made from the Arabic, but they never had a wide circulation. One is dated to 1206, a second was made in Toledo in 1234, and the third appeared sometime before 1250. The Tiburtinus translation remained the standard text throughout the Renaissance. It was printed in Venice in 1484, 1493 and 1519; in Paris in 1519;[1] in Basel in 1533, 1541 and 1551. The 1493 and 1519 editions of Tiburtinus also include the translation of Aegidius.

The first editor of the Greek text of the *Tetrabiblos*, Joachim Camerarius, made a translation directly from the Greek of the first two books; for the latter books

[1] *Quadripartitum iudiciorum, opus Claudii Ptolemei Pheludiensis, ab Joanne Sieurreo brittuliano Bellouacensi perbelle recognitum*, in edibus Ioannis de Porta: Parisiis [1519]. An extremely rare print. We thank Charles Burnett for having examined the exemplar in the British Library in London.

he still used the Arabic-based translation of Tiburtinus (Nürnberg 1535). In 1548, in Louvain, Antonius Gogova published another translation of the *Tetrabiblos*, using Camerarius for the first two books, but complementing them with his own translation of the last two books. And finally, Philipp Melanchthon published a complete translation from the Greek in Basel, in 1553. Until recently, it was generally assumed that one had to wait until the Renaissance to have the first Latin translation from the Greek *Tetrabiblos*. However, already in 1909, A. Björnbo had noticed the existence of a medieval translation directly from the Greek in the Oxford manuscript, *Bodleian Library, Digby 179*, and this information was taken over by Ch. Haskins and Thorndike–Kibre. Inspired by Björnbo's remark, in 1983 Fernand Bossier started examining the Oxford manuscript, alongside his student Luc Anthonis, and came to the conclusion that it was the work of William of Moerbeke.

CHAPTER I
MANUSCRIPT TRADITION OF THE TRANSLATION

Moerbeke's translation is transmitted in two manuscripts, *Ox* and *Ve*, but only one (*Ox*) has the complete text.

1. OXFORD, BODLEIAN LIBRARY, DIGBY 179, ff. 171r-208v *(Ox)*[1]

Manuscript in parchment, containing 208 folia. Text in two columns, each of 35/36 lines, written by one copyist, an English hand of the first half of the 14th century. Before it entered the Digby collection, the manuscript belonged to the famous mathematician and astrologer Thomas Allen (1542-1632). The manuscript contains two different Latin translations of Ptolemy's *Apotelesmatica*.

(a) ff. 1r-170v: <Quadripartitum Ptolomei cum commento Haly>
This is the Arabo-Latin translation of the *Apotelesmatica* by Aegidius de Thebaldis with the commentary of Ibn Ridwān (known in Latin as Haly). The introduction of the translator starts at f. 1r: 'Dixit Egidius de Tebaldis Lombardus de ciuitate Parmensi: scire et intelligere gloriosum est...' The first lemma of Ptolemy's text followed by the commentary of Haly starts on f. 3r: 'Res misori quibus perficiuntur pronosticationes accepte de astronomia maiores et nobiliores sunt due'. In the inferior margin another hand added an alternative translation: 'incipit liber quadripartite Ptholomei Pheludiensis de judiciis astrorum secundum aliam translacionem. Prohemium.' A third hand again noted above 'prohemium': 'et apparet Bredōn'. This alternative translation of the Ptolemy lemmas is found on all following pages until the last lemma in book IV.

According to G. Watson (*Bodleian Library Record* 9 [1976], 215 n. 2) the translation in the margins is in the hand of Simon Bredon. Simon Bredon (c. 1300-1372), a fellow of Merton College in Oxford, had a reputation as physician, mathematician and astronomer. He composed a commentary on Ptolemy's Almagest, which is preserved in two other manuscripts from the Digby collection (168, 178).[2] In his testament made 12 October 1368, Simon bequeathed his manuscripts to Merton College. There is a reference in his will to 'quadripartitum tholomei de iudiciis astronomie' (p. 84). This is most probably the actual Digby 179.

---

[1] See the catalogue of the Digby collection by W.D. Macray, Oxford, 1883, with the additional notes by R.W. Hunt (†) and A.G. Watson (1999).

[2] On Simon Bredon see Talbot (1962-1963), Snedegar (1999), Molland (2006).

Since Björnbo[3] and Haskins,[4] it is generally accepted that Bredon is the author of the alternative translation copied by him in the margins of Digby 179. The incipit is even listed in Thorndike–Kibre 1138. However, this attribution is only based on the first marginal note on f. 3r:

> Pronosticatio per astronomiam fienda dependet a duobus, ab astrologia quadam uniuersali (...) dicens causam primam mendacie quia eam intelligere nesciunt quidam, secundam eo quod propter difficultatem eam addiscere dedignantes dicunt quod est res quam attingere nequit homo. Ego tum ostendam quod de hac scientiam potest homo attingere et quantum proficuum sit in ea.

This is a kind of summary of the first three chapters of the *Quadripartitum*, but certainly not a translation. It looks like the 'prohemium' of a commentator. Only from f. 3v onward does Bredon start copying the alternative translation of Ptolemy:

> Capitulum I. Rerum Sire in quibus est pronosticabilis scientie stellarum perfectio magnas et precipuas duas esse deprehendimus, quarum altera que precedit et est fortior.

It is immediately clear that this 'alia translatio' is nothing but a copy of the translation of Plato of Tivoli. The alternative translation ends on the inferior margin of f. 166r:

> explicit liber alarba Ptolomei in distinctione iudicandi per stellas de futuris et in (*sic*) constructionis et destructionis contingentibus.

It is evident that Simon Bredon did not himself make this alternative translation, or even a revision of an existing translation,[5] but merely copied what he found in another manuscript, namely the translation of Plato of Tivoli. He had thus in this manuscript three different translations of the *Quadripartitum*. Only the short prohemium on f. 3r may have come from him.

(b) ff. 171r-208v: <Liber iudicialium Claudii Ptolomei>

The copy starts with the list of chapters of book I: 'In primo libro Iudicialium Claudii Tholomei ad Sirum insunt ista capitula.' After the list, the proemium begins with: 'Hiis qui instituunt per astronomiam pronosticum finem, o Sire' . Explicit: 'bene utique habebit huic tractatui conuenientem imponere finem. Explicit liber quartus Tholomei' (see image, p. 151). After having completed Ptolemy's trea-

[3] Björnbo (1909) 391.

[4] Haskins (1924) 131-2.

[5] Pace Snedegar (1999) 295: 'While there is no reason to believe that Bredon had the linguistic competence to attempt a new translation, he drafted his own Latin recension, mixing and matching readings from the two translations and making some independent amendments.'

tise, the copyist adds an astrological calculation to determine the month in which one may die. It is evident that he found this text in his model. (See image, p. 152; on this addition, see Appendix I.) Titles of the chapters have been added in red after the completion of the copy. Initials are rubricated.

On ff. 193r-194r (book III, ch. 12 [11]: *De annis uite*), Simon Bredon copied in the margin a commentary on the calculation of the *yleg*: 'cum cognoueris yleg... et Veneri dirigere yleg'. On f. 194v follows another note which ends: 'et similiter albumasar differentia quarta capitulo secundo.' Bredon also added some terms *supra lineam* in the Moerbeke translation, for instance, to explain Moerbeke's translation of ἀφέτας *dimissores* in III.447, 458: + .i. yleg. (on *yleg*, see p. 31)

Bredon also used the inferior margins of ff. 201v-202r to add the Tiburtinus translation of the pseudo-Ptolemaic 'scientia in proiectione radiorum'. Incipit: 'cum proiectorum radiorum planetarum seu stellarum scire uolueris'. Explicit: 'erit locus radiationis equate. Explicit'. On the next page (202v), Bredon adds another note: 'nota quod radiationum alia est... gradus ecleptice habere sunt plures. Explicit.'

There are two longer lacunae in the text. On f. 171vb the scribe only used the first five lines and left the rest of the page unwritten until the end (ca. 32 lines). This omission reaches from I.84 (102°) *fructificationes* to I.121 (151°) *theoria.*[6] The second lacuna is found on f. 172va. Here, only the 24 first lines are written, while the following 13 lines are left unwritten. This omission goes from I.187 (238°) *et uniuersaliter* to I.222 (283°) *contrapassionem inferentium.* The omitted text is, in both cases, of similar size. A possible explanation for two similar lacunae at this distance may be the loss of a bifolium in a quire of its model – for example, the third folium in a quaternion (as Luc Anthonis hypothesized). The copyist must have noticed that something was wrong, as he left the two places open in the text. These omissions are not found in the Venetian manuscript. On f. 195vb, lin. 22 the copyist jumps from III.673 *mediocres* (898° μετρίους) forwards to III.777 *adustione* (1037° κατακαύματος); on f. 197ra, lin. 7, from III.879 *facile acquisitibiles* (1172° εὐπορίστους) back to III.673 *magnitudinibus* (899° μεγέθεσι); on f. 198ra, lin. 30 from III.777 *ferro aut* (1037° σιδήρου ἢ) to III.879 *penitiuas* (1171° μεταμελητικάς); on f. 202va, lin. 21 the text jumps from IV.105 *acceptione* (129° λήψεως) forwards to IV.206 *parum coeuntes* (259° νωχελεῖς), on f. 203vb, lin. 2 from IV.314 *habentibus* (407° ἐχουσῶν) back to IV.105 *et si quidem* (130° κἂν μὲν), and on f. 204vb, lin. 24 from IV.205-6 *Saturno quidem* (259° μὲν Κρόνου) forwards to IV.314 *ad signa* (407° πρὸς ... τὰ ζῴδια). These transpositions are corrected in the margin. All of these

[6] Throughout this introduction, line numbers followed by "°" refer to Hübner's edition of the Greek text, plain line numbers to our edition of the Latin translation.

examples show that the model used for the copy of Moerbeke's translation was in great disorder. It was certainly not the model used for the Venetian manuscript.

2. VENEZIA, BIBL. MARCIANA, LAT. XIV, 242 (= 4295), ff. 1r-12v (*Ve*)[7]

This convolute contains the following texts, written on different paper by different hands:[8]

- I <*Liber iudicialium Claudii Ptolomei*>.
- II Sentence of a tribunal in Venice (dated 10 june 1396) in favour of Alberto di Bartolomeo from Guidecca.
- III A collection of 143 sonnets by Petrarca, dated circa 1400.
- IV A fragment of a register of payments dated 1466-1467.

The translation of Ptolemy only contains book I and II, ch. 1-9, 619 (813°). The text ends abruptly with the words 'studio mathematici circa iudicia particularia' at the end of f. 12v, which is also the end of the first quire (a senio). The copyist added in the inferior margin the catchword 'particularia. Observare' to make the connection with the following quire (II.620), which is now lacking. What we have is only the first quire of a manuscript which may have contained the full text of the *Iudicialia*. We do not know how and why this mutilated Latin text came together with Petrarca's sonnets and the sentence of a tribunal in favour of Alberto di Bartolomeo.

The Ptolemy text is written in two columns of 47 lines by an Italian hand of the late 15th century.[9] The copy has been made with great care. Alternative readings are indicated in the margin (*al'*). Titles of the chapters have been added in red. Open spaces are left for the rubricated initials of each chapter, with indication of the first letter. The copyist was very careful in composing the table of termini in I, ch. 21, and the table *Expositio regionum* in II, ch. 4. On f. 9r he added in the inferior margin a geographic design with the four parts of the inhabited world. In contrast to *Ox*, *Ve* has a clear articulation of the text in paragraphs. At the end

[7] See P. Zorzanello, *Catalogo dei codici latini della Biblioteca nazionale Marciana di Venezia, non compresi nel catalogo di G. Valentinelli,* vol. III, Classe XIV, Trezzano s/N 1985, p. 401-402.

[8] We owe the reference to P.O. Kristeller, *Iter Italicum,* II, p. 269: 'Ptolemy, Iudicialia ad Syrum (i.e. Almagestum)'. Kristeller probably made the wrong identification ('i.e. Almagestum') because he read the notice in the old written catalogue 'Claudii Ptolemaei Almagesti liber primus et secundi pars'. Zorzanello corrected the old notice 'Est autem Ptolemaei Tetrabiblos, s. quadripartitum (non Almagestum ut in cat.)'. He adds 'versione di chi?' (whose translation?).

[9] We thank Robert Wielockx for his meticulous analysis of the handwriting. See the images of f. 1r and 9r on p. 149-150.

of each section there is some space left with a double dash // (probably an indication for a later rubricated paragraph sign). In *Ox*, one only finds occasionally a majuscule letter to indicate a new section. The articulation of the texts into paragraphs in *Ve* makes sense and often corresponds to the paragraphs that Hübner adopted from the edition of Boll. They may have come from the autograph of the translator himself. One finds, indeed, a similar system of division in paragraphs in Moerbeke's autograph of his translations of Archimedes. They are also found in the best manuscripts of the translations of the *Metaphysics* and the *Meteorology.*

3. THE RELATION OF THE TWO MANUSCRIPTS

Both manuscripts are characterized by common errors and thus derive from a common model. This model must therefore be different from the translator's autograph. One finds numerous confusions of letters, such as i/e, a/o, a/e, s/f, c/f, and errors in the interpretation of abbreviations at the endings of words. These errors were probably made in the first copy of the autograph. In all these cases the editors had to make corrections, mostly inspired by a comparison of the translation with the Greek text. Examples:

Book I: qui/que (30, 156, 372, 526; also II.202, 299) — animis] animalibus (157) — ipsa[1,2]] ipso (169) — sit proficua] fit proficua sit (173) — prima] primam (181) — autem] aut (182, 247; or the inverse 893) — calefaciendo] tale faciendo (303) — congregantur] concreantur (337) — bone] bene (372) — disponunt] -untur (430) — Sagittario] sagittariis (438) — Bootem] boetem (464) — Cratere] cartere (495) — campestria] -postria (601) — spirituosus] spumosus (706) — primo] primos (778) — permutata] permutate (792) — diuisionem] diuisiones (844) — sic] sit (859) — qui] quoniam (880) — luna] bona (898) — similitudine] similitudini (940) — undecumque] unumcumque (954)

Book II: consistentias] constantias (14) — preinductioni] preinductis (23) — affectuum] effectuum (47) — capillis] capilli (69) — plantarum] planetarum (72) — oppositam] oppositis (145) — fenicia] fenix (227) — multarum] -orum (290) — notolibice] notalibice (295) — Cyrenaica] Cureniaca (318) — magici] magicis/magis (336) — Kiklades] -kaldes (358) — adicere] addicere/dicere (380) — ascendens[2]] accendens/accendendum (386) — mansorie] mansione (398) — significationum] significatorum (400) — dispositarum] dispositorum (404) — accedentibus] accidentibus (439) — meridiem] medium (481) — natatilibus] natalibus (484) — effectiuus] effectus (516) — extiterint] existerit (530) — contabefactiuos] contabesatiuos/-saturus (573) — efimerinarum] efimerinorum (599) — significationum] significatorum (599).

A further indication that the common model of both manuscripts was not the translator's autograph, but a copy of it, is the fact that both often have a superfluous *et* in the text. There are eleven such instances in the first common part: I.31, 65, 91, 113, 527, 551, 655; II.141, 243, 471, 576. The addition of *et* may be explained by the fact that the copyist of the apographon sometimes added a sign to fill a line. This sign was erroneously interpreted as *et* by the copyist of the common model of the Latin tradition.

A final argument for a common archetype are shared omissions in both manuscripts. In the first part we have only a few minor omissions: I.136 *a*; 400 *que*; 523 *quatuor*; 568 *octo*; II.35 *hoc quidem*; 102 *regionis*; 244 *et*; 333 *qui*; 476 *circa*^{2}; 517 *natura*. In all these places we have added the missing terms because we thought it unlikely that they were omissions made by the translator or omissions in his Greek model. The largest omission is found in II.504-5: *apud medium <celi super terram ... apud> occasum*. It is difficult to explain this omission as originating from the Greek model. *Medium celi* is Moerbeke's standard translation of μεσουράνημα (30 instances). In this passage we only have *medium* without *celi*. The Latin text reads *apud medium occasum*. If the error had occurred in the Greek model, the omission should have started in the middle of the term μεσου/ρανήματι. There are many more, and more important omissions in the second part where we only have *Ox*. See for instance II.745-6, 777; III.101, 117, 413; IV.15, 219, 230-1, 238. Here it is impossible to determine whether they were in the lost model of both manuscripts or are due to greater negligence of the copyist of *Ox* in the work's second part.

In the section where both manuscripts can be compared, *Ve* is a more careful and reliable copy of its model than *Ox*. Given the inferior quality of the copy in *Ox*, and the different anomalies in its text (lacunae, transpositions), which can only be explained by the confused state of its model, it seems as though *Ox* derives from the common model via another copy, whereas *Ve* would be a direct copy of it. The fact that *Ox* is a much older copy than *Ve*, and thus closer in time to the archetype, does not undermine this hypothesis. There are many other cases where a recent manuscript is a direct copy of an older model.

CHAPTER II
WILLIAM OF MOERBEKE TRANSLATOR OF THE *IUDICIALA*

1. ATTRIBUTION OF THE TRANSLATION TO WILLIAM OF MOERBEKE

No translator of the *Iudicialia* is indicated in either of its two manuscripts. There are, however, strong arguments to attribute the translation to William of Moerbeke.

1.1. External arguments

The first consideration which led Bossier and Anthonis to propose Moerbeke as the possible translator is the fact that the translation was used by Henry Bate (for Bate's quotations, see Chapter III). We know that both scholars were in contact with one another from the year 1274, and that Bate had privileged access to certain rare translations of Moerbeke's. Thus, for instance, he is the only person to know Moerbeke's translation of a selection from Proclus' commentary on the *Timaeus*. Bate also had a copy of Moerbeke's translation of Philoponus, *De intellectu* (which even Thomas Aquinas did not know), and he possessed a manuscript of Simplicius' commentary on the *De caelo* in Moerbeke's translation. Although Bate does not identify the translator of the *Iudicialia*, the fact that he quotes such a rare text is an indication that this translation, too, may have come from his learned friend Moerbeke.

There is, however, an even stronger argument pointing to Moerbeke (one not noticed by Bossier and Anthonis, as it happens): the fact that Thomas Aquinas quotes the new translation in his commentary on the *De caelo* II, lectio XVIII, 11:

> Et inde est etiam quod, secundum Ptolomaeum in quadripartito, quod *ea quae* sunt *Saturni attribuuntur ad universalia loca temporum; ea autem quae* sunt *Iovis, ad loca annualium temporum; ea vero quae* sunt *solis et Martis et Veneris et Mercurii, ad loca mensium; transitus vero lunae ad loca diurna*. Coniunctiones etiam superiorum planetarum coaptantur effectibus magis universalibus et permanentibus, secundum astrologos.

Although Thomas still uses the traditional title *Quadripartitum* of the Arabo-Latin translation, this quotation undoubtedly comes from the Greco-Latin translation *Iudicialia*. See in our edition IV.643-9 (849°-59°):

> Attendendum autem et ad superingressus qui fiunt ad loca temporum quasi non contingentes et ipsis coassumptis ad temporum effectus, *hiis quidem igitur qui Saturni ad loca uniuersalia temporum, hiis autem qui Iouis ad loca annualium, hiis uero qui*

> *solis et Martis et Veneris et Mercurii ad loca mensium, hiis autem qui lune transitibus ad loca diurna.* Tamquam uniuersalibus quidem temporum dominatoribus existentibus principalioribus ad perficiendum effectum, particularibus autem cooperantibus aut impedientibus.

The quotation in Thomas is so literal that there can be no doubt that it comes from this translation. One may notice certain small adaptations, such as the addition of the copula *sunt* and the use of the neutrum *ea que* instead of the masculine *hiis qui* [= *eis qui*]. We know that Thomas was working on the commentary of the *De caelo* at the end of his life – in 1272-73, when he was at Naples – and we know that he never completed it. In that commentary, Thomas made extensive use of Moerbeke's translation of Simplicius' commentary on *De caelo*, which he received directly from Moerbeke. If Thomas also used this extremely rare translation of the *Iudicialia*, then he must have likewise known it through the translator himself, William of Moerbeke.[1] Interestingly, Thomas had already quoted the same passage from Ptolemy some time earlier, when writing his commentary on *Metaphysics* XII, but at that time he did not yet know Moerbeke's translation, and used the translation of Tiburtinus. See book XII, lectio IX, 2560-2561:

> Et inde est quod, sicut dicit Ptolemaeus in quadripartito, effectus Saturni coaptantur ad universalia loca et tempora, Iovis ad annos, Martis, solis, Veneris et Mercurii ad menses, lunae autem ad dies. Et inde est quod effectus planetarum apparent in istis inferioribus secundum ordinem eorumdem. Nam primi tres superiores videntur ordinari ad ea quae pertinent ad existentiam rei secundum seipsam: nam ipsa stabilitas esse rei attribuitur Saturno, perfectio autem rei et bona habitudo correspondet Iovi, virtus autem rei, secundum quod se contra nociva tuetur et ea propellit, correspondet Marti.

The relevant passage in Tiburtinus' translation reads:

> Existentiam itaque Saturni in locis temporum uniuersalium, Iouis quidem in locis annorum, solis autem, Martis, Veneris Mercuriique in locis mensium, lune uero in dierum locis obseruabimus.

In this section Thomas does not give a literal quotation, but a paraphrase, and he even expands the testimony with comments that are not in Ptolemy's text (the three effects coming from the three superior planets: *stabilitas esse rei, perfectio rei et bona habitudo, virtus rei*). With this free adaptation of the text it is difficult to demonstrate that Thomas used the Tiburtinus translation, but it nevertheless

---

[1] This is a new important argument in support of the claim that Thomas had a 'privileged relation' with the translator and received from him copies of new translations. See Steel (1986) against R.-A. Gauthier, wo considers this relation as a 'myth'. Standard scholarship adopts Gauthier's view. See for instance Torrell (2005) 177: 'we have no historically trustworthy proof of any such collaboration between William and Thomas'.

seems plausible, since this was the standard translation used by authors in the 13th century.

A literal quotation from another passage is found in the *Summa Contra Gentiles*, III, c. 86:

> Ptolemaeus etiam in Quadripartito dicit: Rursus nec aestimare debemus quod superiora procedant ineuitabiliter ut ea quae diuina dispositione contingunt et quae nullatenus sunt uitanda, necnon quae ueraciter et ex necessitate proueniunt (*cf.* I.259°-63°).

Here, there is no doubt that Thomas quotes from the translation of Plato of Tivoli:

> Rursus, non est extimandum quod superiora super sua significata procedant ineuitabiliter uelut ea quae a diuina dispositione contingunt et quae nullatenus sunt euitanda necnon quae ueraciter et ex necessitate proueniunt.[2]

Given these quotations, one must conclude that when Thomas was working on this part of the *Summa* (around 1265 in Italy), and again when he was writing his commentary on *Metaphysics* XII (in Paris/Naples around 1272-73),[3] he had not yet seen Moerbeke's new translation. Thomas probably encountered it when he returned to Naples in 1272 and started working on *De caelo*. His interest in astrology may also have been stirred by the fact that he was asked to give expert advice on the question of whether it is permissible to draw on astrology. In his short response entitled *De iudiciis astrorum*, maybe dating from his second Parisian regency, Thomas first declares that there is no problem in using 'the judgment of stars' for whatever is related to natural conditions, which depend on the influences of the celestial bodies:

> non est inconueniens, secundum aliquas alias occultiores obseruationes stellarum, circa corporales effectus uti astrorum iudicio.

This argument may have been influenced by Ptolemy's preface to the *Tetrabiblos*, where we find a similar comparison of farmers and seamen forecasting future events. Nevertheless, Thomas insists – as do all Christian authors – that the stars can never determine our voluntary actions.

There may be a reference to Moerbeke's translation of the *Iudicialia* in the inventory of the manuscripts of Gonzalve Gudiel, Archbishop of Toledo, which was made in Viterbo on the 12th of June, 1280. When he stayed at Viterbo, Gudiel

---

[2] All quotations from Tiburtinus' translation come from the 1533 edition (Basel, Johannes Hervagius), which we occasionally collated with the first edition (Venice, 1484, Ratdolt).

[3] On the date of the composition of the commentary on *Metaphysics* XII, see R.-A. Gauthier, *Thomas Aquinas, Quaestiones de Quolibet* (Opera Omnia XXV), 1996, p. 498: 'Parisiis 1271-72 et Neapoli 1272-1273'. It is possible that Thomas finished his commentary on book XII when he was already in Naples.

collected an impressive number of translations by Moerbeke (often directly copied from the translator's autographs) and Michael Scot, as well as an array of other scientific works. Some of these manuscripts are still preserved in the cathedral library of Toledo. There is one reference in Gudiel's list to Ptolemy: 'Item Almagistri de Greco. Item libri Tholomei de judiciis.' The first title probably refers to the Greek translation of the *Almagest* that was made in Palermo around 1160.[4] Interestingly, Thomas Aquinas knew this rare translation and quoted from it, rather than quoting the standard Arabo-latin translation of the Almagest.[5] The second item in Gudiel's entry is more puzzling. The title *De judiciis* is not commonly used for a translation of the *Tetrabiblos* from the Arabic.[6] The fact that this *De judiciis* follows a translation of the Greek *Almagest* makes it plausible that it was itself a translation from the Greek. The copyist of Gudiel's inventory may have adopted 'Tholomei de judiciis' as a short title for 'Claudii Ptolomei iudicialia ad Syrum'.[7] Hence, it is not impossible that Gudiel had a copy of the translation of the Greek *Tetrabiblos* in his precious Moerbeke collection, together with the Greek-Latin translation of the *Almagest* which was known to Thomas. But unfortunately, there is no manuscript corresponding to this description in Toledo – or in any other Spanish library.

That Moerbeke translated the *Tetrabiblos* should come as no surprise, since we know from Bate that Moerbeke shared his interest in astrology. In addition, and probably in the same period – as we will establish – Moerbeke produced a translation of another work by Ptolemy, *De analemnate*, a mathematical treatise that is of importance to astronomy (as it may help to calculate the position of the heavenly bodies in the celestial sphere). *De analemnate* is preserved in the celebrated autograph along with Moerbeke's translations of Archimedes and Eutocius, *Vat. Ottob. lat.* 1850. As the Archimedes translations were finished in 1269, the autograph, which also includes *De analemnate,* cannot have been made much later than that.[8]

---

[4] On this translation see Heiberg (1910), Haskins (1924), 157-164 and 191-193, Thorndike (1958), 34-49 (in particular 36).

[5] See *In De trinitate*, q.6, a.1, v.71-75; 279-287 (ed. Leonina); *cf.* q.5, a.1, v.92; a.3, v.67 and 423; q.6, a.4, v.55 and 286.

[6] Albert refers once to the *Tetrabiblos* as 'scientia de iudiciis'. See *Super Ethica*, III, lectio VII, p. 175,50-51: 'Ptolemaeus dicit in scientia de iudiciis' (same quotation as in Thomas, *Summa contra Gentiles*, III, c.86). However, Albert does not refer to the title of the work, but to the discipline that is the subject of this work. For the title he usually has *Quadripartitum*, sometimes connected with the Arabic title: 'in libro, qui dicitur arabice (h)Alarba, Quadripartitum autem latine'. There is one reference to a 'liber de nativitatibus' in *Physica* II, tr. XVIII, c.2, p. 125,54-55.

[7] See Hernández–Linehan (2004) 488: 'The work of Ptolemy most closely answering to this description is William of Moerbeke's translation of the *Iudicialia ad Syrum*'.

[8] The text has been first edited by Heiberg (1907). We use the edition of Edwards (1984).

All of these arguments clearly lead in the direction of Moerbeke, but it is only an analysis of translation method that can demonstrate that the *Iudicialia* translation is his work. This is also the case, it should be noted, for many other important translations that are not explicitly attributed to Moerbeke.

1.2. Internal arguments: an analysis of the translation method

1.2.1. Particles, conjunctions, prepositions, adverbs

Since the seminal study of Lorenzo Minio-Paluello first appeared,[9] the specific manner in which particles, conjunctions, prepositions and other invariant terms are translated has been used as a strong argument for the attribution of a translation to a particular translator. The introductions to modern editions of Moerbeke's translations, as well as the indices, offer impressive comparative material to determine the characteristic format of his translations.

When we set the Ptolemy *Iudicialia* translation beside this documentary material, it is immediately evident that the way particles, conjunctions, prepositions and adverbs are translated in the *Iudicialia* perfectly corresponds to Moerbeke's usage in his other translations. As we give an exhaustive list of the correspondences between Greek and Latin terms in this edition's indices, it makes no sense to repeat here the material we have collected. We will only mention certain exemplary translations that are characteristic of Moerbeke and that distinguish him from other translators. Some of the terms in the following list are not exclusively characteristic of Moerbeke, since they may also have been used by other translators, albeit in different proportions. For example, Bartholemew of Messina too uses *sine* for ἄνευ, but he mostly prefers *absque* (which is absent in Moerbeke); he uses *propter quod* for διό, but also *unde* (which is absent in Moerbeke); he translates ἔτι as *adhuc*, but more often as *amplius* (the inverse proportion in Moerbeke); τοιοῦτος may be translated as *talis*, but also as *huiusmodi* (which is never found in Moerbeke). When one takes into account not only the terms used, but also their proportion – and no less crucially, the absence of certain terms – we get quite a fair idea of Moerbeke's standard vocabulary.[10] And as we will see, the vocabulary used in the *Iudicialia* translation corresponds to this standard usage.

(a) Particles

ἀλλά sed, οὐ μὴν ἀλλά quin immo; ἄν + *opt.* utique + *futur. indic.*; ὃς ἄν quicumque; ἄν (or ἔαν) si; αὖ rursum; δέ autem (*passim*), uero (*post* autem), -que (6), sed (2), tamen (5), enim (3), etiam (4), *om. raro* (*post* δέ, *post* καί, *ante* οὐκέτι *et al.*); δή itaque (8), utique (6), autem (2), igitur *post* μέν (1), *om. ante* οὖν (1); εἶτα deinde (11), postea (2); ἔτι adhuc

[9] See Minio-Paluello (1947).

[10] For a comparison of the translation method of Bartholomew and Moerbeke see Vuillemin-Diem (2014), Beullens (2014) and Cordonier–Steel (2012) 415-416.

(*passim*), amplius (1), insuper (1); ἤδη iam (*semper*); μέν quidem; μέντοι tamen (40), attamen (2), autem (2), uero (1), quidem (3); οἷον puta (17), ut puta (2), ut (1), uelut (3); οὐκέτι non adhuc (3), non iam (3); πάλιν rursum (57), iterum (30)

Interestingly, the translator always avoids the syntagm *si utique*. This combination of conjunction and particle is often used by other translators (e.g. Burgundio of Pisa and Bartholomew of Messina), but never by Moerbeke (except in the construction *ac si utique*). Also notable are the numerous instances of *scilicet* (57) and *uidelicet* (19) in the translation. In many cases they correspond in the Greek text to the particle τε or τε καί. In other cases the translator adds *scilicet* or *uidelicet* to render the definite article ὁ when it is posited after a noun in an attributive function, or to introduce a noun or a definite article in apposition. In yet other cases, they are introduced by the translator himself. The frequent use of *scilicet* – and less frequent, of *uidelicet* – is characteric of Moerbeke from his earlier to his later translations, though one may find also examples in other translators (e.g., in Burgundio of Pisa). Consider the following examples:

I.755°-7° (563-5) τό τε ἀπὸ τῆς ἐαρινῆς ἰσημερίας πρῶτον δωδεκατημόριον, τὸ τοῦ Κριοῦ, καὶ τὸ ἀπὸ τῆς μετοπωρινῆς, τὸ τῶν Χηλῶν: primum *uidelicet* a uernali equinoctio triginta partium, *scilicet* Aries, quod ab autumpnali equinoctio, quod *scilicet* Chelarum

The addition of *uidelicet* corresponds to τε in the Greek text, the two *scilicet* are added in the sense of 'namely'. But *uidelicet* can also have this function, see:

I.997° (750) ἐν τῷ ὑψώματι τοῦ ἡλίου (τῷ Κριῷ): in exaltatione solis, Ariete *uidelicet*

I.999°-1000° (753) πρώτῳ ζῳδίῳ (τῷ Ταύρῳ): primo animali ... Tauro *uidelicet*

I.1000°-1° (754) τὸ δὲ διάμετρον (τὸ τοῦ Σκορπίου): diametrale autem, Scorpius *scilicet*

II.231° (177-8) τοῦ νοταπηλιωτικοῦ τριγώνου τοῦ τε κατὰ τὸν Ταῦρον : trigonalitatis notafiliotice (id est ...), eius *scilicet* que est secundum Taurum

III.92°-3° (74-6) τὸν μὲν ἀρχαῖον ... τρόπον τὸν κατὰ τὸ συγκρατικὸν εἶδος: antiquum quidem modum ..., illum *uidelicet* modum

IV.692° (527-8)τις εἱμαρμένη μείζων, ἡ τῆς τῶν χωρῶν αὐτῶν: quoddam fatum maius, ipsum *uidelicet* quod regionum ipsarum

(b) Conjunctions

ἅτε uelut; διό propter quod; διόπερ propter quod quidem; ἐάν si; ἐπεί/ἐπειδή quoniam; καθάπερ sicut (6), quemadmodum (5); ὅπως quatinus (5), ita ut (1), ut (1); ὅπως ἄν qualitercumque; ὅταν quando (*passim*), cum (1); ὅτε quando; ὅτι quod (*affirm.*), quia (*caus.*) [but never *quoniam*]; ὡς ut (*passim*), tamquam (29), quasi (4), sicut (4), ita ut (3), puta (2), quomodo (2), uelut (2), ueluti (2), in quantum (1): ὡς ἄν ac si; ὥσπερ sicut (15), quasi (13), quemadmodum (3), tamquam (3), uelut (3), ut (1); ὥστε ita ut (3), quare (1)

(c) Prepositions

ἄνευ sine; ἕνεκεν gratia (11), propter (1); *propter* is rare as translation for ἕνεκεν, but see *De fato* (5x); παρά + acc.: preter (16), propter (3), penes (3), extra (1), ultra (1), ad (1); πλήν excepto (3), uerumtamen (1), nisi (1); πρός + acc.: ad (*passim*), contra (1), usque ad (1), circa (3), secundum (1), quantum ad (2), in + acc. (1), uersus (1), per comparationem ad (3): rare, but see Simplicius, *Cat.* 450,43 and Themistius, DA, 74,85 (παρά + acc.: comparati ad).

One often finds (18x) the construction *quantum ad* to translate εἰς, περί or πρός (*cum acc.*) or a Greek accusative without a preposition, as in II.553°-4° (422) τὰ μεγέθη ... τὸν χρόνον *quantum ad magnitudines ... quantum ad tempus*. Similar usages are frequent in Moerbeke (HA, *Politica, Rhetorica*). Also noteworthy is the frequent use of the preposition *penes* to translate κατά + acc., παρά + acc., περί + acc. There are 77 cases of *penes* in this translation of Ptolemy, and this seems to be a distinctive feature of Moerbeke's translations (in particular when it is a translation of παρά + acc.), around the years 1267-69.[11] Moerbeke uses it occasionally in his translation of Themistius, DA (dated 1267), Proclus' *Elementatio* (dated 1268), in the *Rhetorica* (18x), and frequently in the Archimedes translation (1269) and *De analemnate* (15x), though other Latin equivalents remain dominant.

(d) Adverbs

ἀκολούθως consequenter (9), conformiter (6), sequendo (1); εἰκότως merito; ἑξῆς consequens (-ter) (12), sequens (2), deinceps (3); ὡς ἐπίπαν ut ad omne (dicatur) (30), ut in totum (dicatur), ut in toto, quasi in totum (4), sepius (1), ut (quasi) in omnibus (3), ut in pluribus (3); εὐθύς mox ; ἠρέμα debiliter (30) (very characteristic, never in other translators), remisse (4), quiete (1); ἴσως forte; λοιπόν de cetero (2); μεταξύ intermedius; κατὰ μικρόν paulatim; ὅλως uniuersaliter (14), totaliter (9), omnino (5); ὅμως (δὲ —) tamen; παντάπασιν penitus; πανταχῇ ubique; πανταχόθεν undecumque; κατὰ τὸ παντελές penitus; παντελῶς penitus (4), omnino (1); πάντῃ πάντως omnino penitus; πάντοτε semper; πάντως omnino, modis omnibus; πολλάκις sepe; σχεδόν fere; ὡς ἐπὶ τὸ πολύ ut in pluribus (7), ut in plurimis (1); ὡσαύτως eodem modo (15), consimiliter (1)

1.2.2. Vocabulary

The vocabulary of the *Tetrabiblos* offers an incredible challenge to any translator, medieval or modern. It is not only the technical vocabulary of astrology and astronomy that poses problems, but also the fact that Ptolemy indulges in creating composite terms and uses an impressive number of *hapax legomena* and other rare terms for which Latin equivalents cannot readily be found. Here again, we

---

[11] The preposition *penes* is not found in the translations of Boethius, James of Venice, Burgundio, Robert Grosseteste, *Metaphysica vetus*. However, it is attested in Aristippus' Plato translations, in Nicholas Siculus and Bartholomew of Messina, and in the anonymous translation of the *Rhetorica*.

need not duplicate what is collected exhaustively in the present edition's indices. We will merely give a list of translations that are characteristic of Moerbeke. As regards technical vocabulary, the comparison with other translations proves to be difficult, because the *Tetrabiblos* stands apart in terms of its content (but the same is true of Moerbeke's Archimedes translations). There is, however, Moerbeke's translation of another Ptolemy treatise, *De analemnate.*[12] Following Anthonis, we will first indicate some parallels in ordinary and astrological vocabulary between *De analemnate* and the *Iudicialia* (separated by /).

ἀκριβῶς certitudinaliter, diligenter, examinate / certitudinaliter (3), diligenter (1), κατὰ τὸ ἀκριβές examinate (5); ἀνατολή oriens / oriens; ἀνατολικός: orientalis / orientalis; ἁρμόζω congruo / congruo; βόρειος borealis / borealis; διάστημα distantia / distantia; δῆλον palam / palam; ἔγκλισις declinatio / declinatio; ἔκθεσις expositio / expositio; ἑξῆς consequenter / consequenter; ἐπίσκεψις consideratio / consideratio (19), intentio (1), speculatio (13); εὐπόριστος facile acquisibilis / facile acquisitibilis; ἐφαρμόζω adapto / adapto; ἐφοδεύω tracto / tracto (1), pertracto (2), aggredior (1), expono (1), trado (1); ἔφοδος processus / accessus (3), supergressus (2), uia (2), processus (1), regula (1), ars (1), incursus (1); ἰσημερινός equinoctialis / equinoctialis; ἰσημερία equinox / equinoctium; κεφαλαιωδῶς summatim / summatim; κορυφή uertex / uertex; λῆψις acceptio / acceptio; μεσημβρινός meridianus, meridionalis, equinoctialis[13] / meridianus, meridionalis, equinoctialis; μηνιαῖος mensualis / mensualis; νότιος australis / australis; οἰκεῖος conueniens, proprius / conueniens, proprius; παραλείπω pretermitto / pretermitto; πρόσνευσις adnuitio / adnuitio; προτάσσω premittere / preinstituo (1), premitto (1), preordino (2); πρόχειρος promptus / promptus; ῥᾳδίως de facili / de facili; τεταρτημόριον (τὸ) quarta (pars) / quarta (*sc.* pars), quadratura (2); τοιοῦτος talis / talis; χειμέρινος hyemalis / hyemalis; ὡριαῖος horarius / horarius

De analemnate only offers a limited vocabulary. If we compare the vocabulary of the *Iudicialia* with the corpus of Moerbeke translations insofar as they have been edited, we could establish a much longer list of parallels. The indices of our edition give ample evidence of a vocabulary characteristic of Moerbeke. Here, we only give some striking examples:

αἰγωπός ('goat-eyed') caprinus (oculos). – *Cf.* HA 492a3; GA 779b1.

ἀκμαῖος (ὁ) perfecte etatis (is qui) (*cf.* ἀκμαιότης status I.692°, συνακμάζων perfecti status IV.295°). – *Cf.* GA 788a3 *qui perfecte etatis*; *In Parm.* 324,11-12 *akmazans, id est in perfecta etate.*

[12] Unfortunately the original Greek text is lost. The Greek can only be partially recovered thanks to the palimpsest of the *Ambrosiana* and marginal notes in the autograph. We rely on the index made by Edwards.

[13] *Equinoctialis* is found four times in *De analemnate* as translation for μεσημβρινός: see p. 36,14; 44,1; 45,1; 46,8. The editor is surprised at the use of this Latin equivalent. In his apparatus he notices at the first case in n. 49 'meridionalem scribendum'. For the three other cases he supposes that there was a confusion in Greek between μεσημβρινός and ἰσημερινός: see n. 117; 135; 151.

ἀκριβής (-ῶς) diligens (–ter). – This translation is often used by Moerbeke (but is rare in his later translations): *cf. In Mete., Mete., In De int., Pol., Rhet., In Cat., In De caelo, Themistius.*

ἀκριβῶς certitudinaliter (*cf.* also ἀκριβόω certitudinaliter nosco, certitudinaliter examino). – *certitudinaliter* is a characteristic term of Moerbeke: see GA 738a16; 787a6; 781a15; *Pol.* 1282b5; *In Mete.* 245,37; 357,18; *Cat.* 8b12; *In Cat.* 27,43; 274,86; 294,11; *In De caelo* 504,55; *In Parm.* (7x). See also *Mete.* 363b32 ἀκριβόω *certitudino.*

κατὰ τὸ ἀκριβές: examinate (*adv.*). – *Examinatus*(-*e*) for ἀκριβής (-ῶς) is again a characteristic term of Moerbeke (once however it is found in the *vetus* of *De sensu* 442b15); see *In Cat.* (10x); *Rhet.* 1418b1; *In De caelo*, 152,22; see also: ἀκριβολογέομαι *examinate dissero In Cat.* 257,69; ἀκριβεία *examinatio Rhet.* 1414a16 and *In Cat.* 522,44. Interestingly Moerbeke avoids *examinatus* in his later translations.

ἀνοίκειος inconueniens. – *Cf. In Cat.* (4x); in later translations Moerbeke prefers *incongruus.*

ἀνόητος insensatus. – *Cf.* HA 610b23; 622a3; *Rhet.* 1391a14; *Pol.* 1269a6 (*imp.* + *perf.*); *De fato* 86,93.

ἀφορμή occasio. – *Cf. In Cat.* (7x); Proclus, *Opusc.* (2x); *In Parm.* (3x).

δεινός industrius. – *Cf. Rhet.* 1373a5 (*cf.* 1401b3); *Pol.* 1274a26; *In Parm.* 62,41 71,3.

δεόντως oportune. – *Cf. In Cat.* (2x); *Opusc.* (1x) ; *In Parm.* (6x); Moerbeke often uses *oportunum* to translate δέον.

διακρίνω diiudico. – *Cf. In De int.* 410,70; *In Cat.* 148,51; 266,40; *In De caelo* 247,26. This translation is not found in later works.

δύσις occasus, occidens. – *Cf. Mete., In Mete., In De caelo.*

ἐμφανίζω insinuo. – *Cf. In De int.* 90,49; 94,31; *In Cat.* 202,92; *In Parm.* 50,52; 358,69.

ἐπιβολή adiectio. – *Cf. In Cat.* (2x); Procl., *In Tim.* (1x), *Opusc.* (3x), *In Parm* (32x).

εὐστόχως bene coniecturabiliter. – *Cf.* PA 639a5; *De fato* 110,34; *cf. An. Post.* 89b10 (εὐστοχία: *facilis coniecturatio*, a correction by Moerbeke of James of Venice's *subtilitas quaedam*); *Rhet.* 1412a12 (εὐστόχος *bene coniecturans*).

θησαυριστικός thesaurizatiuus. – *Cf.* HA 488a20 *thesaurizatiua alimenti.*

κατάληψις comprehensio. – *Cf. In Cat.* (1), *In De int.* (5x); *In Parm.* (2x); in *De somno* 458a29 Moerbeke replaces *inceptio* (*tr. vetus*) with *comprehensio.*

μετάβασις transitus. – *Cf. In Mete.* (1x); *In Cat.* (6x); *In Parm.* (18x); *Mete.* (*transitio* 1x).

ὁλοσχερής grossus; -ῶς grosse. – *Cf. In Cat.* 126,67 *grosse*; *In De caelo.* 400,74 *grossior*; Proclus, *Prov.* 18,21; 43,14: ὁλοσχέρεια *grossities.*

ὁρμή impetus. – *Impetus* (never *motus*) is Moerbeke's standard translation.

ὀσφύς lumbi. *Cf.* HA 494b.

παραλλαγή (ἐναλλαγή) uicissitudo (*cf.* ἐναλλάξ uicissim; ἐνηλλαχώς uicissitudinans). – *Cf. In Cat.* 228,89; 231,83; *In Parm.* 400,9. The term *uicissitudo* and related terms are only found in Moerbeke's translations. One finds besides *uicissitudo* also *uicissitudinatio*, the adjective *uicissitudinarius*, the adverb *uicissitudinate* and the verb *uicissitudinor* with participle *uicissitudinatus* (*cf. Mete.* 386a1, PA 661b18 ἐπαλλάττων; HA 501a17 ἀνεπάλλακτος *non vicissitudinatus*). Frequent, but more common, is *uicissim* (mostly as translation of ἀνὰ or παρὰ μέρος).

προσλαμβάνω assumo. – *assumo* or *coassumo* are frequent in all of Moerbeke's translations.

πρόφασις pronuntiatio III.1138° (851), occasio IV.132° (107). – Moerbeke rightly distinguishes between the two meanings of the term: *cf.* Robbins 'purpose – pronouncements' (p. 331/383). For the translation *occasio*, see *De fato* 94,70; *Rhet.* 1371a1,3 *In De int.* 4,55 197,80; *In Parm.* 66,53.

συνεγγίζω appropinquo. – *Cf. In Cat.* (4x); *In Parm.* (2x)

ὠφέλιμος proficuus. *Cf. Rhet.* (11x), *In Cat.* (8x), *In Parm.* (1x).

On διαμετρέω diametrizo, ἐναντίωσις contrariatio, ἐπιστημονικός scientionalis, παθητικός passionum illatiuus, see below, p. 27.

The verb εἶναι is usually translated by *esse*, but sometimes by *existere*; *existens* is of course the evident choice to translate the participle ὤν. All of this is standard translation practice. Distinctive of Moerbeke, however, is the use of the forms of the perfect subjunctive (*fuerit/fuerint*) to translate the Greek subjunctive or optative form of εἶναι (or sometimes of τυγχάνειν): ᾖ, εἴη *fuerit*. In this Ptolemy translation, there are 101 cases of *fueri(n)t*.

Also characteristic of Moerbeke is the transliteration of Greek terms, followed by an explanation. An exhaustive list is found in the *Supplementum latinum* [D] of the indexes. See for instance:

II.325° (245) βορραπηλιωτικῷ boreafiliotice (id est aquilonari simul et orientali)

II.582° (443) προηγήσεις precessiones (id est retrogradationes) (same addition in *Fragmentum Toletanum*, ed. 504,28. Heiberg)

III.72° (60) καταρχήν katarchen (id est optentionem principii)

Another practice which is characteristic of Moerbeke is the translation of composite nouns (with φιλο- πολυ- εὐ-), of which there are many examples in Ptolemy.[14] Thus, one finds in the *Tetrabiblos* no fewer than 65 different terms composed with φιλο-. For some of these composite nouns, the translator could find Latin simple terms. Thus he renders φιλάνθρωπος as *benignus* and *humanus*, τὸ φιλάνθρωπον and φιλανθρωπία as *benignitas*, φιλανθρωποῦν as *amicabilis*, φιλόστοργος as *affectuosus*, τὸ φιλόστοργον as *affectio*, φίλεχθρος as *discors*, -ως as *inimicabiliter*. These cases are however rare. The translator generally opts for a system of periphrastic translations, whereby the prefix φιλο- is rendered by *amor*, *amator* and other related terms. For the translation of substantive nouns or substantified adjectives we find *amor* + gen.: φιληδονία *amor delectationis*, τὸ φιλελεύθερον *amor libertatis*, φιλοκαλία *amor pulcritudinis*, τὸ φιλοκάθαρον *amor puritatis*, φιλαλήθης (σκοπός) *amor ueritatis*. For the translation of the adjectives we have generally *amator* (or *amatrix*) followed by a genitive. See the following extraordinary list:

[14] We summarize here the analysis of Bossier (1994).

amator (*-trix*) *agonum* φιλαγωνιστής, *amator amicorum* φιλόφιλος, *amator antiquorum* φιλάρχαιος, *amator* (-trices) *armorum* φίλοπλος (-οι), *amator artis* φιλότεχνος, *amator astrologie* φιλαστρόλογος, *bonorum amator* φιλόχρηστος, *communitatis amator* φιλόκοινος, *corporis amatrix* φιλοσώματος, *amator dationis* φιλόδωρος, *amator dei*(*-orum*) φιλόθεος, *amator discipline* φιλομαθής, *amator domesticorum* φιλοίκειος, *amator exercitii* φιλαγωνιστής, *amator extraneorum* φιλόξενος, *fistularum amator* φιλόμουσος, *amator gesticulationis* φιλόστροφος, *amator (-trix) glorie* φιλόδοξος, *amator heremi* φιλέρημος, *amator (-trix) honoris* φιλότιμος, *amator iustitie* φιλοδίκαιος, *amatrix laborum* φιλόπονος, *amator libertatis* φιλελεύθερος, *amator litigii* φίλερις, *amator litterarum* φιλογράμματος, *amator malignitatis (malignorum)* φιλόμοχθος, *amator medicinalium* φιλίατρος, *amator monachorum* φιλομόναχος, *amator musice (musicorum)* φιλόμουσος, *amator (-trix) musarum* φιλόμουσος, *natorum amatrix* φιλότεκνος, *amator operationis* φιλοπράγμων, φιλόπρακτος, *amator operis* φιλεργός, *amator ornatus* φιλόκοσμος, *amator (-trix) pecuniarum* φιλοχρήματος, *amator planctus* φιλόθρηνος, *amator populi* φίλοχλος, *amator possessionis* φιλοκτήματος, *amator premii* φίλαθλος, *amator pulchritudinis* φιλοκαλλωπιστής, *amator rationis* φιλόλογος, *amator saltationis* φιλορχηστής, *sanguinis amator* φίλαιμος, *amator sermonis* φιλόλογος, *sociorum amator* φιλέταιρος, *amator speculationis* φιλοθέωρος, *tormentorum amator* φιλοβάσανος, *amator tumultus* φιλοθόρυβος, *uictorie amator* φιλόνεικος: *amatrix uirorum* φίλανδρος. See also privative adjectives ἀφιλότιμος *non amator honoris*, but, ἀφιλότιμος *sine amore honoris*.

Rare constructions are:

amans + inf.: *amans speculari* φιλοθέωρος

amatiuus + gen.: *amatiuus armorum* φίλοπλος, *amatiuus domesticorum* φιλοίκειος, *amatiuus laborum* φιλόπονος, *amatiuus libertatis* φιλελεύθερος

sectator + gen.: *sectator heremi* φιλέρημος

sectans + acc.: φιλόφρων *sapientiam sectans*

delectans in + inf.: *delectans in addiscere* φιλοπεύστης

The large number of periphrastic translations in this translation is astonishing, but it corresponds to Moerbeke's method. One may, for instance, analyse the indices of the *Rhetorica* sub verbis φιλο-. The editor, B. Schneider, indicates for each Greek term the translation of the *Vetus* (v) and that of William (g). Whereas the *Vetus* generally transliterates *philandroteroi, phileraste, philargiria, phileteria, fileterus, philogelos*, etc., Moerbeke translates *amatores uirilitatis, amatores amantium, amor operositatis, amor sociorum, amator sodalium, amator risus*, etc. But even in cases when the *Vetus* gives a periphrastic translation, the anonymous translator does it in a different manner than Moerbeke. Besides *amator* + gen. the former also uses *amans* + gen. and *amicus* + gen. to translate the φιλο-composites, for instance *amantes uiri, amici amoris, amicus risus*. However, as Bossier noticed, the last two constructions are never used by Moerbeke, and this negative rule is confirmed by the *Iudicialia*. The translation of the numerous φιλο-composites in the *Iudicialia* results, therefore, in a good argument for the attribution of the translation to Moerbeke.

Less numerous are the compositions with πολυ- (26 cases). Since there are also composite terms in Latin beginning with *multi-* (*multiplex, multiformis*, etc.), these are obvious correspondences to the Greek terms. However, here again, Moerbeke prefers periphrastic constructions. In most cases he replaces the composite Greek term by a descriptive Latin genitive with *multus*. For instance:

> πολύανδροι *multi uiri*, πολυγύναιοι *multarum mulierum*, πολυΐστωρ *multarum historiarum*, πολυκτήμονες *multarum possessionum*, πολυμερής *multarum partium*, πολυπράγμων *multe actionis, multorum negotiorum*, πολύσπορος *multe dispersionis*, πολύσπερμος *multi spermatis*, πολύτροποι *multorum modorum*, πολυφλέγματος *multi flegmatis*, πολυχρόνιος *multi temporis*, πολύχρυσος *multi auri*.

In other cases we find the adverbial *multum* followed by an adjective or participle or a verb. Thus:

> πολυθρύλλητος *multum notatus, multum diuulgatus, multum historiosus*, πολύκοινος *multum communis*, πολυμήχανος *multum ingeniosus*, πολυχρονέ(ί)ω *multum tardo*, πολύψογος *multum probrosus*.

This practice again perfectly corresponds to the usage of Moerbeke in his other translations.

Finally, there are the numerous composite terms with the prefix εὐ- (about 100 different terms). Here again, we see that the translator prefers periphrastic translations (in 70 different cases), either a descriptive genitive with *bonus* (εὐάγωγος *bone conuersationis*, εὐέκτης *bone habitudinis*, εὐόφθαλμος *bonorum oculorum*) or an adjective or participle preceded by the adverbial *bene* or *facile* (*bene aptus, bene cessibilis, bene coniecturabilis, ... facile accusabilis, facile acquisitibilis, facile adibilis, facile applicabilis, facile auertibilis*). A full list is found in this volume's indices. This practice corresponds, again, to what we find in Moerbeke's translations. In particular, the translations of Greek verbal adjectives ending on -τος as *facile* followed by an adjective on *-bilis* seem to be characteristic of Moerbeke. Similar comments could be made on the translation of composites with the prefix δυσ- where Moerbeke prefers combinations with *difficile* (*difficulter*), *gravis*.

Ptolemy likes to use the definite article τό to substantify adjectives. This construction becomes 'excessivement lourde et pédante' in cases where the article substantifies an adjective in the comparative or superlative form.[15] The translator does not shrink from this philological challenge, as one can see in the following passages:

---

[15] See on this practice Bossier (1994) 113-4.

ἐπὶ μέντοι τὸ εὐμορφώτερον καὶ ἐπιχαριτώτερον καὶ γυναικοπρεπωδέστερον καὶ θηλυμορφώτερον καὶ ἀθυμότερον [εὐχυμότερον Hübner] καὶ τρυφερώτερον

tamen ad formosius et gratiosius et ad muliebrem decentiam magis et magis femininum et minus animosum et tenerosius (III. 909°-11° [681-3])

ἐπὶ τὸ δικαιότερον καὶ ἀνυτικώτερον καὶ τιμητικώτερον καὶ σεμνότερον καὶ θεοσεβέστερον, κατὰ δὲ τὸ ἐναντίον καὶ ἀνοίκειον ἐπὶ τὸ ταπεινότερον καὶ ἐπιπονώτερον καὶ ἀσημότερον καὶ ὠμότερον καὶ μονογνωμονέστερον καὶ αὐστηρότερον καὶ δυσδιαγωγότερον καὶ ὅλως ἐπὶ τὸ δυσκατορθώτερον

ad iustius et ad eminentius operationis et honorabilius et reuerentius et ad maiorem dei cultum. Secundum contrarium autem et inconueniens ad humilius et laboriosius et minus insigne et insocialius et singularioris sententie et austerius et ad id quod peioris regiminis et uniuersaliter ad id quod peius dirigibile (III. 1448°-53° [1103-7])

Another extreme example is III.91°-2° (73-4) where Ptolemy substantifies an adjective connected with a genitive:

ἐπιλογισμὸν περὶ τοῦ δυνατοῦ τῆς τοιαύτης προγνώσεως

ratiocinationem de eo quod possibile est talis pronosticationis

We have only discussed a number of distinctive usages. If, however, one browses through the indices at the back of the volume, one can see that the vocabulary of the *Iudicialia* overwhelmingly points to Moerbeke as translator of this Ptolemaic work. To be sure, there is a technical vocabulary not found in other Moerbeke translations. And what is more, it seems that the *Iudicialia* translator likes more variation in translating the same Greek term than Moerbeke usually does. Yet there are examples of a freer translation style in Moerbeke's oeuvre, for example, the translation of Alexander's *De fato* and of Ammonius' commentary on *Perihermeneias*.

One could, alternatively, try to challenge the attribution to Moerbeke by finding counter-examples. Is there a term (apart from the technical vocabulary related to astronomy) that is never found elsewhere in Moerbeke translations? Or again, is a favourite Moerbeke term lacking here? At IV.346° (266), the translation *supersedens* for ἀντιληπτικός may surprise. However, the Greek term has, in this passage, the rare meaning of 'capable of self-restraint'. In Latin *supersedere* can mean 'to preside over, to be superior to, to restrain or desist from'. The term thus fits excellently in this context. Anthonis points to the use of the verb *(e)uitare* and its derivatives: φυλάσσομαι *uitor* I.337°, ἀμετατρεψία *ineuitabilitas* I.362°, ἀνεμποδίστως *ineuitabiliter* I.347°, ἀφύλακτος *ineuitabilis* I.294° 348°, δυσφύλακτος *ineuitabilis* II.836°: τὸ -ον *difficultas euitationis* I.58°, φυλακτικός *euitatiuus* III.1413°: τὸ -όν *euitatio* I.353°, ὑποσιώπησις *collocutionis euitatio* IV.451°. One finds *euitare* and its derivatives never in Moerbeke's Aristotle translations. But in *De fato* we have for παραβαίνω *euito* (2), ἀπαράβατος *ineuitabilis* (5), -ως *ineuitabiliter* (3). See also Proclus *De providentia*, 2,8. This is not really a counter-argument, then, especially since Moerbeke may have had reasons to prefer here the verb *euitare*. Take the verbal adjective δυσφύλακτος I.58° (50). This term can have two meanings: 'hard to keep, guard' or 'hard to keep off or prevent' (a danger, for instance). As the term here has the second meaning,[16] it would be wrong to use the translation *difficultas seruandi*; *difficultas euitationis* is the right choice. Finally, the term *ineuitabilis* is also used in the Tiburtinus translation (see the quotation from Thomas Aquinas, above, on the question on whether the influences of the stars are 'unavoidable').

Another negative example mentioned by Anthonis is the absence of the term *interimo* as a rendering of ἀναιρέω and its derivatives. In the *Iudicialia* we find *perimo, peremptiuus, peremptorius, peremptio*, but never a form of *interimo*, which is the preferred translation of Moerbeke. We noticed, however, that Moerbeke often uses *perimo* (23) besides *interimo* (44) in his translation of Simplicius, *Cat.* He also uses *peremptiuus* and *peremptio*. But it must be admitted that the complete absence of forms of *interimo* in the *Iudicialia* remains puzzling.

1.2.3. Syntax

> Ptolemy is fond of long, involved sentences and has a number of mannerisms, among them a fondness for the infinitive with the article and an almost Teutonic habit of piling up long strings of modifiers between article and substantive, which often results in sequences of two or even three articles. (Robbins, p. XXIV).

The translator of the *Iudicialia* adopts different strategies to deal with the recalcitrant syntax. We will examine three translation techniques that are characteristic of Moerbeke's method.

(a) Infinitives with article

The translator generally translates τὸ (τῷ, τοῦ) + *inf.* with the construction *hoc quod + verb. finit.*: I.166°-7° (134-5),[17] 1245°-6° (951), III.80° (65), 581° (443), 1545° (1174), IV.767°-8° (581-2); or with *eo quod + verb. finit.*: I.202° (160-1), 396° (311), 897°-9° (677-8), 1255°-6° (959-60), 1260° (962), III.70°-1° (58-9) *eo quod per illud et ea que ante partum possint precognosci*, III.91°-2° (74), III.1013°-15° (760-2) *eo quod lesura ... efficiat et non habeat ..., passio autem ... superueniat*. See also the frequent cases of διά τὸ + *inf.*: *quia + verb. fin.* (28), *propterea quod + verb. fin.* (12) , *propterea quia + verb. finit.* (2), *quia* (2).

[16] Interestingly the Proclus Paraphrase understands the term in the first meaning, which is definitely wrong: *cf.* Robbins, p. 5, n. 3.

[17] In I.166° (134-5) πρὸς τῷ + *inf.* is translated as *cum hoc quod ... accidit + inf.*

(b) Absolute genitive

Ptolemy likes absolute genitive constructions. The translator sometimes renders them with the Latin absolute ablative; in most cases, however, he opts for the construction *cum* + verb in subjunctive. See the following examples:

I.33° (30-1) δύο τῶν μεγίστων καὶ κυριωτάτων ὑπαρχόντων, ἑνὸς μὲν,… δευτέρου δέ: cum duo insint maxima et principalissima: unum quidem… secundum autem

I. 807°-9° (603-4) τῆς ἐκ τῶν τοιούτων διατυπώσεων ποιότητος … δυναμένης προεκτίθεσθαι: cum qualitas que ex talibus occurrit figurationibus possit preexponi

I.1060° (801) δεικνυμένου: cum sit demonstratum

II.927° (703) ὀφειλόντων: cum … debeant

III.651° (490) σελήνης δὲ ἀφιείσης: luna autem cum dimiserit

III.672°-3° (508) τῆς σελήνης ἀφέτιδος οὔσης: cum luna fuerit dimissor

III.736° (551-2) ταύτας δὲ εἰληφότες: has autem cum acceperimus

See also a case of participle in instrumental dative position:

I.1068° (805-6) οὐδ' αὐτοῖς ὡς ἔφαμεν ἀληθοῦς ἐχομένοις σκοποῦ: cum ipsi non habeant ueram intentionem

Sometimes the genitive construction is rendered by *quod* or *quia* + *verb. finit.* See:

I.767°-8° (572-4) οὐ τῶν καταστημάτων … ἀκρατοτέρων, ἀλλ' ἡμῶν ἐγκεχρονικότων … ἀντιλαμβανομένων: *non quod consistentie … sint … magis impermixte, sed quia nos … morati sumus … percipimus.*

The construction *cum* + *verb. finit.* in subjunctive is also used to translate the predicative participle in the nominative. See:

I.1150° (888) οἶκοι ὄντες *domus cum sint*, 743° (553) προτάξαντες cum *preinstituerimus*, II.685° (520-1) αἴτιοι γινόμενοι *cum sint causa*, III.551 [736] εἰληφότες *cum acceperimus*, and further examples in I.871° (653-4), II.808° (615), III.94° (76-7).

(c) Attributes between article and substantive

The absence of the definite article in Latin is one of the major difficulties in translating a Greek text into Latin, in particular a philosophical or scientific text like that of Ptolemy. In some cases, the Latin pronoun *qui* (*que, quod*) takes over the function of the Greek article (see examples in index sub ὁ). We have already seen how the translator deals with infinitives which are introduced by an article. Let us now see how he deals with Ptolemy's 'habit of piling up long strings of modifiers between article and substantive'. The translator adopts a standard practice to translate those phrases: he renders the Greek definite article by the pronouns *ea, eo, hiis, eorum, earum,* followed by the relative *qui* or *que.* After the relative pronoun one may find the copula (*est* or *sunt*), but in most cases it remains implicit. In this way, the translator can connect all the genitives and other attributive specifications inserted between the definite article and the substantive to the substantive they determine. See the following examples:

I.51°-2° (44) τὴν ἀπὸ τοῦ περιέχοντος αἰτίαν ἐμφανιζόντων: **eam que** a continente causam insinuantium

I.66°-8° (56-7) ἐπὶ πᾶσαν τὴν περίγειον καὶ δι' ὅλων μεταβλήτην, τῶν ὑπὸ τὴν σελήνην πρώτων στοιχείων: ad omnem **eam que** circa terram et ad per omnia transmutabilem naturam **eorum que** sub luna primorum elementorum

I.130°-1° (105) τὰς κατ' αὐτὸ τὸ ὑποκείμενον ἀλλὰ τάς γε δυνάμει ποιητικάς: **eas que** secundum ipsum quod subicitur sed uirtute effectiuas

I.134°-5° (108) ἐκ τῆς συγκράσεως πάντων τὸ ἴδιον τῆς ποιότητος διαλαβεῖν: deprehendere **eam que** ex commixtione omnium qualitatis proprietatem

I.226° (178) περί τε τὰς ἐξαρχῆς ἐπιτηδειότητας δυνάμεων: circa **eas que** ex principio ydoneitates potentiarum

I.234° (184) τὴν κατὰ τὸ χρήσιμον ἐπίσκεψιν: **eam que** secundum utile considerationem

I.260°-1° (205-6) ἀπὸ τῆς ἄνωθεν αἰτίας: ab **ea que** desuper causa

I.1220° (932) τοῖς ἐξ ἀρχῆς οἴκοις: **hiis que** a principio domibus

I.1234°-5° (942-3) τῆς κατὰ τὸ ἀνόμοιον τῶν περιεχόντων ζῳδίων κράσεως: **ea que** secundum dissimilitudinem continentium signorum complexione

II.612°-3° (468) τῷ κατὰ τὰ ἑπόμενα κέντρῳ τοῦ τόπου τῆς ἐκλείψεως: **eo qui** secundum sequentia angulo loci eclipsis

II.614°-5° (469) θεωρηθέντων δὲ οὕτως τῶν εἰς τὴν αἰτίαν τοῦ συμπτώματος παραλαμβανομένων ἀστέρων: consideratis autem ita **hiis que** ad causam simptomatis coassumuntur astris

IV.738° (561) τοῖς περὶ τὸ ποιητικὸν αὐτῆς συμβεβηκόσιν: **hiis que** circa efficientiam ipsius accidentibus

IV.224° (177) τοῦ περὶ συναρμογῶν λόγου: **eo qui** de coniugiis sermone

These three translation methods, and in particular the last one, are characteristic of Moerbeke.

1.3. Special character of the translation

All of these indicators confirm the attribution of the translation of the *Iudicialia* to William of Moerbeke. Nevertheless, the translation has some distinctive features. Although it remains basically a verbum-de-verbo translation, William takes much more liberties as translator than he does in his Aristotle translations or in the later translations which are extremely literal, such as his translation of the commentary on the *Parmenides*. This is evident in the greater variation in vocabulary, as we have already noted. Moerbeke often gives interpretative translations, adapting his translation to the context. Take the translation of *pronosticatio* γνῶσις in the title of chapter I, 2 [7° 63°]. At first one might assume that the translator read πρόγνωσις in his model, but other examples in this context, such as γνωσθέντων *preuisorum* I.58° (49) and δυσθεώρητον *difficultas precognitionis* I.57° (48), show that it is the translator himself who chose the term *pronosticatio* as more appropriate here than *cognitio* or *notitia* to indicate the subject of this chapter. A similar example of contextual interpretation is II.519° (395) παρόδους *configurationes* (where *transitus* would be the literal translation). Examples of such 'free' translations are frequent in this treatise. See III.95° (77) ἀκριβοῦν *omnimoda diligentia pertractare*; I.45°-6° (39-40) ὡς ἄν τις φιλαλήθει μάλιστα χρώμενος σκοπῷ *quomodo si quis intendat maxime amori ueritatis*. Instead of *intentione utens* Moerbeke translates *si ... intendat* and also modifies the attributive adjective into *amori ueritatis*. See also I.95° (78) πρὸς τὸ καθόλου τῆς ποιότητος τεταγμένον *ad uniuersalem impressionem qualitatis*; II.122° (95-6) ἰδιότροποι περιστάσεις *singulares proprietates*; IV.115°-6° (94) ψήφους οἰκοδεσποτείας *rationes dignitatis*. Other examples can be found in the Greek-Latin apparatus of the edition where we put between brackets those deviations from the (literal) Greek texts which, in our view, are due to the translator himself.

William does not even hesitate to adapt the structure of sentences to reflect their logical order. See the following examples where we indicate with numbers between brackets the order of words in the Greek text. Take, for instance, the opening paragraph of Ptolemy's preface:

I.34°-7° (32-3) καθ' ὃ τοὺς γινομένους ἑκάστοτε σχηματισμοὺς τῶν κινήσεων ἡλίου τε καὶ σελήνης καὶ ἀστέρων πρὸς ἀλλήλους [τε καὶ τὴν γῆν] καταλαμβανόμεθα
per quod (1-2) motuum solis et lune et astrorum (7-10) factas semper (3-4) ad inuicem (11-12) figurationes (5) comprehendimus (13)

I.40°-2° (35-6) τὸ μὲν πρῶτον ἰδίαν ἔχον καὶ δι' ἑαυτὴν αἱρετὴν θεωρίαν, κἂν μὴ τὸ ἐκ τῆς ἐπιζεύξεως τοῦ δευτέρου τέλος συμπεραίνηται
primum quidem, propriam (1-3) et propter se eligibilem (5-7) habens (4) theoriam (8), etiam si (9) finis (14) qui est ex connexione secundi (11-13) non (10) concludatur (15)

I.134°-5° (108) ἐκ τῆς συγκράσεως πάντων τὸ ἴδιον τῆς ποιότητος διαλαβεῖν deprehendere (7) eam que (4) ex commixtione omnium (1-3) qualitatis (6) proprietatem (5)

Another interesting example is I.174°-80° (139-44), where Moerbeke not only modifies the order of clauses in one sentence, but reorganizes the whole section. Or again, see III.92°-9° (74-9)

> τὸν μὲν ἀρχαῖον τῶν προρρήσεων τρόπον τὸν κατὰ τὸ συγκρατικὸν εἶδος τῶν ἀστέρων πάντων ἢ τῶν πλείστων, πολύχουν τε ὄντα καὶ σχεδὸν ἄπειρον, εἴ τις αὐτὸν ἀκριβοῦν ἐθέλοι, κατὰ τὴν διέξοδον καὶ μᾶλλον ἐν ταῖς κατὰ μέρος ἐπιβολαῖς τῶν φυσικῶς ἐπισκεπτομένων ἢ ἐν ταῖς παραδόσεσι ἀναθεωρεῖσθαι δυνάμενον παραιτησόμεθα διά τε τὸ δύσχρηστον καὶ τὸ δυσδιέξοδον.
>
> antiquum quidem modum predicendi *recusabimus propter usus difficultatem et inexplicabilitatem*, illum uidelicet modum qui secundum speciem optentiuam astrorum omnium uel plurium, cum sit multiplex et fere infinitus, si quis uelit omnimoda diligentia pertractare, et magis possit in particularibus adiectionibus eorum que naturaliter considerantur quam in traditionibus per singula speculari.

As one can observe, Moerbeke inserted the last phrase of the sentence (παραιτησόμεθα διά τε τὸ δύσχρηστον καὶ τὸ δυσδιέξοδον) immediately after the direct object τρόπον = *modum*. To connect the attributive explanation τὸν κατὰ ... with τρόπον he repeated *modum* and added *scilicet*: *illum scilicet modum* There are many examples of such additions, by the translator, of terms which are not found in the Greek text. The following cases can be distinguished: (1) Moerbeke repeats a preposition (*a/ab, ad, circa, de, ex, in, per, propter, secundum*), whereas in the Greek text the preposition is only given once for the different nouns depending on it. (2) Moerbeke repeats a term whereas the Greek text only refers to it implicitly by the use of a definite article (see the example of *modum* above). (3) Finally, Moerbeke adds *suo arbitrio* pronouns, particles, nouns, verbs. There are the standard additions of *est* or *sunt, fiunt, fuerit, scilicet* and *uidelicet* discussed before, and also the additions of pronouns. Other additions are more intrusive. E.g. (added words are in bold):

I.14 **orizontis** angulorum — 363 distantie due **scilicet dies et nox** — 733 propter domum **Martis, scilicet** Scorpionem — 835-6 **in hac tabula** (*sc.* 'termini secundum Egiptios') **tractatur de dispositione terminorum planetarum seu de diuisione** (*cf. adn. gr.-lat.*) — 873-4 numerus collectorum **terminorum** — 876 propter corrosionem **libri** — 890 quia in ordine **quantum ad alia signa** postponuntur — 960 ex eo quod sunt ablatiui **a proprio motu** (*pro* ἀφαιρετικούς) — II.142 ut ex hiis **consequitur** — 172 unde amor libertatis **conuenit** ipsis — 222 qui ibi **morantur** — 224 **proprietas** habitantium regionem **est** incompositio — 385 quemadmodum in **hominis** generatione — 437-9 Particularibus autem remissionibus... **similiter consideratis** (*cf.* Robbins: 'we deduce from') — 460 **Si autem ambo easdem habent rationes** (*cf. adn. gr.-lat.*) — 603 **Per comparationem** autem ad continens — 607-8 et corruptionis que ... **euenit** per hec — 621 regiones et ciuitates quibus simptoma **futurum** significatur — III.3 Prohemium **de ordine sequentis intentionis ad precedentem** — 174 eleuationes **super horizontem** — 204-5 Mars adiungatur soli **uel lune de die** Saturnus autem lune **de nocte** (*sed cf. adn. gr.-lat.*) — 435 gradibus **acceptis** — IV.637-8 et signi facti **abiectione occurrentis** (*sc. anni*) dominum coassumentes (*cf.* Robbins: 'and taking the ruler of the last sign') — 659-60 per **proprietatem** configurationum

A comprehensive list of these additions is found in the *Supplementum latinum* in the Indexes. All of these examples show that Moerbeke's method is in this text much freer, and more *ad sensum*, than in his Aristotle translations and other translations of his later period. Still, as we have argued, this 'freer' translation method is also encountered in Moerbeke's translation of Alexander's *De fato* and of Ammonius' commentary on *Perihermeneias*.

1.4. When was the translation made?

As Fernand Bossier pointed out in a pioneering study,[18] there are some variations in Moerbeke's translation method, which make it possible to situate an undated

[18] See Bossier (1989).

translation in the long period between about 1258 and 1286, when he was active. Having examined how some particles and adverbs are translated, Bossier supposed that the translation of the *Iudicialia* was made around 1264 (p. 291). The evidence Bossier adduces is limited to the fact that Moerbeke no longer uses *multotiens* to translate πολλάκις, as he had done in his earlier translations, but only *sepe* (6 cases). However, as Moerbeke continued using *sepe* in his later translations, this is not a reliable argument against a later date of composition. In fact, there are some indications that the translation was made after 1266. For instance, the fact that Moerbeke translates ὅταν by *quando*, and no longer by *cum* (with one exception), situates the translation after 1266. *Merito* for εἰκότως is not found in early translations such as *Mete.* (there *uerisimiliter*), but it is standard in Simplicius, *Cat.*, in the Archimedes translations and in all later translations. Again, *quatinus* for ὅπως is not found in the early translations. There is also the frequent use of *scilicet* (57) and *uidelicet* (19), and of *penes* (77 cases), which indicates a composition around 1266 (not yet in Simplicius' *Cat.*).[19] There is the use of *contrariatio*, a term coined by Moerbeke, appearing first in Simplicius, *Cat.* (6x).[20] Anthonis pointed to *scientionalis* (2x) as translation of ἐπιστημονικός besides *scientificus* (3x). *Scientionalis* is absent in early translations, is not yet found in Simplicius, *Cat.*, but becomes dominant in the later translations (first attested in Themistius, DA, dated 1267). Bossier also noticed *diametrizo* which renders διαμετρέω, first occurrence of verbs on *-izo*, which later become more en vogue. The translation at III.1018° (764) of παθητικός as *passionum illatiuus* offers another chronological indication. A similar translation is found in Simplicius, *Cat.* (4x), in *Rhet.* (4x), and *Pol.* 1342b3. This notable translation is exclusively characteristic of Moerbeke (no parallels are found in other translators) and seems to be limited to translations of the period around 1266 (date of *Cat.*). The usual translation of παθητικός is *passiuus*.[21] There are, however, other features which exclude a much later date. Thus the translation *de cetero* for λοιπόν is only found in the early translations until Archimedes, but will disappear in the later translations. Another argument against a late translation is λαμβάνω. This verb is generally translated as *accipio* and only once as *sumo*, which becomes the preferred translation later. It is therefore plausible to situate the composition of the translation sometime between Simplicius *Cat.* (1266) and the Archimedes translations (1269). We have undoubtedly a *terminus ante quem*, 1272, the quotation by Thomas in his commentary on *De caelo*.

---

19 On *penes* see above 15.

20 Rare term characteristic of Moerbeke, see G. Vuillemin-Diem, *Metaphysica, Recensio et Translatio Guillelmi de Moerbeka*, A.L. XXV 3.1, p. 221-222, 249.

21 The Latin term *illatiuus* is not uncommon, but is used in the sense of 'inferring, concluding'. Moerbeke uses it to translate οἰστικός in *El. Theol.*, *In Parm.*, *De decem dub.*

2. MOERBEKE: A TRANSLATOR WITH AN INTEREST IN SCIENCE

As we have seen, Ptolemy is a difficult author, writing long tortuous sentences, stretching beyond measure the possibilities of Greek syntax, such as the use of articles and absolute constructions. But to translate the *Tetrabiblos* is also difficult because of the technicalities of its subject, astrology. No good translation of a scientific treatise is possible without an expert knowledge of the science discussed in it. This is true for the translation of Archimedes, of Aristotle, of Galen, of *De analemnate* – and of the present translation. In this section we will examine further examples which display both the linguistic genius of the translator and his expertise in the field.

2.1. Astrological terms

To start with, Moerbeke's translation of the title of Ptolemy's work (*iudicialia*) is surprising. It is probably Ptolemy himself who created the term Τὰ *ἀποτελεσματικά* as his title for the practical part of astronomy, to distinguish it from the theoretical part wherein the movements of the celestial bodies are studied. The ἀποτελέσματα are the effects produced by the celestial bodies in their various configurations on the sublunary world, and in particular, on the human beings born under their influence.[22] Moerbeke of course knew the literal meaning of the term ἀποτελεστικός, which he translates elsewhere as *effectiuus* (5x), *perfectiuus* (1x). Compare ἀποτέλεσμα being translated as *effectus* (15x) or *euentus* (3x), and ἀποτελέω generally as *efficio* (*perficio* 2x, *facio* 1x). In opting for *Iudicialia* to translate the title of Ptolemy's treatise, Moerbeke was probably influenced by the tradition of Arabic-Latin judicial astrology (with treatises 'de iudiciis astrorum').[23] Ptolemy occasionally uses the term γενεθλιαλογία or γενεθλιακός, which Moerbeke translates as *geneaticus sermo* or *sermo natiuitatis*. Moerbeke also uses the term *natiuitas* to translate γένεσις in its astrological sense (*generatio* in this meaning is not evident).

It is of great importance for an astrologer to know the different configurations that heavenly bodies may have towards one another (also called 'aspects' because the stars and planets are supposed to 'look' at one another). Ptolemy distinguishes in chapter I, 14 four aspects, which correspond to four geometrical relations.[24] Some stars are diametrically opposed on a straight line, some are at a triangular

[22] See on this innovation Simplicius, *In Phys.* 293,11-14: οἱ δὲ νεώτεροι (*sc.* Ptolemy) διελόντες τοὔνομα τὴν μὲν τὰς κινήσεις τῶν οὐρανίων ἐπισκοποῦσαν ἀστρονομίαν καλοῦσι, τὴν δὲ περὶ τὰ ἀποτελούμενα ἐξ αὐτῶν διατρίβουσαν ἀστρολογίαν ἰδίως ἐπονομάζουσι.

[23] See above p. 12 and n. 5.

[24] See Robbins, p. 73-4, n. 2.

distance, some have a quartile and some have a sextile opposition. Let us see how Moerbeke interprets this technical vocabulary.

Ptolemy is the first to use the term διαμηκίζω to indicate that two stars stand diametrically opposed in one straight line. Moerbeke has the following translations. Three times (IV.644° 667° 882°) he simply latinizes and writes *diametrizo* (as in his translation of the equivalent διαμετρέω *diametrizo*: see II.672° III.700° 1051° IV.446°). In the other cases he interprets more freely and thus shows that he understands the astronomical meaning of the term:

> *in diametrali situ existo* III.503°, *prospicio ex longinquo* III.1077°, *per longitudinem disto* IV.556°, -ῶν *in maxime remoto aspectu existens* IV.621°, -σας *diametraliter aspiciens* IV.629°, -ῶν τυγχάνω *diametraliter aspicio* IV.681°, -ομαι *longitudinis aspectu impedior* III.1462°. See also for διαμετρέω: -ῶν *diametralis (est)* III.513° 785°: -ήσας *in diametrali aspectu* III.301°, -ούμενος *diametraliter aspectus* III.1044°.

One may notice the use of *aspectus* in these periphrastic translations. *Aspectus* is often found in Latin astrological treatises, but Moerbeke uses it rarely as a technical term to translate one single Greek term. In this translation we find only ἐπιθεώρησις *aspectus* IV.467° 471° and συνεπιθεώρησις *aspectus* IV.626° (*cf.* ἐπιθεωρέω *inspicio* II.1039°, *aspicio* IV.277° 508° 513° 529° 827°). However, Moerbeke takes advantage of this term to render the meaning of the different configurations, not only, as we have seen, in the case of diametrical oppositions, but also in the vocabulary related to the three other geometrical relationships between stars.

> See for the quartile relation, besides the simple transliteration τετραγωνίζω *tetragonizo* (I.817° IV.882°): -ῶν *in quadrato aspectu (existens)* III.300° IV.621°, -σας *aspiciens quadrangulariter* IV.628°, -ῶν τυγχάνω *aspicio quadrangulariter* IV.681°, τετράγωνος *quadratus (sc. aspectus)* I.827° 831° 833°, *in quadrato aspectu* III.279°, *quadrangularis* III.568° 646° IV.450° 648°, *tetragonalis* III.648° 658° IV.275° 526°, οἱ (*sc.* συσχηματισμοί) τετράγωνοι *que (sc. configurationes) ex quadrato* I.837°, τὸ -ον *quadratura* I.829° 868° 909°: κατὰ -ον *quadrangulariter* IV.821° (*cf.* τετραγωνικῶς *quadrangulariter* III.981°). For the triangle relation: τριγωνίζω: -ῶν *in trino aspectu* III.273°; and finally for the sextile: ἑξαγωνίζω: -ῶν *in sextili aspectu (existens)* III.273°, ὁ ἑξάγωνος (συσχηματισμός) *exagonalis* I.868°: ἑξάγωνος ὢν *scilicet in sextili aspectu* III.298°.

One may be surprised to find *longitudinis aspectu impediatur* rendering διαμηκισθῶσιν in III.1462°. However, in this passage Ptolemy is considering how the positions of Mercury and the moon, when 'they are overcome, or surrounded, or held in opposition by unfamiliar stars in injurious aspect', may cause various diseases of the soul: ἐάν τε πρὸς τὸν ἀνατολικὸν ὁρίζοντα καθυπερτερηθῶσιν ἢ ἐμπερισχεθῶσιν ἢ διαμηκισθῶσιν *siue ... superentur aut habitudine inuoluantur aut longitudinis aspectu impediantur* (III.1461°-2° [1114-5]). The beneficial influence of the moon and Mercury is then blocked and impeded by the malificent stars. In fact, stars standing in diametrical opposition of the ascendant often have a harmful influence. Moerbeke's free translation renders the meaning of the text very well. His interpretation is also confirmed by his translation of κατοπτεύων as *aspectu impediens* in the next paragraph, which gives the reason why epileptic children are born: *Martem autem de nocte habent in angulo et aspectu impedientem* (κατοπτεύοντα) (III.1481° [1128]). A less fortunate translation in the above-quoted text is ἐμπερισχεθῶσιν *habitudine inuoluantur.* The verb ἐμπεριέχω means 'to surround' and thus blockade a star. Moerbeke, however, recognized in this form of the verb a derivation of the term σχέσις *habitudo,* which leads to his nonsensical translation.

Another important notion in astrology is the μεσουράνησις or mid-heaven, the star in culmination at the moment of nativity (see Robbins, p. 165, n.1 with reference to Bouché-Leclercq, p. 271). Moerbeke translates μεσουράνημα generally as *medium celi,* μεσουράνησις as *mediatio celi* (in two cases *medium celi*), μεσουρανέω as *celum mediare* or *in medio celi esse,* τὸ -οῦν once as *attingens ad medium celi.* The place opposite to mid-heaven is τὸ ὑπὸ γῆν μεσουρανοῦν *oppositum medii celi sub terra* (I.800°).

The most important factor in predictions is the determination of the ὡροσκόπος, the sign rising at the horizon at the moment of birth, translated by Moerbeke as *ascendens* and *horoscopus.* The term *horoscopus* renders two times ὥρα (usually translated as *hora* or *tempus*).

The horoscope, the superior mid-heaven and the inferior are called the κέντρα of the nativity. When the term κέντρον is used in this astrological sense, Moerbeke consistently translates it as *angulus,* and the adjective ἐπίκεντρος as *angularis* or *in angulo.*[25] He uses *centrum* in I.1248°, but rightly to indicate 'the centre of earth', and *aculeus* in I.567° and III.1027° to indicate the sting of the Scorpion.

ζῳδιακός is always translated as *zodiacus.* Moerbeke also uses the term *zodiacus* for the elliptic circle going through the zodiac: ὁ διὰ μέσων τῶν ζῳδίων, ὁ διὰ μέσων, ὁ διὰ μέσων κύκλος. ζῴδια. The different signs on the zodiac, are generally translated as *signa;* 20 times, however, as *animalia,* which is etymologically correct (ζῴδιον is the diminutive of ζῷον), but confusing. See ζῳδιακῶς *per proprietatem animalis* (IV.590°-1°), or take the title of chapter II, 12 *De particulari animalium natura ad consistentias.* No Latin reader could have thought that this chapter discusses, as Robbins translates, 'the nature of the signs, part by part, and their effect upon the weather' (περὶ τῆς μερικῆς πρὸς τὰ καταστήματα φύσεως τῶν ζῳδίων). Besides, *consistentia* is not a good choice either to refer to weather conditions. *Signum* is also the standard translation of δωδεκατημόριον, which is the 12th part of the zodiac. In some cases, however, Moerbeke opts for *duodecima pars* or adds *duodecim* (XII) to *signa.*

---

[25] Puzzling is the translation of ἐπικεντρότερος *is qui fuerit magis in cuspide* in II.603°. Cuspis (ἀκίς) is the translation of the sting of the arrow in the Sagittarius.

In the determination of a person's nativity the calculation of the length of his life is fundamental. Ptolemy devotes a long and complex chapter (III, 12 *De annis uite*) to this question. In his view, this calculation 'depends entirely upon the determination of the prorogative places and the stars that rule the prorogation, and upon the determination of the destructive places or stars' (Robbins).

> ἤρτηται μὲν γὰρ τὸ πᾶν ἔκ τε τῆς τῶν ἀφετικῶν τόπων διαλήψεως καὶ ἐκ τῆς αὐτῶν τῶν τῆς ἀφέσεως ἐπικρατούντων καὶ ἔτι ἐκ τῆς τῶν ἀναιρετικῶν τόπων ἢ ἀστέρων (III.558°-60° [425-8])
>
> Dependet quidem enim omne ex perceptione dimissoriorum locorum et ex perceptione ipsorum qui dominantur super dimissionem, et adhuc ex perceptione peremptiuorum locorum uel astrorum. (Moerbeke)
>
> Hec autem obseruatio ex loco *alhileg* et eorum qui super *alhileg* potestatem habuerint necnon et stellarum interficientium locis consideratur. (Tiburtinus)
>
> Conuenit enim ut aspiciamus in locis *ylegiorum* et similiter in planetis interficientibus. (Aegidius)

The essential term here is ἀφέτης and ἀφετικὸς τόπος, which Robbins renders with the Latin astrological term 'prorogator' and 'prorogative places'. The prorogator is 'a point on the ecliptic that, traveling at the rate of one degree of oblique ascension a year toward either the descendant or ascendant, determines a person's length of life' (*Encylopedia Britannica*). Moerbeke translates the term ἀφετικός consistently as *dimissorius*, ἀφέτης as *dimissor* and ἄφεσις as *dimissio*. This is a literal translation of the Greek term, which means 'letting go, release'. One wonders, however, if readers of Moerbeke's translation, versed in astrology, would have understood what the term *dimissorius* means here in a technical sense, unless somebody would have told them that it corresponds to the term *yleg* or *alhileg*, which was commonly used in the Arabo-Latin translations. One understands why Simon Bredon felt the need to add in *Ox yleg* above *dimissorius* in III.447 and 458. On should notice *en passant* that Moerbeke never uses the Arabo-Latin terms that had become the standard vocabulary in Latin astrological works through the translations from the Arabic, such as *almugea* and *haiz*.

Μοῖρα in the astrological sense is translated as *gradus* (49x), only in four cases (I.28° 1189° 1191° 1195°) as *pars*, but here again Moerbeke has reasons to deviate from the general usage. In chapter I, 22 Ptolemy is explaining the meaning of μοῖρα, and it would be confusing to have it translated as *gradus*, which is the meaning the 'parts' have after interpretation.

Ἀπόρροια is translated as *refluxus* in I.30° (in the chapter list), 1236° 1244°, ἀπορρέω *refluo* I.1239°, otherwise always as *defluxus*, which is the standard translation of Moerbeke. The translator must have assumed that in chapter I, 24 (*De coniunctionibus et refluxionibus*) ἀπόρροια (ἀπορρεῖν) has a technical astrological meaning, which could not be rendered by *defluxus*. The term refers to the movement apart of two planets after they had come close at the same zenith (see Robbins, p. 112-3, n. 3). The Arabo-Latin translation of Tiburtinus uses *separatio* as a technical term (so also Robbins). In other contexts the term ἀπόρροια refers to the emanations coming from the heavenly bodies: here *defluxus* is the obvious translation: see I.81° II.596° III.525° 580° 923°. However, Moerbeke is not always consistent: he also has *defluxus* where it should be *refluxus*: see II.1020° III.1152° 1201°.

Planets may prevail, dominate, be master of the house or yield to others. Moerbeke here again finds his way through the metaphorical astrological vocabulary of οἰκοδεσποτέω, οἰκοδεσποτεία and related terms usually rendered as *dominari, dominatio, dominium*, καθυπερτερέω as *supero, superemineo, obtineo, preualeo*. We notice στηριγμός *statio*, στηρίζω *sto*, -ζων *stationarius*, which render well the astronomical term 'standing still, being fixed'. Moerbeke knows the standard Latin equivalents of the names of planets, stars and constellations and their elements, such as Φάτνη *Presepe*, Στάχυς *Spica*. For Χηλαί he uses both *Chele* (*Khele*) and *Libra*. But Moerbeke did not understand the term λαμπήνη, which means 'covered chariot' and designs a favourable aspect of planets, and translated it three times as *claritas*, connecting it to λαμπρός.

Moerbeke does not follow the Greek system of referring to the planets with an article followed by a genitive: the (star) of Mercury, of Venus, etc. ὁ τοῦ Ἑρμοῦ ... ὁ τῆς Ἀφροδίτης ... ὁ τοῦ Κρόνου ... ὁ τοῦ Ἄρεως. These constructions are simplified as *Mercurius, Venus, Saturnus, Mars*. In the translation of sections on planets there are often changes of gender. Thus, when Ptolemy speaks of ἀστέρες and uses the corresponding masculine forms in articles, pronouns, adjectives, Moerbeke translates *astra* and correspondingly puts all related forms in neutrum plurale, but, as it happens, he returns in the next sentence to the Greek usage and has the masculine forms. See for instance:

> Aiunt autem astra masculin**a** et feminin**a** fieri et secundum figurationes ad solem: matutin**os** quidem enim existentes et precedentes masculin**os** fieri, uespertin**os** autem et sequentes feminin**os**. (I.457°-61° [355-7])

This change may due to inadvertence of the translator influenced by the Greek model. However, the fact that Moerbeke translates ἀστήρ not only as *astrum*, but also as *planeta* (masculine), may have explained why he tends to deal with *astrum* (*sc. planeta*) as if it were a masculine noun. *Planeta* is of course also used as a translation of πλανήτης (πλανώμενοι are *erratici, errantes, planete*). To translate τὰ φῶτα Moerbeke uses *luminaria*, a term also commonly used in the Arabo-Latin astrological texts.

2.2. Medical terms

Moerbeke is not only confronted with the technical vocabulary of astronomy, but also with difficult medical terms describing all kinds of diseases. We have selected, *unum ex multis*, a long section of chapter IV, 9 where Ptolemy describes what kind of diseases may cause our death, depending on our nativity. The underlined terms in the Latin translation are explained in the commentary that follows.

> ὁ μὲν οὖν τοῦ Κρόνου τὴν κυρίαν τοῦ θανάτου λαβὼν ποιεῖ τὰ τέλη διὰ νόσων πολυχρονίων καὶ φθίσεων καὶ ῥευματισμῶν καὶ συντήξεων καὶ ῥιγοπυρέτων καὶ σπληνικῶν καὶ ὑδρωπικῶν καὶ κοιλιακῶν καὶ ὑστερικῶν διαθέσεων, καὶ ὅσαι κατὰ πλεονασμὸν τοῦ ψυχροῦ συνίστανται· ὁ δὲ τοῦ Διὸς ποιεῖ τοὺς θανάτους ἀπὸ συνάγχης καὶ περιπνευμονίας καὶ ἀποπληξίας καὶ σπασμῶν [καὶ κεφαλαλγίας] καὶ τῶν καρδιακῶν διαθέσεων, καὶ ὅσαι κατὰ πνεύματος ἀμετρίαν ἢ δυσωδίαν ἐπισυνάπτουσιν· ὁ δὲ τοῦ Ἄρεως ἀπὸ πυρετῶν συνεχῶν καὶ ἡμιτριταϊκῶν καὶ αἰφνιδίων πληγῶν καὶ νεφριτικῶν καὶ αἱμοπτυϊκῶν διαθέσεων καὶ αἱμορραγιῶν καὶ ἐκτρωσμῶν καὶ τοκετῶν καὶ ἐρυσιπελάτων καὶ ὀλέθρων, καὶ ὅσα τῶν νοσημάτων κατ' ἐκπύρωσιν καὶ ἀμετρίαν τοῦ θερμοῦ τοὺς θανάτους ἐπιφέρει· ὁ δὲ τῆς Ἀφροδίτης διὰ στομαχικῶν καὶ ἡπατικῶν καὶ λειχήνων καὶ δυσεντερικῶν διαθέσεων ποιεῖ τοὺς θανάτους, ἔτι δὲ καὶ διὰ νομῶν καὶ συρίγγων καὶ φαρμάκων δόσεων, καὶ ὅσα τοῦ ὑγροῦ πλεονάσαντος ἢ φθαρέντος ἀποτελεῖται συμπτώματα· ὁ δὲ τοῦ Ἑρμοῦ διὰ μανιῶν καὶ ἐκστάσεων καὶ μελαγχολιῶν καὶ πτωματισμῶν καὶ ἐπιλήψεων καὶ βηχικῶν καὶ ἀναφορικῶν νοσημάτων, καὶ ὅσα τοῦ ξηροῦ πλεονάσαντος ἢ φθαρέντος συνίσταται.

> [3] Saturnus quidem dominium mortis suscipiens facit fines per egritudines diuturnas et pthises et per reumatismos et contabescentias et rigores febriles et per egritudines splenis et uentris et per ydropises et per matricales dispositiones, et quecumque secundum habundantiam frigidi existant. [4] Iupiter autem facit mortes a squinanchia et peripneumonia et apoplexia et a spasmis et cardiacis dispositionibus, et quecumque <secundum> spiritus immoderationem aut prauitatem odoris copulantur. [5] Mars autem a febribus continuis et emitriticis, et repentinis percussionibus nefreticis et a dispositionibus sanguinem spuere facientibus et a sanguinis fluxibus et aborsibus et partubus et ab erisipelis et aliis molestiis, et quecumque egritudinum secundum incensionem et immoderantiam calidi mortes inferunt. Venus autem per stomaticas et epaticas et per impetigines et dissentericas dispositiones facit mortes, adhuc autem per deuorationes et fistulas et farmacorum dationes, et quecumque humore habundante aut corrupto simptoma efficiunt. [7] Mercurius autem per manias et extases et per melancolias et per casus et per epilempses et per tussiuas et euaporatiuas egritudines, et quecumque sicco habundante aut corrupto [quod] consistunt (IV.592°-613° [454-70])

Here again we can only admire Moerbeke's talent in translating a text where we modern scholars need dictionaries on every line. Moerbeke has almost always found the right Latin equivalent. To be sure, he often takes it easy and just gives Latin transliterations of Greek terms, but this was already an accepted practice in medieval medical handbooks, and it even remains so in contemporary English. See, for instance, in Robbins's translation: pneumonia, phthisis, apoplexy, rheumatism, spasms, nephritic, erisipelas, lichens, etc. In some cases Moerbeke may have used a transliteration to hide his ignorance, as in *febribus emitriticis*. Ἡμιτριταϊκός means here, according to our dictionary, 'half every three days', that is, a 'semi-tertian fever'. With the exception of this term, Moerbeke always finds the right interpretation of the Greek terminology for diseases in Ptolemy's list. Take *contabescentia* (consumption, wasting away of organs) for σύντηξις opposed to φθίσις. *Squinanchia* is the right translation for συνάγχη, which means angina, inflammation of the throat. 'The term was pronounced as Synanke, which in Middle Latin became Quinancia, then squinanchia, and finally Quinsy.'[26] *Impetigines* translates λειχήνων. Where Robbins uses the Latin transliteration of the Greek term, Moerbeke opts for *impetigo*, a scabby eruption on the skin. Puzzling is the translation of νομῶν as *deuorationes*. The term νομός (νομή) has the original meaning of 'pasture', while the variant form νομή can also have the metaphorical meaning, in medical treatises, of the 'spreading' of sores and ulcers on the flesh. By using *deuorationes* Moerbeke may have wanted to indicate some 'devouring' of the flesh, such as ulcers. Robbins translates 'cancers', which is not altogether right. The Greek term coming closest to cancer is κονδύλωμα in III.1097°, which Moerbeke renders very well as *lupus aut cancer*. We could give many more examples of Moerbeke's astonishing expertise in medical terminology, but let us end with two skin deseases: III.1084° (811) *morphea* for ἀλφός (white leprosy), and III.1118° (837) *scrophula* for χοιράς (scrofulous swellings in the glands of the neck). Both terms are still used in contemporary medicine.

2.3. Translation of hapax legomena and rare terms

As the above examples of translations give ample evidence, Moerbeke is a genius in coining new terms to translate rare Greek terms. Ptolemy's treatise with its wealth of rare terms and hapax legomena for describing characters of the soul, diseases of the body or a host of professions offers a rare opportunity for a linguistic genius. We have discussed numerous examples of creative vocabulary production in section 1.2.2, and many more could be derived from the volume's indices. Take III.898° (672) ἀναφαλαντιαίους (bald on the forehead) *subcaluastros*. (When translating *HA* 518a28 Moerbeke did not yet understand a related term, he simply transliterated *anafalanthiasis*.) When confronted with a rare complicated term, Moerbeke often resolves the term into its components. See III.1186° (890) θρασυδείλους *audentes magna mala cum formidine*; III.1247° (941-2) δυσεντεύκτους *difficile precibus acquiescentes* (which is better than Robbins's 'hard to speak with'). Moerbeke kept in his translation the meaning of ἔντευξις as 'petition, intercession for a person'. Or the hapax εὐπεριστόλους *facile uestu ornabiles* (*cf.* περιστέλλω dress, cloth). However, in such a reduction into elements the metapho-

[26] See Nybakken (1959) 11.

rical meaning of a term (in this case 'to be circumspect') risks getting lost.[27] This method may lead to funny examples as IV. 170° (139) παλαιστάς *ad palmam mensuratores*. Παλαιστής means wrestler, but the term παλαιστής is also a variant form of παλαστή 'palm of the hand', 'measure'. Compare HA 606a14, where Moerbeke translates 'aures autem capre palme quatuor digitorum'. He must have supposed the term had the same meaning here, and made of it a profession: 'those who measure according to the palm'.

Let us end with a difficult and intriguing philological problem with sexual connotations.

III.1052°-3° (787-8) εὐνοῦχοι ἢ ἑρμαφρόδιτοι ἢ ἄτρωγλοι καὶ ἄτρητοι γίγνονται
eunuchi aut ermafroditi aut *sine folle* aut sine foramine fiunt

This is the only instance of the term ἄτρωγλοι in the TLG. The term ἄτρητος is more often used, and was also known to Moerbeke because we find it three times in Aristotle's HA, where it is translated as *non foraminosus* (488a25) and *non perforatus* (516a26 b35). In IV.336° Moerbeke translates the same term as *imperforatus*, here as *sine foramine*. The problem is the interpretation of ἄτρωγλοι. The only manuscript *Ox* reads *sine folle*. But what could that mean? *Follis* is a bag, cushion or pillow filled with air. However, ἄτρωγλοι is probably related to τρώγλη, which means according to L-S-J a hole gnawed in clothes, or holes of mice, or holes in a wall (see HA 552b26, translated by Moerbeke as *latibulum*), but also 'the holes of canals in the flesh'. This could also be the meaning of the Latin term *fossa*, literally a trench or ditch. For that reason Bossier proposed to correct the text and write *sine fossa*. However, *fossa* would be a conjecture, not evident from a palaeographical point of view. Moreover, in this interpretation ἄτρωγλος becomes almost a synonym of ἄτρητος. We therefore keep the reading of *Ox*. In medieval Latin, *follis* can also be used to indicate the scrotum; in this sense *sine folle* fits very well in the enumeration of sexual aberrations. Moerbeke probably interpreted the term through the context and not by etymology. Robbins translates nicely: 'the children born have no ducts and vents'.

2.4. *Quandoque bonus dormitat Homerus*

Moerbeke's talents do not make him immune to error. Here are some rare examples of wrong interpretations of Ptolemy's syntax, mainly in dealing with the use of the Greek genitive:

I.50°-2° (43-4) plurimorum et excellentiorum simptomatum, ita euidenter eam que a continente causam insinuantium. – A Greek absolute genitive (τῶν γε πλείστων ... ἐμφανιζόντων) has been wrongly translated as a genitive in Latin.

---

[27] See on this term, Bossier (1994) 131-4.

II.102°-4° (81-2) quia appropinquat zodiaco locus qui secundum uerticem et errantium circa zodiacum astrorum διὰ τὸ συνεγγίζειν αὐτῶν τὸν κατὰ κορυφὴν τόπον τοῦ ζῳδιακοῦ καὶ τῶν περὶ αὐτὸν πλανωμένων ἀστέρων. – The Greek genitive (τῶν πλανωμένων ἀστέρων) is wrongly translated by the genitive in Latin. It should be a dative 'appropinquat zodiaco locus ... et errantibus circa zodiacum astris': 'because their zenith is close to the zodiac and to the planets revolving about it' (Robbins).

See also the example we discuss in the section on Nifo below (p. 47).

3. MOERBEKE AND PLATO OF TIVOLI

Moerbeke is the first to translate the *Tetrabiblos* directly from the Greek. He must have known, however, the translation from the Arabic made by Plato of Tivoli, for this translation was commonly used by scholars in the 13th century. Did Moerbeke find inspiration in this translation, or help in understanding difficult sections, or in rendering technical vocabulary? We have compared large sections of both commentaries and the answer is clearly negative: Moerbeke did not use the Arabo-Latin translation when working on his translation from the Greek. Even in terminology, the difference between both translations is radical. As said before, Moerbeke avoids all hybrid Arabo-Latin astrological terms. To show the difference between the two translations, let us just give a specimen of Tiburtinus' translation, namely the translation of book I, chapter 23. This text can be compared with the Moerbeke translation and the adaption by Agostino Nifo of the same chapter quoted at the end of Chapter III. As is immediately evident, the Arabo-Latin translation is often far removed from the original Greek text, whereas Moerbeke's translation is incredibly accurate and stays close to the original. Nevertheless, as we shall see in the next chapter, Moerbeke's translation never succeeded in replacing Plato's translation, which remained the standard translation until the Renaissance. Scholars were used to it, and honestly, Plato's version reads more easily than the accurate Moerbeke, who held too much to the Greek *littera*.

Capitulum 23 in almugea, idest in visione ad inuicem facie ad faciem et in alchinara, idest splendore et in his similibus[28]
Erraticarum quidem stellarum similitudines in signis sunt velut prediximus. Item quod planete proprie habeant almugeam similiter ostenditur, cum eorum unusquisque aliquam in figura cum sole vel luna societatem habuerit modo qui dicitur almugea; hoc est ut sit inter unumquodque et solem ac lunam ex longitudine quantum est inter ipsius domum et domum solis, aut lune ex longitudine, veluti si Venus esset in sextili aspectu alterius luminaris, ita quod a sole occidentalis vel a luna foret orientalis, esset tunc in almugea, hic est ergo modus almugea uniuscuiusque planetarum cum luminaribus.

[28] We quote the text from the first edition (1484), using for the correction of some typos the edition of 1533.

Hoc autem sequitur alchinara que sit cum fuerit in sua haiz et in eius simili ex locis sibi convenientibus, cumque duas vel plures auctoritate aliquis eorum habuerit, in loco sui gaudii et honori esse dicitur, eorumque vires tunc augmentantur eo quod signis in quibus sunt assimilantur, que vires eorum adiuvant.

Dicuntur etiam gaudere, cum signis in quibus fuerint predicto modo non assimilantur sed planete que fuerint ex eorum haiç ipsorum locis assimilantur. Que similitudo, licet remotissima sit, est tamen aliqua. Cumque in locis extraneis, scilicet in haiz suis haiz contrariis inueniuntur, eorum vires proprie valde decrescunt et tunc sunt eorum complexiones signorum in quibus fuerint complexioni contrarie, quapropter fiet et ex hoc opus ex duabus naturis collectum.

CONCLUSION

Moerbeke's translation of Ptolemy's *Tetrabiblos* is an extraordinary achievement in his outstanding scientific career. This is the first translation directly from the Greek of a major astrological work, which has confronted all scholars since late antiquity with great problems of interpretation. Moerbeke made a translation solidly based upon an excellent Greek model (see Chapter IV), exhibiting his expert knowledge of the subject, creating a wealth of new Latin technical terms, remaining accurate and close to the text, and yet free enough in his interpretation to make Ptolemy's long sentences digestible. As we have shown, Moerbeke worked on this translation in the period preceding his Archimedes translations. If we bring together all that he produced in that short period – Simplicius, *Cat.* (1266), Themistius, DA (1267), Philoponus, DA (1268), Ammonius, *De int.* (1268), Proclus, *El. theol.* (1268), Aristotle, *Rhetorica* (around 1267?), Ptolemy, *Iudicialia* (1268?), Ptolemy, *De analemnate* and Archimedes–Eutocius (1269), plus the monumental Simplicius, *De caelo* (1271) – one is amazed, and wonders how he could have realized all this work in conjunction with his other duties.

CHAPTER III
THE RECEPTION OF MOERBEKE'S TRANSLATION

Moerbeke's translation of the *Quadripartitum* did not attract many readers. As we have seen, the first to have noticed the new translation was Thomas Aquinas, to whom Moerbeke must have communicated a copy when Thomas returned from Paris to Italy in 1272. Henry Bate, a learned friend of the translator, received another copy and similarly made use of it. Finally, we found that Agostino Nifo used Moerbeke's translation when he produced his own Ptolemy translation at the beginning of the 16th century. We have done extensive search in commentaries and astrological works of the later Middle Ages and Renaissance, and no other scholar seems to have noticed Moerbeke's translation. All continue quoting from the Arabo-Latin translation of Tiburtinus. To mention just two examples: Guido Bonatti (13th century) in his *Liber astronomiae* and John Ashedon (14th century) in his *Summa astrologiae* have long quotations and paraphrases of Ptolemy: they all come from Tiburtinus' translation, which remained dominant far into the Renaissance period.[1]

1. HENRICUS BATE

Henry Bate of Mechelen (1246-ca. 1310)[2] had a strong interest in astrology, as is evident from his translations of the works of Abraham ibn Ezra, his construction of an astrolabe and equatorium, his establishment of astronomical tables, and his original astrological autobiography, 'Liber servi Dei de Mechlinia de ducatu Brabantiae super inquisitione et uerificatione natiuitatis certe ex indiciis ac subsequentibus nato post natiuitatem'. His treatise on the astrolabe is dedicated to William of Moerbeke. In the preface, Henry reminds William that when they were both in Lyon – they both attended the Council of Lyon between May and July 1274 – he had told Moerbeke how he had made himself a very useful astrolabe and had promised to write down how he had constructed it and how it could be used for observations and calculations. After his return to Mechelen, Bate immediately

---

[1] We express our gratitude to the Bayerische Staatsbibliothek in München, to the Deutsche ForschungsGemeinschaft, to the Gallica project of the BNF in Paris, to the Biblioteca Virtual del Patrimonio Bibliografico of Spain for having made digital versions of rare editions free accessible via internet. They are of great help for scholars.

[2] On Bate, see Poulle (1972) and the introductions to the editions mentioned in n. 3.

started writing the treatise 'at the request' (*ad petitionem*) of Moerbeke. It was finished on 11 October 1274. In the work's first part, Bate describes in full detail how he constructed the astrolabe from bronze material; and in the second part, he explains how the instrument can be used. As Emmanuel Poulle has pointed out, 'Bate's astrolabe was quite clearly destined for astrological use; designed to facilitate the rapid acquisition of astrological data'. All of his astronomical studies were primarily motivated by a practical astrological interest. Bate's treatise on the construction of the astrolabe not only reveals his friendship with Moerbeke ('tamquam uere amicitie internexu firmo pernexus'), but also indicates that both scholars shared a keen interest in astrology. Bate kept his promise and sent William his 'magistralis compositio astrolabe'. From William, he received – in return (?) – the newly translated Ptolemy. One finds quotations from this Ptolemy translation in three of Bate's works: the *Liber servi Dei* (written in 1280); the introduction to Bate's translation of Ibn Ezra's *De mundo uel celo* (dated 1281); and in Part 19 of Bate's vast scientific-philosophical encyclopaedia, *Speculum divinorum*. This encyclopaedia was completed between the years 1301 and 1305, but Bate had probably accumulated much of his material in the preceding years.[3]

As Bate tells us in the preface of his *Speculum divinorum*, he intends to quote his authorities *secundum litteram* so that their true meaning may be clear and free from all distorting interpretations.[4] Therefore he likes to compare texts when he has access to different translations, and especially when he can compare a translation from the Arabic with a translation made directly from the Greek. Bate knew, of course, the *Quadripartitum* in the translation of Plato of Tivoli, which had become the standard text, but he is pleased to quote besides – and against it – Moerbeke's translation from the Greek (as in *Liber servi Dei*, p. 72,14-15 [quoted below as T.9]: 'alia uero translatio que de Greco melius habet hoc modo'). In the introduction to his translation of Ibn Ezra (see T.10), Bate even quotes three different translations of the same text, giving preference to the translation 'que scilicet im-

[3] The edition of the *Speculum divinorum et quorundam naturalium* is being undertaken by the De Wulf-Mansion Center (Leuven). So far the following volumes have been published. Pars I with introduction (ed. E. Van de Vyver, 1960); Partes II-III (ed. E. Van de Vyver, 1967); Partes IV-V (ed. C. Steel, 1993); Partes VI-VII (ed. C. Steel and E. Van de Vyver, 1994); Partes VIII-X (in preparation); Partes XI-XII (ed. H. Boese and C. Steel, 1990); Partes XIII –XVI (ed. G. Guldentops, 2002); Partes XVII-XIX (in preparation); Partes XX-XXIII (ed. C. Steel and G. Guldentops, 1996). We have checked all published volumes and the provisional edition of the two volumes in preparation. The quotations from the *Quadripartitum* are all found in Part 19.

[4] 'Sermones denique philosophorum et aliorum quorumlibet huic insertos operi, prout secundum litteram iacent in originalibus, ... ut mens illorum et sententia, nullatenus a quoquam depravata nec in aliquo variata, certior habeatur aut clarius pateat auctoribusque suis imputentur singula, sive bene dicta sive male.' (*Ep. Introd.*, p. 4,37-43 ed. E. Van de Vyver).

mediate de Greco interpretata'. Given Bate's interest in astrology and his overall preference for Greco-Latin translations, one might expect more quotations from Moebeke's translation of Ptolemy's *Quadripartitum*. However, as is clear from his own *Nativitas*, Bate seems to have relied mainly on his Hebrew and Arabic authorities (Albumasar, Ibn Ezra, and many others). Still, his Ptolemy quotations are extremely valuable. Not only do they offer a strong external argument for the attribution of this Ptolemy translation to Moerbeke (while providing a *terminus ante quem*), but they also contribute to an evaluation of the text tradition of the translation of Ptolemy's work. We give here a complete survey of Bate's quotations of Ptolemy.

(a) Liber serui dei[5]

T.1. Saturnus autem dissinteriam significat secundum quod dicit Ptolemeus 3° Quadripartiti, capitulo 14 (p. 8,11-12). Vide *Iud.* III.816-7: 'Vniuersaliter autem Saturnus quidem (...) facit (...) dissentericos'.

T.2. Sol igitur in proposito gradum ascendentis aspiciens uitam confirmat. Vnde Ptolemeus, 2° Quadripartiti, capitulo 9°: *Sol quidem enim et luna ordinatores et quasi presides sunt aliorum, cum sint ipsi causa eius quod secundum actum totius et dominationis astrorum, et adhuc fortitudinis et debilitatis dominantium* (p. 15,19-22). Vide *Iud.* II.520-3.

T.3. Item natiuitas uernalis est de qua Ptolemeus in Quadripartito 4° *que a uernali equinoctio ad estiuam uersionem facit boni coloris, bone magnitudinis, bone habitationis, bonorum oculorum, quod plurimum homines in humido et calido* cuiusmodi complexio apparet in hoc nato (p. 21,16-20). Vide *Iud.* III.701-3.

T.4. Vnde Ptolemeus in Quadripartito capitulo 8° secundi: *equinoctialia quidem* signa *sacris et religionibus circa deos significationes insinuant* (p. 26,8-10). Vide *Iud.* II.498-9.

T.5. De quo Ptolemeus 2° [*sic!*] Quadripartiti in fine sic ait: *Attingentes quidem enim uel accedentes ad medium celi maxime uirtuosi sunt, secundo autem quando in ipso orizonte fuerint* etc. (p. 27,1-2). Vide *Iud.* I.962-3.

T.6. Vnde Ptolemeus 3° Quadripartiti capitulo de qualitate anime: *signa quidem igitur tropica,* Mercurium continentia, *communiores faciunt animas popularium et ciuilium rerum concupiscitiuas, adhuc autem amatrices glorie et deo supplices, bene aptasque et boni motus quesitiuas et astrologicas et dominatiuas* (p. 30,24-27). Vide *Iud.* III.873-6.

T.7. Primo *de coniugio,* hoc est *de conuictu uiri et mulieris secundum leges,* ut dicit Ptolemeus, qui lunam principaliter in hoc negotio *considerare* precipit (p. 47,15-16); iterum luna Mercurio applicata domino inquam domus in qua est significat iuxta Ptolemeum mulieres *perspicaces et acutas* siue boni intellectus (p. 48,16-18); Item Ptolemeus luna in angulo *occidentali* significat natum *tarde nubere* (p. 48,23-24). Vide *Iud.* IV.177-9; 190-1.

T.8. *Iupiter* enim, ut dicit Ptolemeus, *dignitates adducat per fidem et prefacturas* [*sic*] *et per sacerdotalia* (p. 65,12-13). Vide *Iud.* IV.34-5.

---

[5] The *Liber servi Dei* is preserved in four manuscripts. We quote the text according to a provisional edition by Van de Vyver (based on two manuscripts), preserved in the De Wulf-Mansion Center in Leuven.

T.9. Et hoc est quod dicit Ptolemeus in Quadripartito capitulo tertio primi inquirens quod 'huius artis maximas uires Egyptii monstrauerunt quoniam in omnibus libris medendi scientiam stellarum pronosticationi coniunxerunt, propterea quod hec huiusmodi cognouerunt nec in eodem hora illa insimul coniungerent ut incantationes et medicinas facerent ad delendum impedimenta presentia seu futura, generalia seu particularia per corpus circumdans accidentia nisi quoniam opinati sunt quod inde prouent<ur>a nec mutentur nec conuertantur.' Alia uero translatio que de Greco melius habet hoc modo: *Egyptii qui maxime talem artis uirtutem produxerunt, coniunxerunt ubique medicinalem cum pronosticatione per astronomiam; non enim umquam utique instituissent in contrarium regimina aliqua et preseruatoria contra dispositiones continentis superuenientes aut presentes uniuersales et particulares, si qua ipsis mutabilitatis et ineuitabilitatis euentorum inesset opinio* (p. 72,7-20). Vide *Iud.* 1.279-84.

(b) De mundo vel celo

Bate added a long introduction to his translation of Ibn Ezra's *De mundo*, also known as *Liber de reuolutionibus annorum*.[6] This translation was completed in 1281. In Bate's introduction, we find two quotations from Moerbeke's translation of Ptolemy. Interestingly, in the first passage Bate first quotes two different translations from the same text made from the Arabic. The second comes from Plato of Tivoli; we have not been able to identify the first.[7] After these two, Bate quotes the translation that had been made 'directly from the Greek'.

T.10. Constat autem quod in Almagesto non inuenitur illud dictum, sed eius contrarium. Igitur si usquam inueniri debeat, hoc esse deberet in Quadripartito, capitulo quidem 11 secundi libri, ubi tractatur de initio anni. Illud autem quod ibi scriptum est, est huiusmodi secundum unam translationem: 'Quia ergo uel magis conueniens nihil naturali rationi propinquius habere obseruatione deprehendi potest hiis quatuor punctis anni principia relinquuntur, nos itaque quid singulis accidat anni temporibus prescire uolentes in eisdem principiis Solis et lune conuentum aut oppositionem utrum scilicet istorum solis ingressum in illius quarte principio preueniat considerare oportet. Si itaque eorundem conuentum aut oppositionem committetur eclipsis efficacior erit significatio. Quod si antequam sol Ar(ietis) principium ingrediatur contingat in totum uer et uernali qualitate. Si ergo illa dilatabitur in cancro estatem in libra autumnum, sed in capricorno hyemem totam occupabit, temporum itaque alternatio et generalis eorum proprietas omnesque eorum status causam a sole specialiter assumunt.' Alia quidem translatio

---

[6] The text is included in the Venice edition (P. Liechtenstein, 1507) of the Latin translations of the works of Ibn Ezra (*Abrahe Avenaris in re iudiciali opera*). Some errors may be due to the edition.

[7] We thank Charles Burnett and David Juste for having checked this passage in the three other translations from the Arabic: they do not have the text quoted by Bate. Charles Burnett also checked the translation in the margins of Tiburtinus' translation in *Vat. Reg. lat.* 1285, to no result. He notices: 'What is curious is that the terminology of Bate's mysterious translation corresponds with Hermann of Carinthia: *conventus* and *oppositio* for conjunction and opposition (other translations have *coniunctio* and *oppositio* or *coniunctio* and *preventio*)'.

sic habet: 'Et quod magis conueniens naturalique rationi propinquius in obseruatione rerum anni deprehendimus, est istorum quasi 4 principiorum obseruatio, necnon coniunctionum et preuentionum solis et lune, que per dicta tempora precedentes prope ipsa fuerint maxime, aut in quibus fuerint eclipses, ita ut per principium quod ex solis essentia in Ar(iete) deprehenditur uernalem qualitatem cognoscimus et per initium ex eiusdem essentia in Cancro deprehendimus qualitatem estiualem, per principium autem quod ex ipsius in Libram ingressu cognoscere qualitatem autumnalem. Initium uero quod per eiusdem introitum in Capricorno accipitur hyemalem qualitatem demonstrat, propterea quod temporum generales qualitates et eorum omnes modi non sunt nisi per solem.' Tertia uero translatio que scilicet immediate de Greco interpretata hec tenet: *Conuenientius autem mihi uidetur quod magis naturale ad annuales considerationes quatuor principiis uti, obseruando propinquissime sibi prius factas* coniunctiones *solis et lune* coniunctiones coniunctionales *aut plenilunares, et harum maxime rursum eclipsatas, ut a principio quidem quod est penes Arietem* ut [*error pro* uer] *quale sit, consideremus, ab eo quod est penes Cancrum estatem, ab eo* autem *quod est penes Libram autumnum, ab eo autem quod est penes Capricornum hyemem. Vniuersales quidem enim temporum qualitates et consistentias sol facit.* Ex his igitur nisi littere uiolentia fiat, non potest elici illud dictum Auenare (f. LXXVI rv). Vide *Iud.* II.688-94.

T.11. Vnde in primo Quadripartiti: *estimo* aut *de ipsa pronosticatione et si non ex toto sit sine errore tamen quod possibile est de ipsa maximo studio dignum uidetur* (f. LXXVII r). Vide *Iud.* I.274-75.

(c) Speculum diuinorum

T.12. Ptolemeus autem in primo Quadripartiti: *estimo,* inquit, *sicut de ipsa prognosticatione et si non ex toto sit errore, tamen* quidem [*quod? cf.* T.11] *possibile est de ipsa maximo studio dignum uidetur.* Vnde *conueniens utique erit neque quoniam fallere quandoque contingit talem prognosticationem, totum ipsius perimere, sed ut in tam magnis et diuinis pronuntiationibus acceptare et amabile existimare quod possibile est* (...); *et sicut de medicis, cum requirunt aliquem de ipsa egritudine et de proprietate infirmi non querelamus interrogantes, ita et hic. Egiptii* nimirum *qui maxime artis* astrologie *uirtutem produxerunt, coniunxerunt ubique medicinalem cum prognosticatione per astronomiam* (Pars 19,9). Vide *Iud.* I.275-6; 164-71; 279-81.

T.13. Vnde Ptolemeus in Quadripartito de configurationibus loquens astrorum super monstruosis huiusmodi: *si Mercurius,* inquit, hiis *testificetur cum hiis quidem* orbationibus nati *responsa dantes* erunt *per hec lucratiui* (Pars 19,10). Vide *Iud.* III.370-371.

T.14. Hinc enim Ptolemeus in 3° Quadripartiti: *Venus,* inquit, *per pronunciationem deorum et responsionum lesuras quidem modo quodam decentes et graciosas instituit, passiones autem per ea que a diis medicamina facile consolabiles.*' In alia uero translatione (...). Deinde posterius: *demoniaci* quidem fiunt, ait, *et humidi capitis, quando malignantes sic se habentes obtineant lunam existentem in apparentia, Saturnus quidem festinantem ad coniunctionem, Mars autem festinantem ad plenilunium*; et infra: *coaptatis autem benificis Ioue et Venere in occidentalibus partibus existentes ipsi. Beneficis autem in orientalibus angulariter, curabiles quidem, facile tamen productibiles in exemplum faciunt passiones; sub Ioue quidem per curas medicinales et aut per regimen diete aut farmaco, sub Venere autem per responsa et auxilia a diis.* Et infra: *secundum demonizationes autem aut humorum molestationes reuelationibus et prophetiis et plagis* *et similibus exemplarium. Singulariter autem et locorum figuram continentium que quidem solis et que Martis ad manias maxime operantur, que autem Iouis et Mercurii ad epylempsias, que autem Veneris ad dei illustrationes et elocutiones.* Secundum aliam uero translationem ... (Pars 19,19). Vide *Iud.* III.851-3; 1131-3; 1137-41; 1147-51.

T.15. Vnde Ptolemeus in 4° sui Quadripartiti capitulo 5: *si locum*, inquit, *operatiuum luna obtineat a coniunctione cursum faciens cum Mercurio, in Sagittario* quidem *et Piscibus in mortuis diuinatores et demonum commotores, in Virgine autem et Scorpione magos, astrologos, pronuntiatores, prescientias habentes, in Libra autem et Ariete et Leone raptos, sompniorum iudices, coniuratiuos. Magnitudinem autem* huiusmodi *operationum ex uirtute astrorum dominantium oportet* accipere. *Orientales quidem enim existentes* planete *aut in angulis faciunt operationes principales, occidentales autem aut decidentes ab angulis secundarias operationes, et a beneficis quidem magistrati magnas et gloriosas et lucrosas et sine offensione et complacentes, a maleficis autem humiles, inglorias, sine lucro et incertas* (Pars 19,19). Vide *Iud.* IV.157-169.

As we can see, the quotations in Bate confirm the quality of the Latin tradition of the *Iudicialia*. Of course, Bate sometimes adapts and shortens the text he quotes. Where we have the two Ptolemy manuscripts, Bate generally confirms the reading of *Ve* against the errors of *Ox*. Where only *Ox* remains as a witness, Bate sometimes offers a better reading, but he also has some errors (some of which may be due to errors in the edition we quote, as in T.10). Most important is T.14, where an omission in *Ox* can be corrected thanks to the Bate quotation.

Confirmation of Ve against Ox: I.275 tamen *Ve* Bate] tantum (*abbr.*) *Ox* 284 inesset *Ve* Bate] inde esset *Ox* 963 accedentes *Ve* Bate] accid- *Ox* 964 fuerint *Ve* Bate] fuerunt *Ox* II.499 circa *Ve* Bate] quibus *Ox*

Confirmation of Ox against Ve: I.275 uidetur *Ox* Bate] uidet *Ve*

Errors in Bate against Ve Ox: I.283 immutabilitatis *Ve Ox*] mutabilitatis Bate

Correct readings in Bate against Ox: III.852 modo Bate] moto *Ox* 853 consolabiles Bate] -lationes *Ox* 874 concupiscitiuas Bate] -scentias *Ox* 876 diuinatiuas *ed.*] dn̄atiuas *Ox*, dominatiuas Bate 1132-3 festinantem ad coniunctionem, Mars autem Bate] *om. Ox* 1133 plenilunium Bate] plenissimum *Ox* 1140 faciunt Bate] *om. Ox* et aut Bate] *inv. Ox* 1147 demonizationes Bate] -ionibus *Ox* IV.157 luna Bate] lumina *Ox* 163 talia] taliū *Ox* 166 principales Bate] et *praem. Ox* 168 a Bate] et a *Ox*

On II.692-3, *Ox* has an omission ('ab eo uero quod est penes Chelas'), which is corrected by the same hand in the margin. As one can see in T.10, Bate quotes the full text, but with one difference: he has *Libram* instead of *Chelas*. Moerbeke uses both *Libra* and *Chele* to render the Greek Χηλαί. Did Bate replace *Chelas* with the more common *Libram*?

Errors in Bate against Ox: II.690 solis] coniunctiones *praem.* Bate coniugationes] coniunctiones coniunctionales Bate pleniluniares] -nares Bate 691 eclipticas] eclipsatas III.702 habitudinis *Ox*] habitationis Bate 703 habentes *Ox*] homines Bate 1141 farmacie *Ox*] farmaco Bate (D), farmato Bate (CDE) 1150 cooperantur *Ox*] operantur Bate IV.162 coniuratores *Ox*] –iuos Bate

2. AGOSTINO NIFO

The Italian humanist Agostino Nifo (1469/70-1538) admired Ptolemy's astrological treatise and often defended him against his critics. In 1513, Nifo published *Ad apotelesmata Ptolemaei eruditiones* in Naples.[8] This is a commentary on the first book of the *Apotelesmatica*, starting from chapter four. Nifo divided his commentary into three books: 1. *De potestate planetarum* (ch. 4-17); *De iurisdictionibus stellarum* (ch. 18-22); 3. *De opulentiis stellarum* (ch. 23-24). Nifo does not use the standard translation of the *Quadripartitum* by Tiburtinus, of which there were already two editions which he could have used (1484, 1493), but rather made his own literal translation of the lemmas of the Greek text (*uerbum uerbo reddens*) because he wanted to avoid the errors attributed to Ptolemy by scholars using the Arabo-Latin translation (*interpres Arabs*). He also relied on the anonymous Greek commentary (*Graecus sine nomine*). Nifo's translation is the first translation from the Greek after Moerbeke's. He did, however, use Moerbeke's version while producing his own translation. To demonstrate Nifo's dependence on Moerbeke, we have analyzed his translation of chapter 23. We first give the Greek text in Hübner's edition, then Moerbeke's version confronted with Nifo's translation. Literal correspondences between the two Latin translations are in italics, while Nifo's adaptations of Moerbeke's text with the use of alternative terms are underlined; his additions are printed in bold.

Περὶ προσώπων καὶ λαμπηνῶν καὶ τῶν τοιούτων.

[1] Αἱ μὲν οὖν συνοικειώσεις τῶν τε ἀστέρων καὶ τῶν δωδεκατημορίων σχεδὸν ἂν εἶεν τοσαῦται. λέγονται δὲ καὶ ἰδιοπροσωπεῖν μὲν ὅταν ἕκαστος αὐτῶν τὸν αὐτὸν διασῴζῃ πρὸς ἥλιον ἢ σελήνην σχηματισμόν, ὅνπερ καὶ ὁ οἶκος αὐτοῦ πρὸς τοὺς ἐκείνων οἴκους, οἷον ὅταν ὁ τῆς Ἀφροδίτης λόγου ἕνεκεν ἑξάγωνον ποιήσῃ πρὸς τὰ φῶτα διάστασιν, ἀλλὰ πρὸς ἥλιον μὲν ἑσπέριος ὤν, πρὸς σελήνην δὲ ἑῷος ἀκολούθως τοῖς ἐξ ἀρχῆς οἴκοις.

[2] λαμπήναις δ' ἐν ἰδίαις λέγονται εἶναι καὶ θρόνοις καὶ τοῖς τοιούτοις, ὅταν κατὰ δύο ἢ καὶ πλείους τῶν προεκτεθειμένων τρόπων συνοικειούμενοι τυγχάνωσι τοῖς τόποις ἐν οἷς καταλαμβάνονται, τότε μάλιστα τῆς δυνάμεως αὐτῶν πρὸς ἐνέργειαν αὐξομένης διὰ τὸ ὅμοιον καὶ συμπρακτικὸν τῆς τῶν περιεχόντων δωδεκατημορίων ὁμοφύλου οἰκειότητος.

[3] χαίρειν δέ φασιν αὐτούς, ὅταν κἂν μὴ πρὸς αὐτοὺς ᾖ ἡ συνοικείωσις τῶν περιεχόντων ζῳδίων, ἀλλὰ μέντοι πρὸς τοὺς τῶν αὐτῶν αἱρέσεων, ἐκ μακροῦ μὲν μᾶλλον οὕτως γινομένης τῆς συμπαθείας, κοινωνούσης δὲ ὅμως καὶ κατὰ τὸν τοιοῦτον τρόπον τῆς ὁμοιότητος, ὥσπερ, ὅταν ἐν τοῖς ἠλλοτριωμένοις καὶ τῆς ἐναντίας αἱρέσεως τόποις καταλαμβάνωνται, πολὺ παραλύεται τῆς οἰκείας αὐτῶν δυνάμεως, ἄλλην τινὰ φύσιν μικτὴν ἀποτελούσης τῆς κατὰ τὸ ἀνόμοιον τῶν περιεχόντων ζῳδίων κράσεως. (1.1211°-35°)

---

[8] *Ad Sylvium Pandonium Boviani episcopum Eutichi Augustini Niphi philothei Svessani ad apotelesmata Ptolemaei eruditiones.* Impressum Neapoli: Per Petrum Mariam de Richis Papiensem, 1513.

Capitulum 23m
De faciebus et claritatibus et talibus

[1] *Conuenientie quidem igitur* astrorum *et* 12 signorum *fere utique* erunt *tot. Dicuntur* autem et proprias habere facies, *quando unusquisque* ipsorum *eandem* conseruat *configurationem ad solem* uel *lunam, quam quidem et* domus ipsius *ad illorum* domos, puta *quando Venus uerbi gratia* sextilem *distantiam* habuerit *ad* luminaria, sed *ad solem quidem uespertina* existens, *ad lunam* autem *matutina,* conformiter *hiis que* a principio domibus. [2] *Dicuntur autem in* claritatibus *propriis* esse et thronis et talibus, *quando* secundum duos aut plures preexpositorum *modorum* conuenientes fuerint *locis in quibus* deprehenduntur, *tunc maxime* uirtute ipsorum *ad* effectum augmentata *propter similitudinem et* cooperationem *proprietatis* conformis signorum.

[3] *Gaudere* autem aiunt *ipsos, quando* etiam si *non ad ipsos* fuerit conuenientia continentium *signorum,* sed *tamen ad eos qui sunt earundem* partitionum; a remotis *quidem magis sic facta* compassione, *communicante* autem *tamen secundum talem modum similitudine.* Quemadmodum quando *in alienis* et *in locis contrarie* partitionis *deprehenduntur, multum dissoluitur* de propria *ipsorum* uirtute, cum efficiat *quandam mixtam naturam* ex ea que *secundum dissimilitudinem continentium signorum complexione.*

Liber tertius qui est de opulentiis stellarum
De faciebus et faculis caeterisque id genus

[1] *Conuenientiae quidem igitur* stellarum *et* dodecatemoriorum *fere utique tot* sunt, sed se quidem proprie uultispicere *dicuntur,* quoties *unaquaeque* earum *eandem ad Solem* aut *Lunam configurationem* seruat, *quam quidem et* eius domicilium *ad illorum* domicilia, ut *quando Venus u.g.* per rationem sexangulam *ad* lumina *distantiam* seruat, uerum *ad Solem quidem uespertina, ad Lunam* uero *matutina,* consequenter *his quae* prius domiciliis. [2] *In* faculis *uero propriis* atque soliis caeterisque id genus *esse dicuntur* **stellae,** *quando* duobus uel pluribus dictorum *modorum* iis ipsis *locis* conueniunt, *in quibus* constituuntur. Potestate **namque** earum *ad* agendum *tunc maxime* adaucta *propter similitudinem et* coactionem unigeniae **ipsorum continentium** dodecamoriorum *proprietatis,* **proprio in solio constitui atque splendere dicuntur.**

[3] *Gaudere* uero *ipsos* dicunt: *quando* quamuis *non ad ipsas* sit *signorum* in quibus continentur similitudo, est *tamen ad eas quae earumdem sunt* factionum, *facta quidem* de longinquo *magis sic* mutua receptione, *communicante tamen* **et** *secundum talem modum similitudine* ipsa, ueluti quum *in alienis* ac *contrariae* factionis *locis deprehenduntur, multum dissoluitur* peculiaris *ipsarum* proprietatis, **aliam** *quandam naturam mistam complexione secundum dissimilitudinem continentium signorum* perficiente.

As can be seen particularly in §§ 1 and 3, Nifo takes Moerbeke's translation as the substrate of his own translation. This is already evident in the first sentence, with the series of particles *quidem igitur* and *fere utique*, which a humanist writer would generally avoid, and the use of *conuenientiae* to translate συνοικειώσεις (see also *conueniunt* replacing *conuenientes fuerint* in 1220°-1°, but in 1228° Nifo replaces *conuenientia* with *similitudo*). Other obvious examples are πολὺ παραλύεται *multum dissoluitur*; ἐν τοῖς ἠλλοτριωμένοις *in alienis*; τῶν περιεχόντων κράσεως *continentium complexione*; φύσιν μικτὴν *mistam naturam*; καταλαμβάνονται (1232°) *deprehenduntur* (but *constituuntur* in 1223°); οἰκειότης *proprietas*. An interesting trace of his use of Moerbeke is Nifo's phrase *uerbi gratia per rationem* to render λόγου ἕνεκεν (1218°). Nifo first writes with Moerbeke *uerbi gratia*, and then adds his own translation *per rationem*, a needless duplication.

The dependence on Moerbeke is not immediately evident because Nifo radically reformulates the Moerbeke version in an attempt to make it more idiomatic classical Latin. Thus, he replaces throughout his translation *autem* by *uero*, *etiam si* by *quamuis*, *quemadmodum* by *ueluti*, *talibus* by *caeteris id genus*, *uirtus* by *potestas*, *domus* by *domicilium*, *luminaria* by *lumina*, *astra* by *stellae*, *conformiter* by *consequenter*, *a principio* by *prius*, *aiunt* by *dicunt*, *a remotis* (a characteristic phrase of Moerbeke's) by *de longinquo*. Moerbeke's *ad effectum augmentata* becomes Nifo's *ad agendum adaucta*. And of course, Nifo avoids Moerbeke's *fueri(n)t*. For the astronomical vocabulary, the humanist prefers more technical terms, thus replacing the ambiguous *signa* with a learned transliteration, *dodecatemoria* (but even so, Nifo once keeps Moerbeke's *signorum* in § 3). To render λαμπηνῶν, he opts for *faculis* ('torches'), which is not better than Moerbeke's *claritatibus*: as noted previously (see p. 32), the correct meaning is 'chariots'. But even with the replacement of so many terms, the structure and rhythm of Nifo's translated sentence remains that of Moerbeke, at least in §§ 1 and 3. See, for instance, *communicante ... tamen secundum talem modum similitudine*.

This being said, it is evident that Nifo was also working with a Greek manuscript of Ptolemy's text, which probably belonged to the γ family. Thus, Nifo adds at the end of § 2, *proprio in solio constitui atque splendere dicuntur*, which corresponds to the addition in Σ βγ ἰδιοθρονεῖν καὶ λάμπειν λέγονται. He adds after *caeterisque id genus* (τοῖς τοιούτοις 1124°) *esse dicuntur stellae*, which corresponds to λέγονται οἱ ἀστέρες in γ. He has *namque* to translate γάρ, which is added in Σ MSγ after τότε (1224°). Nifo also corrects some small omissions in Moerbeke's version. See 1225° adding *ipsorum continentium* corresponding to τῶν περιεχόντων; 1230° adding *et* after *tamen* corresponding to καί; 1234° adding *aliam* corresponding to ἄλλην. Finally, Nifo corrects in the last line, in 1234-5° (942-3) *cum efficiat ... ex ea que complexione*, rendering the Greek absolute genitive (which Moerbeke did not catch) as *complexione ... perficiente*. As he made other interventions in the text, Nifo probably also corrected these small omissions himself. One cannot exclude that he had access to a better manuscript than *VeOx* where those omissions were not present. However, unfortunately, Nifo's adaptations in the text are too intrusive to make use of his testimony in our edition.

There are also long quotations from the *Quadripartitum*, again inspired by Moerbeke's translation, in *De nostrorum calamitatum causis liber* (Venice 1505). In his preface Nifo says that he translated his quotations *e Greco codice Ptholomei* (f. 2v), and there is no doubt that he used a Greek codex to make his translation. Nevertheless, here again, Nifo seems to have had Moerbeke's translation at his disposition, though he rewrites his medieval source radically. See, however, the following quotation, which is certainly an adaptation of Moerbeke's version:

> δεῖ δὴ λαμβάνειν τὴν τῆς ἐκτροπῆς προγινομένην ἔγγιστα συζυγίαν, ἐάν τε σύνοδος ᾖ ἐάν τε πανσέληνος, καὶ τὴν μοῖραν ἀκριβῶς διασκεψαμένους συνόδου μὲν οὔσης τὴν ἀμφοτέρων τῶν φώτων, πανσελήνου δὲ τὴν τοῦ ὑπὲρ γῆν αὐτῶν ὄντος ὄντος ... ἰδεῖν [III.137°-41°]

Moerbeke III.111-3	**Nifo f. 24r**
Oportet itaque *accipere coniugationem* ante *partum* factam *proxime, siue coniunctionalis* sit *siue* pleniluniaris, *et* gradu *diligenter considerato* (*coniunctione* quidem existente *amborum luminarium, plenilunio* autem gradum *super terram* existentis ipsorum) ... uidere	*oportet* igitur *accipere coniugationem* que *proxime partum* precessit, *siue coniunctio* ea fuerit *siue* oppositio [*sic*], *et considerata diligenter* portione, si *coniunctio* fuerit *amborum luminarium* sin *plenilunium* eius quod *supra terram* luxerit intuebimur, (et reliqua)

Let us end with a quotation in *De falsa diluuii prognosticatione* (Augsburg 1520), which reveals again Nifo's dependence on Moerbeke.

> συνίστανται δὲ δυνάμεις ἐπί τε τούτου καὶ τῶν λοιπῶν καὶ διὰ τῶν ἐν τοῖς πρὸς ἥλιον καὶ σελήνην σχηματισμοῖς παρατηρήσεων, ἐπειδήπερ οἱ μὲν οὕτως οἱ δὲ οὕτως τὴν τοῦ περιέχοντος κατάστασιν ἐπὶ τὸ μᾶλλον ἢ ἧττον συντρέποντες φαίνονται. [I.403°-7°]

Moerbeke I.314-7	**Nifo f. 7r**
Constant autem uirtutes *in hoc et in reliquis per obseruationes in configurationibus* ad *solem* et *lunam*, quoniam hii *quidem* isto modo, hii autem illo, *continentis* consistentiam *uidentur* in *maius* aut *minus immutare*	In primo Apotelesmatum alio modo Ptolomaeus inquit: Et deprehenduntur uires *in hac* (*sic*), *et in reliquis per obseruationes* illas, quae fiunt *in figurationibus* quas habet *luna* cum *sole* quando *haec quidem* sic, aliae uero aliter *continentis* habitum secundum *magis* et *minus mutare uidentur*

Much more research is needed on Nifo's interpretation and his use of Ptolemy. It is evident, however, that he deserves a place among the Renaissance translators of the *Quadripartitum*.

CHAPTER IV
THE GREEK TEXT AND THE TRANSLATION

A. THE TRADITION OF THE GREEK TEXT

It is not easy to situate the Greek model of Moerbeke's translation in the tradition of the Greek text. In fact, all extant manuscripts of the *Apotelesmatica*, with the exception of some excerpt-manuscripts (L, y and the *Vat. gr.* 1291, which contains Ptolemy's Handy tables), were copied after the translation was made and their mutual relations are complicated because many corrections were introduced in later manuscripts from a variety of sources, leading to contaminated versions. Much better is the tradition of Ptolemy's major astronomical work, the *Syntaxis mathematica*, of which there still exist two manuscripts of the 9th century, one of which is the *Vat. gr.* 1594 'quo nullum pulchriorem elegantioremque unquam uidi' (Heiberg)[1]. This manuscript now contains the *Syntaxis* with prolegomena, *De apparentiis* (incomplete), *De criterio* and the first book of the *Hypotheses*, but it may originally have included other works of Ptolemy.[2] At present, the oldest preserved copy of the *Apotelesmatica* is the *Vat. gr.* 1038, a manuscript of the end of the 13th or beginning of the 14th century.

1. THE VATICANUS GRAECUS 1038 AND RELATED MANUSCRIPTS

The *Vat. gr.* 1038 (V), a parchment codex of large format, contains in its first part (ff. 1-129v) the mathematical works of Euclid: *Elements* II 8-XV (first quaternion is lost), *Optica vetera*, *Phaenomena* and *Data* (with Marinus' commentary), followed by pseudo-Heron, *De mensuris*; and in its second part (ff. 137-384) a collection of works of Ptolemy: *Syntaxis* (137r-323r); *Psephophoria* with additional material (323v-333v); *De apparentiis* (336v-342r); *De criterio* (342r-345v); *Hypotheses* I (346v-350v); *Apotelesmatica*, mutilated at the end (352r-384v).[3] The codex is written by one and the same professional copyist.[4] Fr. Boll was convinced that the whole Ptolemy collection in V was copied from one and the same model, most probably a manuscript of the 9th century related to *Vat. gr.* 1594 (B) (if not that manuscript

[1] See Heiberg, vol. I, p. IV.

[2] The *Vat. gr.* 1594 had originally 43 quires; of these 2-3 and 37-40 are lost. On this manuscript see Acerbi (forthcoming).

[3] For a description see Heiberg, vol. II, p. XXIV. See also Acerbi, sect. 2.

[4] See Heiberg, *Euclides. Elementa*, vol. V.1, p. VII-VIII: 'totus codex eadem manu eleganti calligraphi exercitati scriptus est compendiis paucis et fere in extremis versibus usurpatis'.

itself). As Boll observed, notwithstanding its many particular errors, V offers a faithful, almost servile copy of its lost model, without any attempt to improve or correct the text: 'eine getreue Kopie einer sehr guten Vorlage'; 'sie geht manchmal bis zur Gedankenlosigkeit'.[5] Boll is confident that, if we abstract from its particular errors, we may find in this copy the state of the text as it was in the 9th century (p. 85). Heiberg, however, concluded from his investigation of the textual tradition of the *Syntaxis* that V used two different models for that text. In the second part of the *Syntaxis,* the copyist of V clearly had B as model and even copied this model scrupulously in the smallest errors: 'cum librarius imperitus omnes minimos errores archetypi religiose servaverit'.[6] However, in the first part of the *Syntaxis,* the copyist used another model, the *Vat. gr.* 184, a bombycin manuscript which can be dated to the years 1269-70.[7] If Heiberg is right, the dependence on the *Vat. gr.* 184 is a strong argument to date V late in the 13th century, if not in the early 14th century. In the minor astrological works, V depends again on B according to Heiberg. However, regarding *De criterio,* its editor, Fr. Lammert, noticed that, though V has many particular readings in common with B, it has also so many stupid errors that it would be difficult to admit that it was copied directly from such a beautiful manuscript as B.[8] These observations should make us cautious in speculating about the lost model of the text of the *Apotelesmatica* in V. This model was certainly not the *Vat. gr.* 1594. Acerbi's recent study shows that the text of the *Apotelesmatica* is too large to fit within the lost quires of that manuscript.[9] Moreover, there are clear indications that V did not copy the *Apotelesmatica* directly from a 9th century manuscript, but from an intermediate model deriving from that source.[10]

Deriving from the same hyparchetype as V are the excerpts in the *Laurentianus gr.* 28,34 (L). This manuscript, copied around 1060-70 in Constantinople, maybe in Psellus' circle, contains an astrological collection that was probably composed in the 9th century.[11] It is the oldest witness of the anonymous commentary on the *Apotelesmatica* (on ff. 28v-58r). The scholar who composed the anthology in the 9th

[5] See Boll (1899) 83-84, *cf.* Hübner, p. XVIII: 'hunc codicem et aetate et qualitate praestantissimum a librario admodum indocto fideliter ex archetypo codicis Vaticani gr. 1549 litteris uncialibus exarato descriptum esse F. Boll docuit'.

[6] Heiberg, vol. II, p. XLVI.

[7] The manuscript can be dated, since an annotation by the hand of one of the main copyists is dated 1269-70. We owe this information to F. Acerbi.

[8] Lammert (1952) V.

[9] See Acerbi (forthcoming).

[10] After the completion of the introduction we discovered that the Oxford manuscript *Bodleian Library,* Auct. T.5.4 (U) was copied from the same model as V, but independently of it. This model was not the hyparchetype ψ, from which also L derives, but an intermediary copy. See addendum to this chapter below.

[11] On this manuscript see Caballero-Sánchez (2013a) 88-92.

century, probably found this commentary in an ancient uncial manuscript in the form of scholia around the Ptolemaic text.[12] He may have copied from the same manuscript also a series of extracts of the Ptolemaic text, which he inserted in his anthology between extracts of Hephaestio, Paul of Alexandria and other astrological authors. On ff. 104r-107r we find in L a first series of extracts from book I of the *Apotelesmatica* with the title Περὶ τῆς τῶν πλανωμένων ἀστέρων φυσικῆς ἰδιοτροπίας καὶ τῶν ἀπλανῶν. Another hand added Πτολεμαίου. This section contains chapter 4 (starting with the concluding paragraph of chapter 3: 380° ἀρξάμενοι), chapter 5 (without title), chapter 8 (without title), chapter 9, chapter 10, chapter 11 (including the opening of what is in Hübner ch. 12): I.380°-445°; 488°-745°. After a series of extracts from other authors, one finds on ff. 146r-152v a new series of extracts, for the most part from Ptolemy: II, ch. 12 (936°-993°), I, ch. 12 (748°-775°); I, ch. 14 (810°-839°); I, ch. 18-19-20 (876°-1018°); I, ch. 21 last part with *tabula secundum Ptolemaeum* (1139°-1185°); follows *tabula secundum Egyptios* (1073°-1087°); I, ch. 22-23-24 (1190°-1212°; 1220°-1226°; 1238°-1265°); follows Paulus, *El. Apot.*, 5, p. 18,1-10 with *tabula*; Ptolem., II, ch. 13 (994°-1061°). A comparison of the excerpts in L with the corresponding chapters in V makes it possible to reconstruct the state of the hyparchetype of V and L (ψ), leaving out the many particular errors of V and L.

Another collection of extracts from the *Apotelesmatica* is found in the *Vindobonensis phil. gr.* 115 (y) of the 13th century. This manuscript contains on ff. 7r-16v excerpts from chapters 7 and 9 and chapters 10 to 14 (incomplete) of the second book. This collection goes back to the same lost hyparchetype (ψ) from which both V and L ultimately derive.[13] This lost hyparchetype had an excellent text with only minor errors. As we shall see, it was also used as model for the Paraphrase of 'Proclus'.

2. THE TRADITION BESIDES V

Besides V, the fragmentary witnesses (L and y) associated with it, and its later copies, one can distinguish according to Boll–Hübner three other classes of manuscripts which, in their view, derive from three models that go back to an archetype (ω), from which also the hyparchetype of V and L (ψ) originated. Class α is represented in the edition by the *Parisinus gr.* 245 (Y), of the 15th century, and the *Norimbergensis gr.* Cent. V, app. 8 (Σ), a manuscript of the late 14th cen-

[12] The commentary is incomplete because of the loss of some quires between quire 7 and 8. On f. 56v the text ends abruptly (book II, p. 63, l.21 in the edition of Wolff) and continues on f. 57r (p. 175, l. 51 Wolff) until the end (p. 179, l. 38). See Caballero-Sánchez (2013a) 90-91.

[13] See Hunger (1961) 226 (information in Hübner, p. XVII, should be complemented). Unfortunately y is only rarely mentioned in the apparatus of Hübner's edition.

tury, that once belonged to Bessarion. Class β is represented by three manuscripts, which were copied in the early 14th century in Constantinople: the *Laurentianus gr.* 28,20 (D), copied some time after 1322 by Macarios, a collaborator of Nicephorus Gregoras,[14] the *Marcianus gr.* 314 (M), written around 1335 by the so-called Metochites-scribe,[15] and the *Monacensis gr.* 419 (S). Finally class γ is represented by three late 14th century manuscripts, *Angelicus gr.* 29 (A), *Vat. gr.* 208 (B) and *Laurentianus gr.* 28,16 (C).

According to the stemma of the editors, the three classes derive *independently* through their respective hyparchetypes (α, β and γ) from the archetype ω. Unfortunately, neither Boll nor Hübner give arguments for this threefold division of the text tradition besides V. As we did not make a new collation of the Greek manuscripts (except L), we had to rely on an analysis of the apparatus of Hübner's edition to understand the tradition. Hübner, however, is not always complete in the apparatus. Often one has to infer indirectly what the reading is of a manuscript that is not mentioned. Nevertheless a study of the wealth of material collected in the apparatus did help us to clarify the complex transmission of the text. On some points, we had to correct the suppositions of the editors. Thus, as we will show, β and γ do not constitute two different traditions deriving *independently* from the archetype ω. The model of subfamily γ depends on a manuscript from the β tradition. We will use the joined sigla βγ to indicate class β together with the subclass γ. Moreover, class α and βγ are often together opposed as one group against V, so that the text tradition is originally not fourfold, as Boll and Hübner thought, but twofold.

2.1. Class βγ

The manuscripts of classes β and γ have so many *Trennfehler*, so many conjectural corrections and other particularities in common, that they must be considered as one group deriving from one common exemplar (β), in which all these characteristics were found. The text in hyparchetype β, as we can reconstruct it from D M and S, seems to be the result of a radical editorial revision. Within this group, the manuscripts of class γ constitute a *subordinated* family, which depends itself upon a manuscript in the β tradition. The text in γ was the result of a new editorial revision. Finally, Σ (a manuscript of group α) often adopts readings of the group βγ and of γ in particular in books III-IV. We first give a list of common readings clearly separating group βγ from the main tradition of the text.

Book I.97° καὶ διχοτόμοις] *om.* βγ 104° πᾶσιν] καὶ πᾶσιν Σ βγ 158° ἐν ἴσῳ] ἐν ἑκάστῳ Σ βγ 194° γὰρ] *om.* βγ 210° ἀναιρεῖν] ἀθετεῖν βγ 216° καὶ ὥσπερ] *inv.* βγ 244° συνεργεῖ] *om.* βγ 249° ἄχρηστον] εὔχρηστον βγ 267° πρώτας] *om.* βγ 281° ἀποβαίνειν ἀνάγκη] -νει

[14] On this manuscript, see Caballero-Sánchez (2013a) 112-115.

[15] On this manuscript, see Caballero-Sánchez (2013a) 115-117.

-γκαίως βγ 295° ὅσοι δυνατοὶ σημειοῦσθαι] οἱ δυνατοὶ σήμερον βγ 306° παρακολουθήσει] ἀκολουθήσει βγ 354° κἂν ὀλίγα] ὀλίγα βγ 379° ἤδη] *om.* βγ 400° τὸ κατὰ ψύξιν] *om.* βγ 402° περὶ τὴν γῆν] περιττῶν βγ 408° *hic* Iovem *codices classis* βγ *inserunt, ut solitum planetarum ordinem restituant* 433° γόνιμα ... ποιητικὰ] *inv.* βγ 441° Ἄρεως] κακοποιοὺς *add.* βγ 497° ἑῷοι VL] ἑῳ μένω Y: ἑῷοι μόνον βγ 503° παμπληθεῖς] παμπόλλους βγ 508° δὲ ὄντος] ὄντος ἀκολούθου 522° Ἄρεως] Διός βγ + Heph. Ep. 523° λαμπαύρας] λαμπαδίας βγ + Heph. Ep. 533° ὀφθαλμῶν] ποδῶν βγ + Heph. + Heph. Ep. 545° ἐν τῇ ὀσφύϊ] ἐπὶ τῆς ὀσφύος βγ + Heph. + Heph. Ep. (*'fort. recte' Hüb.*) 558° Σκορπίου] *om.* βγ + Heph. Ep. 558° ἄκραις αὐταῖς] ἄκροις αὐτοῖς βγ 566° τῆς Ἀφροδίτης] τοῦ Ἑρμοῦ βγ + Heph. Ep. 644° λοιποὶ] *om.* βγ 664° Ἑρμοῦ] Ἄρεως (*per signum*) βγ + Heph. 687° ζῴου] *om.* βγ 816° τρίτον] + ὀρθῆς βγ 879° φύσιν] φυσικὴν αἰτιολογίαν Σ βγ (*cf.* Co) 903° ἀγαθοποιοῦ] ἀγαθοὶ βγ 910° εὐκράτῳ] εὐκρατοτέρῳ βγ 924° ζῳδιακὸς] + κύκλος βγ 944° θηλυκῶν] + ζῳδίων βγ 1163° Ταῦρον καὶ Αἰγόκερων] Καρκίνον βγ 1226° οἰκειότητος] ἰδιοθρονεῖν καὶ λάμπειν λέγονται *add.* Σ βγ 1451° ἀσημότερον] ἀσεμνότερον Σ βγ 1469° διακέκριται] + πως Σ βγ

Book II.33° μὲν] *om.* βγ 38° προδιειληφέναι] διειληφέναι βγ 44° ἐτησίων] ἐνιαυσιαίων βγ 74° *et* 87° συνέχειαν] ἀεὶ *praem.* βγ 77° διαπύρωσιν] διάθεσιν βγ 79° ὑπὸ *om.* βγ 184° πρόσνευσιν *om.* βγ 185° κατ᾽αὐτὸ τὸ] κατὰ βγ 188° οἰκήσεων] οἰκηώσεων βγ 237° Ἄρεως] Κρόνου βγ 273° ἀπλοδίαιτοι] ἁβροδίατοι βγ + Co (*cf.* 289° 344°) 292° καὶ παρατηρητικὸν] *om.* βγ 318° ἐλευθέρας] ἐλεύθερον Σ βγ 324° μὲν] *om.* βγ 329° πλουσιώτατοι] πλουσιώτεροι βγ 331° καὶ] καὶ περὶ βγ 358° πως] *om.* βγ 367° διὸ] *om.* βγ 377° ἀρρενοποιοῦσαι] ἀρρενωποὶ βγ 413° εὐθηνίᾳ] ἐλευθερίᾳ βγ 442° τεθηλυσμένοι] τεθηγμένοι βγ 521° προγινώσκειν] γινώσκειν βγ 568° ἐπιτάσεις] περιστάσεις βγ 576° παρόδων] παρανατελλόντων βγ 617° ἔτυχον ὄντες] τυγχάνουσι βγ + Heph. 648° καρποφηθέντων] ἐτησίων καρπῶν βγ 715° καὶ λοιμικὰ] *om.* βγ 792° ἐναλλοιώσεως] ἀλλοιώσεως βγ 1055° τὰς προσνεύσεις] πορεύσεις βγ

Book III.36° καθολικῶν] καθόλου βγ 43° καὶ] *om.* Σ βγ 81° πρὸς] γοῦν Σ βγ 105° προυποτιθέμενοι] προυποτιθεμένοις βγ 108° σωμάτων] ἀστέρων βγ Σ^2 112° πρῶτον δὲ] καὶ πρῶτον βγ 214° ἐνεργῶς] ἐναργῶς βγ Heph. 265° ἢ] *om.* Σ βγ Heph. 298° ὢν] *om.* Σ βγ Heph. 329° πρακτικῶν] πραγματικὰς Σ βγ 372° λοιπῶν] ὄντων Σ βγ 373° ἔτι] ἐπὶ Σ βγ 417° τόπους] δύο τρόπους Σ βγ 418° δὲ] γὰρ Σ βγ + Heph. 421° πάθωσιν] καθορῶσιν Σ βγ 424° πλείονες] πλείοσι Σ βγ 440° μόνον] ἕνα Σ βγ 477° προσμαρτυρῇ] συμμαρτυρῇ Σ βγ 527° τοῖς ... μέρεσι] ἐν τοῖς ... μέρεσι Σ βγ Heph. 551° ἕκαστα] ἕκαστον Σ βγ 559° αὐτῶν] αὐτῆς Σ βγ 590° φέροντος] ἀφαιροῦντες Σ βγ 640° ὄντων] + ὅλον Σ βγ 641° εἶτα] καὶ Σ βγ 771° τῶν διαστάσεων] τῆς προτέρας (τοιαύτας γ) διαστάσεως Σ βγ 772° ὅσοις] οἷς Σ βγ 775° δὲ] + πάλιν Σ βγ 783° ὑπεροχῆς] ὑπὲρ γῆν Σγ: ὑπὲρ γῆς MS, *cf.* D 824° ο᾽ καὶ νη᾽] ὡρῶν Σ βγ 832°-3° ἡ ... ἀποχὴ ... ὑπέκειτο] αἱ ... ἀποχαὶ ... ὑπέκειντο Σ βγ 886° δασυστέρνους καὶ μεσοφθάλμους] μεγαλοφθάλμους Σ βγ 910° θηλυμορφώτερον] εὐσχημονέστερον Σ βγ 917° λευκοὺς μέν, οὐκ ἐπὶ τὸ εὔχρουν δὲ ὁμοίως, τετανότριχας] *om.* Σ βγ 918° λοξοφθάλμους τε] κοιλοφθάλμους Σ βγ 954° ἢ ἐπὶ τὸ] καὶ Σ βγ 1057° καὶ Λέοντι καὶ Σκορπίῳ καὶ Αἰγοκέρωτι] *om.* Σ βγ 1077° τὰ φῶτα ἢ] μὲν τὰ φῶτα Σ βγ 1078° ὦσιν] *om.* Σ βγ 1122° ἄκρα] ἄγρια Σ βγ 1233° κακογνώμονας] μονογνώμονας Σ βγ Heph. 1320° ἐκστατικοὺς] στασιαστικοὺς ἐκστατικοὺς Σ βγ 1341° γεωμέτρας] φιλογεωμέτρας Σ βγ 1378° ἀσελγαίνοντας] + ἀπεργάζεται Σ βγ 1451° ἀσημότερον] ἀσεμνότερον Σ βγ 1537° νομίμας ὥσπερ] *inv.* Σ βγ

Book IV.44° Διὸς] δύνοντος Σ βγ 97° πολυκτημοσύνη] φιλοκτημοσύνη Σ βγ 109°-110° ἤ ... ἔχων] ἔχη Σ βγ 117° φάσιν] + ἑῴαν Σ βγ 154° παρασχίστας] περιχύτας Σ βγ Heph. 218° καταρριψοκινδυνίας] ῥιψοκινδυνίας γ Heph. (-νευσίας Σ β) 246° αὐτῶν] αὐτὸν Σ βγ 285° οὖν] *om.* Σ βγ 296° *et* 298° τριγωνικοῖς] τριγώνοις Σ βγ 227° αὐτῶν] αὐτὴν Σ βγ 305° τῷ] ὁ Σ βγ + Heph. 326° ὦσι τόποις] + καὶ προσώποις Σ βγ + Heph. Co 395° δὲ καὶ] τε βγ 448° δωδηκατημορίοις] *om.* Σ βγ 619° ὅταν] + ἢ βγ 643° ἢ τετραπόδοις] *om.* βγ 665° ἐπισήμοις] ἐπισήμως βγ 657° καὶ ταύρῳ] ἢ κενταύρῳ βγ 663° καὶ συμπτωμάτων] ἢ συρμάτων βγ, *cf.* Heph. 803° καταληφθεὶς] κατακλίθη (-η) βγ 888° ἐὰν μὴ μόνον] *om.* βγ

This is an impressive list, and not even a complete one. One may notice that, besides copyist's errors, there are many conjectures, attempts to correct what was a problematic reading in the archetype. Some conjectures are excellent and are rightly adopted by the modern editors. Moreover, the scholar responsible for the editing of the text also occasionally used the indirect tradition of Hephaestio to correct Ptolemy's text. An interesting case is found in chapter 9 of book I, where we can compare the text of Ptolemy with both the *Apotelesmatica* of Hephaestio and its *Epitome* IV. We noticed that βγ often adopt readings of the Byzantine *Epitome* IV whereas the text in α (Y) corresponds to the complete version of Hephaestio's *Apotelesmatica*.

I.522° Ἄρεως VY Heph. Procl. (*deest* L)] Διὸς βγ Heph. Ep. 523° καλούμενος δὲ Λαμπαύρας V L Procl.] καλούμενος δὲ Λαμπαδίας βγ Heph. Ep.: *om.* Y Heph. 558° Σκορπίου V LY Heph. Procl.] ζυγοῦ Σ: *om.* βγ Heph. Ep. 566° τῆς Ἀφροδίτης α L Heph.] τῆς Ἑρμῆς V: τοῦ Ἑρμοῦ βγ Heph. Ep. 568° ἑπομένη L M S γ Heph. Ep., ἐπιμένη V] λεγομένη α Heph.: *om.* D 574° ἐφαπτίσι V L βγ Heph. Ep. Procl.] πτέρυξι Σ: πτέρηξιν Y: πτέρυξι ἤτοι ἐφαπτίσι Heph.

The examples show that, whereas α (or Y alone) usually follows the *Apotelesmata* tradition of Heph., the editor of the hyparchetype of βγ corrected the text according to *Epitome* IV.[16] This editor not only used the *Epitome*, but also had access to the anonymous Commentary. In fact, the leading manuscripts of β (D M S) also contain the text of the commentary in a particular text tradition.[17] Given the quality of the corrections and the use of Hephaestio and the commentary, one must conclude that the editor of the βγ version was a scholar interested in astrology, and not just a copyist.

Within this group the manuscripts of class γ constitute a particular subfamily deriving from a model that corrected even more radically the text from the β tradition, collating it again with a manuscript of Hephaestio. According to Pingree the model of this class was a recension made by the Byzantine astrologer John Abramius, who was active around 1370-1390.[18] See for instance the following particular readings of γ:

16 See other examples in Hübner, p. XXIX.

17 See Caballero-Sánchez (2013a) 104ff. ('la recension ω').

18 See Pingree (1971) 202.

I.464° δυσὶν] *om.* γ 489° *after* πλανωμένων *addition in* γ 491° ἀνατολῆς] συνόδου γ + Co 496 κρύψεως] συνόδου γ 524°-5° οἱ δὲ λοιποὶ … τοῦ Ἄρεως] *om.* Σ γ 1193° τουτέστι … ἥμισυ] *om.* γ II.77° ἐμφανίζοντα] τοιαύτην *praem.* Σ γ 109° ὑπολάβοι] + διὰ τοῦτο Σ γ 1043° ἐπιτάσεις] ἢ ἀνέσεις *add.* Σ γ 1053° φώτων] φάσων Σ γ III.560° καὶ ἔτι … ἀστέρων] *om.* γ 598° κληροῦ] τῆς τυχῆς *add.* γ 591°-3° *very different text in* Σ γ *with addition of scholia* 594°-9° *long section considered as inauthentic in* γ 752° ἔγγιστα] *om.* γ 770° ἡ ἀρχὴ τῶν] ὁ ἀφετικὸς τόπος Σ γ (*om.* Y β) 780° ἑπτακαιδεκάκις — ποιήσωμεν] *different longer text in* γ 871° *addition of a gloss in* γ 1034° ὡς εἴπομεν] *om.* γ IV.227°, 230°, 257° *marginal addition from* Heph. *in* γ 255° λαμβάνουσι] ποιεῖ λαμβάνειν τὰς γυναῖκας γ 320°-3° *long omission in* γ

John Abramius used a manuscript from the β tradition as model for his own copy. It may have been D (or a manuscript related to it) as γ shares with D many particularities. See, for instance, III.736°, 1170°-1°, 1181°, 1184°, 1240°, 1246°, 1263°, 1267°. D is in itself a particular case. This manuscript, which was copied by a collaborator of Nicephoros Gregoras, is not a pure representative of β and often goes with V (and even Y). Because of its relation to D the γ class too often goes with V or (and) Y against M and S. To conclude, for the constitution of the text, γ has no independent authority, though it may occasionally offer an intelligent correction.

2.2. Class α

The sigle α indicates in the apparatus the consensus between Y and Σ, which is mostly found in the first two books, for Σ joins γ from book III on, as can be seen in many examples listed above. The text in Σ is also contaminated by the indirect tradition. Thus, we find additions and corrections that come either from the anonymous Commentary or from the Proclus Paraphrase. See, for instance, I.677°, a longer version with additions similar to the Commentary; II.62° κλίματα] ἔθνη Σ *cf.* Co; 1112°-5° long addition *ex Heph.* Σ *mg.*; III.561°, 587°, 600° *extra titulos add. ex Proclo.* Even if Σ may occasionally have a good correction, it is an unreliable witness of the text tradition when it stands alone. The best representative of α (and in the apparatus often the only one) is Y, which, however, is disfigured by errors of all kinds. Nevertheless, this manuscript comes closest to V in the tradition: 'partim confirmat, partim supplet, nonnumquam immo corrigit' (Hübner). In a few cases it is the only manuscript that preserves the authentic text. For instance in I.1021°-1022° where all other manuscripts have an omission, and at the end of the treatise where V is lacking.

2.3. Bipartition of the text tradition

α and βγ are often together opposed to V. This is no surprise, as V has many particular errors they do not share. There are, however, cases where α βγ stand as a group against V, when V is not alone, but is confirmed by L, Proclus and the

Moerbeke translation, and offers a reading adopted by the editor. We shall give below a full list of these cases, when we examine the place of Moerbeke's translation in the tradition (see B.1.3). Contrary to the hypothesis of Boll–Hübner, there seems to have been originally not four, but two different branches in the text tradition, each with distinctive features, one represented by V, the other by α βγ.

3. THE INDIRECT TRADITION

The most important indirect witness of Ptolemy's work is the Egyptian astrologer Hephaestio of Thebes (born 380 A.D.), who composed a handbook of *Apotelesmatics* in the early 5th century, in which long excerpts from Ptolemy are integrated. There also exist shorter versions of this work, the *Epitomai*, published by Pingree in the second volume of his edition. The fourth *Epitome*, which is transmitted in four 14th century manuscripts, is particularly important for parallels with Ptolemy's text. Hephaestio is undoubtedly the oldest indirect witness of Ptolemy's text. However, his authority on the Ptolemaic text must be qualified. First, Hephaestio rewrites, shortens, and adapts the Ptolemaic text. Second, there is undoubtedly reciprocal contamination between manuscripts of Hephaestio and manuscripts of the direct Ptolemy tradition ('alterum altero contaminatum esse', Hübner, p. XXVI). The fact that the scholar who made the astrological compilation, which is found in L, mixed chapters of Ptolemy and Hephaestio (and other late antique astrologers dependent on Ptolemy) is a strong indication that Hephaestio's text was used and read together with Ptolemy's work. Interestingly, some variant readings in the manuscripts of Ptolemy's text are also found in the manuscripts of Hephaestio's text. As we have seen above, there are evident connections between readings in βγ and in the fourth *Epitome*. The editors of Ptolemy and Hephaestio concluded from the parallel variation in the Hephaestio tradition that the division of the Ptolemy tradition into families (ψ α βγ) originated in late antiquity.[19] But the occurrence of similar variations in the tradition of both works may also be due to reciprocal contamination of manuscripts in the Byzantine period. Nevertheless Hephaestio remains an important authority for the establishment of Ptolemy's text, as his work may in principle give us access to Ptolemy's work as it was read in the early 5th century.

Also of great importance for the edition is the paraphrase attributed to Proclus. We will continue to designate it as 'Proclus', though there are no arguments to attribute this work to the Lycian philosopher. In its oldest manuscript, the *Vat. gr.* 1453, from the 10th century, the work is anonymously transmitted with the title Τῶν ἀσαφῶς εἰρημένων Πτολεμαίῳ καὶ δυσπαρακολουθήτως ἐν τῇ αὐτοῦ

[19] See Pingree in his edition of Hephaestio: 'Textum codicum Ptolemaei Apotelesmaticorum in classes a Boll–Boer cognitas post fere annum 415 diremptam esse concludendum est' (p. VIIs.). See also Hübner, p. XXVI-XXVIII.

τετραβιβλίῳ ἐπὶ τὸ σαφέστερον καὶ εὐπαρακολούθητον μεταχείρησις. A later hand added in the margin the attribution to Proclus, which is also found in recent manuscripts and in the edition of 1635. The title describes very well the intention of its author, namely, to provide a more readable and understandable version of the often obscure and tortured phraseology of the Alexandrian master. Whoever the author may have been, he was himself an outstanding expert in astrology, as is also clear from the additional tables he inserted. In many ways, the paraphrase is an excellent guide for our understanding of Ptolemy's arguments, and it deserves a modern critical edition. Because the paraphrase remains very close to Ptolemy's text, and its oldest manuscript is from the 10th century, it has an important authority in the edition of Ptolemy's text. However, several caveats must be made. In the first place, when we (including Boll and Hübner) quote 'Proclus', we rely upon the Renaissance edition of this text, not upon a collation of the Vatican manuscript itself. Secondly, one may never forget that 'Proclus' rewrites and simplifies the Ptolemaic text, which undermines its authority for this edition, in particular in the case of omissions. Thirdly, the 'Proclus' text was used in the Byzantine period to improve and correct the direct transmission of Ptolemy's text. To use 'Proclus' to confirm a particular reading in the Ptolemaic text tradition may, therefore, be a circular argument. Thus Σ and even more S are contaminated by so-called Proclus. The insertion of the conclusion of the Proclus Paraphrase to supply the missing conclusion in the βγ group is the most obvious example of this contamination (see Appendix II). Nevertheless, the paraphrase is an important authority on the state of the text of the *Apotelesmatica* before the 10th century. It generally confirms the readings of V (and L), although never their particular errors. The model used for the paraphrase thus seems to attest the same text tradition that was also present in ψ, from which V originates. (See the list of examples below).

Besides the Proclus Paraphrase, there is a commentary, which is tentatively attributed to Proclus in the edition of Wolff (Basel 1559). The oldest witness of this commentary is the 11th century *Laurentianus* (L), which we presented above. The commentary goes back to late antiquity and may have been composed in the 4-6th century, though there are no arguments for an attribution to Proclus. The anonymous commentator introduces the different chapters and sections of the treatise and gives comments on problematic passages of the text. For the edition of the Ptolemaic text, the commentary has only a limited importance, namely when it quotes lemmas of the text, which are explained. Moreover, the transmitted text is often corrupted. Probably, the copyist of the model of L found the commentary in the form of scholia in an ancient manuscript of the *Apotelesmatica*.[20]

[20] For an exhaustive study of the text tradition of the commentary and an edition with Spanish translation of the first 37 scholia corresponding to the prologue of the *Tetrabiblos*, see Caballero-Sánchez (2013b).

B. THE GREEK MODEL OF MOERBEKE'S TRANSLATION

In Moerbeke's translations, it is in most cases clear which element or which variant of the Greek tradition he read. His translation method makes it possible to retrieve, for an essential part, the text of the Greek manuscript he used. This is also the case in Moerbeke's translation of the *Tetrabiblos*, which was not known to the editors of the Greek text. It is therefore an important new witness to the Greek tradition, especially because all the manuscripts of this work of Ptolemy, with the exception of the excerpt manuscripts (L, y, *Vat gr.* 1291), are more recent than the translation, which was probably made around 1268. We have seen that the main representatives of the direct text tradition are V (+ L when present), family α and family βγ. Moreover, we noticed that α + βγ often stand as a group against V, not only in cases were the copyist of V made an obvious error, but also when V has a better reading adopted by the editors of the Greek text. We are now prepared to examine two important questions. Where in this tradition do we have to situate the Greek model (G)[21] that Moerbeke used for his translation? What contribution could Moerbeke's translation make to the authoritative Greek edition of Hübner, who did not know it? We will discuss the first question in the following section, leaving the second for Chapter V.

1. PRINCIPAL CHARACTERISTIC OF G: A REPRESENTATIVE OF BRANCH ψ

When we compared the Latin translation line for line with Hübner's Greek edition, we noticed that Moerbeke must have had a manuscript with a very good text, as we find few particular errors in the translation that could be explained by errors in its Greek model. There are certainly a series of longer omissions, but they may also be due to accidents in the transmission of the Latin text. It also became clear that, of all collated manuscripts, the translation follows most closely the text tradition represented by the late 13th century *Vat. gr.* 1038 — *codex et aetate et qualitate praestantissimus* — together with the excerpt manuscript L, 10th century, when present. For this reason, in the Greek-Latin apparatus we compare the Latin text with V and with the edited Greek text. Taking into account Moerbeke's translation method, we mention in the apparatus all cases where the Latin text does *not* agree with V or/and the edited Greek text, but we do not mention all cases where the Latin text corresponds to a reading of V, when this reading — alone, with some, or with all others — has been adopted in the edited text. However, to answer the questions about the place of Moerbeke's Greek model in the text tradition and its possible contribution to the Greek edition of the Ptolemy's text, we could not

[21] The siglum G means in our whole edition the translation or the translator, or its Greek model. From the context it is always clear, which of the three meanings is to understand.

rely solely upon the limited information provided in the Greek-Latin apparatus. Therefore, we decided to present in Chapter V, as a complement to that apparatus, a comprehensive list of all readings that indicate how the Greek model translated by Moerbeke is related to the different traditions of the Greek text, such as we know from Hübner's edition[22]. In the following discussion we shall often refer to that list. (For the meaning of the symbols that precede variant readings, see below p. 95).

1.1. G agrees with V (L) alone against variants of all other witnesses

Of all the Greek manuscripts cited in Hübner's edition, the translation follows, when we exclude omissions (which might be due to errors in the Latin tradition), not only closely, but often exclusively the text tradition represented by V (and L when present) against all other manuscripts. This means that G agrees with V (L) when V or V + L are *alone* (or only together with one or two of the contaminated witnesses, Σ, γ or D) to present the edited text against variants of α, β, γ. In many cases, the Proclus Paraphrase agrees with this tradition. In the following list we give only the line numbers of Hübner's edition: the passages are cited *in extenso* in Chapter V. When G + V (L) are accompanied by one of the contaminated witnesses, the line numbers are given in italics:

Book I.4° 6° 7° 8° 63° *78*° *224*° 279° 288° *321*° 383-4° 400° 470° *508*° 523° *598*° *651*° 753° 778° 796° 841° *864*° 909° *959*° 1028° *1134*° 1138° 1167° **Book** II.*8*° *18*° 31° 38° *116*° *150*° 175° 205° 229° 233° 235° 270° 293° *311*° 313° 330-1° *376*° 378° *435*° 444° *447*° *498*° 506° 512° 513° 524° 525° 615° *669*° *684*° 759° (576) 759° (577) 917° 931° **Book** III.2° 3° 4° 5° 18° 23° 101° 117° *164*° 232° 259° 347° 394° *425*° 567° 823° 884° 1096° 1139° 1223° 1326° *1380*° 1427° 1478° 1486° 1489° 1504° **Book** IV.3° 4° 13° *40*° 153° 331° 352° 396° 469° *497*° 644° 698° 699° 770° 782-3°— *inde ab* 797 *deficit* V

1.2. G is dependent neither from V nor L

G does not have the numerous particular errors of V. This can easily be observed in our Greek-Latin apparatus: the individual readings of V, which are not confirmed by G, are indicated. When edited readings are confirmed by G and not by V, V's variants are almost always in the company of other witnesses. This immediately rules out a direct dependence of G from V (which would also be difficult for chronological reasons). See the following examples from a section where L is pre-

[22] In cases of omissions and additions we also checked the 12th century Arabo-Latin translation of Plato of Tivoli (Tib.) in the edition of Johannes Hervagius (Basel, 1533) and asked advice about the Arabic translation (Arab.), which was made in Bagdad in the 9th century in the circle of Ḥunayn b.Isḥāq. We acknowledge our gratitude to Bojidar Dimitrov (collaborator within the *Ptolemaeus latinus* project), who checked for us the relevant passages in the edition in preparation by Keiji Yamamoto.

sent (the readings before the square bracket are here, as elsewhere, the readings adopted in Hübner's edition).

BOOK I

566° (436) τῆς Ἀφροδίτης α L + G (Veneri)] τῆς Ἑρμῆς V: τοῦ Ἑρμοῦ βγ

597°(*sic*) (455) λίνου L M S γ + G (lino)] αἴνου V D, *cf.* α

604° (460) οἱ δὲ (ὑπὸ τῆν οὐρὰν) L D γ + G (Que autem sunt)] ἡ δὲ V Y

814° (608) δύο ὀρθὰς γωνίας L α + G (duos rectos angulos) Procl.] ὀρθὰς ἴσας γωνίας V: ὀρθὰς δύο γωνίας DM γ, *cf.* S

845° (633) τε καὶ καταφέρεσθαι L βγ + G (et descendunt) Procl.] *om.* V α

On the other hand, in the sections of the *Apotelesmatica* where L is present, the model of G is not dependent from L: it does not have the deviant readings of L, even when they are significant or suggestive. Some examples:

BOOK I

383° (300) πρώταις ταῖς V + G (primas, *sc.* uirtutes)] πρώτης τῆς βγ: πρὸ τῆς Y: πρώτης ἀψάμενοι τῆς L

384° (300) ἀστέρων δυνάμεσιν V A + G (uirtutes siderum)] δυνάμεως L, *cf. cett.*

384° (300) ἡλίου τε καὶ σελήνης V *cett.* + G (solis scilicet et lune)] *om.* L

390° (305) τῶν κατὰ τὰς ὥρας μεταβολῶν V *cett.* + G (transmutationum secundum tempora)] τῶν ὁρῶν μεταβολὰς L

560° (431) ἐν μέσαις V *cett.* + G (in mediis)] ἐν τῷ μέσῳ τῶν Χηλῶν L

611° (465) καλούμενος V *cett.* + G (uocata)] λεγόμενος L

623°-624° (473) οἱ δὲ — τοῦ Ἑρμοῦ V *cett.* + G (Clare autem que sunt ... Mercurio)] *om.* L Y

918° (691) πως V *cett.* + G (aliqualiter)] πρὸς L

1019° (769) περὶ ὁρίων διαθέσεως V D S γ + G (De terminorum dispositione) Procl. (60,-4)] περὶ τῶν ὁρίων τῶν πέντε ἀστέρων L: περὶ (τῶν) ὁρίων α M: περὶ ὁρίων Co (39,-5)

1186°-7° (905-6) γίνεται —Ἑρμοῦ ος′ V βγ + G (fiunt —Mercuri 76)] *om.* L α

G is therefore an excellent witness of branch V (L), and, in a stemmatic perspective, G descends, independently from V and L, from the hyparchetype ψ.

1.3. Bipartition of the tradition: G with V (L) alone against common readings of α βγ

When we examine branch V (L) + G in relation to the other two branches, α and βγ, we observe — always distinguishing readings that are 'adopted' or 'rejected' (not adopted) in the edition of Hübner — a number of readings which oppose V (L) + G to a common reading of α βγ, both in adopted and in rejected positions. This evidence makes us conclude, as we have already supposed above (IV A), that there is a fundamental bipartion of the text tradition: V (L) + G (almost always confirmed by Procl.) on one side, and α + βγ on the other. This conclusion does not depend on the existence of G. Even without the translation, one could reach the same conclusion. However, the testimony of the translation makes the bipartition more evident, as we have now a supplementary witness to confirm readings of V (and L, Procl.).

As common readings of α βγ we also count in the following series some instances where one of the members of α βγ has a derived mistake, or where Σ or γ — known for heavy contamination — are exceptions.

1.3.1. **G *with* V (L)*: adopted readings against α βγ***

There are about 60 cases, where G + V (L) + Procl. (when available) present the edited text against a *common* reading of α βγ rejected by the editor:

BOOK I

31° (29) α′ V + G (capitulum primum)] *om.* α βγ

224° (177) τοιαύτης V Σ + G (tali)] αὐτῆς Y βγ

383°-4° (300) πρώταις ταῖς ... δυνάμεσιν V + G (primas uirtutes)] πρὸ τῆς ... δυνάμεως Y: πρώτης τῆς ... (*om.* δυνάμεως) β BC; *cf.* A, *cf.* L

400° (313) κατὰ ψῦξιν V L + G (secundum infrigidationem), *cf.* Tib. (suaque frigiditate)] *om.* α βγ

470° (365-6) ἐν αὐτῇ V + G (in ipsa)] κατ' αὐτὴν α βγ (*Robb.*)

598° (455) ὁ ... λαμπρὸς V Σ + G (clara ... que) Procl. Heph. *cod.* A (οἱ... λαμπρὸς *cod.* U)] οἱ ... λαμπροὶ Y βγ Heph. Ep.

651° (491-2) περὶ τὸ στόμα V L γ + G (circa os) Procl.] ἐπὶ τοῦ στόματος α β Heph.

753° (562): τῶν ... παρόδων V L + G (motuum) Procl.] τὴν ... πάροδον α β, *cf.* γ

778° (581) ἓξ δὲ V + G (sex autem) Procl.] τὰ δὲ ἴσα α βγ

796° (593-4) τὸ ἀπηλιωτικώτερον V + G (magis orientale) Procl.] τὸν ἀπηλιώτην α βγ

841° (629) ὁμοίως V + G (similiter)] *om.* α βγ Heph. ('*fort. recte*')

864° (649) ἃ V Σ + G (que)] τὰ μετὰ Y βγ

909° (685) τοῦ τετραγώνου V L + G (quadrature)] τετράγωνον α βγ (*Robb.*)

1028° (777) ἐπεὶ V + G (quoniam)] ἐπὶ Y βγ (*om.* Σ)

1134° (876) (μόλις) ἡμῖν (*ante* δυνάμενον) V Σ + G (uix nobis poterat)] ἡμῖν *post* δυνάμενον (1135°) Y βγ

1138° (878) διασεσωσμένης V + G (quia stabat)] διασεσωσμένοις Y (-ων Σ: -ην βγ)

1167° (900) τοῖς ἑξῆς V L + G (consequentibus, *fort.* ἑξῆς *legit*) Procl.] τοῖς ἐφεξῆς α βγ

1226° (936) ὁμοφύλου L + G (conformis), *cf.* -φίλου V] -φυοῦς α βγ

BOOK II

31° (27) ὅπερ V + G (quod) Procl.] ὃ καὶ αὐτὸ α βγ Arab.

175° (137) τὰ μὲν V (*Robb.*) + G (que quidem)] τὰ (μὲν *om.*) α βγ (*Boll*)

205° (159) τὸν ... συσχηματισμὸν V + G (configurationem)] τὸ τοῦ (*om.* γ) κατὰ τὸν (τὸν *om.* D) σχηματισμὸν α βγ

229° (176) τὰ ... κείμενα V + G (que iacent ...) Procl.] (τὰ *om.*) ... κείμενα α βγ (*Boll Robb.*)

235° (180-1) ἀπεφάντησαν V + G (dicti sunt)] ἀπέβησαν α βγ, *cf.* Arab.

293° (223) πέντε V + G (quinque) Procl.] *om.* α βγ Arab.

313° (237) φιλάνθρωποι V + G (humani) Procl.] φιλανθρωπότεροι α βγ

444° (332) ἑσπερίου V + G (uespertina), *cf.* Procl.] ἑσπέριον α βγ

506° (385) γενέσεως V + G (hominis generatione)] γενέσεων α: τῶν γενέσεων βγ

525° (399) γὰρ τότε V + G (enim tunc) Procl. Arab.] *om.* α βγ

759° (576) ποτίμων V + G (potabilium), *cf.* Procl.] ποταμίων α βγ

759° (577) ἐπὶ γῆς V + G (super terram), *cf.* Procl. ἐκ τῆς γῆς] *om.* α βγ Arab.

818° (622) ἀστέρες V D γ + G (astra) Procl.] *om.* α M S Heph. Arab.

917° (696) ἰδιοτρόπους V + G (proprias) Procl.] ἰδιοτροπίας α βγ

BOOK III

3° (3) α΄ προοίμιον V + G (prohemium ... cap. 1^{m})] *om.* α βγ

4° (4) β΄ V + G (cap. 2^{m})] α΄ D γ: *om.* Σ M S: [Y]

5° (5) γ΄ V + G (cap. 3^{m})] β΄ Y D γ: *om.* Σ M S

23° (24) γενεθλιαλογίαν V + G (geneaticum sermonem) Procl.] γενεθλιαλογικὸν α βγ ('*fort. recte*')

101° (81-2) ὁλοσχερὲς V + G (uniuersalem habitum)] ὁλοσχερέστερον α βγ, *cf.* Procl. (καθολικώτερον)

117° (94) κατ' αὐτὴν τὴν (σύγκρισιν) V + G (penes fetus concretionem)] αὐτὴν *om.* α βγ, *div. mod. pro* κατ' ... τὴν

232° (183) βραδύτεροι V + G (tardiora) Procl.] βράδιον α βγ

347° (269) μέγα V + G (magnum) Procl.] *om.* α βγ

394° (305) λόγον V + G (rationem) Procl.] τρόπον α βγ

425° (329) κυΐσκεσθαι V γ + G (procreari) Heph.] τίκτεσθαι α β, *cf.* Procl.

823° (617) τῶν συγκρινομένων κέντρων V + G (ab angulis comparatis) Procl.] τῷ συγκρινομένῳ κέντρῳ α βγ Heph.

884° (661-2) τῇ μὲν μορφῇ V + G (forma quidem)] τὴν μὲν μορφὴν α βγ, *cf.* Procl.

1096° (821) κρύπτων τόπων V + G (occultorum locorum), *cf.* Heph.] ἢ *praem.* α βγ

1139° (852) ἐπιχάριτα V + G (gratiosas), *cf.* Procl.] ἐπιχαρῆ Σ βγ (ἐπιχάρος Y)

1326° (1009) ἐλεητικούς V + G (miseratiuos), *cf.* Procl.] ἐλεγκτικούς βγ (ἐλεκτικούς Σ: *om.* Y)

1427° (1088) θεωρητικούς *hic* V + G (speculatiuos) Procl., *cf.* Tib.] *post* εὐστόχους (1428°) Σ M S: *om.* Y D γ

1478° (1127) ἢ[1] V + G (aut) Procl. Heph.] *om.* α βγ

1489° (1135) προειρημένον V + G (predictum), *cf.* Procl.] προκείμενον α βγ

1504° (1145) ἀκαταστασίαις V + G (inquietudinibus), *cf.* Procl.] ἀκατασχεσίαις α βγ Heph.

BOOK IV

3° (3) α΄ προοίμιον V + G (Prohemium ca. p^{m})] *om.* α βγ

4° (4) β΄ V + G (ca. 2^{m})] α΄ Y D γ (*om.* Σ M S) *et sic deinceps*

40° (36) καὶ[1] V S + G (et[2]) + Procl.] ἢ α D M Heph. [γ]

153° (122) ζωγράφους V + G (pictores) Procl.] στρατιώτας Σ D M γ: *om.* Y S Heph.

Ab 327° *deficit* Σ

331° (255) κεντρωθέντες V + G (in angulis existentes), *cf.* S] -θέντος Y (-θὲν G°: -θέντα D M γ)
469° (360) ἢ τὸ ἔγγιστα V + G (uel propinquissime), *cf.* Procl.] ἔγγιστα (*om.* ἢ τὸ) Y βγ
497° (382) πραγμάτων V AB + G (negotia) Procl.] γραμμάτων Y βC
644° (492) ἢ τετραπόδοις V + G (aut quadrupedibus) Procl.] τετραπόδοις Y: *om.* βγ
699° (533) γεννώμενον V + G (nascitur) Procl.] γενόμενον Y βγ (*Robb.*)
770° (584) ἡλικίαν λαχὼν V + G (etatem sortiens), *cf.* Procl.] *inv.* Y βγ
Ab 797° *deficit* V

This list reinforces the close relation of the lost model of Moerbeke's translation with V, as was already evident from the passages listed in section 1.1, which refer to readings where V (L) + G support the edited text not only against *common* readings of α βγ, but also against different readings *inside* α βγ — a longer series.

1.3.2. **G** ***with*** V (L)***: rejected readings against α βγ***

The close connection of G with V and its privileged place within the text tradition ψ is confirmed by instances where G shares the rejected reading of V (L) against a common reading of α βγ adopted in the edited text. We again have more than sixty adopted readings of α βγ as a whole, against the rejected readings of V (L) + G. In this part we add the symbols which refer to questions or comments in Chapter V, see p. 95.

BOOK I

(-?) 319° (251) μὲν α βγ Procl.] *om.* V + G
(-?) 435° (338) τοῦ ψυχροῦ ... τοῦ ξηροῦ (Y M S γ)] *inv.* V L D + G (siccum ... frigidum) Procl. (*Robb.*)
(-) 444° (345) ὡς (ἀμφότερα) Y βγ, *cf.* Procl.] *om.* V L + G (ambo)
(-) 497° (387) ἑῷοι μόνον βγ, *cf.* Co le (25,1), *cf.* ἑῷοι μένω Y)] ἑῷοι V L + G (matutini)
(-?) 570° (438) περὶ α βγ] ἐπὶ V L + G (in) Procl.
(-?) 570°-1° (438-9) ὁ ... ἔχει βγ Heph., *cf.* ἐν ... ἔχει Y] οἱ ... ἔχουσιν V + G (que ... habent) Procl. (*Boll*)
(-?) 587° (449) ἱματίῳ (χύματι Σ) α βγ Heph.] μετώπῳ V L + G (fronte) Procl.
(-) 785° (586) ἔτι τὸ α βγ, *cf.* Tib. (et etiam)] *om.* V + G Procl.
(-*) 820° (612) μιᾶς α βγ] γωνίας V L Procl. + G (anguli) (*Boll*)
(-?) 853° (640) ἀλλήλοις μέρη α βγ] τά τινα ἀλλήλοις μέρη V + G (inuicem partes quasdam)
(-?) 863° (649) ἁπλῶς α βγ] *om.* V + G, *cf.* Procl.
(-?) 869° (652) καὶ α βγ Procl.] τὰ V + G (que)
(-?) 896°-7° (676-7) Αἰγόκερως καὶ ὁ Ὑδροχόος α D S γ, *sed* ὁ *om.* α, *cf.* Tib.] Ὑδροχόος καὶ ὁ Αἰγόκερως L M + G (Aquarius scilicet et Capricornus) Procl., *cf.* V
(-?) 939° (708) συνίσταται α βγ, *cf.* Tib. (colligitur)] γίνεται V L + G (fit) Procl.
(-?) 973° (732) μόνῳ α βγ] μὲν V L + G (quidem) (*Robb.*)
(-) 1061° (801) ἐν α βγ Procl.] ἔτι V + G (adhuc)

(-?) 1094° (844) πρῶτος α βγ] πρῶτον V + G (primo) Procl.

(-) 1103° (851) ἥδε α βγ] *om.* V + G Procl.

(-?) 1133° (875) περισσῆς α βγ] πολλῆς V + G (multa) (*Boll*)

(-?) 1135° (877) ὑποτυπῶσαι α βγ] ὑποδεῖξαι καὶ τυπῶσαι V + G (ostendere et exprimere) (*Boll*)

(-?) 1257° (960-1) δυτικοὺς καὶ ἀφαιρετικοὺς α βγ Procl.] *inv.* V L + G (ablatiui ... et occasiui)

BOOK II

(-*) 43°-5° (38-9) οἷαί (*sc.* περιστάσεις?) εἰσιν αἱ τῶν ἐτησίων (ἐνιαυσιαίων βγ) ὡρῶν κατὰ τὸ μᾶλλον καὶ ἧττον ἀλλοιώσεις (ἐναλλοιώσεις M S) α βγ] οἷαί εἰσιν αἱ τῶν ἐτησίων ὡρῶν καὶ ἡ κατὰ τὸ μᾶλλον καὶ ἧττον ἀλλοίωσις V (*Boll*) + G (quales sunt que temporum anni et que secundum magis et minus alteratio), *cf.* Procl., *cf. Robb.* (... καὶ ... ἀλλοιώσεις)

(-?) 113° (90) ἀπαλώτεροι α S (*cf.* ἁπλούστεροι D M γ)] ἀπαλώτεροι μᾶλλον V G (teneriores magis)

(-) 183° (142-3) (ὡς ἐκ τούτων ἕκαστον τῶν τεταρτημορίων) ... συνοικειοῦσθαι α βγ] (...) συνοικειοῦται V + G ([ut ex hiis consequitur: quelibet quartarum] ... coaptatur); *cf.* Procl. (88,3-6: Ἐκ τούτου οὖν φανερὸν, ὅτι ἕκαστον τῶν τεταρτημορίων ... συνοικειοῦται) .

(-) 199° (154) Βασταρνία Ἰταλία Γαλλία α βγ Arab.] *om.* V + G Procl. Heph.

(-) 216° (168) Βασταρνία α βγ] βαρσανία V + G (Barsania) Procl.: *om.* Heph.

(-?) 229° (176) Κύπρος α βγ] κύπρου V + G (Cipri) Procl. (*Robb.*)

(-?) 246° (189) Κύπρον Σ βγ] κύπρου V Y + G (Cipri) Procl.

(-) 305° (231) πανουργότεροι δὲ α, *cf.* Co le (πανοῦργοι τε βγ)] *om.* V + G Procl.

(-) 350° (263) Λυδία α βγ Arab.] λυδία λυκία V + G (Lidia Licia) Procl.

(-*) 410° (307) Καρχηδο(ω)νίαν α βγ] Καρχηδόνα V + G (Carchidonem) Procl. (*Robb.*)

(-?) 478° (*in tabula*) παράλια Μικρᾶς Ἀσίας α βγ] μικρὰ Ἀσία V + G (Parua Asia) Heph. (*'haud scio an recte?' Hüb.*)

(-*) 505° (383) τοῦ ζῳδιακοῦ κύκλου α βγ] τοῦ ζῳδιακοῦ V + G (zodiaci) Procl.

(-?) 666° (507) καὶ *ante* περὶ α βγ] *om.* V + G (*Boll*)

(-?) 729° (553) λαβὼν α βγ] λαχῶν V + G (sortiens)

(-) 762° (579) ἐκ τῆς ... καταφλέξεως α βγ Heph. (*Apot. et Epit.*)] ἐκ τῆς (!) ... καταφλέξεων V + G (ex inflammationibus)

(-) 958° (728) σεισμοποιὰ καὶ ἀχλυώδη α βγ Heph. (1,1,62) Arab.] *om.* V L + G Procl.

(-?) 1035°-7° (792-3) διὰ τοῦ ... συσχηματισμοῦ α βγ] διὰ τοῦ ... συσχηματισμῶν (συσχημῶν V) V L + G (per id quod ... configurationum: διὰ τὸ ... συσχηματισμῶν)

BOOK III

(-*) 309° (242) (θανάτους ...) αἰφνιδίους (... ποιεῖ) α βγ] (...) αἰφνιδίως (...ποιεῖ) V + G (mortes ... efficit repentine ...) Procl.

(-*) 513° (394) διαμετρῶν α βγ] διάμετρος ᾖ V + G (diametralis sit), *cf.* Procl.: διάμετρος Heph.

(-?) 749° (562) ὁ τῆς ἀρχῆς τοῦ Κριοῦ α βγ] ἡ ἀρχὴ τοῦ Κριοῦ V + G (principium Arietis) Procl. (*Robb.*)

(-*) 824° (617) προκειμένων α βγ, *cf.* Procl. (προστιθέντος)] προυποκειμένων V + G (presuppositarum)

(-?) 1192° (896-7) ὀργανικάς α βγ] ὀργανικάς τε καὶ V + G (organicasque et) Procl.

(-?) 1209° (910) καὶ[2] α βγ] ἢ καὶ V + G (aut etiam)

(-) 1284° (974) παντοποιούς βγ, *cf.* Y, *cf.* πάντα ποιοῦντας Procl.] παντοίους V + G (omnimodo): [Σ]

(-) 1314° (998) εὐεπιβούλους, ἐπιεικεῖς, ἐπάνδρους, νικητικοὺς α βγ Arab.] *om.* V + G Procl.

(-?) 1394° (1063) τῆς ψυχῆς α βγ] *om.* V + G Procl.

(-*) 1516° (1154) ἐστὶν α β] *om.* V γ + G Procl. ('*fort. recte*')

(-?) 1517° (1154-5) ἡ ... κατὰ τοῦτο (Y: κατ' αὐτὸ Σ βγ) πάλιν τὸ ἐξαίρετον θεωρουμένη α βγ Heph.] ἡ ... πάλιν ἐξαιρέτως θεωρουμένη (κατὰ τοῦτο ... τὸ *om.*) V + G (que ... rursum precipue consideratur), *cf.* Procl.

(-?) 1566° (1187) εὐπροσκοπώτερον α βγ] εὐπροκοπώτερον V + G (facile profectibilius) (*tacite Boer*)

BOOK IV

(-?) 37° (35) πίστεων α βγ] πίστεως V + G (fidem) (*Robb.*)

(-?) 99° (82) καὶ[3] α βγ Procl.] ἢ V + G (aut)

(-) 124° (100) ἐν α βγ Procl.] καὶ ἐν V + G (et in)

(-?) 191° (150) τῶν πράξεων α βγ] τῆς πράξεως V + G (operationis)

(-?) 262° (209) λέγομεν α βγ, *cf.* Procl.] λέγω μὲν V + G (Dicamus: λέγωμεν)

(-?) 271° (214) συμβαίνῃ α βγ] συμβαίνει V + G (accidit)

325° *hic desinit* Σ

(-?) 409° (315) ὀλίγων Y βγ] ὀλίγων ὄντων V + G (paucorum existentium)

(-) 506° (388) ἀλλοτριωθέντες Y βγ, *cf.* Procl.] ἀλλοτριωθέντων V + G (alienatis)

(-*) 517° (396) συμφώνως Y β Heph., *cf.* συμφώνους γ] συμφωνῶσι V + G (concordes fuerint) ('*fort. recte, recepit Robb.*')

(-?) 556° (425-6) διαμηκίσωσιν (-ζωσιν Y, -κοίσωσιν S) α βγ] διαμηκίζουσιν V + G (per longitudinem distant)

(-?) 572° (439) ἐκ τῆς τῶν ... ἐπεμβάσεων ποιότητος Y βγ Heph.] ἐκ τῆς τῶν ε' πλανωμένων ... ἐπεμβάσεων ποιότητος V + G (ex qualitate accessuum quinque planetarum ...), Procl.: ἐκ τῆς τῶν <τῶν> ε' πλανωμένων ... ἐπεμβάσεων ποιότητος (*Boer Robb.*)

(-*) 783° (593) τῷ βίῳ Y βγ] ἐν βίῳ V + G (in uita) Heph. ('*haud scio an recte*' *Hüb.*)

797° (602-3) *hic desinit* V

From the two series (1.3.1 and 1.3.2), and in particular from the rejected readings, we may conclude, as we have already supposed, that there were originally two main branches in the tradition, both deriving independently from an archetype ω: on one side V (L) + G + Procl., on the other side α βγ. That V (L) + G + Procl. derive from an hyparchetype ψ is evident from section 1.3.2, where α βγ support together the edited text against the rejected readings of V (L) + G (+ Procl.) in more than sixty cases. To be sure, other editors (Boll, Boer, Robbins) adopted some of the readings of V (L) + G, which are rejected by Hübner; and even Hübner himself is sometimes in doubt over whether to accept or reject a reading; and we, too, have questions regarding some of the rejected readings (see below under 1.3.3 and the numerous cases introduced with (-?) in Chapter V below). Nevertheless, there remain clear cases of negative readings in ψ, such as omissions, where the α βγ branch seems to have preserved the authentic text, e. g. II.199°, 305°, 958°; III.1314°. From these cases we may conclude that there existed an hyparchetype ψ, as Boll and Hübner postulated (to explain the connection between V, L and y). A similar case can be made for the second branch of the tradition. The common rejected readings of α βγ in 1.3.1 are sufficiently numerous and obvious to postulate an hyparchetype for α βγ. (See for example, besides several minor omissions, the rejected readings of α βγ in I.778°; II.205°, 235°, 759°; III.394°, 1326°; IV.497°, 699°.) This means that branch α and branch βγ ultimately depend on a common hyparchetype φ, distinct from ψ. Boll and Hübner did not hypothize such an hyparchetype. In their view, α, β and γ derive directly, and independently from the archetype ω, and specific readings of these branches were already found in the Ptolemaeic text used by Hephaestio. One must concede that the hypothesis of an hyparchetype φ, common to α and βγ, is less evident than that of ψ, because in contrast to ψ, different corrections and contaminations entered *en masse* into the manuscripts of the families α, β and γ, which makes their consensus in opposition to ψ rare. This may explain why fewer particular features and obvious errors are found to distinguish the two branches of the tradition.

1.3.3. Some questions

If the distinction between the two branches in the tradition is obvious, it is less clear how we have to evaluate this bipartition. There are nearly no cases of evident copyist errors (real 'Trennfehler') distinguishing both branches. The different readings in the two traditions often seem to be both possible, though one could be the correction of the other. Let us have a closer look at some of the rejected readings of V (L) + G that oppose group α βγ (1.3.2). Some of these rejected readings of V (L) + G may have been the primitive readings in the archetype of the whole tradition (ω), but amended by the editor of the hyparchetype of α βγ:

I.435° (338) τοῦ ψυχροῦ ... τοῦ ξηροῦ Y M S γ] *inv.* V L D + G (siccum ... frigidum) Procl. (*Robb.*). — 'Two of the four humors are fertile and active, the hot and the moist ... two are destructive and passive, the dry and the cold': this is the reading of VL + G + Procl., but, if one would respect the exactly correspondent order to 'hot and moist,' one should write the inverse: the cold and the dry, which is the reading of α β(-D) γ. There are two similar examples of inversion in book I, in the same distribution of witnesses:

I.896°-7° (676-7) (τὰ διάμετρα ζῴδια τοῦ τε Καρκίνου καὶ τοῦ Λέοντος ὅ τε) Αἰγόκερως καὶ ὁ Ὑδροχόος α D S γ, *sed* ὁ *om.* α] Ὑδροχόος καὶ ὁ Αἰγόκερως L M + G (Aquarius scilicet et Capricornus) Procl.: Ζυγὸς καὶ ὁ Αἰγόκερως V [the copyist of V made an error in reading the astrological sign of Aquarius]. — 'The signs opposite to Cancer and Leo, namely Capricorn and Aquarius': this is the strictly right order. The reading of (V) L + G, 'Aquarius ... Capricornus' is not false, nobody would misunderstand it, it is only somewhat careless. One may understand why it was corrected. The divergences in α βγ could be a sign of a correction.

I.1257° (960-1) (ἀνατολικοὺς ... καὶ προσθετικοὺς ...) δυτικοὺς καὶ ἀφαιρετικοὺς α βγ Procl.] (...) ἀφαιρετικοὺς καὶ δυτικοὺς V L + G (ablatiui ... et occasiui). — Again, two pairs of adjectives stand in opposition. It seems that there are again two redactions: one a correction of the other. δυτικοὺς καὶ ἀφαιρετικοὺς of α βγ is the right order which corresponds to ἀνατολικοὺς καὶ προσθετικοὺς in 1256°. But it could also be a correction of the reading of V L + G, which is not false, but only less logically ordered. The author of the Proclus paraphrase may have corrected the order for the same reason as the editor of hyparchetype φ (unless the latter depends on Procl.).

These three minor examples with the same distribution of witnesses indicate that some meticulous scholar wanted to introduce more order in the transmitted text. This also seems to be the case in a number of syntactical/grammatical differences which are found in the same distribution in the manuscripts:

IV.271° (214) συμβαίνῃ α βγ] συμβαίνει V + G (accidit). – The indicative in V + G could well be the reading of the archetype (ω), wich was changed into the conjunctive by someone who wanted the form of the verb (after ὅταν) grammatically more correct.

IV.556° (425-6) διαμηκίσωσιν (D M γ: -ζωσιν Y, -κοίσωσιν S) α βγ] διαμηκίζουσιν V + G (per longitudinem distant). — Again, the same difference (indicative / conjunctive) as in IV.271°, and with the same witnesses. Notice the divergences in α βγ, which seem to indicate corrections in the archetype (ω).

There are six further instances in book III § 14 of the same differences between an adopted conjunctive and a rejected indicative which depend directly or indirectly on ὅταν, but with somewhat changing witnesses; see III.1210° 1212° 1215°(2) 1217° 1220° in the list. Notice that, in all six instances, G translates the rejected indicative, always attested by Heph., either with V (3) or against V (3), but together with S/β (6) and α (5).

Let us examine three other cases where the editor of the hyparchetype φ seems to have introduced corrections in grammar and syntax:

II.181°-3° (142-3) ὡς ἐκ τούτων ἕκαστον τῶν τεταρτημορίων ... τριγώνοις συνοικειοῦσθαι α βγ] ὡς (*om.* V, *habet* G) ἐκ τούτων ἕκαστον τῶν τεταρτημορίων ... τριγώνοις συνοικειοῦται V + G (ut ex hiis consequitur: quelibet quartarum ... trigonalitatibus coaptatur); *cf.* Procl. (88,3-6 Ἐκ τούτου οὖν φανερὸν, ὅτι ἕκαστον τῶν τεταρτημορίων πρὸς δύο καὶ ἀντικείμενα τρίγωνα συνοικειοῦται). — Was the anacolouthon in V, which Moerbeke read as well as Procl., simply a copyist's mistake? Could the change of the construction at the end of a six-line argument not be the primitive reading in the common archetype, and the infinitive in α βγ be a grammarian's correction? Moerbeke (as Procl.) read in V συνοικειοῦται, and he tried to solve the anacoluthon by inserting *consequitur*. Procl. tried to improve the sentence in a similar way. One can see why the editors of the Greek text prefer the reading of α βγ, but consider the same situation in the following examples:

II.1035°-6° (792-3) διὰ τοῦ καθ' ἑκάστην (-ον Y: -ου Σ G° S) τῶν τοιούτων (τούτων τῶν Y: τούτων Σ) ... συσχηματισμοῦ α βγ] διὰ τοῦ καθ' ἑκάστην τῶν τοιούτων ... συσχηματισμῶν (συσχημῶν V) V L + G (per id quod ... configurationum: διὰ τὸ ... συσχηματισμῶν). — On the side of V L + G there is again an anacolouthon, which makes the sentence harsh. On the side of α βγ, there are hesitations, only MD γ have the adopted text: a sign that the common archetype was corrected?

IV.718° (548) ἐφαρμόζωμεν γ, Y (-μώζομεν), β (-μόζομεν) Heph. (191,2: ἐφαρμόσωμεν)] ἐφαρμόζειν V + G (adaptare): συνεφαρμόζειν Procl. — It is difficult to construe the infinitive ἐφαρμόζειν with ἀναγκαῖον προϋποτίθεσθαι καὶ σκοπεῖν 712°, because of ἀλλὰ (716°), which corresponds to ὅπως μὴ (712°). ἐφαρμόζειν is again an anacolouthon at the end of a long sentence. It could well be the reading of the common archetype (if not the author's), while the conjunctive, which induced hesitations in the witnesses, could be a correction (from Heph.).

In all three cases we find in α βγ similar attempts to set right a complex syntactical construction. We add another case, where the reading of V, confirmed by G, points to a mistake (or an ambiguous correction) in the common archetype, but to which α and βγ reacted differently without restoring the right text:

II.22-23° (21-22) (συνάψομεν ...) τὰ καθ' ἕκαστα τῆς τῶν ... ἐμπιπτόντων πραγματείας *coni. Hüb.*] τὰ καθ' ἕκαστα τῆς πραγματείας (!) τῶν ... ἐμπιπτόντων πραγματείας V + G (particularia negotii de incidentibus ... in operando)] καθ' ἕκαστα τῶν ... ἐμπιπτόντων τῆς πραγματείας Y: τὰς (*sc.* πραγματείας) καθ' ἕκαστον τῶν (τῆς D) ... ἐμπιπτόντων πραγματείας Σ βγ. — In V we find πραγματείας two times. This must have also been the case in the model of Moerbeke, who tried to make sense of the second πραγματείας by translating it as *in operando*. The copyist of the archetype (ω) might have intended to write πραγματείας immediately after the article τῆς, to which it is connected, instead of having it after ἐμπιπτόντων, forgetting to delete it after ἐμπιπτόντων. In any case, there was a problem in the archetype with τῆς and the widely separated πραγματείας: Y and Σ βγ tried to solve the textual problem in different ways, without getting back to the original text, which is coniectured by Hübner. This case illustrates the difficulty mentioned above: the supposed hyparchetype φ is, in this case, characterised only negatively, i.e. by the fact of not having copied the problematic second πραγματείας. De facto, we are again confronted with a division into three branches, as α and βγ make other conjectures.

In the following cases, the opposition between the two branches is more evident. Here again we notice that a rejected variant of V, which is confirmed by G, seems not to be a simple copyist's error, but may have been the original reading in the archetype ω, corrected by the ancestor of α βγ.

I.820° (612-3) μιᾶς ὀρθῆς α βγ] γωνίας ὀρθῆς V L Procl. + G (anguli recti) (*Boll*). — Both readings are possible, but μιᾶς ('two parts of one right') could be an improvement of γωνίας in order to be even more precise and to adapt the expression exactly to the preceding μίαν ὀρθὴν (818°). But γωνίας ὀρθῆς is already completely precise: 'two parts of a right angle'.

I.939° (708) καὶ συνίσταται βορρολιβυκόν α βγ, *cf.* Tib. (colligitur)] καὶ γίνεται βορρολιβυκόν V L + G (et fit borolibica) Procl. — Both verbs are possible, and γίνεται is not an error. Was συνίσταται a correction, in order to assimilate the verb to the following three expressions in the same context (συνίσταται νοταπηλιωτικὸν. ... ἀπηλιωτικόν, ... λιβυκὸν 954° 967° 980°)?

I.973° (732) κατελείφθη μόνῳ λοιπῷ ὄντι α βγ] κατελείφθη μὲν λοιπῷ ὄντι V L + G (... quidem ...) (*Robb.*). — Neither μὲν nor μόνῳ is found in Proclus; μόνῳ however is found in the lemma of the commentary (38,9) and the commentator tries to explain it: this fourth triangle is assigned to Mars alone, whereas the other triangles are assigned each to two luminaries. The addition of μόνῳ is not necessary; it is more precise, but redundant. The μὲν, on the contrary, is adapted to the context, because it corresponds to the following συνοικοδεσποτοῦσι δὲ (975°), which introduces the co-rulers.

I.1133° (875) περισσῆς α βγ] πολλῆς V + G (multa) (*Boll*). — Both readings are perfectly possible: see on the one hand Iamblichus, *De com. math. sc.* 59,25 (περιττῆς δεόμενα ἀποδείξεως), on the other hand Philoponus, *In De gen. et corr.* 154,28 (πολλῆς δεόμενον ἀποδείξεως) and Chrysostomus (PG 51, p. 38, l.21). The reading περισσῆς may be preferable, because it is more precise, but πολλῆς is not a copyist's error, and it is difficult to imagine that περισσῆς, the stronger and more precise expression, was corrected into πολλῆς.

I.1135° (877) ὑποτυπῶσαι α βγ] ὑποδεῖξαι καὶ τυπῶσαι V + G (ostendere et exprimere) (*Boll*). — This is the only occurrence of this hendiadys in the TLG. Could the reading of α βγ be a correction in order to contract and simplify?

II.505° (383) τοῦ ζῳδιακοῦ κύκλου α βγ] τοῦ ζῳδιακοῦ V + G (zodiaci) Procl. — The addition of κύκλου is not necessary, *cf.* τοῦ ζῳδιακοῦ in 501°. The text which follows is not about the quality of a circle, as in the context of I.684°.

One may conclude from the examples we have examined that V (L) + G (+ Procl.) — i.e. branch ψ — may often have preserved, perhaps not the 'better', but at least the primitive reading against a seemingly 'better' corrected reading of branch φ. One should of course not generalise this hypothesis. There are other cases — in particular omissions, as in II.958° and in III.1314° — where it seems that the tradition of α βγ has preserved the original text against V + G (+ Procl.).

2. G WITH OTHER WITNESSES AGAINST V (L)

So far we have only discussed the cases — and these are the majority — where G is in conformity with V. We concluded that G is an excellent representative of ψ. However, there are strong indications that G had been occasionally corrected. It has several readings corresponding to readings found in the φ-branch, especially in β(γ), and also some double readings. Hübner adopted many of these readings, deviant from V, in the edited text. Even the rejected readings are meaningful variants.

2.1. Relations of G with readings of (Σ) βγ against V Y (or V α)

First of all, there are cases where G is in consensus with (Σ) βγ, against V Y or V α. In some cases D, which is often contaminated with V, does not follow β.

2.1.1 Accepted readings of (Σ) βγ + G

BOOK I

135° (108) τὸ ἴδιον Σ βγ + G (eam que ... proprietatem)] ... ἴδιον (*om.* τὸ) V Y

213° (168) πάντα Σ M S γ + G (omnia), *cf.* Procl.] μὴ *praem.* V Y D

230° (181) ἡ συμβίωσις Σ M S γ + G (conuictus)] εἰ συμβίωσις V Y D

inde ab 235° *usque ad* 509° *deficit* Σ

260° (205) ἅπαντα βγ + G (omnia) Heph. (*post* νομίζειν), *cf.* Procl.] ἕκατα V: ἕκαστα Y (*Robb.*): [Σ]

286° (226) ἴδοι βγ + G (uidebit)] ἴδη V: εἴδη Y: [Σ]

845° (633) τε καὶ καταφέρεσθαι βγ + G (et descendunt) Procl.] *om.* V α

848° (636) ἐκείνων βγ + G (illa), *cf.* Procl.] ἐκεῖνον V Y: -νο Σ

1072-1084° (811-834) *tabulam habet* (*sed sine summa integra*) G: *tabulam habent* (*cum summa integra*) Σ β(-S) Procl.] *om.* V (*spatio relicto*) Y[1] S, *suppl.* Y[2]

BOOK II

190° (149) αὐτὸν Σ βγ + G (ipse)] αὐτῶν V: *om.* Y

302° (229-30) ἔχει δὲ βγ + G (habent autem) Procl.] ἔτι δὲ V: ἔχῃ Y: ἔχοντος Σ

333° (252) οἰκείων βγ + G (domesticis) Procl.] οἰκειωτάτων V α (‘*fort. recte*’)

509° (386-7) καθ’ ὧν βγ + G (in quibus)] καθ’ ἣν V: *om.* α

520° (396) μὲν βγ + G (quidem) Procl.] μέντοι V: μέν τι α

636° (484) νηκτοῖς Σ βγ + G (natatilibus), *cf.* Procl.] νυκτοῖς V Y

666° (507) πόστον δὲ Σ βγ + G (quotam autem) Heph. (L), *cf.* Procl.] πόσὼν δὲ (*sic*) V: μέσον Y

834° (634) καθολικῶν Σ βγ + G (uniuersalium) Procl.] κολικῶν V Y

1079° (825) ἅλως Σ D S γ + G (halo) Procl., *cf.* Y] ἄλλως V M Heph.

1100° (842) ἅλως Σ βγ + G (aree) Procl. Heph. (A)] ἄλλως V Y Heph. (LP)

BOOK III

57° (49) ἐκείνῳ βγ + G (illi) Procl.] ἐκεῖνο V α

77° (63) αὐτῷ Σ βγ + G (ipsi), *cf.* Procl.] αὐτὸ V Y

88° (71) τῷ ποιητικῷ βγ + G (factiuo) Procl.] τὸ ποιητικόν V α

106° (85) οὗ Σ βγ + G (ad quod), *cf.* Procl.] οὐ V Y

141° (114) ὄντος βγ + G (existentis) Heph., *cf.* Procl.] ὄντως V: ὄντας Y: ὄντων Σ

149° (119) οὗτος Σ βγ + G (iste) Heph. + Procl.] οὕτως V Y

200° (156) ταυτολογῶμεν Σ βγ + G (idem loquamur)] ταυτολογοῦμεν Y: ταῦτα (*i.* ταὐτὰ) λέγωμεν V

241° (191) δεῖ Σ M S γ + G (oportet) Heph.] δὴ V D: γὰρ Y

241° (191) τὰ Σ + G (que) Heph. (AP) Procl., *cf.* 244° *infra sub* 3.2] τὰς M S Heph. (C): *om.* V Y D γ (‘*haud scio an recte*’)

342° (266) ἀπομερίζωμεν Σ βγ + G (distribuamus)] -ζομεν V Procl.: -ζόμεθα Y
346° (269-70) οἷόν τε Σ M S γ + G (possibile est), *cf.* Procl. (δύναται)] οἴονται (*sic*) V Y D
381° (295) μὲν Σ M S + G (quidem) Heph.] *om.* V Y D γ
388° (299) ἔτι πολυπραγμονοίη Σ M S + G (amplius ... uoluerit percunctari) Heph.] ἐπιπολυπραγμονοίη V Y D γ (*Boer Robb.*)
458° (353) ὁ τῆς σελήνης Σ βγ + G (locus lune)] οἱ τῆς σελήνης V G° Procl.: ἡ τῆς σελήνης Y
468° (360) ἢ πιθήκων Σ βγ + G (uel de simiis) Heph.] *om.* V Y Procl. (*Robb.*)
520° (399) μάλιστα Σ βγ + G (maxime) Procl. Heph.] *om.* V Y
554° (423) οὗτος Σ βγ + G (iste) Procl.] οὕτως V Y
580° (442) ἐπειδήπερ Σ βγ + G (quoniam quidem)] ἐπείπερ V: ὅπερ Y
652° (491) ἀναιρεῖ Σ βγ + G (perimit) Heph.] *om.* V Y Procl.
833° (621) μεσουρανήματος Σ β + G (a medio celi) Heph.] οὕτως V γ, *cf.* Y
885° (662) εὐεκτικοὺς Σ βγ + G (mediocris pinguedinis) Procl.] εὐεκτίας V Y
889° (665) σπινώδεις Σ βγ + G (subtiles), *cf.* Procl.] σπινώδης V Y
1008° (757) εὐωνύμων πάντων Σ M S + G (sinistrorum omnium) Procl. Heph.] εὐωνυμωτάτων V Y D γ
1209° (910) ἐπίδοξον Σ βγ + G (super gloria)] ἑτερόδοξον V Y
1210° (911) ἄλλους Σ βγ + G (aliis) Heph.] ἀλλήλους V Y
1291° (981) νηλεεῖς βγ + G (duros)] μηλίους V Y: [Σ]
1348° (1028) θεοπροσπλόκους βγ + G (deo supplices) Procl., *cf.* Y] θεοπροσπόλους Σ: θεοπλόκους V

BOOK IV

48° (43) κτήσεις Σ βγ + G (possessiones), *cf.* Proclus] κτίσεις V Y
189° (148) οἷς ἂν Σ βγ + G (quibuscumque)] οἷς ἐὰν V Y Heph.
266° (211) τοῖς εἰρημένοις Σ βγ + G (dictis) Procl.] τῆς εἰρημένης V Y
277° (219) ἡδείας Σ β B C + G (delectabiles) Heph., *cf.* Procl. (καλὴ)] ἰδίας V Y: *def.* A
308° (237) ἡδείας Σ D M γ + G (delectabiles) Heph., *cf.* Procl.] ἰδίας V Y S
326° (251) ἐν θηλυκοῖς ... τόποις Σ βγ + G (in femininis ... locis) Procl. Heph.] ἐν θηλυκῶν ... τόποις V Y
453° (348) σπονδὰς βγ + G (compactiones) Heph., *cf.* Procl.] σπουδὰς V Y
461° (354) οἱ M S + G (loca), *cf.* οἱ ... τόποι Procl. Heph.] ἡ V Y D γ
706° (538) προθέμενος βγ + G (proponens)] μὴ *praem.* V Y
706° (538-9) δέον βγ + G (quod oportunum erit)] οὓς δέον V Y

One should first notice that there are so many rejected readings and even real mistakes of V + Y (II.636° 834° III.1209° 1210° 1291° IV.48° 266° 453°). This cannot be a coincidence. It seems that there is a contamination between V and Y, which means that singular mistakes or false conjectures came by correction from one into the other, probably from V into Y, a manuscript which has integrated many foreign influences. In these cases, the adopted reading which G shares with (Σ) βγ against V could be simply the reading of the hyparchetype ψ. This hypothesis is confirmed by the fact that the Proclus paraphrase almost always goes with G + βγ. In other cases (e.g. III.468° 520° 652° 1008°) as well as in the next part (rejected readings, 2.1.2) there should have been correction in the model of G or in its ancestor. Secondly, one should notice that in nearly 20 cases of consensus between G + βγ we have also a consensus with Heph.

2.1.2. *Rejected readings of (Σ)* βγ + G

BOOK I

(-?) 244° (192) συνεργεῖ V Y Procl.] *om.* βγ + G: [Σ]

(-?) 248° (195) τὰ μείζω V Y, *cf.* τὰ μείζονα Procl.] τὸ μεῖζον βγ + G (maius): [Σ]

(-?) 272° (214) ἐκ πυρώσεων V Y, *cf.* Procl.] ἐκ -εως βγ: ἐκπυρώσεως *legit* G (adustionis): [Σ]

(-?) 281° (221-2) ὅσα δὲ V Y] ἃ δὲ βγ, *sed* συμπτώματα *add.* β *post* ἔχει; ἃ δὲ συμπτώματα Procl.: symptomata autem que G

(-?) 314° (248) θεραπευόντων V Y] -πευσόντων βγ: curaturis G

(-) 441° (342) Ἄρεως V L Y, *cf.* Tib.] κακοποιοὺς *add.* βγ + G (maleficos) Procl.

(-) 682° (511) διὰ τὸ τὸν V Σ] διὰ τὸ τότε τὸν β B C + G (quia ... tunc): διὰ τότε τὸν A: διά τε τὸν Y

(-) 1055° (798) μὴ V Y] μηδὲ Σ βγ + G (neque)

BOOK II

(-) 228° (175-6) κυκλάδες V α Procl.] κυκλάδες νῆσοι βγ + G (Ciclades insule)

(-*) 277° (211) ἐκ τῶν μητέρων α Procl., Co le] ἐκ τῶν μητέρων αὐτῶν βγ + G (ex matribus ipsorum): ἐκ τῶν ἀμητέρων V

(-*) 337° (254) ἀνατολικῶν συσχηματισμὸς *coni. Hüb.*] ἀνατολιῶν συσχηματισμὸς V (*Robb.*): ἀνατολικὸς συσχηματισμὸς βγ + G (orientalis configuratio): ἀνατολικὸν σχηματισμὸν Y: κατὰ ἀνατολικὸν συσχηματισμὸν Σ

(-*) 402° (301) ὡς α] ὥστε βγ + G (ita ut): *om.* V

(-) 442° (330) σαθροὶ V α, *cf.* Procl.] θρασεῖς βγ + G (audaces)

(-) 553° (421) τῶν τε ἡλιακῶν τῶν αὐτῶν V Y Procl., *cf.* Arab.] τῶν τε ἡλιακῶν καὶ τῶν σεληνιακῶν τῶν αὐτῶν β: τῶν τε ἡλιακῶν δηλαδὴ καὶ τῶν σεληνιακῶν Σ γ: solaribus et lunaribus *post* eclipsibus (551°/420) G

(-?) 600°-1° (458-9) τοὺς ... ἔχοντας V Y Procl.: τὰς ... ἔχοντα (*sic*) Σ] τὸν ... ἔχοντα βγ + G (illum ... qui ... habet) Heph.

(-?) 672° (511) ἐπὶ τὸ ἥμισυ α] περὶ τὸ ἥμισυ βγ + G (circa dimidium) Heph.: πρὸς τῷ ἡμίσει V

BOOK III

(-) 5° (6) περὶ μοίρας ὡροσκοπούσης *cf. cett.*] *tit. add.* quomodo oportet accipere horam ca. 4^{m} G (πῶς δεῖ λαβάνειν τὴν ὥραν?): ἐν ᾧ καὶ περὶ τοῦ πῶς δεῖ λαβεῖν τὴν ὥραν γ: περὶ τοῦ πῶς δεῖ λαμβάνειν ὥραν M: *cf. infra* 136°

(-?) 87° (71) κατὰ V Y + Procl.] καὶ κατὰ Σ βγ + G (et secundum) (*Robb.*)

(-?) 136° (109) εὑρισκομένης V α] *tit. add.* πῶς δεῖ λαμβάνειν ὥραν D M + G (Quomodo oportet accipere horoscopum, capitulum 4^{m}): πῶς δεῖ λαβεῖν τὴν ὥραν *ut titulum add.* Co (90,30) *cf.* γ, *mg.* S^{2}

(-?) 157° (124) ἐγγὺς εἶεν τῷ ἀριθμῷ V Y D γ, *cf.* Procl.] ἐγγὺς ἔχοιεν τὸν ἀριθμὸν S + G (propinquum habeant numerum), Co le (92,2): ἔχοιεν τὸν ἀριθμὸν ἐγγὺς Σ M

(-*) 187° (147) περὶ ξενιτειῶν V Y D] περὶ ξενιτείας Σ M S γ + G (de peregrinatione) Procl. Heph. (*Robb.)*

(-) 271° (213) καὶ[1] V Y Heph.] ἢ Σ βγ + G (uel)

(-?) 298° (233-4) ἑξάγωνος ὢν V Y] ἑξάγωνος Σ βγ Heph. + G (scilicet in sextili aspectu)

(-) 412° (317-8) πάλιν ὄντες V Y D Heph.] *inv.* Σ M S γ + G (existentes iterum)

(-*) 418° (323) δὲ V Y] γὰρ Σ βγ + G (enim) Co le, *cf.* Heph.

(-*) 440° (339) μόνον V Y Heph.] ἕνα Σ βγ + G (unum) Procl.

(-*) 476° (366) μηδὲ εἷς V] μηδεὶς Σ βγ + G (nullus) Heph.: μηδεὸ Y

(-*) 490 (377) λόγου V Y D γ] *om.* Σ M S + G, *cf.* Procl. Heph.

(-?) 539° (412) ἀνατολὴν V Y γ] ἢ τὴν ἀνατολὴν Σ β Procl. + G (uel ortum)

(-?) 580° (442) κακοῖ V (κακοὶ Γ: κα'κεῖ G°: κακῇ Y), *cf.* Procl. (βλάπτει)] *om.* Σ βγ + G

(-) 630°-631° (474-5) τοῖς μέχρι τῆς καταδύσεως τοῦ ἀφέτου συναγομένοις V Y Procl. Heph.] τοῖς ὑπὸ τοῦ ἀφέτου συναγομένοις μέχρι τῆς καταδύσεως τοῦ ἀφέτου Σ βγ + G (collectis a dimissore usque ad occasum dimissoris) G

(-?) 675° (510) ἀναλελυμένος V Y Procl. Heph.] βοηθούμενος D + G (auxiliatus): βοηθούμενος ἢ ἀναλελυμένος Σ M S γ

(-?) 1210° (910) ἔχωσι V Y γ] ἔχουσι Σ β + G (habent) Heph.

(-) 1397° (1065) καλοζήλους V Y D γ] κακοζήλους Σ M S + G (mali zelatores) Procl.

(-) 1548° (1176) σαθροὶ V Y Procl.] θρασεῖς βγ + G (proni) Heph., *cf.* θάρσεις Σ, *cf. supra* II.442°

BOOK IV

(-?) 39° (36) (διὰ) φιλικῶν V, *cf.* Y (φυληκῶν)] (διὰ) φιλίων G° (*cf.* Heph. φιλιῶν) + G (per amicitias): (ἐκ) φιλίας Procl.: (διὰ) φιλῶν Σ βγ

(-*) 330° (253-4) καὶ τῆς (*om.* D) κάλπιδος καὶ τῶν ὀπισθίων (τοῦ Λέοντος) V Y D γ] καὶ τῶν ὀπισθίων M S Heph. (*Robb.*) ('*fort. recte*') + G (et in natibus et in posterioribus *dupl. lectio Guill.*)

Ab 797 *deficit* V

As we have seen in the preceding section (2.1.1), some of the accepted readings shared by G with βγ against V Y do not necessarily prove that the model of G was corrected from βγ. G could merely have been, in these cases, a better representative of ψ against deviant readings of V + Y. But other cases show an undeniable connection of the Greek model used by Moerbeke with the (Σ) βγ-family. Significant examples are: additions (accepted by Hübner) such as III.468° 520° 652° 1008° (all with Heph., see above) against the omissions in V Y (+ Procl.). The hypothesis that G was corrected from βγ is confirmed by the list of shared readings rejected by Hübner: additions (rejected) such as I.281° 441° II.228° 553° III.630°, variant readings such as II.442° III.1548° (both σαθροί/θρασεῖς), III.1397° (καλοζήλους/κακοζήλους). The most important case is III.136°. Here, Moerbeke adds, inside chapter 3 (Περὶ μοίρας ὡροσκοπούσης *De gradu horoscopi*), this extra title: 'Quomodo oportet accipere horam' (Πῶς δεῖ λαμβάνειν ὥραν). This section is counted as chapter 4, and the numbers of the chapters in the tabula of Moerbeke's translation are changed accordingly. We find the same title and attempts to change the tabula in βγ and the same title as lemma in the Commentary, but nothing similar in V and α. The addition of this title is quite understandable: it separates the first part of chapter 3, where the general difficulty with regard to the horoscopic point and the inadequacy of several instruments are exposed (§§1-2), from the second part (§§3-4), where a specific reasoning method of finding the horoscopic point is given. The addition of this title could come from an ancient scholion.

The repeated consensus of G with (Σ) βγ against V is evident, but it is also limited, if one considers the mass of divergent readings and common mistakes of that family listed above, which are all absent in G. How should we explain the presence of the (Σ) βγ readings in G? Could it be that Moerbeke used a second manuscript to correct the Greek model he had taken as the basis of his translation? We know that Moerbeke often collated a second Greek manuscript, sometimes also a commentary. This hypothesis is, however, in this case improbable. In fact, Moerbeke himself tells us that he found at the end of the text an addition in 'his Greek model' (see Appendix 1): 'Que sequntur *in greco exemplari* subiniuncta reperi ...'. We therefore cannot suppose that it was the translator who used a second Greek manuscript, but we have to admit that some Greek scholar — perhaps the same person who also added that astrological note — had introduced by correction specific readings from βγ before Moerbeke made his translation. This is also evident in cases where Moerbeke translates a double reading, that of V and that of the (Σ) βγ family. We discuss the principal cases in the following section.

2.1.3. Double readings in G

I.237°-9° (187-8) τί ἂν εἴη συμφορώτερον (...) τῆς τοιαύτης προγνώσεως V Y, *cf.* Procl.] τί ἂν εἴη (...) τῆς τοιαύτης γνώσεως (προγνώσεως γ) σπουδαιότερον M S γ: quid utique erit magis conferens [= VY] (...) tali cognitione studiosius [= βγ] G: *om.* D: [Σ]. — Besides the version of VY there was another version of the same sentence: instead of συμφορώτερον at the beginning (VY) βγ has σπουδαιότερον at the end. Both expressions are found in the Ptolemaic vocabulary of the *Apotelesmatica*, and both are possible. The version of V Y is certainly the better one. Moerbeke's translation combines both and is redundant. But it shows that, in his model, two versions of the tradition had been integrated, probably by correction.

I.1117° (862) γίνεται τὸ πᾶν τξ´ V D γ Procl.] ὅμου τριάκοντα ἑξήκοντα M S: fit totum simul 360 G: γίνεται (-ονται Σ) μοῖραι τξ´ α. — The hesitation in the tradition indicates correction. Moerbeke's reading is redundant, but shows that in his model (or in an ancestor of his model), ὅμου was a correction for τὸ πᾶν, written *supra lineam* and misunderstood as an addition.

II.185° (145) κατ' αὐτὸ τὸ μέρος Σ] κατὰ μέρος βγ: κατ' αὐτὸ τὸ μερικὸν V Y, *cf.* Procl.: secundum ipsam partem partialem (κατ' αὐτὸ τὸ μέρος μερικὸν) G. — Moerbeke translates both readings of the tradition. His model probably had the reading of V Y (see κατ' αὐτὸ) with a correction, μέρος *super* μερικὸν, which could be taken as an addition. See also a longer comment in the list we give in Chapter V.

II.1090° (834) ἢ χλωρὰ (χλόρα V) V Procl.] ἢ ὠχρὰ α γ L Heph. Lyd.: ἢ ὠχρὰ ἢ χλωρὰ D M: ἢ χλωρὰ ἢ ὠχρὰ S + G (uel uiridis uel rubicunda). — Moerbeke had both readings (ὠχρὰ / χλωρὰ) in his model, as is the case in the β-branch. Correction in ω?

III.265° (209) μηδενὸς ἢ τῶν κακοποιῶν V Σ β Co le (99,11), *cf.* Procl.] μηδενὸς τῶν ἀγαθοποιῶν γ Heph., *etiam* Co (99,13-14: μηδενὸς τῶν ἀγαθοποιῶν ... ἢ μόνον τῶν κακοποιῶν): μηδενὸς τῶν κακοποιῶν Y: a benefico nullo uel a maleficis G (μηδενὸς τῶν ἀγαθοποιῶν ἢ τῶν κακοποιῶν). — In the passage which immediately precedes, G confirms a reading of V Y against Σ βγ. Here, it seems that G combines — and certainly not through contamination by γ — both terms which are present in the Greek tradition. Ptolemy's text is difficult and condensed. ἀγαθοποιῶν is found already in Heph. and also in the explication of the Commentator. Was there an addition *supra lineam* already in the common archetype?

III.594° (453) θέλει τε (τε *om.* Y) καὶ δύναται V Y] ἐστί τε καὶ δύναται Σ D M (*desunt* S γ): est et uult (*fort.* ἐστί τε καὶ θέλει *legit*) G. — The two different readings of the Greek manuscript tradition (ἐστί and θέλει) are present together in the model of G, whereas the common reading δύναται is missing. The version without δύναται, which Moerbeke translated, is probably the result of a correction *supra lineam*, but gives a very good text. Note that these lines (594°-596°), excluded by Robbins as a scholion, were explained by Hephaestio (II 11,25) as an allusion of Ptolemaeus to a theory of Nechepsos and Petosiris about a calculation of the *sors fortunae*. It is to this kind of calculation that the certainly late ancient note present in the Greek model of Moerbeke (Appendix 1) is related.

The double readings of G which stem from two traditions always present meaningful alternatives. There are no examples of clearly aberrant readings, of which there are numerous examples in βγ. The same is valid for most of the βγ-readings + G listed above (2.1.2). The correspondences with the βγ-family cannot come from one of the existing manuscripts of the β-family; they are of older origin. We noticed in the first part of the series (2.1.1) that the consensus with βγ against V Y is often confirmed by Hephaestio (nearly 20), and this also holds for the second part, that is, for the 'rejected' readings (about 12 cases). These readings, which were already found in Ptolemy's text as read by Hephaestio, may have been added by a selective correction from an ancestor of the βγ-family into Moerbeke's model, which was fundamentally a very good representative of ψ.

2.1.4. Questions about some rejected readings

As we have seen in the list presented in 2.1.2, G goes with (Σ) βγ (essentially G + β) in a reading rejected by Hübner in around 40 cases. This does not, however, mean that we have here a list of common 'Trennfehler' of G + (Σ) βγ. In fact, in many cases the adopted and rejected readings are alternative versions, one more primitive and the other corrected. Take, for instance, two cases in which the version of

(Σ) βγ + G is rejected by Hübner as an 'omission' against a reading that is present in V Y.

I.244° (192) συνεργεῖ V Y Procl.] *om.* βγ + G. — The verb συνεργεῖ fits well, but is not really needed here, as is evident from the parallel formulations in 237° and 240°.

The second case is even more significant. In book III, chapter 11, § 4, Ptolemy speaks about the places or the signs which are *not* to be considered for the determination of prorogation: 'Of the part above the earth it is not fitting to consider <as prorogative places> either the sign that is disjunct from the ascendant or that which rose before it, called the House of the Evil Daemon' (τοῦ κακοῦ δαίμονος). There are two versions of the immediately following argument, versions that differ at two places:

III.580° (442) ἐπειδήπερ (ἐπείπερ V: ὅπερ Y) κακοῖ V (κακῇ Y, *cf.* Γ G°, *cf.* Procl. βλάπτει) τὴν ἐπὶ τὴν γῆν ἀπόρροιαν ... θολοῖ τε (μὲν Y: γὰρ Procl.) καὶ ὥσπερ ἀφανίζει τὸ ἀναθυμιώμενον ἐκ τῶν τῆς γῆς ὑγρῶν V Y, *cf.* Procl.] ἐπειδήπερ τὴν ἐπὶ τὴν γῆν ἀπόρροιαν ... θολοῖ καὶ ὥσπερ ἀφανίζει τὸ ἀναθυμιώμενον ἐκ τῶν τῆς γῆς ὑγρῶν Σ βγ + G (quoniam quidem astrorum que in ipso defluxum ad terram ... obfuscat et quasi exterminat quod euaporat ex humiditatibus). — The combination of V and Y (+ Γ and G° present in this section) attests that κακοῖ was in their common ancestor. The verb is also indirectly confirmed by Proclus (βλάπτει). Still, it is not required in the sentence; θολοῖ suffices. In Σ βγ + G, it is not found. Moreover, in 582°, there is no trace of the particle τε (μὲν Y: γὰρ Procl.) after θολοῖ, which connects in V the two verbs κακοῖ and θολοῖ. This particle is useless if there is only one verb. If one adopts the reading of Σ βγ + G, the text would be this: 'Of the part above the earth it is not fitting to consider either the sign that is disjunct from the ascendant or that which rose before it, called the House of the Evil Daemon, because it obscures (troubles, makes turbid) the emanation of the stars in it to the earth ... and, as it were, destroys the exhalations from the moisture of the earth'. The fact that both, κακοῖ and τε, are absent in Σ βγ + G is probably neither coincidental nor due to a double deletion or a double omission. There is no hesitation in the seven manuscripts of Σ βγ + G, but in the redaction of V Y (Procl.), there are variations in respect of the two words (see Hübner's apparatus). We suppose that Σ βγ + G present here a first reading of the common archetype which was enlarged by an addition of the two words in an ancestor of V or Y, in order to render Ptolemy's argumentation clearer and more precise.

But even if our interpretation is right — namely, that two versions are often opposed as a primitive redaction versus a corrected one — this does not mean that the primitive version is always to be found in the same group. Take the following example:

I.441° (342) τὸν τοῦ Κρόνου καὶ τὸν τοῦ Ἄρεως ὡς τῆς ἐναντίας φύσεως ποιητικούς V L Y, *cf.* Tib.] τὸν τοῦ Κρόνου καὶ τὸν τοῦ Ἄρεως κακοποιοὺς ὡς τῆς ἐναντίας φύσεως ποιητικούς βγ + G (Saturnum autem et Martem maleficos tamquam contrarie nature factiuos), *cf.* Procl. — The word κακοποιοὺς is added in βγ, in the model of Moerbeke, and in the Proclus Paraphrase (ὁ δὲ τοῦ Κρόνου καὶ ὁ τοῦ Ἄρεως, τῆς ἐναντίας ὀνομίσθησαν, καὶ κακοποιοί 28,12-14). But κακοποιοὺς is already implicitly expressed in the ὡς τῆς ἐναντίας φύσεως ποιητικούς clause. It was added as a counterpart to ὡς ἀγαθοποίους (438°) in order to make the sentence more explicit. This may also be the reason why Procl. (whose paraphrase usually corresponds to V L) added it.

III.675° (510) ἀναλελυμένος V Y Procl. Heph.] βοηθούμενος D + G (auxiliatus): βοηθούμενος ἢ ἀναλελυμένος Σ M S γ. — This is a case with two alternative terms. The combination V + Y + Heph. + Procl. supports ἀναλελυμένος. But the verb ἀναλύω is unexpected: it is nowhere else used by Ptolemy in this text. The reading βοηθούμενος, integrated in Moerbeke's model and in D (β-branch), and which figured as double reading in the rest of the β-family, was 'probably an explanatory gloss' (Robbins). Indeed, a very good explanatory gloss: βοηθέω or βοήθεια are used in the same paragraphs (14 and 15) and in the same sense for 'helping', i.e. the fact that an affliction due to a maleficient planet is diminished or 'released' by the help (βοηθοῦντος) of a beneficient planet (*cf.* 664°/500 βοηθοῦντα = *adiuuantia*; 842°/631 βοηθεῖσθαι = *per auxilia subleuatam*).

2.2. G: relations with α

Besides the readings of β there are also some readings of α, of Y (when Σ is absent or contaminated), of α M S or of α βγ against V (L) in Moerbeke's Greek model. Even if the consensus with these manuscripts is only occasional and rare, there is enough evidence again that it cannot just be a coincidence, but must be explained through a correction of Moerbeke's Greek model by a manuscript of the φ-branch.

2.2.1. G *with adopted readings of* α, Y, α M S, α βγ

BOOK I

51° (43) ὁλοσχερεστέρων Y + G (excellentiorum) Heph., *cf.* Procl.] ὁλοσχερῶν V Σ βγ

152° (122) τοῦ Y + G (quod)] τούτου V Σ βγ (*Robb.*)

807° (603) τῆς ... ποιότητος α + G (cum qualitas)] τοῖς ... ποιότητος V: καὶ ὡς ἡ ποιότητος M S γ, *cf.* D

885° (667) ὂν α βγ + G (existentem)] *om.* V L

914° (688-9) τὸ πλεῖστον *ante* ἀφίσταται α M S + G (secundum plurimum *ante* distat)] *post* ἡλίου (915°) D γ: *om.* V L Procl.

1140° (879) τοῦτον α M S + G (istum)] + *titulum* περὶ ὁρίων ὡς (V: ὁ D γ) Πτολεμαῖος V D γ Procl.: ὁ δὲ τῶν ὁρίων τῆς ὅλης ἐπιβολῆς κατὰ τὸν Πτολεμαῖον τοῦτον ἔχει τὸν τρόπον L (*initium excerpti*)

BOOK II

339° (256) Ματιανὴν α + G (Matianem)] μαντιανὴν V γ Procl. (*cf.* Heph.): μαντειανὴν β

426° (319) βορραπηλιώτην α βγ + G (boream et orientem)] ἀπηλιώτην V Procl.

607° (464) τόπῳ α βγ + G (loco)] χρόνῳ V Procl. (*Robb.*)

970° (739) τρεπτικὸν α βγ + G (uersatiuum) Heph.] τροπικὸν V L, *cf.* Procl. τροπῆς

1048° (801) τινὶ α M S + G (alicui)] τι V L D γ

1076° (823) καὶ α M S + G (et) Procl.] *om.* V D γ

1103° (845) τι α M S + G (aliquam) Heph.] *om.* V D BC Procl., *cf.* A

BOOK III

392° (302) περὶ ... θηλυκῶν α C + G Procl. Heph.] + γενέσεως V D: + γενέσεων AB

500° (384) μὲν α + G (quidem) Heph., *cf.* M S] *om.* V D γ

579° (442) ὃ καλεῖται Y D γ + G (quod uocatur)] καὶ καλεῖται V: ὃ καὶ καλεῖται Σ βγ: ὅπερ καλεῖται Procl.

623° (468-9) ἀλλὰ καὶ α M S + G (sed etiam)] + τὸν V D γ

680° (513-4) ἑκάστης μοίρας α + G (uniuscuiusque gradus) Procl.] ἑκάστη μοίρα V: -στας -ρας βγ

711° (533) ἕκαστον (*sc.* ἡμικύκλιον) α M S + G (unumquodque) Procl. Heph.] ἕκαστος V D γ

817° (612) τῆς ἀρχῆς Y + G (principium) Procl.] *om.* V Σ βγ

836° (626) διαστάσεων Y γ + G (distantiarum)] -εως V Σ β

842° (631) προειρημένον α M S + G (predictum) Procl.] εἰρημένον V D γ

977° (734) τὸ προδῦνον Y + G (id quod preoccidit) Co le, *cf.* Procl.] δῦνον V M: ἡγούμενον Σ γ, *cf.* D G° S

1025° (769) συνάπτῃ δέ Y + G (et coniungatur), *cf.* β Procl.] -πτει δέ V G°; *cf.* Σ γ

1091° (817) καὶ ῥευματώδεις α M S + G (et reumaticos) Procl.] *om.* V D γ Heph.

1104° (827) ἐπιτάσεις α M S + G (intensiones*), *cf.* Procl.] ἐπιστάσεις V D γ

1332° (1015) θηλυψύχους α M S + G (muliebres animo) Heph., *cf.* Procl.] μεγαλοψύχους V D γ

1335° (1016) ἐπιμώμους Y + G (probrosos) Procl.] ἐπιβώμους V: φιλομώμους ὑπομώρους Σ M S: φιλομώρους D γ

BOOK IV

203° (159) μάντεις, θύτας α M S + G (diuinatores, immolatores) Heph., *cf.* Procl.] μαντευτὰς V D γ

738° (561) τοῖς Y S B C + G (hiis) Heph.] τῆς V D M A

We should add here some instances, where inside α there are different mistakes, e.g. II.1051° (803) ἀμπώτεις β AB + G (effusiones) Co le (83,22), *sim.* Σ C, *cf.* Y (ἐμπτώσεις)] ἀμπώτις (*sic*) V: -της L

2.2.2. *G with rejected readings of α,* Y, *α* M S

BOOK I

(-*) 81° (67) ἡμῖν V βγ] *om.* α + G (*Robb.*)

(-?) 600° (456) Περὶ τῶν βορειοτέρων τοῦ ζῳδιακοῦ (*tit.*) V L D γ Procl. Heph. (*cod.* A, *cf. adn.*)] *om.* α M S + G Heph. (U) (*Robb.*)

(-?) 638° (481) Περὶ τῶν νοτιωτέρων τοῦ ζῳδιακοῦ (*tit.*) V L D Procl., *cf.* γ] *om.* α M S + G Heph. (U)

(-*) 690° (518) ὑγρασίας V D γ] ὑγρᾶς οὐσίας L α M S + G (humide substantie) Procl.

(-?) 1025° (775) μὲν V βγ] μὲν γὰρ α + G (quidem enim) (*Robb.*)

(-?) 1086° (837) μοῖραι γ] μὲν Y + G (quidem) Procl.: μὲν ἔτη V D M: [Σ S]

(-*) 1238° (946) καθόλου V βγ] καὶ καθόλου δὲ α + G (Et uniuersaliter autem): καθόλου δὲ L

BOOK II

(-?) 727° (552) λιμοῦ V βγ Procl.] λοιμοῦ α + G (pestilentiam)

(-*) 773° (588) περὶ δὲ τὰ τοῦ ἀέρος πνεύματα V D γ] περὶ δὲ τὰ τοῦ ἀέρος -μάτων α M S + G (circa aerem autem uentorum)

(-) 778° (591) ἔτι V βγ] ἐν α + G (in)

(-*) 857° (652) πίθων Σ βγ Heph.] πιθείων V (*cf.* Procl. πιθείας): πιθίων Y + G (pithiarum)

(-) 946° (719) τὴν πλειάδα V L βγ Procl. Heph.] τὰς πλειάδας (πλοιά-Y) α + G (Pliades)

(-?) 1075° (822) καλούμενα V βγ Procl.] λεγόμενα α + G (que dicuntur) Heph.

(-?) 1088° (833) ὅλον V βγ Procl.] *om.* α + G Heph. ('*fort. recte*')

(-*) 1117° (853) ὡς ἐπίπαν V D γ, *cf.* ὡς ἐπιπολύ Procl.] πάντοτε α M S + G (semper) Heph.

(-*) 1135 (866) γενεθλιακὸν (εἶδος) V Σ β Heph.] γενεθλιαλογικὸν (εἶδος) Y γ + G (geneatici sermonis species) (*Robb.*)

BOOK III

(-*) 9° (10) γενέσεων V D S γ] *om.* Y M + G ('*fort. recte*'), *cf. supra* III.392°

(-*) 53° (46) τέξεων V D, *cf.* γ] γενέσεων α M S + G (generationum) Procl.

(-*) 188° (147) τελευταῖος δὲ V] καὶ τελευταῖος α βγ + G (et ultimus) Procl. Heph. (*Robb.*) ('*fort. recte*')

(-*) 208° (163) τούτῳ V D γ] αὐτῷ α M S + G (ipsi) Heph.

(-*) 300° (235) καὶ V D γ] ἢ καὶ Y + G (uel etiam): ἢ Σ M S Procl. Heph.

(-*) 476° (366) ἀγαθοποιῶν ἀστέρων V D γ Procl.] ἀγαθοποιῶν α M S + G (beneficorum) Heph.

(-*) 484° (371) ὁ (τοῦ) Ἑρμοῦ V D γ Procl.] *om.* α M S + G ('*fort. recte*') Heph.

(-*) 592° (451) ὡροσκόπον V D] ἀνατολικὸν ὁρίζοντα α M S + G (orientalem horizontem) Heph. (120,22) Procl.

(-) 674° (509) διὰ V βγ Heph.] ὑπὸ α + G (a) Procl.

(-*) 1212° (912) ἀτιμώρητοι V] εὐτιμώρητοι α βγ + G (de facile punibiles) (*Robb.*)

(-*) 1217° (916) διατελῶσι V D M γ] διατελοῦσι α S + G (perficiunt) Heph.

(-*) 1220° (918) τυγχάνωσιν V γ] τυγχάνουσιν α β + G (existunt) Heph.

(-*) 1365° (1041) φιλοτέκνους V Σ βγ] φιλοτέχνους Y + G (artis amatores) Procl.; *cf.* 1404°

BOOK IV

(-?) 198° (155) ἢ Σ M S γ] καὶ Y + G (et) (*Boer*): *om.* V D

(-*) 225 (177) τῶν ... συμβιώσεων V D γ Heph.] τῆς ... συμβιώσεως Y + G (conuictu) Procl.: τῆς (τῶν M) ... συμβιβάσεως Σ M S

(-*) 252° (199) συνάπτων V D γ] *om.* α M S + G Procl.

(-*) 336° (258) αὐλικοὺς V βγ (Co le 157,37) (*Robb.*)] αὐλίσκους Y + G (canales habentes) Co (τὸν αὐλίσκον 157,38): *om.* Procl. Heph.: [Σ]

(-?) 518° (396) ἐναντίως V Heph.] ἐναντίους Y βγ + G (contrarias)

(-?) 719° (549) ἐπιβολὴ V βγ Procl.] τις ἐπιβολὴ Y + G (quedam adiectio): [Σ]

797° (602-3) *hic desinit* V

Some items of the first series (adopted readings) probably only show that the model of G was a better copy of ψ against mistakes of V, or against particular readings which V shared with D and γ, e.g. II.339° III.1335°. That there was a strong contamination between V D γ (Σ) is obvious in III.1047° (785): Αἰγοκέρωτι] + καὶ τοῖς νεφελοειδέσι συνάπτουσα ὀφθαλμὸν πηροῖ σελήνη V D, *simil.* Σ γ. — In most of these cases, G does not confirm V. Moreover, even Procl. does not confirm them. But in the first two books, when the adopted reading of G stands against common readings of V and L, one may presume that the model of G, which belongs principally to ψ (V L), was corrected (I.885° 914° 1140° II.970° 1048°).

A real connection can be seen in the rejected readings. There are several minor syntactical additions: I.1025° 1238° IV.719° where α + G or Y + G stand against V βγ, and another one, more problematic, in III.300° where Y + G stand against V D γ *and* against Σ M S. An effective correction in the model of G from a manuscript of the α-branch is certainly shown by II.946°, where the reading of α + G is opposed to V *and* L: τὴν πλειάδα V L βγ Heph.] τὰς πλειάδας (πλοιά- Y) α + G (Pliades). Moerbeke should have seen the plural, because otherwhere he translates the singular by a singular. But this is also a minor case. There are two omissions in α (M S) + G against V βγ or V Dγ of one significant word: I.81° (ἡμῖν) IV.252° (συνάπτων). A more substantial connection is also obvious from certain important variants, II.727° (λιμοῦ / λοιμοῦ), 1075° (καλούμενα / λεγόμενα), 1117° (ὡς ἐπίπαν / πάντοτε), III. 53° (τέξεων / γενέσεων), 592° (ὡροσκόπον / ἀνατολικὸν ὁρίζοντα), 1365° (φιλοτέκνους / φιλοτέχνους), IV.336° (αὐλικοὺς / αὐλίσκους). In all these cases, the model of G or its ancestor seems to have been corrected or contaminated from a manuscript with readings present in α or (only) Y. Here again, however, we have the impression that there are often two possible readings which both fit in the same context, so that one could have been the correction of the other. See, for example, the two cases of omissions (ἡμῖν and συνάπτων) and two cases of oppositions in vocabulary:

I.80° (67) (ἥ τε σελήνη πλείστην ὡς περιγειοτάτη διαδίδωσιν) ἡμῖν ἐπὶ τὴν γῆν τὴν ἀπόρροιαν V βγ] (...) διαδίδωσιν ἐπὶ (ὑπὸ Y, *cf.* Σ) τὴν γῆν τὴν ἀπόρροιαν (*om.* ἡμῖν) α + G (Luna uero, tamquam propinquissima terre, exhibet super eam plurimum defluxum) (*Robb.*). — '(The moon, too, as the heavenly body nearest the earth,) bestows her effluence most abundantly upon mundane things'. In this sentence ἡμῖν is not necessary, it is even somewhat strange. Was ἡμῖν cancelled by α + G, or, on the contrary, was it added by V βγ?

IV.252° (199) (καὶ [*sc.* ἥλιος] ἐν μὲν μονοειδεῖ ζῳδίῳ τυχὼν ἢ ἑνὶ τῶν ἑῴων ἀστέρων) συνάπτων (μονογάμους) V D γ (*sed cf. var. lect. in adn. Hüb.*)] (...) συνάπτων *om.* (...) α M S + G (et [*sc.* sol] in signo quidem unius specie existens aut in uno matutino planetarum, semel nubere). — 'If the sun is in a sign of a single figure or applies to (συνάπτων) one of the planets, she makes them men of one marriage'. In this context, the word συνάπτων is unnecessary from a syntactical point of view (see τυχὼν), but it fits well into Ptolemy's system of relations; see the parallel place about the moon IV.233°: ἐὰν (*sc.* σελήνη) μὲν ἐν μονοειδεῖ ζῳδίῳ ἢ καὶ ἑνὶ τῶν ἀστέρων συνάπτουσα τύχῃ. However, the absence of συνάπτων is very well attested by α and two members of β, and it would be very strange to suppose that this word was deleted by correction in the model of G. Was it perhaps an addition in the common archetype?

III.53° (46) τέξεων V D, *cf.* γ (ἕξεων)] γενέσεων α M S Procl. + G (generationum). — This is the only occurrence of τέξις in the whole text. Ptolemy generally uses γένεσις, which Moerbeke translates as *generatio* and later in the text as *natiuitas*. In his model Moerbeke certainly had γενέσεων, as had Proclus. τέξεων is a conjectural gloss, explaining τῶν ἀνθρωπίνων γενέσεων. It is noteworthy that there have been such glosses in V (and especially in V D together, see below) which worked their way into the text.

III.592° (451) (ὃν ἔχει λόγον καὶ σχηματισμὸν ὁ ἥλιος πρὸς τὸν) ὡροσκόπον V D (*Boll Robb. Hüb.*)] (...) ἀνατολικὸν ὁρίζοντα α M S + G (orientalem horizontem) Heph. (120,22) Procl., *cf.* γ. — 'Whatever relation and aspect the sun bears to the 'horoscope', this the moon will have to the Lot of Fortune and so it may be as it were a lunar horoscope (καὶ ᾖ ὥσπερ σεληνιακὸς ὡροσκόπος)'. Reading the whole sentence and especially the continuation of the argument, one may ask if the expression ἀνατολικὸν ὁρίζοντα, which is very well attested and confirmed by G, was not the primitive redaction and the reading ὡροσκόπον of V D a correction in order to explicate and to assimilate the terms of the comparison? All Greek editors, however, judged this case inversely.

The readings G has in common with α or Y against V are rarer than those G shares with βγ. Moreover, they are often readings in α (respectively Y) connected with some manuscripts of β. Moreover, we have here again, both in accepted as in rejected readings, a significant connection with Hephaestio, mostly when α is together with β or with the manuscripts M S of β. Of all the readings in which G deviates from its main source ψ, the larger half agrees with the edited text, and those rejected by the editor offer a coherent text, which may often be preferable, see in Chapter V all cases preceded by (-*). For these reasons we do not believe that G was contaminated from different origins. We rather suppose that a Byzantine scholar occasionally corrected G with the use of a manuscript of the φ-branch. This manuscript was certainly older (and better) than the existing manuscripts of the φ-branch, and may even precede the distinction between α and βγ.

3. G DIFFERENT FROM THE WHOLE TRADITION

3.1. Two important cases

There are instances where Moerbeke's translation — or better, its model — allows us to understand the diversity in the whole tradition. The best example is in book I.994°-6° (749-50):

(ὅπου γὰρ τὸ θερμὸν αὔξεται,) μειοῦται ἐκεῖ τὸ ψυχρόν, καὶ ὅπου τὸ ψυχρὸν αὔξεται, ἐκεῖ μειοῦται τὸ θερμόν *Hüb.*

(Vbi enim calidum augetur,) minuitur frigidum, et ubi frigidum augetur, minuitur calidum G

(...) ἐκεῖ τὸ ψυχρὸν μειοῦται, καὶ ὅπου τὸ ψυχρὸν αὔξει, ἐκεῖ τὸ θερμὸν μειοῦται Procl.
(...) ἐκμειοῦται ἐκεῖ τὸ ψυχρόν, καὶ ὅπου τὸ ψυχρὸν αὔξεται, ἐκμειοῦται ἐκεῖ τὸ θερμόν V
(...) ἐκεῖ μειοῦται τὸ ψυχρόν, πάλιν ὅπου τὸ ψυχρὸν αὔξεται, μηοῦτε ἐκεῖ τὸ θερμόν L
(...) μειοῦται ἐκεῖ τὸ ψυχρόν, καὶ ὅπου ἐκεῖνο μειοῦται, τὸ ψυχρὸν (θερμὸν Y) αὔξεται (-ξάνει Σ) α
(...) μειοῦται ἐκεῖ τὸ ψυχρόν (τὸ ψυχρὸν ἐκεῖ D), καὶ ὅπου ἐκεῖνο μειοῦται, τὸ ψυχρὸν αὔξεται β
(...) μειοῦται τὸ ψυχρὸν ἐκεῖ, καὶ ὅπου ἐκεῖνο μειοῦται, αὔξεται τὸ ψυχρὸν γ
(*cf.* Tib. (...) frigiditas remittitur, et ubi frigiditas intenditur, calor remittitur)

G does not have the two instances of the superfluous ἐκεῖ, which are found in the Greek tradition and in Proclus. They had likely been added (*supra lineam*?) in the archetype, as is evident from hesitations in the manuscripts. V starts in both cases with ἐκμειοῦται, a non-existent form, probably because the copyist misread ἐκεῖ μειοῦται (as has L), and added ἐκεῖ again after the verb (as in α β); γ again has ἐκεῖ on another place. The second ἐκεῖ disappeared entirely in the α βγ tradition: they all read (with some variations) ἐκεῖνο μειοῦται, τὸ ψυχρὸν αὔξεται instead of τὸ ψυχρὸν αὔξεται, μειοῦται ἐκεῖ. One may explain this divergent reading as follows. In their common model, the second ἐκεῖ was read as ἐκεῖνο, and once this error was made, the inversion followed. Only G has the correct text (which is confirmed by the Arabic tradition). We would thus delete the two instances of ἐκεῖ.

The second instance is a case where a rejected and somewhat corrupted reading of one β-manuscript against an omission in all others is surprisingly confirmed, in its correct form, by Moerbeke's translation. It may be an authentic reading. Its absence in the rest of the tradition is easily explained:

III.1246° (941-2) μεταμελητικοὺς — εὐλαβητικούς *habent* V D γ Procl.] *habet et* μελητικοὺς (*sic*) *add.* S: *habet* (penitiuos — reuerentiales) *et* futuricios (μελλητικοὺς) *add.* G: μεταμελητικοὺς *habent, sed sequentia verba* (αὐστηρούς — εὐλαβητικούς) *om.* α M (*fort. per hom.*). — The term μελλητικός ('inclined to delay') is rare. It is used once in Vettius Valens' *Astrology*, I, 3, 115. This term was certainly found in Moerbeke's model. With its negative signification, it fits very well into the context. Either it was present as an additional gloss in an older state of the transmission of the text or, more probably, the rare term is simply authentic: V D γ omitted it (in the orthography of S it seemed to be a duplication of μεταμελητικοὺς), and for α M it was the cause of an omission by homoioteleuton (μεταμελητικοὺς ... μελ<λ>ητικοὺς).

3.2. G agrees with coniectures of Hübner

In a certain number of cases, Moerbeke's translation confirms Hübner's conjecture against readings of the whole tradition. One finds examples of this in the list given in Chapter V (marked by the symbol §). See I.21° 190° 392° 1092° 1197 II.219-20° 472° 501° 781° 812° 889° III.244° 677° 736° 954°. One cannot exclude the possibility that Moerbeke himself made some conjectures when translating. In other cases, it is more probable that he encountered the correct reading in his Greek model, for example:

BOOK I

21° (18) ὁμοίως *coni. Boll Hüb.* + G (similiter)] ὁμοίων *codd.*

190° (152) κατακρατεῖ *coni. Boll Hüb.* + G (optinet), *cf.* Procl.] κατακρατεῖν *codd.*

392° (307) τοσούτῳ *coni. Hüb.* + G (tanto), *cf.* Tib. (tanto)] τότε V L: *om. cett.* Procl.

1092° (841) αὐτοὺς *coni. Boll Hüb.* + G (ipsos)] αὐτοῖς V β, *cf. cett.*

BOOK II

219°-20° (170) Ἀπουλία καὶ Σικελία καὶ Γαλλία *coni. Hüb.* + G (Apulia et Sicilia et Gallia)] σικελία καὶ γαλλία V D: γαλλία καὶ σικελία α M S: γαλλία καὶ ἀπουλία καὶ σικελία γ

472° (355) Ματιανή *coni. Robb. Hüb.* + G (Matiane)] ματτιανή α: μαντιανή βγ: γαμπανή V

501° (381) καθ' ὧν *sec.* Heph. *coni. Hüb.* + G (in quibus)] μεθ' ὧν Σ βγ: καθὸν V: καθ' ὃ Y

781 (593-4) ὡς ἂν ᾖ *coni. Boll Robb. Hüb.* + G (ac si esset)] ὡς ἀνεὶ V: ὁσανεὶ Y: ᾧ ἂν ᾖ Σ βγ

BOOK III

244° (193) περιεχόμενα Q (*Marc. gr.* 594) Heph. + G (circumdata)] πῶς ἐχόμενα V: περιεχόμενοι Σ βγ, *cf.* Y

736° (551) ταύτας δὲ εἰληφότες *coni. Hüb.* + G (has autem cum acceperimus)] ταύτας δὲ ἃς εἰλήφαμεν V Y (*Boer*): ταύτας γὰρ διειληφότες Σ M S: ταύτας διειληφότες D γ, *cf. Robb.*

The authentic conclusion of book IV (wrongly rejected by Hübner as 'altera operis conclusio' (C2°-11°, see Appendix II) is a special case. For the edition of this conclusion Hübner could only rely on the α-family, represented by Y G° Φ, as the manuscript V is mutilated at the end and the family βγ adopted the conclusion of the Proclus Paraphrase. The translation is undoubtedly the best witness of the authentic conclusion. It confirms five conjectures of Hübner/Robbins (symbol §) against the three manuscripts of α. In one case one must propose with G a new correction in the text (C2°). In two cases G confirms the reading adopted by Robbins instead of a coniecture by Hübner (C8° C9°). In another case the translator seems to have himself made a correction in order to solve a construction problem (C4°). For such a small section of the text (only seven lines) this is an impressive list, which again shows the quality of G.

(§) C2° (678) φυσικαῖς *coni. Robb. Hüb.* + G (naturalibus)] δυσικαῖς Y G° (Φ *non citatur*)

(-*) C2° (678) χρηματείας *coni. Robb. Hüb.*] χριμαντείας Y: χρωματίας G° Φ: πραγματείαις *recte legit* G (negotiis)

(§) C4° (679) πολύχους *coni. Hüb.* + G (multi sudoris)] -χρουν Y: -χροαν G° Φ: -χόως *coni. Robb.*

(?) C4° (680) ἐνθάδε] relinquendum *praem.* G. — Moerbeke probably added this verb himself because there is a problem in C2°-5° with the construction of the two accusatives τὰς δὲ κατὰ μέρος ἐπιβολὰς ... τὸ διεξοδικὸν, which cannot depend on the same verb ὑποληπτέον. Moerbeke translates: 'particulares autem adiectiones ... relinquendum, hic autem ... pertractationem ... accipiendum'. Robbins and Hübner propose another conjecture, adding κατὰ before τὸ διεξοδικὸν.

(§) C6°-7° (681-2) <κατὰ> τὴν (τῇ G°) ... εὐστοχίαν Y G° (*Hüb.*)] τῇ ... εὐστοχίᾳ *coni. Robb.* + G (facili coniectura)

(-*) C8° (682) ἐφαρμόζεσθαι *coni. Hüb.*] ἐφαρμό(ω)ζειν Y G° Φ (*Robb.*) + G (adaptare)

(-*) C9° (682) δύναται *coni. Hüb.*] δυναμένου Y G° Φ (*Robb.*) + G (potentis)

(§) C10° (684) καλῶς *suppl. Robb. Hüb.* + G (bene)] *om. codd.*

(§) C10° (684) ἂν ἔχοι *coni. Robb. Hüb.* + G (utique habebit)] ἀνέχει G° Φ: ἀνέχῃ Y

3.3. Glosses in G

Although the Latin translation occasionally confirms modern conjectures, it does not substantially complement the Greek text by providing words or passages omitted in the whole Greek tradition. There are, nevertheless, several additions in the Latin text which were probably not supplied by Moerbeke, but translated from his Greek manuscript. The following additions, not present in the Greek, give an intelligent complement to the argument, and may come from a gloss in G:

I.454° (353) ἀρρενικούς *codd.*] propter siccitatem (διὰ τὴν ξηρότητα) masculinos G

II.240° (184) φιλαγωνισταὶ] amatores exercitii propter Mercurium G

II.602° (460) *ante* προκρινουμένου τοῦ τῆς ἐκλείψεως κυρίου] Si autem ambo easdem habeant rationes (εἰ δὲ ἀμφότεροι τοὺς αὐτοὺς ἔχοιεν λόγους?), preferimus dominum eclipsis G. — See for the whole passage 599°-602° the comments in Chapter V.

II.1079°-80° (825-8) ἢ ἄλως ἔχων (*sc.* ἥλιος) περὶ αὑτὸν καθ' ἓν μέρος ἢ ἐξ ἀμφοτέρων τῶν μερῶν παρήλια νέφη ... χειμώνων καὶ ὑετῶν ἐστι δηλωτικός] uel halo habens circa se unam uel etiam duas, aut secundum unam partem aut etiam ex ambabus partibus nubes parelias ... tempestatum et pluuiarum est significatiuus G.

III.259° (204-5) ἢ τῇ σελήνῃ Κρόνος D γ] ἢ τῇ σελήνῃ Κρόνος τῇ σελήνῃ V: ὁ δὲ Κρόνος δὲ τῇ σελήνῃ Y: Κρόνος δὲ τῇ σελήνῃ Σ M S Heph. (*Boer*): uel lune de die, Saturnus autem lune de nocte G. — The multitude of different readings in the whole Greek tradition indicates a problem in the archetype (ω). The reading of V with the repeated τῇ σελήνῃ may come from a copyist error in the archetype (see a similar situation in II.22°-23°). G also had the double σελήνῃ, but with an introduction of δὲ (Κρόνος δὲ) from a reading of the β branch and Hephaestio, prefered by Boer. Yet the words *de die* (ἡμέρας) and *de nocte* (νυκτὸς) are found in no other witness. They give sense to the duplication. It is probable that they have been added by a Greek scholiast who intervened in Moerbeke's Greek model. It could well be the same who added the astrological calculation after the text of the *Tetrabiblos*, which concerns precisely the division of nativities in day and night (see Appendix I)

3.4. The omission of Ptolemy's table of terms

After having praised the qualities of Moerbeke's Greek model, we should not remain silent about one disappointing feature: it does not contain the famous and somehow enigmatic 'Ptolemaic table of terms' in book I, ch. 21, § 28-29.[23] Hübner edited this table from L, the *Vat. gr.* 1291 (z) and the Proclus Paraphrase. It is also present in α D M, but absent in V and S. However, in V, spaces have been left open and in S, lines have been sketched to insert the table. It should be noticed that V (just as Y¹ and S) equally omits the Egyptian table in I, ch. 21, § 9-10, but with the same indications (open space, etc.). Did Moerbeke forget to translate the Ptolemaic table? This is improbable, considering how accurately Moerbeke works: he even transcribed the table in the additional astrological note he found in his manuscript (see Appendix I). Nor is it plausible that the copyist of the model of the Latin translation omitted it, for, in spite of his many errors, he faithfully transcribed the table of regions and the table in the additional note at the end. If it is not a copyist's mistake, then we have to suppose that Moerbeke did not find the table in his Greek model. Interestingly, Moerbeke also did not translate the sentence at the end of § 27, which announces the table (I.1171°: ἔστι δὲ καὶ ἡ τούτων τῶν ὁρίων ἔκθεσις τοιαύτη). This sentence is present in the whole tradition (including V). One cannot exclude the possibility that Moerbeke himself decided not to translate that sentence, when (as we believe) he did not find the table which that sentence announces. It is furthermore interesting that Moerbeke did not find the title ὅρια κατὰ Πτολεμαῖον in his model, though it is present in nearly the whole tradition (V L α D M Procl., *cf.* γ), and omitted only in S.[24] One cannot imagine that Ptolemy would himself have added such a title ('the terms according Ptolemy'). Why should Ptolemy have used this title attributing the invention of the terms to himself, when he presented so many details about the discovered manuscript so as to show that the system was not 'his' but rather of 'ancient origin'? The title must have been added later in the tradition; Robbins and Hübner therefore rightly delete it as inauthentic. Surprisingly again, the conclusion after the table (§ 30), which gives the sums of the terms assigned to each of the five planets, is not found in z L α Procl., but is present in V βγ Tib. and in Moerbeke's translation, which gives the complete indications with β: 'Fiunt autem et horum ex collectione Saturni gradus 57, Iouis 79, Martis 66, Veneris 82, Mercurii 76'. Moerbeke's accurate translation of these sums,[25] which is also present in the Arabo-Latin translation of Plato of Tivoli, is yet another indication that he did not omit the table himself, but simply did not find it in his Greek model. Could it be that Ptolemy only gave the instruction for a table in his text (§§ 22-27), along with the result (§ 30: the total sums of the terms assigned to each planet), but never added the table itself? Perhaps some

---

[23] See on Ptolemy's doctrine of the terms and its reception Heilen (2010).

[24] In the Hervagius edition of Tiburtinus' translation one finds the title *Termini Ptolomaei seu autoris ignoti*. This title is not yet in the first edition (1484).

[25] Moerbeke did not read (translate) the last addition of these sums (360): ὁμοῦ τξ′ D M γ] γίνονται τξ′ V S: *om.* G, *deficit* (§ 30 *om.*) L α.

scholar in late antiquity, starting from the obscure instructions and the total sums and using the information in Hephaestio, composed the table with the numbers and added the title ὅρια κατὰ Πτολεμαῖον. In this hypothesis, the absence of the table in G may not represent a lacuna, but rather give us an indication of the original state of the Ptolemaic text.

But is the presence of the table in L and in Procl. not an indication that the table was also found in the hyparchetype ψ? Let us first see how L presents the table. On ff. 149v-150r, after having copied chapter I, 20, the copyist does not continue with the beginnings of chapter 21 but skips a long section (1020°-1138°) and starts copying from 1139° on — that is, the introduction of the system of the terms found 'in the ancient manuscript'. He rephrases Ptolemy's neutral introduction (ἔχει δὲ οὖν ὁ τύπος τῆς ὅλης αὐτῶν ἐπιβολῆς τὸν τρόπον τοῦτον) to attribute what follows to Ptolemy himself: ὁ δὲ τῶν ὁρίων τύπος τῆς ὅλης αὐτῶν ἐπιβολῆς κατὰ Πτολεμαίον τοῦτον ἔχει τὸν τρόπον. There follows the whole section 1140°-1172°. However, here again the copyist takes some liberties, for after ζῳδίων (1147°) he adds a long digression on the order of the five planets, which is not found in any manuscript. After 1172° comes the table with the title ὅρια κατὰ Πτολεμαίον, but without the concluding calculation (1186°-1188°). Immediately thereafter comes — without transition — the table according to the Egyptians, which is found much earlier in the text (1072°-1084°). As is clear, the copyist of L (or rather the scholar who composed the 9th century anthology copied in L) made an idiosyncratic selection from the text of chapter 21, even adding a certain amount of new material. The selection has as title περὶ τῶν ὁρίων τῶν πέντε ἀστέρων. It is possible that this scholar did not find the table in his manuscript of Ptolemy's *Apotelesmatica* but copied it from another source: this may have been the *Vat. gr.* 1291 (z).

What, then, about the Proclus Paraphrase? The author of the Paraphrase makes of section 1140°-85° a separate chapter with its own title, περὶ τῶν ὁρίων ὡς Πτολεμαῖος. This title, again, cannot be authentic. Interestingly, however, the same title is found in V and D γ (replacing ὡς by ὁ), most probably by contamination with the Proclus Paraphrase. As Heilen has shown, in the oldest manuscript of the Paraphrase (10th century) the terms in the table correspond to what we find in Hephaestio (where the values are not yet given in a table). The same copyist added other values, however, which correspond to what we find in L and z and all the later manuscripts. In the paraphrase, there is no concluding calculation of the several sums, as we find it in V βγ Tib. + G. Is it possible that the author of the paraphrase composed himself the table using the information he found in Hephaestio?[26] But why then did he write the alternative values in the table? Is it possible that he compared his table with z?

[26] Heilen (2010) noticed 'striking correspondences between the values [in the table in the Paraphrase] and the Ptolemaic table of Terms as transmitted by Hephaestio, Chapter

Whatever the explanation one may wish to adopt, one cannot simply conclude from the presence of the table in L and in Procl. that the hyparchetype ψ had the table *secundum Ptolemaeum*.

CONCLUSION

For his translation of the *Apotelesmatica*, Moerbeke had at his disposition an excellent exemplar of the Greek text. It belonged to the text tradition ψ, which is represented by V (and L) and the Proclus Paphrase, but does not depend on them. There are no indications in the translation of individual errors in the Greek text he read (except omissions, which, however, may be explained through errors in the Latin tradition). Moerbeke's translation thus helps to determine the original state of ψ by eliminating the particular errors of V (and L). Given its quality and the absence of particular errors, G could have been the hyparchetype ψ itself, if there were not found in it a number of readings stemming from the other tradition, α and/or βγ. As Moerbeke did not use a second Greek manuscript, these elements must have been introduced in his model by correction before he made the translation. Still, the Greek text at the basis of Moerbeke's Latin translation was not the result of a simple contamination from different sources. Some scholar using a manuscript from the φ-branch occasionally corrected the original text, which was an excellent copy of the ψ tradition. These corrections were probably made by substitution (*in rasura*) or by superscript, which explains the double readings. That the scholar intervening in G made intelligent corrections is confirmed by the fact that many corrected readings have also been adopted by the Greek text's modern editors. Even the readings which are not adopted make sense within Ptolemy's argument. The quality of these deviant readings, the fact that they combine specific elements of both families, α and β, and the great number of readings also present in Hephaestio, all serve to indicate that the corrector found these readings in a manuscript of branch φ preceding the division between α and βγ. The presence in G of specific readings of φ does not exclude the possibility that Moerbeke's model could have itself been the hyparchetype ψ. In fact, the Proclus Paraphrase and the astrological anthology of L (which represent the pure ψ tradition of the text) could have been made before the corrections were made in G. As regards V,

---

1.1.150'. There are, however, three cases of disagreement between Hephaestio and the Paraphrase, which seem to exclude that the Paraphrase draws on Hephaestio himself (see p. 85-86 notes 152-154). Therefore, Heilen concludes that 'both Hephaestio and the author of the paraphrase drew on the same (now lost) branch of manuscript tradition.' However, the disagreement may also be explained by errors in the transmission of the Paraphrase and of Hephaestio.

this manuscript was copied around 1300, after Moerbeke's translation was made. Besides, V was not directly copied from ψ (see Addendum), but from a more recent copy of it. This intermediary copy also could have been made before the corrections were introduced in G. As we have seen, even the absence in G of the table of the terms *secundum Ptolemaeum* does not necessarily undermine the supposition that G was identical with ψ. The hyparchetype ψ was undoubtedly a manuscript from the 9th century, since the Proclus Paraphrase (of which the oldest manuscript is from the 10th century) depends upon it, while the excerpt-manuscript L depends on an anthology made in the 9th century. Even if G was not ψ, it must have been closely related to it (as it is the best representative of the ψ tradition), and would probably have been copied in the 9th century.

That Moerbeke found in his Greek manuscript, placed after Ptolemy's text, an astrological calculation, which he translated with the title 'quo mense morietur quis in omni natiuitate' (see Appendix I), is another argument for an ancient date of his Greek model. This note, which is found in no other Ptolemaic manuscript, probably dates back to late antiquity. It uses Egyptian names for two months, and it presents a method of calculation that is connected with questions about the *sors fortunae* also found in Hermetical astrological texts of late antiquity. This note could have been appended by the same scholar who made the corrections to Ptolemy's text. One cannot exclude, however, that the note was found in the original state of ψ itself. We do not know whether V had this note, as this manuscript is mutilated at the end. The Proclus Paraphrase does not help us any further, as its author was only interested in explaining Ptolemy's text. If the note was found in ψ, it could have been copied directly from the ancient model of this hyparchetype.

It is well known that Moerbeke had access to an astonishing number of important manuscripts of the 9th century. He translated Alexander of Aphrodisias' *De fato* from the *Marc. gr.* 258, which he personally owned: the manuscript preserves Moerbeke's mark of possession and other inscriptions by his hand. Moerbeke translated (or revised) *Cael., Phys., Gen., Meteo.* and the *Metaphysics* (probably also *Hist. an.*) from the Aristotelian codex *Vindobonensis gr.* 100 (+ *Par. suppl. gr.* 1156), which he had for about ten years at his disposition. He used the *Marc. gr.* 226 (Simplicius *In Phys.*) for his translations of the *Physics*, and he had access to the *Marc. gr.* 196 (Olympiodorus, Damascius). He translated Archimedes and Eutocius from a manuscript of the same period, now lost, but, mentioned in the catalogues of the library of Boniface VIII in 1295 and 1311. Moerbeke translated the text of Ptolemy's *De analemnate* — only partially preserved in a single palimpsest of the 6th century — from a rare manuscript (now lost), similarly mentioned in the papal catalogue. The questions regarding where he encountered these precious manuscripts (whether in Greece or the south of Italy), and under what circumstances, and how he managed to acquire some of them while using others for a considerable period of time, are still open.[27] But it should come as no surprise that Moerbeke had a 9th-century Greek manuscript of our Ptolemaic text.

ADDENDUM

The recent study of Raúl Caballero-Sánchez on the manuscript tradition of the anonymous commentary on the *Tetrabiblos* drew our attention to Oxford, *Bodleian Library*, Auct. T.5.4, a parchment manuscript, copied at the end of the 13th century, which presents Ptolemy's *Tetrabiblos* surrounded by the anonymous commentary. Hübner refers to this manuscript (U) in his presentation of the *Vaticanus gr.* 1038 (V) and some related manuscripts, which are 'necessitudine quadam' connected with V, but also contaminated with readings of β, in particular in the last two books (p. XVIII). As Hübner did not use U in his edition and even considered it in his stemma as a copy deriving from V, we did not consider it in our collation of the Latin translation with the Greek text tradition. Our edition and introduction were already completed, when we had the chance of examining the Oxford manuscript in a digital version provided by Caballero-Sánchez. A collation of this manuscript immediately revealed that U does not derive from V. For the text of the *Tetrabiblos*, the copyist used two different manuscripts. The change occurs at the transition from f. 48v to f. 49r. In the second part, starting with ἄν in line II.683°, U is related particulary to M and S and certainly belongs to β, as it shares all its particularities. In the first part (I-II.683°), U is undoubtedly related to V. It has many readings and even lacunae in common with V, but does not have particular errors of V. U even has in book I, ch. 21, the tables of terms 'according to the Egyptians' and 'according to Ptolemy', which are in V only indicated by an open space. U has its own errors, which are not found in V, but fewer than V. A later hand, using a manuscript from another tradition (the ms. used for the second part?) corrected some of the errors. The conclusion is obvious: U (in its first part) and V both derive from a same manuscript, now lost. As V and U also share with L (when excerpts are present) readings against α βγ, their common model derives from the hyparchetype ψ, from which also the anthology L (both in the excerpts from the *Tetrabiblos* and in the commentary) and the Proclus Paraphrase derived. The testimony of U definitely shows that V was not directly copied from the 9th century hyparchetype, but derives from it via an intermediary copy, In our view, the manuscript U should have been used in the edition of the Greek text to discriminate particular readings of V and to determine with greater precision the state of the text in ψ. As our edition was completed before we collated the manuscript, it was impossible to integrate the readings of U in the apparatus and in the introduc-

[27] For a recent survey of the discussion see Ronconi (2012).

tion. However, the collation of U did not make us modify our views on the place of G in the text tradition presented in this chapter, namely that G derived directly from ψ — or could have been ψ itself — with corrections by a Byzantine scholar, made before Moerbeke translated the text.

Examples of readings of U and V against the adopted readings in book I

In what follows, we present a list of readings that substantiate the above conclusion. We selected the chapters where we have also the excerpts in L. The reading adopted by Hübner stands always before the square bracket. In every entry the readings of all mss. cited by Hübner are present, either explicitly (as variants after the square bracket) or implicitly (when a ms. has the reading of the edited text). We mention G whenever the Latin terms allows us to decide which Greek reading Moerbeke translated.

I.392° τοσούτῳ *coni. Hüb.* (G)] τότε VUL 397° τοῦ^{1}] τῷ VU: τὸ L 403° δυνάμεις] αἱ δυνάμεις VUL 425° ἀναπωτικός (G)] ἀεὶ ποιητικὸς U: ἀεὶ ποτικὸς V: ἀνειπωτικὸς L 435° καὶ *ante* παθητικά YM(G)] τε V: τε καὶ ULSDγ 436° ψυχροῦ ... ξηροῦ] *inv.* VULD(G) 436° πάλιν] *om.* VULMS(G) 444° ὡς] *om.* VUL(G) 497° μόνον] *om.* VUL(G) 501° δύσεως *cett.*] *om.* V: κρύψεως U(G) 508° δέοντος *coni. Kroll*] δὲ ὄντος VULY(G), *cf.* βγ 522° τῆς] ὁ τῆς VUL (*Robb.*), *cf.* Σ 533° ποδῶν] ὀφθάλμων VULD^{1}α (G) 537° τῇ σελήνῃ Y] τῆς σελήνης VULβγ 542° Ἄρεως (G)] Διὸς VU [L] 565° σπονδύλοις γ (G)] σφονδύλοις VULYβ 570° περὶ] ἐπὶ VLU(G) 570° ὁ ... ἔχει] οἱ (ὁ U) ... ἔχουσι VUL(G), *cf.* Y 587° ἱματίῳ] μετώπῳ VUL(G), *cf.* Σ 595° λίνου L(G)] ἀινου VU 600° *et* 638° *titulum hab.* VULD Heph.] *om.* αMS(G) 670° οἱ λαμπροὶ α] ὁ λαμπρὸς VUβγ(G) [L] 672° *titulum hic hab.* αMS] *hab. post* παρατηρήσεως VULDγ(G) 680° τὸ μᾶλλον] μᾶλλον VUL, *cf.* Y 686° *(sic)* τὸ] *om.* VUL Procl. 687° τὴν] *om.* VUD 688° ἔαρος] ἀέρος VULYDγ(G) 689° διὰ τὸ (G)] διατάττουσι VU 728° θερμαῖς (G)] θερμασίαις VU 734° *titulum hic* αDγ] *post* δυνάμεις (745) VU(G): *om.* L 739° ἀναλόγως ULαβγ(G)] ἀνάλογον V 740° αὐτὰ LΣβγ(G)] αὐτὸ VU, *cf.* Y 748°-50° θερινῆς ... τῆς (G)] *om. per homoiteleuton* VU 763° ἰσημερινοῖς (G)] ὡς μέσοις τοῖς τε τροπικοῖς καὶ (*seq. lac. 7 litt.*) *add.* V: μέσοις τοῖς τε τροπικοῖς καὶ τοῖς (*lac. 4 litt. ante et post*) *add.* U (*sed del. alia manus*) 769° αὐτῶν] *om.* VUL(G) 772° τε] *om.* VUL 772° καὶ ἰσημερινῶν] *om.* VUL(G) 814° δύο ὀρθὰς Lα(G), *cf.* DMγ] ὀρθὰς ἴσας VU 816° τρίτον LY, *cf.* βγ] τρίτην VU(G) 820° μίας] γωνίας VUL(G) 826° τὴν] τὸ VUL 826° διάμετρον (G), *cf.* γ] *om.* VU(*add. supr. lin. alia manus*) 828° τὴν^{2} (G)] τε *add.* VU 837° τετράγωνοι α] διάμετροι ULβγ(G): τρίγωνοι V κατὰ διάμετρον] τετράγωνοι VULβγ(G) 879° βορειότατα] βορειότερα VULDMγ(G) 885° ὄν (G)] *om.* VL, *ante corr.* U 893° οὖν] *om.* Uβ(G): διὰ V: δὴ L, *cf.* Y *et* Σ 896° Αἰγόκερως ... Ὑδροχόος UαDSγ] Ὑδροχόος ... Αἰγόκερως LM(G): Ζυγὸς ... Αἰγόκερως V 898° τοῦ τὸν (G)] τοῦτον VU (τῶν L), *cf. cett.* 912° καὶ ὁ Ταῦρος (G)] εὖ Ταῦρος VU 914° τὸ πλεῖστον (G), *cf.* Dγ] *om.* VUL 924° αὐτῷ *Hüb.*] ἑαυτῷ LΣβγ(G): αὐτῷ VUY 924° τριῶν (G)] *om.* VU *cf. cett.* 930° τῷ Διὶ καὶ ἡλίῳ Dγ (G)] δι (*sequente lacuna*) V, *om.* UDγ, *cf. cett.* 930° παρὰ τὴν αἵρεσιν (G)] τὴν αἵρεσιν V: παρ' αἵρεσιν U 931° οἰκοδεσποτείαν (?G)] συνοικοδεσποτείαν VUL 935° μὲν] *om.* VULβ(G) 937° τε] *om.* VUL 939° συνίσταται] γίνεται VUL(G) 946° αὐτοῦ UL] αὐτῷ V 948°-9° ἡ δὲ Παρθένος πρὸς τῷ ἰσημερινῷ, ὁ δὲ (G)] *om.* V: ὁ δὲ Αἰγόκερως πρὸς τῷ χειμερινῷ *om. ante corr.* U 955° αὐτὸ νοταπηλιωτικὸν (G)] αὐτὸν ὄντα ἀφηλιωτικὸν

VL, *ante corr.* U (*dub.*) 959°,960° καὶ (G)] *om.* VU 970° τῇ (G)] τὴν VU *cf.*Σ 973° μόνῳ] μὲν VUL(G) 982° συνοικοδεσποτείαν Lβγ] οἰκοδεσποτείαν VUα 994°-6° *textus in* U *ut in editione Hübner, cf. supra* 3.1. 998° ὡσπερανεὶ (G)] ὥσπερ ἂν εἴη VU *cf. cett.* 1140° τοῦτον (G)] *post* τοῦτον *titulum habent* περὶ ὁρίων ὡς Πτολεμαῖος VULDγ 1163° Ταῦρον L(G)] *om.* V[1]Uαβγ, *add. manus recens mg.* V 1188° ὁμοῦ DMγ] γίνονται VUS: *om.* (G): [Lα] 1193° μοίρας (G)] μόρια VU [L] 1207° συγχρῆσθαι Lα(G), *cf.* βγ] συγχωρῆσθαι VUβ 1226° ὁμοφύλου L (G): -φυοῦς αβγ] -φίλου VU 1247° ἐστι αγ (G)] ἔτι VUβ: ἔτη L 1253° τὰς γωνίας (G)] τῶν γωνιῶν VUDγ 1254° συσχηματίσμων βγ, *cf.* Y] σχηματισμῶν VUL(?G) 1257° δυτικοὺς καὶ ἀφαιρετικοὺς] *inv.* VUL(G) 1264° μὴ (G)] *om.* VU 1265° ἀδύναμοι (G)] ἀδύνατοι VULDMγ.

Examples of errors of V not present in U: I.493° (long omission); 499° (omission); 500°; 632°; 769°; 889°; 895° (2x); 897°; 905°; 906°; 910°; 912°; 919°; 927°; 929°; 931°, 947°; 948° (omission); 960°; 967°; 970°; 977°; 979°; 993°; 1005°; 1029°; 1031°; 1036°; II.275° (omission); 276° (omission).

The copyist of V often makes errors in reading astronomical signs. An interesting case is the frequent confusion between Ἑρμῆς and Ἀφροδίτη in V, which is not found in U. The reason of the confusion is probably the fact that the copyist of the model put a circumflex upon the astronomical sign for Aphrodite, when it had the position of a genitive. An inexpert copyist could mistake this genitive form for the sign of Hermes [☿] which is differentiated from that of Aphrodite [♀] by an extra scroll upon the sign to indicate his helmet. See for instance I.556°; 566°; 947°; 977°; 1096°; 1100°. The copyist of U does not make these mistakes. He shares however the uncertainties of V in reading the abbreviation of ἀνατολή:

I.461° ἀνατολῆς *post corr.* U, ἀνατολικοῦ V II.56° ἀνατολὰς *om.* V *cum spatio*, τῶν περὶ τὰς ἀνατολὰς *om.* U *cum spatio, add. alia manus* II.108° ἐπειδὴ τὰς ἀνατολὰς *om.* V *cum spatio, om.* U *cum spatio, add. alia manus* II.118° ἀνατολι *cum spatio* V, τῷ ἀνατολικῷ *add. alia manus* U.

Other examples of omissions common to U and V are: I.1013°; II.159; II.170° ὅλης ... κατὰ *om.* V *cum spatio, add. alia manus* U.

As is evident from these examples, V and U share a large number of errors and omissions, which must have been in the model from which both were copied around 1300. They also share with L a number of particular readings, which must have been in the hyparchetype ψ. When there is a consensus VUL, the reading generally corresponds to Moerbeke's translation. The discovery of this new partial representative of the ψ tradition reduces the authority of V, which can no longer be seen as a direct copy of the lost hyparchetype, and enhances the value of G, which was a direct copy of the hyparchetype, if not the hyparchetype itself.

CHAPTER V
THE CONTRIBUTION OF MOERBEKE'S TRANSLATION TO THE EDITION OF THE GREEK TEXT

In this chapter, we will examine how Moerbeke's translation may contribute to the assessment, and occasionally to the improvement, of Hübner's authoritative edition of Ptolemy's Greek text. As we demonstrated in the preceding chapter, Moerbeke's Greek model (G) is the most reliable representative of the ψ-branch of the Greek tradition, which Hübner only knew through V (and occasionally through L). The additional testimony of G makes it possible to reconstruct with greater certitude the state of this archetype in the 9th century and to eliminate the numerous particular errors of V. When G does not confirm a reading of V, that reading must generally be rejected. There are, for example, a number of places where Hübner follows V, when it agrees with D and γ but is not confirmed by G. In our view, these cases are problematic. Moreover, a new collation of the Proclus Paraphrase made clear that this Paraphrase depends upon the same archetype, as it generally shares the readings of G + V. The Paraphrase is probably not much older than its oldest manuscript, the *Vat. gr.* 1453 (10th century). The testimony of G thus definitively establishes the existence of a hyparchetype ψ and allows us to distinguish this branch of the tradition from the other tradition represented by α and βγ.

An analysis of the adopted and rejected readings of both branches revealed a bifurcation of the tradition, which should not be understood as a division resulting from errors in the two hyparchetypes. It rather seems to us that there were two redactions at the origin of the tradition, one more primitive, and the other introduced to explicate an obscure or difficult text, to restore really or seemingly lost elements, to render the text more precise where it seemed to be too general, to eliminate grammatical difficulties, and to make the text more fluid and coherent by minor additions and assimilations of certain expressions. The distinction between the two redactions does not strictly correspond to the distinction of the two text branches: the 'primitive' text is sometimes in one branch, sometimes in the other, and sometimes only in certain members of one or the other. This was probably the reason why Boll and Hübner supposed that there had been four branches at the origin — V, α, β and γ — of which specific readings were known already at the time of Hephaestio. Moreover, as we have seen, G also has a number of corrections stemming from the ϕ tradition (shared with β, or α, or α + some β), and not found in V (L) and Procl. These 'deviant' readings of G are not simply contaminations from the other family, but rather indicate that some scholar may have introduced valuable readings from a manuscript of the ϕ-branch preceding the division between α and βγ.

What, then, can Moerbeke's translation contribute to the edition of the Greek text? As can be seen in this edition's apparatus, Moerbeke's translation confirms most of the adopted readings in the Greek edition and almost never offers new material (such as words or passages omitted in the direct tradition). However, as previously stated, the presence (or absence) of G makes it possible to determine with greater precision the state of the text in ψ, and to distinguish more clearly the two traditions of the text. In fact, the major difficulty for an editor is to decide for one of the two traditions. As Hübner declares in his preface: 'Quia neuter semper verum praebet [V Y], unoquoque loco pensandae sunt utriusque lectiones. Praesertim illis locis, ubi Y cum Hephaestione contra V consentit, dignus est qui respiciatur' (p. XXXII). Whereas Boll usually prefers V 'ceteris antiquiori', and Robbins rather follows Y and other manuscripts of the α branch, Hübner puts both V and Y on the same level and only decides between them after having 'weighed' ('pensando') the divergent readings, often following the authority of Hephaestio. We noticed that the Latin translation is often the best arbiter when difficult editorial choices must be made.

Hephaestio is undoubtedly an important witness of the early state of the text in the 4th century. However, he often freely adapted his source. Moreover, it is possible that the Ptolemy manuscript he used had already incorporated a number of variants that never entered the direct tradition of the *Apotelesmatica*. Finally, the text tradition of Hephaestio is itself problematic, and some recent manuscripts may have been contaminated with the Ptolemy tradition (just as many recent manuscripts of Ptolemy have been corrected and contaminated by Hephaestio). Therefore, the testimony of Hephaestio is itself often ambivalent and can be made use of in two ways: one can invoke Hephaestio's authority so as to defend a particular reading, or one can reject that reading on the argument that it is due to a contamination by Hephaestio. To avoid this sort of circular argumentation, one should consider the different relationships of the text's subfamilies to the Hephaestio tradition. We apply the following principles:

1) When Heph. confirms a reading of only βγ against all other manuscripts and against G, we consider this to be generally a contamination of βγ. There are indeed in βγ numerous examples of contamination originating from Heph. (see IVA *supra*). This does not mean that such a reading must always be rejected. However, if it is accepted, it is on the authority of Heph.
2) The same principle holds when Heph. confirms readings of only α (or Y) against all other manuscripts and against G.
3) When Heph. confirms against V + G a reading present in α βγ, the situation is less clear. One cannot exclude the possibility that here again we have a conta-

mination. It is, however, more plausible that, in such cases, this branch of the tradition preserved an authentic reading which is not present in the ψ tradition.

4) When Heph. confirms a reading of V (L, Procl.) + G, this is almost always the reading to be adopted. In fact, as there are no indications that the manuscripts of family ψ were contaminated by Heph., their agreement with Heph. is a strong confirmation of the authenticity of their readings.

5) And finally, when G confirms a reading of Heph., whatever the other witnesses may be, it is almost always the reading to be adopted.

In this chapter, we shall present (in the order of the text) a list of the most important readings that illustrate how G is situated in the text tradition, and shall comment on passages where the translation may improve the edition of the Greek text. These comments should not be taken as criticisms of Wolfgang Hübner's impressive edition. The editors of the Latin text do not have Hübner's expertise in astrology, and their comments are often only questions or are meant to show how this new document could help to make some choices of the Greek editor more consistent. The list should be used as a complement to our Greek-Latin apparatus. In all examples given below, the reading adopted by Hübner is put first before the square bracket. The examples are preceded by the following symbols:[1]

(++) = G confirms a reading of V (L) that is adopted by Hübner in his edition.
(+) = G confirms against V a reading of another subfamily or particular manuscript adopted by Hübner
(§) = G confirms a conjecture of Hübner (or another editor).
(-) = G confirms a reading not adopted by Hübner.
(-*) = the reading confirmed by G seems better to us: we propose to modify Hübner's edition.
(-?) = the reading confirmed by G seems also possible.

Most of the items prefixed by (-*) (-?) are commented. Some have already been discussed in the preceding chapter, and in those cases we refer the reader back to that chapter.

1. BOOK I

V (α =YΣ β =DMS γ =ABC)

(-*) 1° (1) Κλαυδίου Πτολεμαίου Σ βγ] *om.* V Y + G, *sed cf.* 4°-5°

(-*) 2° (1) τῶν πρὸς Σύρον ἀποτελεσματικῶν D M] τετράβιβλον καὶ ἕτερον βιβλίον τοῦ Ἀντιώχου ... Y: μαθηματικῆς τετραβιβλίου συντάξεως Σ, *cf.* S γ] *om.* V + G, *sed cf.* 4°-5°

[1] In some cases we also checked the Arabo-Latin translation of Plato of Tivoli (Tib.) and asked advice about the Arabic translation (Arab.). See Chapter IV, n. 22.

(-*) 3° (1) βιβλίον α΄ Σ D, *cf.* Mγ] *om.* V Y S + G, *sed cf.* 4°-5°. — The complete title is only found in D. It seems that the archetype ω had no general title. All elements of the general title (1°-3°) are in the title of the chapters' index preceding book I in V S + G (4°-5°). As for the general title, see Hübner, p. xxxvi-xxxix.

(++) 4°-5° (2-3): Τάδε ἔνεστιν ἐν τῷ α΄ βιβλίῳ κεφάλαια Κλαυδίου Πτολεμαίου τῶν πρὸς Σύρον ἀποτελεσματικῶν V S + G (In primo libro — insunt ista capitula), *cf.* γ] ὁ πίναξ σὺν θεῷ τῆς βίβλου πτολεμαίου Y: *indicem omnino om.* Σ M

(++) 6° (4) α΄ προοίμιον V + G (Prohemium primum) Procl.] προοίμιον S: *om.* α Dγ

(++) 7° (5) β΄ V + G (2m) Procl.] α΄ Dγ: *om.* α M S

(++) 8° (6) γ΄ V + G (3m) Procl.] β΄ Dγ: α΄ Y: *om. cett. Et sic pro numeris deinceps*

(-*) 12° (10) ἡμερινῶν ... νυκτερινῶν Y Procl.] *inv. cett.* + G (De nocturnis et diurnis) (*Boll*) ('*forte recte, quia antiquis nox diem praecedebat cf.* 3,589°, *sed v.* 1,465°' *Hüb.*). — G confirms the reading of all manuscripts against Y. Y seems to have adapted all titles in the list to make them conform to the chapter titles (thus also in I.21°, 22°, 23°, 27°, 28°). The list in Procl., which was composed by the Renaissance editor from the chapter titles, has no authority. Write νυκτερινῶν ... ἡμερινῶν.

(§) 21° (18) ὁμοίως *coni. Boll Hüb.* + G (similiter)] ὁμοίων *codd.*

(++) 31° (29) α΄ V + G (capitulum primum) Procl.] *om.* α βγ

(++) 31° (28) Προοίμιον V βγ + G (Prohemium) Procl.] *om.* α

(+) 51° (43) ὁλοσχερεστέρων Y + G (excellentiorum) Heph., *cf.* Procl.] ὁλοσχερῶν V Σ βγ

(++) 63° (54) β΄ V + G (Ca. 2m) Procl.] α΄ Dγ: *om.* α M S

(++) 78° (66) τὸν V S ('*scilicet* τόπον' *Hüb.*) + G (locum qui est)] τῶν, τῷ, τὸ, τῆς *cett.*

(-*) 81° (67) ἡμῖν V βγ] *om.* α + G Procl. (*Robb.*). — See above, IV B 2.2.2.

(+) 84° (70) αὐτῆς α S γ + G (ipsius), *cf.* Procl.] αὐτοῖς V M D

(+) 130° (104) αὐτῶν Y β + G (ipsorum) Procl.] αὐτὰ V Σ: αὐτὰς γ

(-) 130° (105) τὰς φύσεις βγ Procl.] φήση Y: φύσει V Σ: *om.* G

(?) 135° (108) συγκρίσεως (Y)] συγκράσεως *cett.* Procl.: commixtione G. — The testimony of G (*commixtione*) is of no great help, as the same Latin term may be used for both Greek forms, as we can see in cases where there is no variation among the manuscripts. For σύγκρασις: *complexio* I.299° 334° 722° II.793°; *commixtio* II.682° 699° 806° 809° III.109° 419° 877° 1225°. For σύγκρισις: *concretio* III.48° 118°, *comparatio* III.743°. In the following cases we have σύγκρισις as dominant form (σύγκρασις only in γ): *complexio* I.243° 369° (-κρα- in Yγ), *commixtio* I.185°, or σύγκρασις as dominant form: *commixtio* I.135°; *complexio* I.143°. We will not mention such cases hereafter.

(+) 135° (108) ... τὸ ἴδιον Σ βγ + G (eam que ... proprietatem)] ... ἴδιον (*om.* τὸ) V Y

(+) 152° (122) τοῦ Y + G (quod)] τούτου V Σ βγ (*Robb.*)

(+) 182° (145) συμπτωμάτων Y M S + G (simptomatum) Procl.] συναπτομάτων V Dγ: *om.* Σ

(§) 190° (152) κατακρατεῖ *coni. Hüb.* + G (optinet), *cf.* Procl.] κατακρατεῖν *codd.*

(-*) 193° (154) (οὐ μικρὰς ποιοῦνται) τὰς περὶ τὰ (συνιστάμενα παραλλαγάς) Σ (*Robb.*)] τὰς περὶ τὰς Y: περὶ τὰ V βγ + G (non modicas faciunt circa ea que constituuntur uarietates). — The addition of τὰς (as in α) in Robbins and Hübner is not needed in the construction οὐ μικρὰς ποιοῦνται. *Cf. Synt.* I,1, p. 428,25.

(-*) 197° (156) οἱ γεννώμενοι βγ] οἱ γενόμενοι V Σ + G (qui generantur): οἱ γενάμενοι Y, *cf.* Procl. (*Robb.*). — The reading of βγ fits well in this context, but is certainly a conjecture.

The argument is about (human) animals generated from seeds, hence *γεννώμενοι*. We propose to write *οἱ γενόμενοι* with V Σ (Y) + G.

(+) 213° (168) *πάντα* Σ M S *γ* + G (omnia), *cf.* Procl.] *μὴ praem.* V Y D

(-?) 218° (171) *μεμψόμεθα* V Σ *Boll, Robb.*] *μεμφόμεθα* Y *βγ* + G (querelamus) Procl. — The future form is not required. See next item.

(-*) 218° (171) *λέγουσιν* Σ (*Robb.*)] *λέγοντες* V Y *βγ* + G (interrogantes) (*Boll*). — The third person plural dativ *λέγουσιν* of Σ, preferred by Robbins and Hübner is certainly a better reading, as it fits to *ἰατροῖς* and *ὑποτιθεμένοις*, but it is obviously a conjecture of Σ. The archetype ω read the nominative *λέγοντες*, which is somehow redundant after *μεμφόμεθα*: 'we do not object saying', as it is not explained what 'we are saying'. One understands why Moerbeke translates not literally 'dicentes', but 'interrogantes'. In 220°, 'non subsannandum' may again be a free interpretation of the infinitive *μὴ ἀγανακτεῖν* where *δεῖ* should be added.

(+) 221° (174) *γ′* (*ante titulum*) V + G (*post titulum*) Procl.] *β′* D *γ*: *ι′* Y: *om.* Σ M S

(++) 224° (177) *τοιαύτης* V Σ + G (tali)] *αὐτῆς* Y *βγ*

(+) 230° (181) *ἡ κτῆσις* Y Σ S *γ* + G (possessio), *cf.* Procl.] *ἡ κτίσις* V D M (*corr.* D[2])

(+) 230° (180) *ἡ συμβίωσις* Σ M S *γ* + G (conuictus), *cf.* Procl.] *εἰ συμβίωσις* V Y D

V (Y β =DMS γ =ABC)

(-) 237°-9° (187-8) *συμφορώτερον* (…) *τῆς τοιαύτης προγνώσεως* V Y, *cf.* Procl., *cf.* Tib. (lucretur)] (…) *τῆς τοιαύτης γνώσεως* (*προγνώσεως γ*) *σπουδαιότερον* M S *γ*, *cf.* D: magis conferens (…) tali cognitione studiosius (*συμφορώτερον … τῆς τοιαύτης γνώσεως σπουδαιότερον*) G. — See above, IV B 2.1.3.

(-?) 244° (192) *συνεργεῖ* V Y Procl.] *om.* *βγ* + G. — The verb *συνεργεῖ* fits well, but is not really needed here, as is evident from the parallel formulations in 237° and 240°. It would be more consistent with the editorial choice e.g. in 230° and 260°, to follow here also the reading of *βγ* confirmed by G. This entry and the three following (248°-260°) may show that the readings which G (principal representative of ψ) shares with *βγ* (subfamily of φ) are not ordinary contaminations but due to an intelligent choice.

(-?) 248° (195) *τὰ μείζω* V Y, *cf.* *τὰ μείζονα* Procl.] *τὸ μεῖζον* *βγ* + G (maius). — *τὰ μείζω* could be a correction. Y is contaminated by V. See 213° 260°.

(++) 249° (196) *ἄχρηστον* V Y + G (inutilitatem)] *εὔχρηστον* *βγ*

(+) 260° (205) *ἅπαντα* *βγ* + G (omnia), *cf.* Heph., Procl.] *ἕκατα* V: *ἕκαστα* Y (*Robb.*)

(-?) 272° (214) *ἐκ πυρώσεων* V Y, *cf.* Procl.] *ἐκ -εως* *βγ*: *ἐκπυρώσεως legit* G (adustionis). — The singular is the *lectio difficilior*: the plural could be an adaption to the following plural.

(++) 279° (220) *ὅσων μὲν* V + G (quorumcumque quidem)] *ὅσον μὲν* Y: *ὧν* β: *ἐφ' ὧν γ*

(-?) 281° (221) *ὅσα δὲ* V Y] *ἃ δὲ* *βγ*, *sed συμπτώματα add.* β *post ἔχει*; *ἃ δὲ συμπτώματα* Procl.: symptomata autem que G. — Probably *συμπτώματα* was added *supra lineam* to explain the reference of *ἅ*.

(+) 286° (226) *ἴδοι* *βγ* + G (uidebit)] *ἴδη* V: *εἴδη* Y

(++) 288° (228) *δὲ καὶ* V + G (autem et)] *τε καὶ* β: *καὶ* Y *γ*

(-?) 314° (248) *θεραπευόντων* V Y] *θεραπευσόντων* *βγ* + G (curaturis)

(-?) 319° (251) *μὲν α* *βγ* Procl.] *om.* V + G. — *μὲν* could be a correction to make a correlation to *δὲ* in 335°.

(++) 321° (252) προφυλάσσεσθαι V D γ + G (precauendi) Procl.] φυλάσσεσθαι Y M S Heph.

(+) 341° (267) ἀμετρίαν Y M S γ + G (intemperantiam) Procl.] μετρίαν V D

V (Y β =DMS γ =ABC, *excerpta* L)

(++) 383° (300) πρώταις ταῖς ... δυνάμεσιν V + G (primas uirtutes ...)] πρὸ τῆς ... δυνάμεως Y: πρώτης ἀψάμενοι τῆς ... δυνάμεως L: πρώτης τῆς ... (*om.* δυνάμ-) β B, *cf.* A

(§) 392° (307) τοσούτῳ *coni. Hüb.* + G (tanto), *cf.* Tib. (tanto)] τότε V L: *om. cett.* Procl.

(++) 393° (308) πλεῖστον V L D γ + G (plurimum) Procl.] πλέον Y M S

(++) 400° (313) κατὰ ψῦξιν V L + G (secundum infrigidationem), *cf.* Tib. (suaque frigiditate)] *om.* α βγ

(-?) 436° (338) ψυχροῦ ... ξηροῦ Y M S γ] *inv.* V L D + G (siccum ... frigidum) Procl. (*Robb.*). — See above, IV B 1.3.3.

(-*) 436° (339) πάντα πάλιν (Y D γ)] πάντα V L M S + G Procl. — πάλιν is not needed here. It is a redactional improvement to resume the first πάντα.

(-) 441° (342) τὸν τοῦ Κρόνου καὶ τὸν τοῦ Ἄρεως ὡς τῆς ἐναντίας φύσεως ποιητικούς V L Y, *cf.* Tib.] τὸν τοῦ Κρόνου καὶ τὸν τοῦ Ἄρεως κακοποιοὺς ὡς τῆς ἐναντίας φύσεως ποιητικοὺς βγ + G (Saturnum autem et Martem maleficos tamquam contrarie nature factiuos), *cf.* Procl. — See above, IV B 2.1.4.

V (Y β =DMS γ =ABC)

(-) 444° (345) ὡς Y βγ, *cf.* Procl.] *om.* V L + G

(-?) 454° (353) ἀρρενικούς *codd.*] propter siccitatem (διὰ τὴν ξηρότητα?) masculos G. — A good addition in the model of G, *cf.* 452° διὰ — ὑγρῷ, and 456°. Scholiast?

(-) 469° (365) τεθηλυσμένον μᾶλλον Y Dγ] feminina facta *hic, et* magis *post* ipso (αὐτῇ) 478° (365) G: τεθηλυσμένον V M S. — μᾶλλον is found in Y Dγ at this place, where it corresponds to μᾶλλον in 467°. It is omitted in V M S. In G it is present, but adapted to another place. The adverb was probably added in the archetype either at the wrong place or without a precise localisation.

(++) 470° (366) ἐν αὐτῇ V + G (in ipsa)] κατ' αὐτὴν α βγ

V (Y β =DMS γ =ABC, *excerpta* L)

(-) 497° (387) ἑῷοι μόνον βγ + Co le (25,1)] ἑω μένω Y: ἑῷοι V L + G (matutini)

(-*) 508° (396) ἑξῆς δέοντος *coni. Hüb. secundum Kroll*] ἑξῆς δὲ ὄντος V L Y + G (Cum autem sit consequens): ἑξῆς ὄντος ἀκολούθου βγ. — The correction ἑξῆς δέοντος is not needed. One should keep the reading of V L Y, which is confirmed by G. The phrase ἑξῆς δὲ ὄντος is found elsewhere in Ptolemy: see *infra* IV. 224°, *Syntaxis*, I,1 (p. 216.3,18), I, 2 (p. 250,3). The editor of βγ added ἀκολούθου, maybe influenced by the Proclus Paraphrase (ἀκόλουθόν ἐστι). Write ἑξῆς δὲ ὄντος.

(++) 508° (396) ἀστέρων V L γ + G] *om.* Y β Procl.

V (α =YΣ β =DMS γ =ABC, *excerpta* L)

(++) 523° (407) Λαμπαύρας V + G (Lampauras) Procl.] λαμπαίρας L: λαμπαδίας βγ Heph. Ep.: καλούμενος δὲ Λαμπαύρας *om.* Y Heph. — The term λαμπαύρας is attested in a similar context by Heron of Byzantium (Γεωδαισία 11,64), an author of the 10th century, who may depend on the same model as V + G and Procl. The term λαμπαδίας is attested in Heph. Ep. and Joh. Lydus and seems preferable. But βγ may have corrected the text from Heph.

(-?) 533° (414) ποδῶν βγ Heph., *cf.* Tib. (pede)] ὀφθαλμῶν V L D^1 α + G (oculis) Procl. (*Robb.*) — The reading of the better manuscripts is ὀφθαλμῶν: it seems to have been the reading of the archetype ω. The editor of βγ introduced ποδῶν from Heph.; this reading is also found in Tib. and may have been in the primitive text of Ptolemy. On the reasons why Hübner prefers it, see note at 533°.

(-*) 560° (431) ἐν μέσαις V *cett.* + G (in mediis)] ἐν τῷ μέσῳ τῶν χηλῶν L

(+) 566° (436) τῆς Ἀφροδίτης α L + G (Veneri) Heph. Procl.] τῆς Ἑρμῆς V: τοῦ Ἑρμοῦ βγ

(+) 568° (437) ἑπομένη L M S γ + G (sequens) Heph. Ep., *cf.* ἐπιμένη V] λεγομένη α Heph.: *om.* D

(-?) 570° (438) περὶ α βγ] ἐπὶ V L + G (in) Procl. — Both prepositions are used.

(-?) 570°-1° (438-9) ὁ ... ἔχει βγ Heph.] οἱ ... ἔχουσιν V + G (que ... habent) Procl. (*Boll*): ἐν ... ἔχει Y

(-?) 587° (449) ἱματίῳ Y βγ Heph.] χύματι Σ^{t}: ἱματίῳ $\Sigma^{mg.}$: μετώπῳ V L + G (fronte) Procl. — The connection between the 'left hand' and the 'cloak' of Aquarius is made by Ptolemy in *Syntaxis* VIII, 1, p. 118-120; ἱματίῳ is confirmed by Heph. (both in *Apot.* and in the *Epitome*). There is no instance found in the astrological literature of the μέτωπον of Aquarius. But also the reference to the 'cloak' of Aquarius is rare, and in fact seems to be limited to this case. The addition καὶ τῷ ἱματίῳ (or μετώπῳ) is not found in the Arabo-Latin translations. The reading μετώπῳ of V L + G Procl. — hyparchetype ψ — is not simply a copyist's error. But it is also not a correction. Could ἱματίῳ be an improvement by the editor of φ, taken from Heph.?

(+) 597°<*sic, cf. Boll*> (455) λίνου L M S γ + G (lino) Procl.] αἴνου V D: λίνῳ α Heph. (*Boll*)

(+) 598° (455) ὁ ... λαμπρὸς V Σ + G (clara ... que) Heph. (A)] οἱ ... λαμπροί Y βγ Heph. Ep.

(-?) 600° (456) Περὶ τῶν βορειοτέρων τοῦ ζῳδιακοῦ (*tit.*) V L D Heph. (*sed cf.* A), *cf.* γ] *om.* α M S + G Heph. (U) (*Robb.*). — See 638°.

(+) 604° (460) οἱ δὲ (ὑπὸ τὴν οὐρὰν) L D γ + G (Que autem sunt) Procl.] ἡ δὲ V Y, *cf.* M S

(-?) 638° (481) Περὶ τῶν νοτιωτέρων τοῦ ζῳδιακοῦ (*tit.*) V L D Heph. (A), *cf.* γ Procl.] *om.* α M S + G Heph. (U) (*Robb.*). — Both titles (600° and 638°) are not translated by Moerbeke, and presumably his model did not have them. Both repeat the immediately following lines. They are not found in the preceding table of chapters. Notice the conformity of G with α M S against V L D Procl. in both places.

(+) 651° (491-2) περὶ τὸ στόμα V L γ + G (circa os) Procl.] ἐπὶ τοῦ στόματος α β Heph.

(-?) 670° (501) οἱ ... λαμπροὶ α Heph.] ὁ ... λαμπρὸς V L βγ + G (que ... est ... clara) Procl. — οἱ ... λαμπροὶ correction of ὁ ... λαμπρὸς? The reverse is not plausible.

(-*) 672° (501) Περὶ τῶν ὡρῶν τοῦ ἔτους *post* Ἑρμοῦ α M S] *post* παρατηρήσεως (674°) V L D γ + G (De temporibus anni *post* obseruationem) Co (25,11) Procl. — Write the title after παρατηρήσεως with V L D γ + G, the Commentary and Procl. This seems to be the original position. The sentence αἱ μὲν τῶν ἀστέρων — παρατηρήσεως is the conclusion of chapter nine.

(-?) 682° (511) διὰ τὸ τὸν V L Σ, *cf.* Y] διὰ τὸ τότε τὸν βBC + G (quia tunc), *cf.* A

(-) 688° (516) ἔαρος M S Procl.] ἀέρος V L Y D γ + G (aeris). — The reading ἀέρος is a copyist's mistake, rightly corrected in some recent manuscripts. In 691° (518) G makes the same error.

(-*) 690° (518) ὑγρασίας V D γ] ὑγρᾶς οὐσίας L α M S + G (humide substantie) Procl. — There is no reason to follow V D γ. The syntagm ὑγρᾶς οὐσίας is found in three other places in this treatise I.449°, 456°, II.83°, whereas this case would be the only instance of ὑγρασίας. We propose to write ὑγρᾶς οὐσίας. *Cf.* 788°.

(-*) 734° (555) *titulum habent post* ποιότητος (733°) α D γ] *post* δυνάμεις (745°) V L + G Co Procl. — Write title after δυνάμεις with V L + G.

(++) 753° (562) τῶν ... παρόδων V L + G (motuum) Procl.] τὴν ... πάροδον α β: τῆς παρόδου γ

(-) 769° (574) αὐτῶν Σ D S γ, *cf.* Procl. (αὐτοῦ)] *om.* V L Y M + G. — αὐτῶν is certainly necessary here. Was it added to the text of the archetype ω?

(-*) 772° (576) καὶ ἰσημερινῶν α γ] *om.* V L β + G Procl. Tib. (*Boll*). — The fact that καὶ ἰσημερινῶν is found in α γ and not in β indicates that it may be a later addition. One may understand why the addition was made. The four δίσωμα signs are situated in the zodiac between, on the one hand, the four στερεά and, on the other hand, respectively the τροπικά and the ἰσημερινά. To say that they are between the στερεά and τροπικά is only right for Sagittarius and Gemini; the two other signs, Virgo and Pisces, are between στερεά and ἰσημερινά. If one wants to be precise, one should write μεταξὺ τῶν στερεῶν τε καὶ τῶν τροπικῶν <καὶ τῶν στερεῶν τε καὶ τῶν ἰσημερινῶν>. The correction of α γ only increases the confusion.

V (α =YΣ β =DMS γ =ABC)

(++) 778° (581) ἓξ δὲ V + G (sex autem) Procl.] τὰ δὲ ἴσα α βγ

(-) 785° (586) ἔτι τὸ α βγ, *cf.* Tib. (et etiam)] *om.* V + G Procl.

(+) 788° (588) τῶν ὅλων α M S + G (totorum) Procl.] *om.* V D γ

(++) 796° (593-4) τὸ ἀπηλιωτικώτερον V + G (magis orientale) Procl.] τὸν ἀπηλιώτην α βγ

(+) 806° (602) τὰ τοιαῦτα V γ + G (talia) Procl.] τὰς τοιαύτας β: τὰ τοιαῦτα καλέσαντες α

(+) 807° (603) τῆς ... ποιότητος α + G (cum qualitas)] τοῖς ... ποιότητος V: καὶ ὡς ἡ ποιότης βγ, *sed cf.* D

V (α =YΣ β =DMS γ =ABC, *excerpta* L)

(+) 814° (608) δύο ὀρθὰς γωνίας L α + G (duos rectos angulos) Procl.] ὀρθὰς δόο γωνίας D M γ: ὀρθὰς ἴσας γωνίας V; *cf.* S

(-*) 820° (612-3) μιᾶς ὀρθῆς α βγ] γωνίας ὀρθῆς V L Procl. + G (anguli recti) (*Boll*). — See above (IV B 3.3.)

(-*) 837°-8° (627) οἵ τε (τε *om.* Σ) τετράγωνοι καὶ οἱ κατὰ διάμετρον Σ] οἵ τε τετράγωνοι καὶ οἱ διάμετροι Y: οἵ τε διάμετροι καὶ οἱ (αἱ L: *om.* M S) τετράγωνοι L βγ + G (que ex opposito et que ex quadrato), *cf.* Tib.: οἵ τε τρίγωνοι καὶ οἱ τετράγωνοι V. — There is no reason to combine particular readings of Σ and Y. Write οἵ τε διάμετροι καὶ οἱ τετράγωνοι with L βγ + G.

V (α =YΣ β =DMS γ =ABC)

(++) 841° (629) ὁμοίως V + G (similiter)] *om.* α βγ Heph. Procl. ('*fort. recte*')

(+) 845° (633) τε καὶ καταφέρεσθαι βγ + G (et descendunt) Procl.] *om.* V α

(+) 848° (636) ἐκείνων βγ + G (illa), *cf.* Procl.] ἐκεῖνον V Y: -νο Σ

(-?) 853° (640) ἀλλήλοις μέρη α βγ] τά τινα ἀλλήλοις μέρη V: inuicem partes quasdam G

(-*) 857° (643) ἰσοχρόνως ἀποτελεῖσθαι] ἰσοχρόνος (Γ: -νων Y; -νια Σ G°, -να Procl.) ἀποτελεῖσθαι α, *cf.* Heph.: ἰσοχρόνως ἀποτελεῖν V βγ + G (in equali tempore perficit) (*Boll*). — We propose to follow the same manuscripts V βγ + G (*Boll*) for the whole expression. ἀποτελεῖν *lectio difficilior,* ἀποτελεῖσθαι correction?

(-?) 863° (649) ἁπλῶς α βγ] *om.* V + G, *cf.* Procl.

(+) 864° (649) ἃ V Σ + G (que)] τὰ μετὰ Y: τὰ μετὰ τοῦ βγ

(-*) 866° (651) ἔτι τε καὶ V] ἔτι δὲ καὶ Σ + G (adhuc autem et): καὶ *cett.* — Write ἔτι δὲ καὶ, characteristic of Ptolemy.

(-) 869° (652) καταλαμβανόμενα καὶ Σ] καταλαμβανόμεθα τὰ V + G (deprehendimus que): καταλαμβάνομεν καὶ Y: λαμβανόμενα καὶ βγ. — There are again two redactions, both of which are possible: the reading of Σ chosen by Boll, Robbins, Hübner; and the reading of V, confirmed by Moerbeke. Moerbeke translates the medial form καταλαμβανόμεθα construed with an accusative (τὰ … γινόμενα δωδεκατημορίων) rightly by an active form (*deprehendimus* que fiunt … *signa*), as he did in a similar context at I.35°-7° (… σχηματισμοὺς καταλαμβανόμεθα: *figurationes comprehendimus*). The version of V + G, which is the *lectio difficilior,* fits well in the context of the chapter: ἔτι δὲ καὶ τῶν ἐκκειμένων τεσσάρων συσχηματισμῶν (τοῦ τε διαμέτρου …) κατὰ τὸ παντελὲς ἀμέτοχα καταλαμβανόμεθα τὰ ἤτοι δι' ἑνὸς ἢ διὰ πέντε γινόμενα δωδεκατημορίων. The other manuscripts present the verb in different ways and replace τὰ by καὶ. The reading of V + G could well be the version of the archetype ω, while the other could be different attempts to correct the text.

(-*) 879° (663) βορειότατα α S] βορειότερα V L D M γ + G (borealiora) + Procl. — The comparative form is better attested, and fits better.

(+) 885° (667) ὂν α βγ + G (existentem)] *om.* V L

(-?) 893° (674) μὲν οὖν γ, *cf.* Procl.] μὲν διὰ V: μὲν δὴ L: μὲν δὲ Y: μὲν γὰρ Σ: μὲν β + G (quidem). — Reading of γ result of correction.

(-?) 896°-7° (676-7) (τὰ διάμετρα ζῴδια τοῦ τε Καρκίνου καὶ τοῦ Λέοντος ὅ τε) Αἰγόκερως καὶ ὁ Ὑδροχόος α D S γ, *sed* ὁ *om.* α] ὅ τε Ὑδροχόος καὶ ὁ Αἰγόκερως L M + G (Aquarius scilicet et Capricornus) Procl.: Ζυγὸς καὶ ὁ Αἰγόκερως V [The copyist of V made an error in reading the astrological sign of Aquarius]. — See above, IVB 1.3.3.

(++) 899° (678) συσχηματισμὸν V + G (configuratio), *cf.* Procl.] συσχηματισμῶν L Y β: συσχηματισμοῦ γ

(++) 909° (685): τοῦ τετραγώνου V L + G (quadrature)] τετράγωνον α βγ (*cf. Robb.*)

(+) 914° (688) τὸ πλεῖστον (*ante* ἀφίσταται) α M S + G (secundum plurimum *ante* distat)] *post* ἡλίου (915°) D γ: *om.* V L (Procl.)

(+) 928° (700) ἐκ α βγ + G (ex)] γὰρ *add.* V L, *cf.* Procl.

(-*) 935° (705) προηγουμένως μὲν α γ] προηγουμένως V L β + G (principaliter). — The position of μὲν after προηγουμένως may be due to a later correction.

(-?) 939° (708) συνίσταται α βγ, *cf.* Tib. (colligitur)] γίνεται V L + G (fit) + Procl. — See above, IVB 1.3.3.

(-*) 946° (712) αὐτοῦ Y L βγ] αὐτῷ V + G (sibi): αὐτῶν Σ. — It seems better to follow V, confirmed by G: dative with συνοικοδεσποτέω (see 975°).

(++) 959° (721) τὸ δὲ τρίτον τρίγωνον ὅ ἐστι V L D γ + G (Tertium autem trigonum quod est)] τὸ δε τρίτον τρίγωνον ἐστι M S: τρίτον δὲ τρίγωνον α

(-?) 973° (732) μόνῳ α βγ] μὲν V L + G (quidem) (*Robb.*). — See above, IVB 1.3.3.

(-*) 994°-6° (749-50). — See above, IVB 3.1

V (α =YΣ β =DMS γ =ABC)

(+) 1019° (769) διαθέσεως V D S γ + G (dispositione) Procl. (60,-4)] τῶν πέντε ἀστέρων L: *om.* α M Co (39,-5)

(+) 1021° (771) ἐστιν V D γ + G (est)] *om.* α M S

(-) 1021°-2° (771) ὁ πρὸς τὰς τῶν οἴκων ὡς ἐπίπαν κυρίας α] *om.* V βγ + G, *cf.* Procl. — The explanation of the Egyptian system is only found in α, which is surprising. In fact, it is for stemmatic reasons difficult to assume that it is a common error by omission in ψ and βγ. Is it, then, a conjectural addition in α? Some scholar may have added it, because the addition seems necessary to parallel the explanation of the Chaldaean system, which follows in the next line: ὁ μέν ἐστιν Αἰγυπτιακὸς (ὁ πρὸς τὰς τῶν οἴκων ὡς ἐπίπαν κυρίας), ὁ δὲ Χαλδαϊκὸς (ὁ πρὸς τὰς τῶν τριγώνων οἰκοδεσποτείας). However, the style of the additional explanation is characteristic of Ptolemy (*cf.* the phrase ὡς ἐπίπαν). Furthermore, the addition was certainly known to the commentator, who quotes it in a shortened form as ὡς ἐπίπαν κυρίας. The commentator explains the sense of κυρίας and why it is ὡς ἐπίπαν. As he says, the Egyptian system is not entirely based upon the government of the houses (ταῖς τῶν οἴκων οἰκοδεσποτίας), but also uses the triplicities and the exaltations. And finally, this explanation is also found in the Arabic translation of the text. See Tib.: 'At qui Aegyptius dicitur, maxime a domorum dominis accipitur, Chaldaicus autem a dominis triplicitatum assumitur'. One therefore has to admit the authenticity of the additional explanation. The explanation for this may have been found in the margin of the archetype ω. But on that hypothesis, it is again difficult to explain why neither ψ nor βγ have it. Maybe we have to admit, after all, that it was a conjecture? In fact, an intelligent scholar could have easily construed it using the commentary. He could find the last part in the quoted lemma ὡς ἐπὶ πᾶν κυρίας and complete it using the explanation of the lemma and the following parallel explanation of the Chaldaean system (ὃ πρὸς τὰς τῶν τριγώνων); <ὁ πρὸς τὰς τῶν οἴκων > ὡς ἐπὶ πᾶν κυρίας.

(-?) 1025° (775) μὲν V βγ] μὲν γὰρ α + G (quidem enim) (*Robb.*)

(+) 1028° (777) ἐπεὶ παραδείγματος ἕνεκεν V + G (quoniam exempli gratia)] ἐπὶ παραδείγματος δὲ ἕνεκεν Y, *div. err. cett.*

(-) 1048° (792) ὁ αὐτὸς ἀριθμὸς α] ὁ ἀριθμὸς (*om.* αὐτὸς) V β + G (numerus): [γ]

(+) 1049° (792) ἂν V S + G (utique)] ἐὰν D M: *om.* α: [γ]

(-*) 1055°-6° (798) ἐγγιζούσης τῆς ἀληθείας V M S γ] ἐγγιζούσῃ τῆς ἀληθείας *coni. Boll* + G (propinque ueritati, *fort.* ἐγγιζούσῃ *legit et perperam per dat. loco acc. redd., cf.* 1055° συνισταμένῃ = constitute): ἐγγιζομένου τῆς ἀληθείας D: ἐγγὺς οὔσῃ τῆς ἀληθείας α (*Robb.*): ἥτις οὐδὲ ὅλως ἐγγίζει τῇ ἀληθείᾳ Procl. — See Hübner, p. xvii. Moerbeke probably read ἐγγιζούσῃ τῆς ἀληθείας in his model, but erroneously kept the dative in Latin as he also did at 1055° συνισταμένῃ = *constitute*, whereas the Greek construction requires a Latin accusative to make the connection with 'traditionem'. Notwithstanding these Graecisms, the translation may point to the right reading ἐγγιζούσῃ τῆς ἀληθείας, *cf.* Procl. ἥτις οὐδὲ ὅλως ἐγγίζει τῇ ἀληθείᾳ. In the sense of 'approaching', the verb ἐγγίζω can be construed with a genitive. The reading of α, adopted by Robbins, is a conjecture.

(-) 1061° (801) ἐν α βγ, Procl.] ἔτι V + G (adhuc)

(+) 1072°-84° (811-34) *tabulam habet* Σ Y^{2} β(-S) γ + G Procl.] *om.* V (*spatio relicto*) Y^{1} S

(-?) 1086° (837) μοῖραι γ] μὲν Y + G (quidem) Procl.: μὲν ἔτη V D M: [Σ S]. — μοῖραι is a conjecture of γ, as is the addition of ἔτη in V D M. The reading of Y, confirmed by G and Procl., could be at the origin of both corrections.

(-*) 1087° (838) γίνεται ὅμου τξ΄ γ] γίνονται τξ΄ V D + G (fiunt 360) Procl.: *om.* Y M [Σ S, *inde a* 1085°] — The accepted reading is a conjecture in γ. See also 1117°.

(-*) 1092° (841) διαβάλλειν V D γ] ἐπιβαλεῖν/ἐπιβάλλειν α M S + G (scire). — Moerbeke has probably translated ἐπιβάλλειν *ad sensum* with *scire* (κατανοεῖν Procl., *comprendre* Charvet, *understand* Robbins). Tib. translates *memoriae commendari*. The reading of V D γ seems inappropriate here. Write ἐπιβάλλειν.

(§) 1092° (841) αὐτούς *coni. Boll Hüb.* + G (ipsos)] αὐτοῖς V β, *cf. cett.*

(-?) 1094° (844) πρῶτος α βγ] πρῶτον V + G (primo) Procl.

(-) 1103° (851) ἤδε α βγ] *om.* V + G Procl.

(-*) 1113° (860) συνάγεται V D γ] συνάγονται α M S + G (Colliguntur) (*Robb.*)

(-?) 1117° (862) τὸ πᾶν V D γ Procl.] ὅμου M S: μοῖραι α: totum simul (τὸ πᾶν ὅμου?) G. — See above, IVB 2.1.3.

(-?) 1130° (873) τάς τε τῶν προειρημένων γενέσεων μοιρογραφίας β, *cf. diversimode* α γ, *cf.* Tib.] τάς τε τῶν γενέσεων προειρημένας μοιρογραφίας V + G (predicte circa natiuitates graduum descriptiones): [Procl.]

(-?) 1133° (875) περισσῆς α βγ] πολλῆς V + G (multa) (*Boll*). — See above, IVB 1.3.3.

(-*) 1134° (876) ἀδιάγνωστον βγ] ἀδιάσωστον V α + G (non salua): [Procl.]. — Intelligent correction in βγ. The archetype (ω) had probably the reading of V α + G.

(+) 1134° (876) ἡμῖν *hic* (*post* μόλις) V Σ + G (uix nobis poterat)] ἡμῖν *post* δυνάμενον (1135°) Y βγ

(-?) 1135° (877) ὑποτυπῶσαι α βγ] ὑποδεῖξαι καὶ τυπῶσαι V + G (ostendere et exprimere) (*Boll*). — See above (IVB 1.3.3.)

(++) 1138° (878) διασεσωσμένης V + G (que salua erat)] διασεσωσμένοις Y: -μένων Σ: -μένην βγ

V (α =YΣ β =DMS γ =ABC, *excerpta* L)

(+) 1140° (879) τοῦτον α M S + G (istum)] *titulum* περὶ ὁρίων ὡς (ὡς V: ὁ D γ) Πτολεμαῖος *add.* V D γ Procl., ὁ δὲ τῶν ὁρίων τῆς ὅλης ἐπιβολῆς κατὰ τὸν Πτολεμαῖον τοῦτον ἔχει τὸν τρόπον L (*initium excerpti*). — G confirms the absence of the title. *Cf.* 1171°.

(-*) 1148° (886) προτασσομένου γ (*sic in textu Hüb., Boll*)] προτασσομένων V β Procl. + G (preordinatis) (*sic lemma in adn. Hüb.*): προτασσόμενοι α (*Robb.*): προτασσόμενος L. — We would write προτασσομένων with V β Procl. + G, and translate 'whereas again in order, those that have two lordships each are preferred to the one which has but one' (Robbins, modified). *Cf.* Tib. 'quicunque uero duas haberet in signo potestates habenti unam potestatem in ordine preponebatur'.

(-*) 1163° (898) Ταῦρον καὶ <Παρθένον καὶ> Αἰγόκερων] Καρκίνον καὶ Αἰγόκερων α: *om. in textu* V^1: Ταῦρον καὶ Αἰγόκερων *mg.* V^2: Καρκίνον βγ: Ταῦρον L + G (Taurum) (*tacite Robb.*), *cf.* Tib. — We would write Ταῦρον.

(-) 1171° (903-4) ἔστι δὲ καὶ ἡ τούτων τῶν ὁρίων ἔκθεσις τοιαύτη *codd.*] *om.* G. — See above, IVB 3.4.

(+) 1173° (903-4) ὅρια κατὰ Πτολεμαῖον *excludit Hüb.*: *om.* S + G] *habent* V L α D M, *cf.* γ. — See above, IVB 3.4.

(-) 1174°-85° (903-4) *tabulam ex* L, *Vat. gr.* 1291 *et Procl. receptam* (*var. lect. ex* α D γ) *ed. Hüb.*] *om.* V (*spatio relicto*) S (*lineis tantum adumbratis*): *om.* G. — See above, IVB 3.4.

(+) 1186°-7° (905-6) γίνεται — Ἑρμοῦ ος′ V βγ + G (fiunt —Mercurii 76)] *om.* L α. — See above, IVB 3.4.

(-) 1188° (906) ὁμοῦ τξ′ D M γ] γίνονται τξ′ V S: *om.* G: [L α]. — See above, IVB 3.4.

(§*) 1197° (915) μὲν οὐ ... καὶ οὐ *coni. Hüb.*] μὲν οὖν ... καὶ οὐ V α: μὲν ... καὶ οὐ ... β: quidem igitur neque ... neque (μὲν οὖν οὔτε ... οὔτε) G: οὔτε ... οὔτε Co le. — Write μὲν οὖν οὔτε ... οὔτε with (α) + G?

(-) 1197° (915) πιθανὸν ... φυσικὸν] *inv.* G (naturalem ... probabilem), *an suo arbitrio*?

(-*) 1238° (946) καθόλου V βγ] καὶ καθόλου δὲ α + G (Et uniuersaliter autem): καθόλου δὲ L. — The addition of the particle δὲ is expected at the beginning of this paragraph; the καὶ is unexpected. Write either καὶ καθόλου δὲ or καθόλου δὲ.

(+) 1246° (952) παρόδους V L Σ D γ + G (transitus) Procl.] *om.* Y M S

(+) 1247° (953) ἐστι α γ + G (est) Procl.] ἔτι V β: ἔτη L

(+) 1253° (958) τὰς γωνίας L α M S + G (angulos) Procl.] τῶν γωνίων V D γ

(-?) 1257° (960-1) (ἀνατολικοὺς ... καὶ προσθετικοὺς) ... δυτικοὺς καὶ ἀφαιρετικοὺς α βγ Procl.] (...) ἀφαιρετικοὺς καὶ δυτικοὺς V L + G (ablatiui ... et occasiui). — See above, IVB 1.3.3.

(+) 1265° (966) ἀδύναμοι α S + G (impotentes) Procl.] ἀδύνατοι V L D M γ (*Boll*). — G translates the reading of α S Procl.; ἀδύνατοι would be *impossibiles, cf.* Index.

2. BOOK II

V (α =YΣ β =DMS γ =ABC)

(++) 8° (7) τῶν ἐκλείψεων V D + G (eclipsium) Procl.] *om.* Y M γ, *cf.* II.512°

(-?) 13° (12) ἐκλείψεως V Σ D S γ] ἐκλείψεων Y M + G (eclipsium) Procl., *cf. tit. cap.* 10 (839°). — Moerbeke read ἐκλείψεων. Nowhere else has he corrected the reading of the titles in the index to conform to the titles in the text. -εων and -εως are often variants in the Greek tradition. Both readings are possible.

(++) 18° (17) Διαίρεσις τῆς καθολικῆς ἐπισκέψεως V D γ + G (Diuisio totius intentionis) Procl.] *om.* α M S

(-?) 19° (19) μὲν δὴ α Co le] μὲν οὖν β: μὲν V γ + G (quidem). — The combination μὲν δή, frequent elsewhere in Ptolemy (26x in corpus in TLG), occurs only once in our whole text (II.1132°: quidem igitur *sic* G). The readings of α Co le and β are probably divergent corrections. See I.893° (674), the same case. Following Boll (*Studien*, p. 204 sqq.), this second book was written earlier than the others, and Ptolemy used another source; the style might be different.

(-*) 19° (19) προεκτεθειμένων α M S] ἡμῖν *add.* V D γ + G (nobis). — The combination προεκτεθειμένων + ἡμῖν is also found in *Synt. math.* I,2 p. 293,13; 429,13. We suggest to add ἡμῖν with V D γ + G, *cf. supra* 8°. ἡμῖν may have been eliminated in α M S because it is somehow redundant with ἡμῖν in the next line (21°). It is also absent in the Arabic version. In turn at I.81°, ἡμῖν is present in V βγ and absent in α G. We thought it was not necessary there. Notice the confirmation by G at both places — here the presence of the pronoun, there its absence.

(-) 22°-4° (21-2) τὰ καθ′ ἕκαστα τῆς τῶν ... ἐμπιπτόντων πραγματείας *coni. Hüb.*] τὰ καθ′ ἕκαστα τῆς πραγματείας (!) τῶν ... ἐμπιπτόντων πραγματείας V + G (particularia negotii de incidentibus ... in operando): καθ' ἕκαστα τῶν ... ἐμπιπτόντων τῆς πραγματείας Y: τὰς καθ' ἕκαστον τῶν (τῆς D) ... ἐμπιπτόντων πραγματείας Σ βγ. — See above, IVB 1.3.3.

(++) 31° (27) ὅπερ V + G (quod) Procl.] ὃ καὶ αὐτὸ α βγ + Arab.

(++) 38° (33) προδιειληφέναι V + G (prepertractasse)] περιδιειληφέναι α: διειληφέναι βγ

(-*) 43°-5° (38-9) οἷαί εἰσιν αἱ τῶν ἐτησίων (ἐνιαυσιαίων βγ) ὡρῶν κατὰ τὸ μᾶλλον καὶ ἧττον ἀλλοιώσεις (ἐναλλοιώσεις M S) α βγ] οἷαί εἰσιν αἱ τῶν ἐτησίων ὡρῶν καὶ ἡ κατὰ τὸ μᾶλλον καὶ ἧττον ἀλλοίωσις V (*Boll*) + G (quales sunt que temporum anni et que secundum magis et minus alteratio), *cf.* Procl. — This is a very interesting case. G confirms the reading of V, which is the *lectio difficilior,* adopted by Boll (and partially by Robbins). In this section, Ptolemy distinguishes two types of prognostic investigation: one considers 'the greater and more periodic conditions (περιστάσεις)', the other 'the lesser and more occasional' (οἷαί εἰσιν αἱ τῶν ἐτησίων... etc.). In the reading of V + G, *two* conditions are considered in the second case: the <conditions> of the seasons (if one adds περιστάσεις), and the variation of intensity (more or less) which has an effect on storms and winds, fertility, etc. In α βγ, only one condition is considered: the changes in intensity (of temperature) of the seasons, which have an effect on storms, winds, fertility, etc. The problem with the reading of V + G is that it is difficult to take the first condition as 'occasional'. The annual rhythm of the seasons is nothing special: but a warm winter or a cold summer is, and this also has an effect on the other meteorological conditions. In fact, there is only one condition, but Ptolemy may have formulated it in a complicated hendyadis. One may understand why the text was corrected in α βγ. The author of the Proclus Paraphrase must have read the text as in V + G. He offers an excellent interpretation, rendering this difficult passage clear: ὡς ἐπὶ τῆς μεταβολῆς τῶν καιρῶν καὶ τῆς κατὰ τὸ μᾶλλον καὶ ἧττον γινομένης ἐναλλαγῆς αὐτῶν πρός τε χειμῶνας...

(-*) 46° (40) ἢ εὐφορίας τε καὶ α] καὶ εὐφορίας καὶ βγ: εὐφορίας τε καὶ V + G (fertilitasque et)

(-?) 82° (67) δαψιλεστέρας V M S (*cf.* Procl. πλείονος)] δαψιλεστέρως Σ G° γ + G (habundantius): δαψηλεσταίρος Y. — The adverbial form is not required. G may have translated with an adverb as this is easier.

(-*) 84° (68) ἀναπινομένης Σ] ἀναπουμένης V: ἀνυπουμένης Y: ἀνικμουμένου DM: ἀνικνουμένης S γ: (substantia humida ... a nullo) exalata (calido) G. — G translates *exalata,* which corresponds to ἀναπνουμένης or ἀναπνεομένης (*cf.* ἐξικμαζομένης Procl.); this form explains the erroneous forms in V and in Y. We propose to write ἀναπνουμένης with G (V Y).

(-*) 106° (85) καθόλου δὲ πάλιν βγ] καθόλου πάλιν V + G (Vniuersaliter rursum): καὶ τούτων δὲ πάλιν α (*Robb.*). — There are three different versions of this passage. The problem concerns καθόλου and δὲ. καθόλου was certainly in the original text, because what follows is a general statement about the connection of the 'north' with 'masculine' features. καθόλου πάλιν, attested in V + G, may have been the primitive version of the text. However, the expression καθόλου πάλιν is syntactically unusual, though not impossible (there are two other instances in Ptolemy). In βγ the particle δὲ was added to indicate the beginning of a new step in the argument. However, the expression καθόλου δὲ is rather rare in Ptolemy (besides this case, there is only one instance in the *Apot.* and two in the *Syntaxis*). When καθόλου as adverb is followed by a particle, it is generally μέν (roughly 30 instances in Ptolemy, most of them in book II). This may be an indication that the particle δὲ was added here by conjecture. A more radical conjecture is found in subfamily α. To connect the beginning of the sentence syntactically with the end of the preceding sentence, καθόλου was changed into τούτων δὲ πάλιν. The editor of the subarchetype *α* most probably was influenced by Procl., who paraphrases πάλιν δὲ τούτων. We propose to accept the primitive version as in V + G.

(-?) 113° (90) ἁπαλώτεροι α S (*cf.* ἁπλούστεροι D M γ)] ἁπαλώτεροι μᾶλλον V + G (teneriores magis). — The addition of a second μᾶλλον, confirmed by G, is redundant, and may have been deleted for that reason in α βγ. The opposite explanation, that it was intentionally introduced in ψ, is not plausible. Moreover, Ptolemy does not shy away from redundancies.

(++) 116° (93) δὴ V D + G (itaque)] δὲ α M S: *om.* γ

(-*) 117° (93) (νυκτερινὸν) δοκεῖ καὶ θηλυκὸν (καὶ εὐώνυμον) α βγ] (...) δοκεῖ κλίμα θηλυκὸν (...) V (*Robb.*): (...) δοκεῖ κλίμα καὶ θηλυκὸν (...) *coni. Boll* + G (nocturnum uidetur esse clima et femininum); *cf.* Procl. (84.2-4: σελήνη νυκτερινὸν ποιεῖ (*sic*) ἐκεῖνο τὸ κλίμα καὶ θηλυκόν. — G confirms the reading of V, followed by Robbins. It even corresponds to the conjecture of Boll (although Moerbeke could have added the *et* himself). There was perhaps a confusion of κλίμα and καὶ in the archetype, due to a *supra lineam* correction of καί, introduced to make the sentence more fluid, and κλίμα was substituted by καὶ. It would be difficult to explain how κλίμα entered in as an addition or correction to the text. The sentence is perfectly intelligible: 'for this reason it (τοῦτο = this part of the world) appears to be a nocturnal clime, feminine and left-handed, in contrast with the Orient' (see Robbins, p. 127). We propose to follow, with Boll and Robbins, the reading of V + G (without καὶ).

(-*) 119° (107) *titulus ante* ἤδη (121°) α] *ante* τεσσάρων δὴ (137°) V βγ + G Procl. Co (*Robb.*). — We suggest to write, following V βγ + G, the title before τεσσάρων δὴ (137°).

(-?) 124° (97) τοῖς V] αὐτοῖς α βγ + G (ipsis). — Unclear: G may have translated the definite article with *ipsis.*

(-*) 127° (99) διὰ ὕψος βγ] ὕψος V α + G (eleuationem). — We suggest to delete διὰ (a stylistic, unnecessary conjecture).

(++) 150° (119) καὶ V D γ + G (et)] *om.* α M S

(++) 175° (137) τὰ μὲν V (*Robb.*) + G (que quidem)] τὰ α βγ (*Boll*)

(-?) 177° (139) ὥσπερ V (*Robb.*)] ἥνπερ Y βγ + G (quam quidem) (*Boll*): ἧπερ Σ. — Both readings are possible but, if ἥνπερ were adopted, this would be the only instance of ἥνπερ in Ptolemy.

(-?) 181°-3° (143) (ὡς ἐκ τούτων ἕκαστον τῶν τεταρτημορίων) ... συνοικειοῦσθαι α βγ] (...) συνοικειοῦται V + G (ut ex hiis consequitur: quelibet quartarum ... coaptatur); *cf.* Procl. (88,3-6: Ἐκ τούτου οὖν φανερὸν, ὅτι ἕκαστον τῶν τεταρτημορίων ... συνοικειοῦται). — See above, IVB 1.3.3.

(-?) 185° (145) κατ' αὐτὸ τὸ μέρος Σ] κατὰ μέρος βγ: κατ' αὐτὸ τὸ μερικὸν V Y: secundum ipsam partem partialem (κατ' αὐτὸ τὸ μέρος μερικὸν) G. — Ptolemy points out that the parts of the quarters situated in the center have an opposite inclination, whereas the other parts are in harmony with the general inclination of the quarter: τῶν μὲν ἄλλων μερῶν πρὸς τὴν καθόλου πρόσνευσιν ἐφαρμοζομένων, τῶν δὲ περὶ τὸ μέσον πρὸς τὴν κατ' αὐτὸ τὸ μέρος ἀντικειμένην. The second part of the sentence posed problems, as is shown in the diversity of the manuscript tradition: (τὸ) μέρος βγ Σ versus τὸ μερικὸν V Y. If we adopt the reading of βγ Σ (as do all modern editors), then τὸ μέρος refers back to τὸ μέσον. The reading of V Y is more difficult, but could be explained in opposition to καθόλου. The parts in the middle have a particular inclination, not the general inclination of the quarter. Procl. read the text as in V Y, but found it difficult to explain. See his Paraphrase: πρὸς δὲ τὸ μερικὸν καὶ ἀντικείμενον τοῖς μέσοις ἐφαρμοζομένων τῶν κατ' αὐτὸ μερικῶν ἀντικειμένων μερῶν. Moerbeke has both readings of the tradition, probably because μέρος had been added above μερικὸν. See above IVB 2.1.3.

(+) 190° (149) αὐτὸν Σ βγ + G (ipse)] αὐτῶν V: *om.* Y

(-) 199° (154) Βασταρνία Ἰταλία Γαλλία α βγ Arab.] *om.* V + G Procl. [Heph.] — With these three names the general enumeration of countries in this paragraph (§ 13) is complete and corresponds to the two more specific enumerations which follow in § 15 and ch. 5 §§ 2-4.

(++) 205° (159) τὸν ... συσχηματισμὸν V + G (configurationem)] τὸ τοῦ κατὰ τὸν (τὸν *om.* D) σχηματισμὸν α β: κατὰ τὸν συσχηματισμὸν γ

(-) 216° (168) βασταρνία α βγ] βαρσανία V + G (Barsania) Procl. (90,5): *om.* Heph. — The geographical term βασταρνία may have been devised by Ptolemy to indicate the region inhabited by the Bastarnae, a German tribe, between the Carpathian Mountains and the river Dnieper. There is no trace of Barsania in ancient sources (*cf.* 473°). The term βαστερνία (*sic*) occurs once in Heph. (19.21), but in another context, and it is not attributed to Ptolemy. The same opposition between βασταρνία and βαρσανία in the same groupings is also found in the table at 473°. It is strange that the unknown name βαρσανία is found two times in V, G and Procl. Is this an error in their archetype ψ? In the Arabo-Latin translation we find *Bestrania*, which seems to correspond to βασταρνία.

(§) 219°-20° (170) Ἀπουλία καὶ Σικελία καὶ Γαλλία *coni. Hüb.* + G (Apulia et Sicilia et Gallia) Procl. (*sic*: 90,10-11); *cf.* Heph. 1,1,84 (Γαλλία Απουλία)] σικελία καὶ γαλλία V D: γαλλία καὶ σικελία α M S: γαλλία καὶ ἀπουλία καὶ σικελία γ

(-) 228° (175-6) κυκλάδες V α Procl.] κυκλάδες νῆσοι βγ + G (Ciclades insule). — *Cf.* 475° (358) (*in tabula*). It is not certain that G read νῆσοι here. Moerbeke may have added *insule* when translating.

(-?) 229° (176) (τὰ παράλια τῆς Μικρᾶς Ἀσίας καὶ) Κύπρος α βγ] Κύπρου V + G (Cipri) Procl. (*Robb.*), *cf.* 246° (189) (περὶ ...) Κύπρον α βγ] Κύπρου V + G (Cipri) Procl. (*Robb.*) — 477° (*in tabula*) Κύπρος *codd.* + G (Ciprus) — 478° (*in tabula*) παράλια Μικρᾶς Ἀσίας α βγ] Μικρὰ Ἀσία V + G Heph. — In these four instances we have a clear opposition V + G Procl. <—> α βγ. In the first group (V + G Procl.) we have a less precise, somewhat negligent formulation, which connects 'the coastal regions' to *both* Asia Minor *and* Cyprus; the second (α βγ) is more precise, and separates the coastal regions of Asia Minor from Cyprus. It would be more consistent with the editor's choice in the next item (229°) to accept the reading Κύπρου of V + G + Procl. (*Robb.*) at both places (229° 246°).

(++) 229° (176) τὰ ... κείμενα V + G (que iacent ...) Procl.] (τὰ *om.*) ... κείμενα α βγ (*Boll Robb.*)

(-*) 232° (179) συνοικοδεσποῦντας V] συνοικοδεσπότας α βγ + G (condispositores). — V against *ceteri* + G has no authority: probably one of the many particular errors of V.

(++) 233° (179) ἔτι V + G (adhuc)] *om.* α βγ

(-*) 234° (180) τὰς χώρας ἐκείνας Σ βγ] τὰς χώρας V Y Procl.: regiones has G. — Moerbeke did perhaps not read ἐκείνας (which he would translate by *illas*, *cf.* Index), but added *has* of his own accord. A demonstrative pronomen is more fluid, but still unnecessary, and the witnesses of V and Y (*lectio difficilior*) should be preferred, see 295°.

(++) 235° (180-1) ἀπεφάνθησαν V + G (dicti sunt)] ἀπέβησαν α βγ (*deest in* Procl.). — Both readings are possible. ἀπέβησαν can be used in the sense of 'to turn out to be so and so' and constructed with an adjective or participle, as is the case here. The passive aorist form ἀπεφάνθησαν is only found in some Patristic texts and later in Byzantine authors..

(-) 240° (184) φιλαγωνισταὶ] + propter Mercurium G. — Addition in Moerbeke's Greek model? Confusion with 243°?

(-?) 246° (189) Κύπρον (Σ βγ)] κύπρου V Y + G (Cipri) Procl. — See 229°.

(++) 248° (190) καθάριοι V D γ + G (mundi) Procl.] φιλοκαθάρ(ι)οι α M S

(-*) 255° (195) νομίμοις βγ] νόμοις V α + G (legibus) Procl. (κατὰ νόμους) (*Robb.*). — We propose to write νόμοις on the authority of V α + G Procl. In the parallel passages of νόμιμος in book II, there are no differences in the tradition.

(-?) 266° (202-3) (καὶ τὰς φύσεις τῶν ἐν αὐτοῖς ἀκολούθως ἄν τις εὕροι) τοῖς ὑπὸ τῶν οὕτως οἰκοδεσποτησάντων ἀποτελουμένας Σ βγ (-μενα Y)] (...) τοῖς ὑπὸ τῶν οὕτως οἰκοδεσποτησάντων ἀποτελουμένοις V + G ([et naturas que in ipsis utique assequentes quis inueniet] hiis que efficiuntur a sic dominantibus). — The version of V + G is grammatically more plausible than the version of (Y) Σ βγ, which Boll and Hübner preferred. Even Robbins, who prints the same Greek text, seems to follow the version of V + G in his translation: 'Therefore one would find that the nature of their inhabitants (τὰς φύσεις τῶν ἐν αὐτοῖς) conform (ἀκολούθως/ἀκολούθους) with the temperaments (τοῖς ἀποτελουμένοις) governed by such rulers (ὑπὸ τῶν οὕτως οἰκοδεσποτησάντων)'. The form -ας, -α may come from an unnecessary correction.

(++) 270° (206) ἔτι δὲ θερμοὶ V + G (Adhuc autem calidi) Procl.] θερμοὶ δὲ ἔτι α: θερμοὶ δὲ βγ

(++) 273° (208) ἁπλοδίαιτοι V Y + G (simplicis diete) Procl. (ἁπλῶς διάγοντες)] ἁβροδίατοι Σ βγ Co le — The reading of V Y + G + Procl. ἁπλοδίαιτοι is a rare term (besides this, only one occurrence in Psellus), but seems to fit better as a characteristic of Saturnus. Moreover, the Arabic translation confirms it: 'their management [is] a simple management'. On the other hand, ἁβροδίατοι fits better in the general context, and is explicitly defended by the commentator in a note. Maybe because of the commentary, Σ βγ adopted it as a conjecture.

(-*) 277° (211) ἐκ τῶν μητέρων α Procl., Co le] ἐκ τῶν μητέρων αὐτῶν βγ + G (ex matribus ipsorum): ἐκ τῶν ἀμητέρων V. — The addition of the reflex pronomen is preferable in the context. Was it omitted by homoioteleuton (see 277° αὐτῶν ... αὐτῶν)?

(-*) 280° (214) τὰ μὲν ἄλλα περὶ τὰς στολὰς καὶ κόσμους β] τἆλλα μὲν τὰς περὶ τὰς στολὰς καὶ κόσμους V: ad alia quidem que circa stolas et ornatus G (τἆλλα μὲν τὰ [*sic*] περὶ τὰς στολὰς καὶ κόσμους): τὰ μὲν ἄλλα τὰ περὶ τὰς ἀνατολλὰς κόσμου Y (G°): *cf.* γ Σ Procl. *in adn. Hüb.* — G confirms the reading of V without the individual mistake (τὰς instead of τὰ) of V. With respect to the variants and corrections in the second part of the passage, especially the errors of Y(G°) Σ γ, (*cf.* adn. Hüb.), it would be more consistent to follow the text of (V) G (branch ψ) in the whole sentence: τἆλλα μὲν τὰ περὶ τὰς στολὰς καὶ κόσμους.

(+) 291° (221) βαβυλωνίαν α βγ + G (Babiloniam)] βαβυλῶνα V Procl.

(++) 293° (223) πέντε V + G (quinque) Procl.] *om.* α βγ Arab.

(-*) 295° (224) τὰς χώρας ἐκείνας Σ γ] τὰς χώρας V Y β + G (regiones). — The addition of ἐκείνας is a conjecture which is in conformity to other passages, see 234° (180), but still unnecessary (*e.g.* 343°). We would follow V Y β + G and delete ἐκείνας.

(+) 302° (229-30) ἔχει δὲ βγ + G (habent autem) Procl.] ἔτι δὲ V: ἔχῃ Y: ἔχοντος Σ

(++) 305° (231) συναλλακτικώτεροι V + G (magis commutatiui) Procl.] συναλλακτικοί α βγ

(-) 305° (231) πανουργότεροι δὲ α, *cf.* Co le] πανοῦργοι τε (τὲ) βγ: *om.* V + G (Procl.). — The first comparative in 305° is confirmed by V + G + Procl. (hyparchetype ψ) against the positive in α βγ. The second comparative, which is attested in α and the commentary, is lacking in ψ, perhaps due to homoioteleuton.

(++) 311° (236) τε εἰσιν V D γ + G (sunt)] *om.* α M S

(++) 313° (237) φιλάνθρωποι V + G (humani) Procl.] φιλανθρωπότεροι α βγ

(++) 330°-1° (250) περὶ τὰ θεῖα καὶ μάγοι V + G (circa diuina et magi) Procl.] καὶ μάγοι περὶ τὰ θεῖα α B C: καὶ μάγοι καὶ περὶ τὰ θεῖα β A

(+) 333° (252) οἰκείων βγ + G (domesticis) Procl.] οἰκειωτάτων V α (*'fort. recte'*)

(-*) 337° (254) ἀνατολικῶν συσχηματισμὸς *coni. Hüb.*] ἀνατολιῶν συσχηματισμὸς V (*Robb.*): ἀνατολικὸς συσχηματισμὸς βγ + G (orientalis configuratio): ἀνατολικὸν σχηματισμὸν Y: κατὰ ἀνατολικὸν συσχηματισμὸν Σ. — We propose to follow the reading of βγ, which is confirmed by G.

(+) 339° (256) Ματιανὴν α + G (Matianem)] μαντιανὴν V γ Procl. (*cf.* Heph.): μαντειανὴν β

(-) 350° (263-4) Λυδία Κιλικία α βγ] Λυδία Λυκία Κιλικία V + G (Lidia Licia Cilicia) Procl. — Λυκία is present neither in the table (495°), nor in the Arabic translation. It is probably a copyist's error in ψ (resemblance of the names?).

(-*) 367° (275) διὰ τοῦτο α] διὸ V + G (propter quod): *om.* βγ. — In the whole of Ptolemy's text, διό occurs only in this third chapter of book II, but it is rather frequent here (6x: 252° 291° 304° 346° 367°* 383°), and it always carries the same sense (translated by *propter quod*). The same applies to διόπερ (12x in book II, 1x in book I). We propose to write διὸ; the reading of α is probably a correction.

(+) 376° (283) φεύγουσαι Σ + G (fugientes) Procl.] φύγουσαι V: φεύγουσι(ν) Y βγ

(++) 378° (284) τὰ θηλυκὰ πάντα V + G (feminina omnia), *cf.* Procl.] τὸ θῆλυ α: τὸ θῆλυ πᾶν β, *cf.* γ

(-*) 380° (285-6) ἀπογυμνοῦσαι (ἀπογυμνοῦσι δὲ M S γ: ἀποτέμνουσι D) ταῦτα τὰ μέρη α βγ] ἀπογυμνώσει τούτων τῶν μερῶν V + G (denudatione harum partium). — The choice of V + G (ἀπογυμνώσει ...) would be more consistent, see the choice of ἀποκοπῇ in 378°.

(-*) 402° (301) ὡς α] ὥστε βγ + G (ita ut): *om.* V. — ὡς nearly always has a comparative signification in this text. We propose to write ὥστε.

(-*) 410° (307) Καρχηδο(ω)νίαν α βγ] Καρχηδόνα V + G (Carchidonem) Procl. (*Robb.*). — Καρχηδο(ω)νίαν seems to be a correction in the model of α βγ in order to assimilate this form to two other instances in the text (*cf.* 391° and 471°). We would retain, with V + G Procl. and Robbins, Καρχηδόνα.

(+) 426° (319) βορραπηλιώτην α βγ + G (boream et orientem)] ἀπηλιώτην V Procl.

(+) 431° (323) οἰκοδεσποτείας Σ M S γ + G (dominatione) Procl.] οἰκοδεσπότας V Y D

(++) 435° (325) ἔθεσι καὶ νομίμοις V D γ + G (consuetudinibus et legitimis)] *inv.* α M S

(-) 442° (330) σαθροί V α, *cf.* Procl.] θρασεῖς βγ + G (audaces). — Σαθροί occurs three times in the whole text: II.442° III.1277° III.1548°. In all three places, we find two different readings in the manuscripts: σαθροί/θρασεῖς. The two other passages have:

III.1277° (969) σαθρούς V Y Procl + G (confractos)] θρασεῖς βγ Heph.

III.1548° (1176) σαθροὶ V Y Procl.] θρασεῖς βγ + G (proni) Heph., *cf.* Σ. — σαθροὶ is uniformly confirmed by Procl., and θρασεῖς is confirmed two times by Heph. Moerbeke read θρασεῖς with βγ at II.442° and III.1548° (with βγ and Heph.), and read σαθρούς with V at III.1277°. Moerbeke had corrections from βγ in his Greek model.

(++) 444° (332) ἑσπερίου V + G (uespertina), *cf.* Procl.] ἑσπέριον α βγ

(++) 447° (334) Διδύμοις V M S + G (Geminis) Procl. Arab.] ἰχθύσι α D γ (*Boll*)

(-*) 459°-60° (346-7) *Titulus ante* Αἱ μὲν οὖν συνοικειώσεις (461°/343) D] *post* (464°/345) τρόπον V + G (modo): *post* τὸν τρόπον τοῦτον 468° (351) M γ Procl. (p. 104-5) Co (p. 62,4), α *non citatur.* — Write title post τρόπον 464° (345).

(++) 469°-97° (352-73) *tabulam habent* V M S + G] *ordinem pervertit* D, planetas *add.* γ: *om.* α

(§) 472° (355) Ματιανή + G (Matiane)] ματτιανή α: μαντιανή βγ: γαμπανή V

(-) 473° (356) Βασταρνία βγ] Barsania (*sic*) (βαρσανία) G: *om.* V: [*deest* α]. — See 216°.

(-*) 475° (358) Κυκλάδες νῆσοι βγ (α?)] Κυκλάδες V + G (Kiklades) Heph. Procl. ('*fort. recte*') — We would follow here V + G Heph. Procl. See 228°.

(-?) 478° (360) παράλια Μικρᾶς Ἀσίας α (?) βγ Tib.] μίκρα Ἀσία V + G (Parua Asia) Heph. Procl. ('*haud scio an recte*?'). — Here again, we would follow V + G Heph. Procl. See 229°.

(-) 492° (373) Νασαμωνῖτις α D S γ Heph.] νασαμωνιτική V M + G (Nassamonitica) Procl. — The form νασαμωνιτική is not attested elsewhere.

(++) 498° (379) γίνονται χῶραι οβ′ V M + G (Regiones omnes sunt 72) Procl. Arab.] *om.* α D S γ. — See Robbins, p. 159, n. 1.

(§) 501° (381) καθ' ὧν Heph. + G (in quibus)] μεθ' ὧν Σ βγ: καθὸν V: καθ' ὃ Y

(-*) 505° (383) τοῦ ζῳδιακοῦ κύκλου α βγ] τοῦ ζῳδιακοῦ V + G (zodiaci) Procl. —See above, IVB 1.3.3.

(++) 506° (384) γενέσεως V + G (hominis generatione)] γενέσεων α: τῶν γενέσεων βγ

(+) 509° (386-7) καθ' ὧν βγ + G (in quibus)] καθ' ἣν V: *om.* α

(++) 511° (389) *titulum hic praebent* V D M γ + G Procl., *cf.* S] *antecipant* (498°) α

(++) 512° (389) τῶν ἐκλείψεων V + G (eclipsium) Procl. (*cf. tit.* II.8°)] *om.* α βγ Co

(++) 513° (391) προεπεσκεμμένων V + G (preconsideratis)] *diversimode cett.*

(++) 515° (392) τῶν ... λαμβανομένων V + G (eorum qui accipiuntur)] τὰς τῶν (τῶν τὰς) ... λαμβανομένων β: τῶν... λαμβάνομεν Y G°: τὰς τῶν (τῶν τὰς) ... λαμβανομένας Σ γ

(+) 520° (396) μὲν βγ + G (quidem)] μέντοι V: μέν τι α. — See also II.563°.

(++) 525° (399) γὰρ τότε V + G (enim tunc) Procl. Arab.] *om.* α βγ

(-*) 533° (406) τοῦ μὲν οὖν (*om.* Σ) πρώτου α] περὶ μὲν οὖν τοῦ πρώτου βγ + G (De primo quidem igitur): περὶ τὸ μὲν οὖν τοῦ πρώτου V. — The expression διάληψιν ποιεῖσθαι is generally constructed with περὶ + genitive, and not with the simple genitive. It would be more consistent with the former choices to follow (V) βγ + G.

(-) 537° (408) τῶν κατ' αὐτὸν τριγώνων V β Procl., *cf.* Y Σ] τῷ κατ' αὐτὸν τριγώνῳ γ + G (trigonalitati ipsius) (*Boll*)

(-) 553° (421) τῶν τε ἡλιακῶν τῶν αὐτῶν V Y Procl., *cf.* Arab.] τῶν τε ἡλιακῶν καὶ τῶν σεληνιακῶν τῶν αὐτῶν β: τῶν τε ἡλιακῶν δηλαδὴ καὶ τῶν σεληνιακῶν Σ γ: solaribus et lunaribus *post* eclipsibus (551°/420) G. — Difficult passage: (1) the addition of τε after τῶν seems to indicate that another adjective should be connected. Hence the conjecture τῶν σεληνιακῶν in βγ and in G, but this makes τῶν αὐτῶν difficult: it disappears in γ and in G; (2) to which noun is τῶν τε ἡλιακῶν connected? It can only be ἐκλείψεων in 551°; therefore Moerbeke put *solaribus et lunaribus* after *eclipsibus* (551°/420). We believe, however, that Hübner rightly keeps the text as it is transmitted by V Y and confirmed by Procl. and the Arabic translation. The particle τε connects, not two adjectives (as in the correction of βγ), but two reasons given by Ptolemy 'inasmuch ... and since the same solar eclipses do not ...' (Robbins).

(-?) 557° (424) κέντρα V D γ Procl.] τά τε κέντρα α (*cf.* MS) + G (et angulos)

(-*) 563° (429) τῶν μὲν *coni. Hüb.*] τῶν μέντοι α + G (tamen) (*Robb.*): καὶ τῶν μέντοι V: τῶν βγ: καὶ τῶν *Boll.* — We would write μέντοι with V α + G. For similar uses of the particle μέντοι: I.1069°, 1104°

(+) 590°-1° (451) τὸ πρὸ G° + G (qui est ante)] τὸ πρώ Y: τοῦ πρὸ βγ: πρὸ Procl.: *om.* V Σ

(-*) 592° (453) πλείστους V βγ] πλείους α + G (plures), *cf.* Procl. (πλείονας). There are 30 instances of πλείους in Ptolemy, but of πλείστους this is the only example, and it is problematic. We thus propose to write πλείους.

(-*) 595° (454) ἔγγιστα βγ] ἔγγιστα καὶ V α + G (propinquissimas et) Procl. (*Robb.*).

(-*) 599°-602° (458-9) — In order to understand the particularities of the Greek and Latin tradition we present the whole passage in its different forms. The introductory clause is the same (εἰ δὲ μὴ ὁ αὐτὸς εὑρίσκοιτο τῆς τε ἐκλείψεως καὶ τοῦ κέντρου κύριος):

(...) ἀλλὰ δύο, τούς ... τὰς πλείους ἔχοντας ... συμπαραληπτέον, προκρινομένου τοῦ τῆς ἐκλείψεως κυρίου *Hüb.*

(...), δύο (δύο δὲ Y) τούς ... τὰς πλείους ἔχοντας ... συμπαραληπτέον, προκρινομένου τοῦ τῆς ἐκλείψεως κυρίου V Y Procl., *Robb.* (δύο τοὺς *connectens*)

(...) ἀλλὰ δύο, τὸν ... τὰς πλείους ἔχοντα ... συμπαραληπτέον, προκρινομένου τοῦ τῆς ἐκλείψεως κυρίου βγ, *cf.* Heph.

(...) ἀλλὰ δύο, τὸν τὰς ... πλείονας ἔχοντα ... ἑλόμεθα, προκρίνοντες τὸν τῆς ἐκλείψεως κύριον Heph.. *cf.* βγ

(...), duorum (*ad sensum*) illum assumendum, qui ... plures habet coaptationes ... Si autem ambo easdem habeant rationes, preferimus dominum eclipis G.

The version V Y Procl. in the way Robbins read it (with the break not after, but before δύο) seems to present the primitive form of the text: 'However, if the same planet is not found to be both, lord of the eclipse and of the angle, we must take together (συμπαραληπτέον) the two which have the greatest number of familiarities to either one of the regions, giving preference to the lord of the eclipse'. In the version of βγ, which is probably contaminated by Hephaestio, ἀλλὰ is added before δύο, and, instead of the plural τούς ... ἔχοντας, one finds the singular accusative τὸν ἔχοντα. The translation of this version will, then, be as follows: 'However, if the same planet is not found to be both ..., but two, we must take the one which has the greatest number of familiarities to either one of the regions, giving preference to the lord of the eclipse'. The problem with this version is that the conclusion of the reasoning (one must give preference to 'the lord of the eclipse') is then already anticipated in τὸν ἔχοντα συμπαραληπτέον. For if we take the one which has the greatest familiarities, this can only be 'the lord of the eclipse'. It makes no sense to say that we prefer that lord, if we had already found the one of two. Surprisingly, the Latin text combines the characteristic of V (Y) in the beginning (no ἀλλὰ or δὲ) with the second significant element of β Heph. (τὸν ἔχοντα *illum qui habet*).

However, to continue immediately after δύο with τὸν ἔχοντα is difficult: 'we must take together the two, the one which has ...'. One understands why Moerbeke translates δύο as if it were a genitive, to connect it to the following τὸν ἔχοντα: *duorum illum assumendum*. Moreover, he adds — before the conclusion — the logical clause which is missing: 'Si autem ambo easdem habeant rationes preferimus dominum eclipsis'. The formulation of the conclusion is similar to what we read in Hephaestio, but it could still be Moerbeke's free translation. Did Moerbeke *suo ingenio* add the premise 'si autem ambo easdem habeant rationes' (εἰ δὲ ἀμφότεροι τοὺς αὐτοὺς ἔχοιεν λόγους), or had it been introduced to his Greek model by a scholiast (the same who introduced the variant τὸν ἔχοντα into his Greek model)? We do not know. As can be seen, Hübner himself combined elements of both text traditions, keeping ἀλλὰ δύο from βγ and the plural τούς ... ἔχοντας from V Y Procl. + G. In our view, it would better to consistently follow V Y Procl.

(-?) 606°-7° (464) αὐτῷ τῷ ἐκλειπτικῷ τόπῳ Σ] αὐτὸ τῷ ἐκλειπτικῷ τόπῳ Y: πρὸς τῷ ἐκλειπτικῷ τόπῳ M + G (cum loco ecliptico): τῷ ἐκλειπτικῷ τόπῳ D S γ: αὐτῷ τῷ ἐκλειπτικῷ χρόνῳ V Procl. (*Robb.*). — G confirms τόπῳ with α βγ against χρόνῳ (V Procl. *Robb.*). The other problematic term is αὐτῷ in V Σ, αὐτὸ in Y and πρὸς in M (and in G?), which is lacking in the other β-members and in γ. The reason for these variations concerning αὐτῷ is likely an ancient correction.

(++) 615° (470) συνεπισκεψόμεθα V + G (conspeculabimur) Procl.] ἐπισκεψόμεθα α βγ Heph.

(-*) 615° (470) καὶ Σ Procl.] κατὰ Y: *om.* V βγ + G Heph. ('*fort. recte*'). — καὶ may have been added in Σ from the Proclean Paraphrase. Delete καὶ.

(+) 636° (484) νηκτοῖς Σ βγ + G (natatilibus) Procl.] νυκτοῖς V Y

(+) 656° (500) ἐθισμῶν βγ + G (consuetudinum) Heph. (A)] ἐθισμένων V (*cf. Robb.*): ἐθή(ί)μων α Heph. (L)

(-?) 666° (507) καὶ α βγ] *om.* V + G (*Boll*)

(+) 666° (507) πόστον δὲ Σ βγ + G (quotam autem), *cf.* Heph. (L), Ep.] πόσὼν δὲ V: πόσον δὲ Procl. Heph. (A): μέσον Y

(++) 669° (508) σχέσεις V Σ + G (habitudines)] σχέσις Y βγ

(-?) 672° (511) ἐπὶ τὸ ἥμισυ α] περὶ τὸ ἥμισυ βγ + G (circa dimidium) Heph.: πρὸς τῷ ἡμίσει V. — The reading (correction?) of α is confirmed by a similar formula in *Synt.* 1,2 (p. 181,3): ἐκτεμόντες ἐπὶ τὸ ἥμισυ.

(-?) 677° (515) Τέταρτον δὲ V βγ Procl.] Τέταρτον α + G (Quartum). — δὲ correction?

(++) 684° (520) διατάκται καὶ V D + G (ordinatores et), *cf.* Procl.] διατοικτικοὶ Y: διατακτικοὶ G° Σ: διατέτακται καὶ M S γ

(-*) 690° (524) καθ' ἕνα ἕκαστον Y] καθ' ἕκαστον V βγ + G (per singulos): καθ' ἓν ἕκαστον Σ. — We propose to write καθ' ἕκαστον, with V βγ + G.

(-?) 727° (552) λιμοῦ V βγ Procl.] λοιμοῦ α + G (pestilentiam)

(-?) 729° (553) λαβὼν α βγ] λαχῶν V + G (sortiens)

(-*) 756° (574) περὶ δὲ (δὴ Procl.) θάλασσαν D γ Procl, *cf.* Heph.] περὶ θάλασσαν δὲ M S: πάλιν ἐν θαλάσσαις α: περὶ δὲ θάλασσας V + G (circa maria autem). — We would write περὶ δὲ θάλασσας, with V + G.

(++) 759° (576) ποτίμων V + G (potabilium), *cf.* Procl.] ποταμίων α βγ (see also II.723°)

(++) 759° (577) ἐπὶ γῆς V + G (super terram), *cf.* Procl. (ἐκ τῆς γῆς)] *om.* α βγ Arab.

(-) 762° (579) ἐκ τῆς … καταφλέξεως α βγ Heph. (*Apot. et Ep.*)] ἐκ τῆς (!) … καταφλέξεων V + G (ex inflammationibus)

(-*) 773° (588) περὶ δὲ τὰ τοῦ ἀέρος πνεύματα V D γ] περὶ δὲ τὰ τοῦ ἀέρος πνευμάτων α M S: circa aerem autem uentorum G. — VD γ have πνεύματα, but as we have seen in Chapter IV, this trio is suspect and almost never offers an authentic reading when it is not confirmed by G. We suppose that it is a conjecture, introduced so as to have a noun after τὰ. Moerbeke read with the other manuscripts the genitive plural, connecting it with εὐκρασίας (*uentorum temperantias*). He is right in his free interpretation of περὶ δὲ τὰ τοῦ ἀέρος as *circa aerem*, i.e., 'concerning the (states) of the air'.

(-) 778° (591) ἔτι V βγ] ἐν α + G (In)

(§) 781° (593) ὡς ἂν ᾖ *coni. Boll Robb. Hüb.* + G (ac si esset)] ὡς ἀνεὶ V: ὁσανεὶ Y: ᾧ ἂν ᾖ Σ βγ

(-?) 792° (602) οἰκείως α D γ] οἰκείας V M S + G (conuenientes). — Was οἰκείως a grammatical correction of οἰκείας? *Cf.* 183° 1035° IV 718°

(+) 795° (604) ἀνακυκλήσεως β + G (reuolutionis), *cf.* α (ἀνακυκλώσεως)] ἀνακλήσεως V γ

(+) 809° (616) ὑπομνηματίζειν α γ + G (describere)] -ζόμενον V D S, *cf.* M

(+) 811° (617) νοουμένους D γ + G (consideratas)] -μενος V: -μένας M S: -μένου α

(§) 812° (618) ἐπὶ τῇ … ἐπιβολῇ *codd.*] τῇ … ἐπιβολῇ *coni. Boll* + G (sollicitudini). — Delete ἐπὶ with Boll and G. Simple dative seems better with καταληφθείη.

(++) 818° (622) ἀστέρες V D γ + G (astra) Procl.] *om.* α M S Heph. Arab. — See 1010°.

(+) 834° (634) καθολικῶν Σ βγ + G (uniuersalium)] κολικῶν V Y

(§) 843° (642) ἅλων Σ^{pc} + G (halo)] ἁλώνων βγ: ἄλλων V Σ^{ac}, *cf.* Procl.: ἄθλων Y

(-*) 857° (652) πίθων Σ βγ Heph.] πιθείων V (*cf.* Procl. πιθείας): πιθίων Y + G (pithiarum). — πιθίων is the *lectio difficilior.* It is a technical term for a jar-shaped comet, see Philoponus, *In Meteor.* 92,34 (LSJ). Write πιθίων with Y + G.

(+) 885° (672) ἤδη Σ γ + G (iam)] εἴδη V Y, *cf.* β

(-?) 886° (673) ὡς προηγουμένῳ α] προηγουμένων V D: προηγούμενον M S: προηγουμένως γ Procl.: principali (προηγουμένῳ) G. — Many different variants at the same place: problem in the archetype? Προηγουμένῳ (α + G) is certainly right. ὡς could be a syntactical correction.

(-*) 886° (673) ἂν V βγ] *om.* α + G ('*fort. recte*'). — Moerbeke almost never omits to translate ἂν (see Index). Probably he did not read it. Delete ἄν.

(§) 889° (676) ἀρχῇ *coni. Boll Hüb.* + G (principio)] ἀρχὴν *codd.*

(-*) 910° (691) ἔσται βγ] ἔστι V α + G (sit). — Write ἔστι.

(++) 917° (696) ἰδιοτρόπους V + G (proprias) Procl.] ἰδιοτροπίας α βγ

(-*) 917° (696) περὶ ἀνέμων M S] ἀνέμων V α D γ + G (uentorum) Procl. — Write ἀνέμων.

(-*) 925° (703) *tit. post* Προεκτεθῆναι δὲ ὀφειλόντων α] *post* εἰπεῖν (dicere) (935°/709) V β (*cf.* S) γ + G Procl. (*Robb.*). — Write title after εἰπεῖν (*dicere*) with V β (*cf.* S) γ + G Procl. *Cf. supra* 119° and 459°-60°.

(++) 931° (705) συνοικείωσιν V + G (coaptationem)] συνοικειώσεως βγ Procl.: οἰκειώσεις α

V (α =YΣ β =DMS γ =ABC, *excerpta* L)

(-) 946° (719) τὴν Πλειάδα V L βγ Heph. Procl.] τὰς πλειάδας (πλοιά- Y) α + G (Pliades)

(-?) 948° (720) κεραυνῶν α M S Heph., *cf.* Procl., *cf.* Tib.] κεραυνώδη V L D γ + G (tonitruosa). — According to the TLG the adjective κεραυνώδης is an hapax, found only once in a Byzantine astrological treatise. The consensus of V L D γ + G is an indication that it was found in the hyparchetype ψ, and maybe even in the archetype ω. Is it a copyist's error, by attraction with the preceding πυρώδη? Was it corrected by α from Heph. into κεραυνῶν, which fits with the following ἀστραπῶν ποιητικά?

(-) 958° (728) καὶ σεισμοποιὰ καὶ ἀχλυώδη α βγ Heph. Arab.] *om.* V L + G Procl.

(+) 970° (739) τρεπτικὸν α βγ + G (uersatiuum) Heph.] τροπικὸν V L, *cf.* Procl. (τροπῆς)

(-?) 1008° (774) καὶ V α β C (*Boll*)] ἢ L A B + G (aut) (*Robb.*)

(-*) 1010° (774) ἀστέρων α M S Co, Arab.] *om.* V L D γ + G (*Robb.*). — *Cf.* 615°. It would be more consistent, here, to follow the reading of V(L) + G.

(-?) 1035°-6° (792-3) διὰ τοῦ … συσχηματισμοῦ α βγ Procl.] διὰ τοῦ … συσχηματισμῶν (συσχημῶν V) V L + G (per id quod … configurationum: διὰ τὸ … συσχηματισμῶν). — See above, IVB 1.3.3.

(+) 1048° (801) τινὶ α M S + G (alicui)] τι V L D γ

(+) 1051° (803) ἀμπώτεις β A B + G (effusiones) Co le (83,22), *sim.* Σ C, *cf.* Y] ἀμπώτις (*sic*) V: -της L

V (α =YΣ β =DMS γ =ABC)

(+) 1066° (815) μὲν² α M S + G (quidem) Heph.] *om.* V D γ Procl.

(-*) 1074° (822) κυκλουμένας y (*Robb.*), *cf.* κυκλούμενον Heph. (A P)] κοιλούμενον V β + G (profundum): κοιλούμενος γ (*Boll*, *cf.* Arat. 828 κοῖλος): καλλούμενον Y: καλούμενον G° Heph. (L): κλωμένας Σ, *cf.* Procl. — κυκλουμένας is a simplifying conjecture in y, which was also adopted in some manuscripts of Heph. The authentic form is undoubtedly κοιλούμενον (*sc.* κύκλον). The parallel in Aratus (825-830), as noticed by Boll, is undeniable:

Εἰ δ' αὔτως καθαρόν μιν ἔχοι βουλύσιος ὥρη,
δύνοι δ' ἀνέφελος μαλακῇ ὑποδείελος αἴγλῃ,
καί κεν ἐπερχομένης ἠοῦς ἔθ' ὑπεύδιος εἴη.
Ἀλλ' οὐχ ὁππότε κοῖλος ἐειδόμενος περιτέλλῃ,
οὐδ' ὁπότ' ἀκτίνων αἱ μὲν νότον αἱ δὲ βορῆα
σχιζόμεναι βάλλωσι, τὰ δ' αὖ περὶ μέσσα φαείνῃ·

The ancient scholiast explains that the apparent solar cavities or hollows are not present in the sun, but are rather due to an optical illusion (αἱ ἐν τῷ ἡλίῳ κοιλότητες οὐκ εἰσὶν αὐτοῦ, φαντασίαι δέ εἰσι τῆς ὄψεως κατ' ἐπιπρόσθησιν ζοφεροῦ ἀέρος). Write κοιλούμενον with V β(γ) + G.

(-?) 1075° (822) καλούμενα V βγ Procl.] λεγόμενα α + G (que dicuntur) Heph.

(+) 1076° (823) καὶ α M S + G (et) Procl] *om.* V D γ

(+) 1079° (825) ἅλως Σ D S y + G (halo) Procl., *cf.* Y] ἄλλως V M Heph.

(?) 1080° (826) (ἅλως …) περὶ αὐτὸν *codd.*] (halo …) circa se unam uel etiam duas aut G (unam – aut *glossa graeca*?)

(-?) 1088° (833) ὅλον V βγ Procl.] *om.* α + G Heph. ('*fort. recte*')

(-?) 1090° (834) ἢ χλωρὰ (χλόρα V) V Procl.] ἢ ὠχρὰ α γ Heph. Lyd.: ἢ ὠχρὰ ἢ χλωρὰ D M: ἢ χλωρὰ ἢ ὠχρὰ S + G (uel uiridis uel rubicunda). — See above, IVB 2.1.3.

(-*) 1097° (840) ὕπωχροι V D S, *cf.* M] ὑπόχλωροι α + G (uirides) Procl.— To be consistent with 1090°, it is better to write ὑπόχλωροι.

(+) 1100° (842) ἄλως Σ βγ + G (aree) Heph. (A) Procl.] ἄλλως V Y Heph. (L P)

(+) 1103° (845) τι α M S + G (aliquam) Heph.] *om.* V D B C Procl., *cf.* A

(-*) 1117° (853) ὡς ἐπίπαν V D γ, *cf.* ὡς ἐπιπολύ Procl.] πάντοτε α M S + G (semper) Heph. — Both expressions are found in Ptolemy. The consensus V D γ is suspect. πάντοτε was probably the original reading (see Heph.), with ὡς ἐπίπαν being a correction, intended to reduce the absolute claim ('generally foretell' v. 'always foretell').

(+) 1127° (860): εὐδίαν δὲ βγ + G (serenitatem autem) Heph., *cf.* α (καὶ εὐδίαν)] εὐδίας δὲ V.

(-?) 1135° (866) γενεθλιακὸν (εἶδος) V Σ β Heph.] γενεθλιαλογικὸν (εἶδος) Y γ + G (geneatici sermonis species)

3. BOOK III

V (α =YΣ β =DMS γ =ABC)

(++) 2° (2) Τάδε ἔνεστιν ἐν τῷ γ´ βιβλίῳ V + G (Hec sunt capitula libri tertii), *cf.* D γ] *om.* (*cum indice*) α S

(++) 3° (3) α´ προοίμιον V + G (prohemium ... cap. 1m)] *om.* α βγ

(++) 4° (4) β´ V + G (cap. 2m)] α´ D γ: *om.* Σ M S: *cum capite sequenti conflans* Y

(++) 5° (5) γ´ V + G (cap. 3m)] β´ Y D γ: *om.* Σ M S

(-) 5° (6) περὶ μοίρας ὡροσκοπούσης V Y D] *tit. add.* (4m) quomodo oportet accipere horam G (πῶς δεῖ λαμβάνειν τὴν ὥραν?): *add.* ἐν ᾧ καὶ περὶ τοῦ πῶς δεῖ λαβεῖν τὴν ὥραν γ: περὶ τοῦ πῶς δεῖ λαμβάνειν ὥραν M: *cf. infra* 136° et IVB 2.1.2

(-*) 9° (10) (Περὶ ἀρρενικῶν καὶ θηλυκῶν) γενέσεων V Σ D S γ] *om.* Y M ('*fort. recte*') + G. — The omission of γενέσεων by Y M in the title of chapter eight is confirmed by G, and is, as Hübner notes, perhaps authentic. In any case, this common omission is not accidental: the same situation is found in the effective title of the chapter 392° (302). There, Hübner adopts the reading of α C Heph., as confirmed by G and Procl., against the addition of γενέσεων (-εως) by V γ. In both items, the reading without the addition of γενεσεῶν certainly presents the authentic title: the text of chapter eight speaks of ἀρρενογονία and θηλυγονία (*procreatio*), not of γένεσις.

(++) 18° (19-20) α´ προοίμιον V + G (Prohemium capitulum primum) Procl., *cf.* D γ] *om.* M S: περὶ σπορᾶς καὶ ἐκτροπῆς (*ex l.* 4 *trahit*) α

(++) 23° (24) γενεθλιαλογίαν V + G (geneaticum sermonem) Procl.] γενεθλιαλογικὸν α βγ ('*fort. recte*'). — For the two different Greek words, see Moerbeke's translation vocabulary in the Index.

(+) 44°-5° (38) τὰς ... συμβαινούσας Σ γ + G (scilicet superuenientes)] τὰς ... συμβαίνειν V Y β

(-*) 53° (46) τέξεων V D, *cf.* γ] γενέσεων α M S + G (generationum) Procl. — See above, IVB 2.2.2.

(+) 57° (49) ἐκείνῳ βγ + G (illi) Procl.] ἐκεῖνο V α

(+) 59° (51) κατ' αὐτὸν α D M γ + G (secundum tempus illud), *cf.* κατ' ἐκεῖνον τὸν καιρὸν Procl.] κατ' αὐτῶν V S

(+) 77° (63) αὐτῷ Σ βγ + G (ipsi)] αὐτὸ V Y

(-?) 87° (71) κατὰ V Y + Procl.] καὶ κατὰ Σ βγ + G (et secundum) (*Robb.*). — See the choice in 57° 77° 88° 106°.

(+) 88° (71) τῷ ποιητικῷ βγ + G (factiuo) Procl.] τὸ ποιητικόν V α

(-?) 97° (78) δυναμένων α γ (*Robb.*)] (εἴ τις ... ἐθέλοι ... καὶ μᾶλλον) δυνάμενον V β + G (si quis uelit ... et magis possit) (*Boer*). — Follow V β + G?

(++) 101° (81-2) ὁλοσχερὲς V + G (uniuersalem habitum)] ὁλοσχερέστερον α βγ, *cf.* Procl. (καθολικώτερον)

(+) 106° (85) οὗ Σ βγ + G (ad quod), *cf.* Procl.] οὐ V Y

(++) 117° (94) κατ' αὐτὴν τὴν σύγκρισιν V + G (penes fetus concretionem) Procl.] κατὰ τὴν σύγκρισιν Y γ; *cf. cett.*

(-*) 121° (98) Ἐπειδὴ V D γ Heph.] ἐπεὶ δὲ α M S + G (Quoniam autem) Procl.

(-?) 136° (109) εὑρισκομένης V α Procl.] *titulum add.* πῶς δεῖ λαμβάνειν ὥραν D M + G (Quomodo oportet accipere horoscopum, capitulum 4): πῶς δεῖ λαβεῖν τὴν ὥραν *ut titulum add.* Co (90,30) *cf.* γ, S² *mg.* — See above, IVB 2.1.2.

(+) 141° (114) ὄντος βγ Heph. + G (existentis), *cf.* Procl.] ὄντως V: ὄντας Y: ὄντων Σ

(-*) 141° (114) κατὰ Σ M S γ] κατά τε V Y + G (et secundum) (*Robb.*). — Write κατά τε and put bracket after ὄντος, not after ἐκτροπῆς.

(+) 149° (119) οὗτος Σ βγ + G (iste) Heph. Procl.] οὕτως V Y

(+) 152° (120) δωδεκατημορίῳ β + G (signo) Procl.] -μόριον V α γ

(-?) 157° (124) ἐγγὺς εἶεν τῷ ἀριθμῷ V Y D γ, *cf.* Procl.] ἐγγὺς ἔχοιεν τὸν ἀριθμὸν Σ M S + G (propinquum habeant numerum) Co (92,2). — Reading of V Y D γ a stylistic correction?

(++) 164° (130) γενεθλιαλογίας V D γ + G (geneatici sermonis)] γενεθλιαλογική α M Co Procl.: [Σ]. — *Cf.* 23°.

(-*) 187° (147) περὶ ξενιτειῶν V Y D] περὶ ξενιτείας Σ M S γ + G (de peregrinatione) Heph. Procl. (*Robb.*)

(-*) 188° (147) τελευταῖος δὲ V] καὶ τελευταῖος α βγ + G (et ultimus) Heph. Procl. (*Robb.*) ('*fort. recte*')

(+) 200° (156) ταυτολογῶμεν Σ βγ + G (idem loquamur)] ταῦτα λέγωμεν V, *cf.* Y

(++) 203° (160) πράξεων V Y D γ + G (operationibus) Procl.] πράξεως Σ M S Heph.

(-*) 208° (163) τούτῳ V D γ] αὐτῷ α M S + G (ipsi) Heph.

(-?) 222° (175) ὅταν κοσμικῶς μὲν V Y βγ] κοσμικῶς μὲν, ὅταν Σ + G (mundialiter quidem, quando) Procl. — Is the reading of V Y βγ an adaptation of the text to correspond to 217°? Or did Moerbeke change himself the order of the words?

(++) 232° (183) βραδύτεροι V + G (tardiora) Procl.] βράδιον α βγ

(-*) 233° (184) *tit. ante* ὁ.. προηγούμενος α M S Heph. Procl.] *tit. post* λόγου (236°/186) V D γ + G (*post* sermone) Co (95,28). — Write title *post* λόγου. *Cf.* titles book I, ch. 12 (736°/555), book II, ch. 3 (121°/109), ch. 4 (459°/346), ch. 12 (925°/709).

(++) 234° (184) τύπος V Y D γ + G (modus) Heph. (A)] τόπος Σ M S Heph. (P)

(+) 241° (191) δεῖ Σ M S γ Heph. + G (oportet)] δὴ V D: γὰρ Y

(+) 241° (191) τὰ Σ + G (que) Heph. (AP) Procl.: τὰς M S Heph. (C)] *om.* V Y D γ ('*haud scio an recte*')

(§) 244° (193) περιεχόμενα Q (*Marc. gr.* 594) + G (circumdata) Heph. Procl.] πῶς ἐχόμενα V: περιεχόμενοι Σ βγ, *cf.* Y

(-?) 259° (204) ἢ τῇ σελήνῃ Κρόνος D γ] ἢ τῇ σελήνῃ Κρόνος τῇ σελήνῃ V: ὁ δὲ Κρόνος δὲ τῇ σελήνῃ Y: Κρόνος δὲ τῇ σελήνῃ Σ M S Heph. (*Boer*): τῇ δὲ σελήνῃ ὁ Κρόνος Procl.: uel lune de die, Saturnus autem lune de nocte G. — See above, IVB 3.3.

(++) 259° (205) ἢ μὴ V + G (uel non), *cf.* Heph.] μήπως Y (*Boer*): ὑπὸ Σ βγ

(++) 265° (208) ᾖ VY + G (fuerit)] *om.* Σ βγ

(-) 265° (209) μηδενὸς ἢ τῶν κακοποιῶν V Σ β Co le (99,11)] μηδενὸς τῶν ἀγαθοποιῶν γ Heph., Co (99,13): μηδενὸς τῶν κακοποιῶν Y: a benefico nullo uel a maleficis G (μηδενὸς τῶν ἀγαθοποιῶν ἢ τῶν κακοποιῶν). — See above, IVB 2.1.3.

(-) 271° (213) καὶ[1] V Y Heph.] ἢ Σ βγ + G (uel)

(++) 280° (220) μόνον V Y S + G (solum) Heph. (A P)] μόνους Σ D M γ Heph. (*cod.* C)

(++) 288° (227) τὸν πατέρα V Y S + G (patrem) Heph. Procl.] τοὺς πατέρας Σ D M γ

(-?) 298° (233-4) ἑξάγωνος ὢν V Y] ἑξάγωνος Σ βγ Heph. + G (scilicet sextili aspectu). — The addition of ὢν makes the text more fluid, but is unnecessary (see parallel without ὢν at I.279°)

(-*) 300° (235) καὶ V D γ] ἢ καὶ Y + G (uel etiam): ἢ Σ MS Procl. Heph.

(-*) 309° (242) (θανάτους ...) αἰφνιδίους (... ποιεῖ) α βγ] (...) αἰφνιδίως (...) V + G (mortes ... efficit repentine ...) Procl. — The adverbial reading of V + G Procl. is the *lectio difficilior.* The adjective of α βγ could have been a grammarian's correction.

(§) 316° (247) νομῶν *Par. suppl. gr.* 597 + G (deuastationes) Heph. (C)] νόμων V Y Heph. (AP): νόσων Σ M S γ, *cf.* D

(++) 330° (258) τρόπου V γ Procl. + G (modi), *cf.* Y] τόπου Σ β

(+) 342° (266) ἀπομερίζωμεν Σ βγ + G (distribuamus)] -ζομεν V Procl., *cf.* Y

(+) 346° (269) οἷόν τε Σ M S γ + G (possibile est)] οἴονται V Y D

(++) 347° (269) μέγα V + G (magnum) Procl.] *om.* α βγ

(+) 381° (295) μὲν Σ M S + G (quidem) Heph.] *om.* V Y D γ (*Boer Robb.*)

(+) 388° (299-300) ἔτι πολυπραγμονοίη Σ M S + G (amplius ... uoluerit percunctari) Heph.] ἐπιπολυπραγμονοίη V Y D γ (*Boer Robb.*)

(+) 392° (302) περὶ ... θηλικῶν α C + G (de ... femininis) Heph. Procl.] + γενέσεως V D: + γενέσεων A B; *cf.* 9°

(++) 394° (305) λόγον V + G (rationem) Procl.] τρόπον α βγ

(-) 412° (317) πάλιν ὄντες V Y D Heph.] *inv.* Σ M S γ + G (existentes iterum)

(-*) 418° (323) δὲ V Y] γὰρ Σ βγ + G (enim) Co le Procl., *sim.* Heph.

(-*) 419° (324) τοῦτο τὸ σύμπτωμα V] τὸ τοιοῦτον σύμπτωμα α βγ + G (talis simptoma), *cf.* Heph.

(-*) 419° (324) παρὰ V] περὶ α βγ + G (circa)

(+) 425° (329) κυΐσκεσθαι V γ + G (procreari) Heph.] τίκτεσθαι α β Procl. — Moerbeke probably translated κυΐσκεσθαι (see Index). The reading of α β is a simplifying correction.

(-*) 440° (339) μόνον V Y Heph.] ἕνα Σ βγ + G (unum) Procl. (ἕν)

(-) 449° (347) μὲν γὰρ V] μὲν Y: γὰρ Σ βγ + G (enim). — This may be a consensus *per accidens,* as Moerbeke does not always translate μὲν before γὰρ.

(+) 458° (353) ὁ τῆς σελήνης Σ βγ + G (locus lune) Heph. Procl.] οἱ τῆς σελήνης V G°; *cf.* Y

(-*) 464° (357) μηδενὸς V Procl. Heph.] ἀλλὰ μηδενὸς α βγ + G (sed nullo) (*Robb.*). — Write ἀλλὰ μηδενὸς. See next item.

(-*) 465° (358) ἀλλὰ τῶν κακοποιῶν *secundum* Procl. *coni. Hüb.*] τῶν κακοποιῶν Σ M S: τῶν δὲ κακοποιόν Y (*cf. Robb.*): *om.* V γ + G Heph. (*Boer*), *cf.* Tib. — The reading of the archetype (ω) in 464°-5° could have been simply: ἀλλὰ μηδενὸς μαρτυροῦντος τοῖς φωσὶν ἀγαθοποιοῦ. α M S tried, like Proclus, to make the argument more explicit. The addition is not necessary. We propose to delete ἀλλὰ τῶν κακοποιῶν.

(+) 468° (360) ἢ πιθήκων Σ βγ + G (uel de simiis) Heph. (πιθήκων 107,17)] *om.* V Y Procl. (Tib.). — This reading shows that Moerbeke translated elements of the Greek tradition which are missing in the V (Y) Procl.-tradition. See above, IVB 2.

(-) 476° (366) μηδὲ εἷς V] μηδεὶς Σ βγ Heph. + G (nullus): μηδεὸ Y

(-*) 476° (366) ἀγαθοποιῶν ἀστέρων V D γ Procl.] ἀγαθοποιῶν α M S + G (beneficorum) Heph. — The addition of ἀστέρων is not needed, see κακοποιοί (475°).

(-*) 484° (371) ὁ τοῦ Ἑρμοῦ γ] ὁ Ἑρμοῦ V D: *om.* α M S + G (*'fort. recte'*)

(-*) 490° (377) λόγου V Y D γ] *om.* Σ M S + G, *cf.* Heph., Procl. — The addition of λόγου is redundant, see λόγου in the preceding line 489°.

(+) 500° (384) μὲν α + G (quidem) Heph., *cf.* M S] *om.* V D γ

(-*) 513° (394) διαμετρῶν α βγ] διάμετρος ᾖ V + G (diametralis sit), *cf.* Heph.

(+) 520° (399) μάλιστα Σ βγ + G (maxime) Heph. Procl.] *om.* V Y

(-?) 539° (412) ἀνατολὴν V Y γ] ἢ τὴν ἀνατολὴν Σ β + G (uel ortum) Procl.

(+) 554° (423) οὗτος Σ βγ + G (iste) Procl.] οὕτως V Y

(++) 567° (433) τὰς V + G (qui dicuntur) Procl.] τοὺς Y: τε Σ βγ (*Robb.*), *cf.* Heph. (A)

(+) 579° (442) ὃ καλεῖται Y D γ + G (quod uocatur)] καὶ καλεῖται V: ὃ καὶ καλεῖται Σ M S: ὅπερ καλεῖται Procl.

(+) 580° (442) ἐπειδήπερ Σ βγ + G (quoniam quidem)] ἐπείπερ V: ὅπερ Y

(-?) 580° (442) κακοῖ V (κακοὶ Γ: κακεῖ G°: κακῇ Y), *cf.* Procl.] *om.* Σ βγ + G. — See above, IVB 2.1.4.

(-*) 592° (451) ὡροσκόπον V D (*Boll Robb.*)] ἀνατολικὸν ὁρίζοντα α M S + G (orientalem horizontem) Heph. (120,22) Procl., *cf.* γ. — See above, IVB 2.2.2.

(-?) 594° (453) θέλει τε (τε *om.* Y) καὶ δύναται V Y] ἐστί τε καὶ δύναται Σ D M (*desunt* S γ): est et uult (*fort.* ἐστί τε καὶ θέλει *legit*) G. — See above, IVB 2.1.3.

(-) 597° (455) κἀκεῖνος V α D S γ] κἀκείνως M + G (et illo modo)

(+) 623° (468) ἀλλὰ καὶ α M S + G (sed et)] + τὸν V D γ

(-) 630° (474) καὶ V Y Heph. Procl.] ἢ Σ βγ + G (uel)

(-) 630°-1° (474-5) τοῖς μέχρι τῆς καταδύσεως τοῦ ἀφέτου συναγομένοις V Y Heph. Procl.] τοῖς ὑπὸ τοῦ ἀφέτου συναγομένοις μέχρι τῆς καταδύσεως τοῦ ἀφέτου Σ βγ + G (collectis a dimissore usque ad occasum dimissoris). — This is again an example of correspondences between G and the group Σ βγ in the third book.

(-) 646° (487) ἐπὶ V Y γ Procl.] ἀπὸ Σ (ἐπὶ *ss.*) β + G (ab) Heph. (*cod.* P)

(-?) 647° (488) ἑξάγωνον G° Σ βγ Heph.] ἑξαγώνων V Y + G (sextilibus) Procl. (*Boer Robb.*). — The accusative form seems to be a correction from Heph.

(+) 652° (491) ἀναιρεῖ Σ βγ + G (perimit) Heph.] *om.* V Y Procl.

(-) 674° (509) διὰ V βγ Heph.] ὑπὸ α + G (a) Procl.

(-?) 675° (510) ἀναλελυμένος V Y Heph. Procl.] βοηθούμενος D + G (auxiliatus): βοηθούμενος ἢ ἀναλελυμένος Σ M S γ. — See above, IVB 2.1.4.

(§) 677° (511) αἱ ... <μοῖραι> *suppl. Boer Hüb.* + G (gradus)] αἱ *codd.* — Perhaps Moerbeke added *gradus* himself.

(+) 680° (513-4) ἑκάστης μοίρας α + G (uniuscuiusque gradus) Procl.] ἑκάστη μοίρα V: -στας -ρας βγ

(-) 684°-7° (517) ὁ τοῦ — τοὺς ἰσημερινοὺς χρόνους (*cett.*)] *om. per hom. in* Y *et* G. — Probably the same omission *per accidens*.

(+) 711° (533) ἕκαστον (*sc.* ἡμικύκλιον) α M S + G (unumquodque) Heph.] ἕκαστος V D γ

(-*) 731° (548) ἐπειδὴ V Procl. Heph.] ἐπεὶ δὲ Σ βγ + G (quoniam autem): ἐπὶ δὲ Y

(+) 734° (550) χρόνους Y + G (tempora) Heph. Procl.] *om.* V Σ β: *alio loco* γ

(§) 736° (551) ταύτας δὲ εἰληφότες (*coni. Hüb.*) + G (has autem cum acceperimus)] ταύτας δὲ ἃς εἰλήφαμεν V Y (*Boer*): ταύτας γὰρ διειληφότες Σ M S: ταύτας διειληφότες D γ, *cf. Robb.*

(-*) 749° (562) ὁ τῆς ἀρχῆς τοῦ Κριοῦ α βγ] ἡ ἀρχὴ τοῦ Κριοῦ V + G (principium Arietis) Procl. (*Robb.*). — It is better to follow V + G Procl. with the *lectio difficilior*. The reading of α βγ may be a correction. See, however, 767°, 777° (indication Hüb.) and the next item: G is perhaps not a reliable witness in both places, because Moerbeke may have simplified in translating.

(-?) 750° (562) ὁ τῆς ἀρχῆς τῶν Διδύμων *codd.*] principium Geminorum G

(+) 764° (572) καὶ α M S + G (et[1])] τοῦ V D γ

(§) 770° (577) ἡ ἀρχὴ τῶν Διδύμων *sec. Procl. coni. Boll Hüb.* + G (principium Geminorum)] ἡ ἀρχὴ τῶν V: ὁ ἀφετικὸς τόπος Σ γ: *om.* Y β

(+) 817° (612) τῆς ἀρχῆς Y + G (principium) Procl.] *om.* V Σ βγ

(++) 818° (612) συμμεσουρανήσεων V Σ β BC + G (commediationum)] μεσουρανήσεων Y A Heph.

(+) 820° (614) ἀπέχει Σ S γ + G (distat)] ἀπέχῃ V Y D M

(++) 823° (617) τῶν συγκρινομένων κέντρων V + G (ab angulis comparatis) Procl.] τῷ συγκρινομένῳ κέντρῳ α βγ Heph.

(-*) 824° (617) προκειμένων α βγ, *cf.* Procl. (προστιθέντος)] προϋποκειμένων V + G (presuppositarum). — The double prefix προϋπο- is used several times III.106 135 IV.712 790: see also προϋφέστηκα IV.691 and προϋποτυπόομαι IV.788.

(-*) 828° (620) χρόνων Σ γ] δύο V: *om.* Y β + G (*cf.* Heph. Procl.). — The precision χρόνων is not needed; we propose to follow Y β + G.

(-?) 828° (620) ἤτοι V Y D γ] καὶ ἤτοι Σ M S + G (et aut) (*Robb.*). — Is it better to write, as Robbins, καὶ ἤτοι?

(+) 833° (624) μεσουρανήματος Σ β + G (a medio celi) Heph. Procl. (199.20), *cf.* Y] μεσουρανοῦντος V γ

(+) 836° (626) διαστάσεων Y γ + G (distantiarum)] -εως V Σ β

(+) 842° (631) προειρημένον α M S + G (predictum) Procl.] εἰρημένον V D γ

(-) 872° (652) ἐπόντας Σ βγ] ἔποντας V Y + G (consequentes)

(++) 884° (661-662) τῇ μὲν μορφῇ V + G (forma quidem), *cf.* Heph.] τὴν μὲν μορφὴν α βγ, *cf.* Procl.

(+) 889° (665) σπινώδεις Σ βγ + G (subtiles), *cf.* Procl.] σπινώδης V Y

(-?) 890° (666) ὑπορρύθμους Σ βγ Heph.] ὑπορίθμους V Y: paucorum pilorum (ὑπότριχας) G. — The reading translated by Moerbeke fits well with the context: a conjectural correction by a scholiast?

(+) 918-9° (689) λοξοφθάλμους ... αἰγωποὺς (αἰγώπους) Y + G (obliquorum oculorum ... caprinos oculos) Procl.; *cf.* ληξοφθάλμους ... αἰγόπλους V] κοιλοφθάλμους ... αἰγόποδας βγ, *cf.* Σ

(-?) 943° (708) παθεινούς *coni. Hüb.*, *cf.* Procl. (νοσερούς)] παθηνούς V (*Robb.*): παθινούς Y: σπανούς Σ γ: σπανθινούς M S: *om.* D: amplarum spatularum (σπαθηνούς?) G. — Moerbeke seems to have read a term similar to what is found in M S (β), *cf.* Σ γ.

(§) 954-5° (716-7) εὐρυθμότερον καὶ ἀρρυθμότερον *coni. Boer Hüb.* + G (meliorem proportionem et deteriorem)] *inv.* Y G°: καὶ εὐρ- καὶ ἀρρ- S: εὐρυθμότερον V Σ M γ: ἀρρυθμότερον D

(-) 963° (723) ὡς V Y D S γ] καὶ Σ M + G (et)

(-*) 967° (726) συνεπικίρναντας V, *cf.* Procl. (-κιρνᾶν)] συνεπικρίνοντας α βγ + G (comparantes): συνεπικρίναντας Heph. (P)

(+) 977° (734) τὸ προδῦνον Y + G (id quod preoccidit) Co le, *cf.* Procl.] δῦνον V M: ἡγούμενον Σ γ, *cf.* D G° S

(-*) 996° (749) ἐπιδείξασθαι V] δέξασθαι Y: ἐπιδέξασθαι βγ + G (suscipere) (*Robb.*). — Write ἐπιδέξασθαι.

(+) 1008° (757) εὐωνύμων πάντων Σ M S + G (sinistrorum omnium) Heph. Procl.] εὐωνυμωτάτων V Y D γ

(+) 1025° (769) συνάπτῃ δὲ Y + G (et coniungatur), *cf.* β Procl.] -πτει δὲ V G°; *cf.* Σ γ

(-*) 1034° (775) τῷ ἡλίῳ Σ βγ Heph.] τῷ ἡλίῳ ὄντες V Y + G (ad solem existentes) Procl. (*Boer Robb.*)

(+) 1047° (785) Καρκίνῳ Y M S + G (Capricorno) Procl.] + καὶ τοῖς νεφελοειδέσι συνάπτουσα ὀφθαλμὸν πηροῖ σελήνη V D, *simil.* Σ γ

(+) 1091° (817) καὶ ῥευματώδεις α M S + G (et reumaticos) Procl.] *om.* V D γ Heph.

(++) 1096° (821) τόπων V + G (locorum), *cf.* Heph.] ἢ *praem.* α βγ

(+) 1104° (827) ἐπιτάσεις α M S + G (intensiones*)] ἐπιστάσεις V D γ

(++) 1139° (852) ἐπιχάριτα V + G (gratiosas), *cf.* Procl.] ἐπιχαρῆ Σ βγ: ἐπιχάρος Y

(+) 1153° (864) αὐτῆς α γ + G (ipsius) Procl.] αὐτῶν V D; *cf.* M S, Heph.

(§) 1155° (866) ποιοίμεθα Σ + G (faciemus)] -ούμεθα V Y γ (*Boer*), *cf.* β

(-?) 1192° (896) ὀργανικάς α βγ] τε καὶ *add.* V + G (organicasque et) Procl.

(-?) 1209° (910) καὶ² α βγ] ἢ καὶ V + G (aut etiam)

(+) 1209° (910) ἐπίδοξον Σ βγ + G (super gloria*)] ἑτερόδοξον V Y

(-?) 1210° (910) ἔχωσι V Y γ] ἔχουσι Σ β + G (habent) Heph. — This is the first instance of six differences between conjunctive and indicative forms in this and the following paragraph (1010° 1212° 1215°(2) 1217° 1220°). They depend directly or indirectly upon ὅταν (*quando*) in 1207° (908), omitted by Hephaestio ('*forte recte*'). The archetype probably had the indicative form. The conjunctives were probably introduced as grammatical corrections. See above, IVB 1.3.3.

(+) 1210° (911) ἄλλους Σ βγ + G (aliis) Heph.] ἀλλήλους V Y (*Robb.*)

(-*) 1212° (912) ἀτιμώρητοι V] εὐτιμώρητοι α βγ + G (de facile punibiles) (*Robb.*)

(-?) 1212° (912) γίνωνται γ] γίνονται α β + G (fiunt) Heph.: *om.* (*spat. rel.*) V. — See 1210°.

(-*) 1215° (914) χαίρωσι γ] χαίρουσι V α β + G (gaudent) Heph. — See 1210°.

(-?) 1215° (914) εὐφημῶνται D M γ] εὐφημοῦνται V α S + G (bonam famam habent) Heph. — See 1210°.

(-*) 1217° (916) διατελῶσι V D M γ] διατελοῦσι α S + G (perficiunt) Heph. — See 1210°.

(-*) 1218° (916) αὐτό τε τὸ πρᾷον Y] αὐτὸ τοῦτο τὸ πρᾷον βγ + G (hanc ipsam mansuetudinem): αὐτὸ τοῦτο (*vel* τοῦ τὸ) πρᾷον V: αὐτὸ τὸ πρᾷον Σ Heph. Procl.

(-*) 1220° (918) τυγχάνωσιν V γ] τυγχάνουσιν α β + G (existunt) Heph. — See 1210°.

(-*) 1246° (941) μεταμελητικοὺς — εὐλαβητικούς *habent* V D γ Procl.] *habet et* μελητικοὺς (*sic*) *add.* S: *habet* (penitiuos — reuerentiales) *et* futuricios (μελλητικοὺς) *add.* G: μεταμελητικοὺς *habent, sed sequentes* (αὐστηροὺς — εὐλαβητικούς) *om.* α M (*omissio fort. per hom., cf.* S *et* G): [Σ]. — See above, IVB 3.1.

(+) 1271° (964) πρὸς συμβιώσεις Y M S + G (ad conuictus) Procl.] ἐνσυμβιώσεις V D, *cf.* γ: [Σ]

(-?) 1283° (974) πάμπαν ἀπίστους βγ] καναπιστάς V Y: turbatores G: *om.* Procl.: [Σ]. — The reading of βγ is a correction of the unintelligible form transmitted in V Y. Moerbeke may have read the same form and guessed what the meaning could be. It is also possible that he read κυκητάς, as at III.1416° (which he translated as *turbatores*).

(-) 1284° (974) παντοποιούς βγ, *cf.* Y, *cf.* Procl. (πάντα ποιοῦντας)] παντοίους V + G (omnimodo): [Σ]

(+) 1291° (981) νηλεεῖς βγ + G (duros)] μηλίους V Y: [Σ]

(-) 1299° (988) ἐλευθερίους α β Procl.] *om.* V γ + G

(-) 1299° (988) δικαίους α M S Procl.] *om.* V + G: *post* σεμνούς D γ

(-) 1314° (999) εὐεπιβούλους, ἐπιεικεῖς, ἐπάνδρους, νικητικοὺς α βγ Arab.] *om.* V + G Procl.

(-*) 1318° (1002) μονοτόνους M S Procl., *cf.* Σ] μονοπόνους V Y D γ + G (singulariter laborantes) (*Boer*): *om.* Heph. — μονοπόνους (hapax) fits better in the context.

(-) 1321° (1004) ἐμπατάκτους V Y] ἐπιτακτικοὺς εὐπατάκτους βγ, *cf.* Σ: ἐμπράκτους Procl. + G (operosos) (*Robb.*). — Similar diversity at 1389°, but there Moerbeke renders the term ἐμπατάκτους correctly (*promptos percutere*). ἐμπάτακτος is an hapax in Ptolemy ('thunderstruck', 'stupid' LSJ).

(++) 1326° (1008) ἐλεητικούς V + G (miseratiuos), *cf.* Procl.] ἐλεγκτικούς βγ: ἐλεκτικούς Σ: *om.* Y

(+) 1330° (1011) φιλοικείους S + G (domesticorum amatores), *cf.* V Y] φιλονείκους Σ M γ, *cf.* D

(+) 1332° (1015) θηλυψύχους α M S + G (muliebres animo) Heph.] μεγαλοψύχους V D γ

(-*) 1333° (1015) καταγυναίους V] κατὰ γυναίων Y: καὶ γυναίους Σ βγ: κακογυναίους Procl. + G (in malum muliebres) (*Robb.*)

(-?) 1334° (1016) λοιδόρους, μοιχούς Procl. (*coni. Robb. Hüb.*)] λοιδόρους V D B C (*cf.* A) + G (maledicos): μοιχούς α M S (*Boer*). — There are two different terms, but no double reading in the direct tradition. G confirms the reading of V D B C. Write λοιδόρους?

(+) 1335° (1016) ἐπιμώμους Y + G (probrosos) Procl.] ἐπιβώμους V: φιλομώμους ὑπομώρους Σ M S, *cf.* D γ

(+) 1348° (1028) θεοπροσπλόκους βγ + G (deo supplices) Procl.] θεοπλόκους V; *cf.* Y *et* Σ

(-*) 1365° (1041) φιλοτέκνους V Σ βγ] φιλοτέχνους Y + G (artis amatores) Procl. — *Cf.* 1404°. φιλοτέχνους fits better in this context.

(++) 1369° (1043) διακριτικούς V Y D γ + G (diiudicatiuos) Procl.] ἀδιακρίτους Σ M S (*Boer*)

(-?) 1372° (1045) ἀδιαφόρους α M S] *om.* V D γ + G Procl. (*Boer*). — Delete ἀδιαφόρους?

(++) 1380° (1052) δεινούς V S γ + G (industrios)] δειλούς Σ D M Procl.: [Y]

(-?) 1394° (1063) τῆς ψυχῆς α βγ] *om.* V + G Procl. — The lack of τῆς ψυχῆς may be an omission in hyparchetype ψ, but more probably it was added in the other branch of the tradition to make this passage conform to the beginnings of paragraphs 10, 20, 28, 36.

(-) 1397° (1065) καλοζήλους V Y D γ] κακοζήλους Σ M S + G (mali zelatores) Procl. — The term κακόζηλος (of bad taste) does not fit here. καλόζηλος is probably an intelligent correction, but it is also a hapax. The Arabo-Latin translation has *ualde zelotypus* (πολύζηλος?).

(-?) 1397° (1065-6) φιλοτέκνους *coni. Hüb.*] καὶ φιλοτέχνους V D + G (et artium amatores): φιλοτέχνους α M S γ Procl., *cf.* Tib. — Is the conjecture necessary?

(++) 1404° (1071-2) φιλοτέχνους V γ + G (artium amatores) + Procl.] φιλοτέκνους Y: χειροτέχνους Σ β, *cf.* 1365°

(-) 1414° (1079) τῆς ἐναντίας V α β] τῶν ἐναντίων γ + G (contrariis). — The plural corresponds to the introductory condition (ἐπὶ μὲν ἐνδόξων διαθέσεων 1403°-4°). Was it an assimilation (*cf.* 1259° 1291° 1331° 1371° ...), or did Moerbeke replace the singular with the plural?

(++) 1427° (1088) θεωρητικούς *hic* V + G (speculatiuos) Procl., *cf.* Tib.] *post* εὐστόχους (1428°) Σ M S: *om.* Y D γ

(++) 1435°-6° (1094) αὐτὴ ἡ V D γ + G (ipsa)] ἡ αὐτῆς α M S

(-*) 1446° (1102) δὲ Σ M S Heph.] δέ πως V Y D γ + G (autem aliqualiter)

(++) 1478° (1127) ἤ¹ V + G (aut) Heph. Procl.] *om.* α βγ

(++) 1486° (1132) ἐπὶ φάσεως οὖσαν *coni. Hüb.* + G (existentem in apparentia), *cf.* ἐπιφάσεως οὖσαν V, *cf.* Procl.] ἐπιφάσεως ἰοὺς ἂν Y: *diversimodo cett.*

(++) 1489° (1135) προειρημένον V + G (predictum), *cf.* Procl.] προκείμενον α βγ

(§) 1497° (1140) (διὰ) ... διαιτητικῆς (ἀγωγῆς) *coni. Boer Robb. Hüb.* + G (per regimen diete), *cf.* διαιτικῆς V] διαγητικῆς Y: ἰατρικῆς *cett.*

(-?) 1502° (1144) συνεχείαις (*codd.*)] constrictionibus (συναγχίαις *or* συνάγχαις?) G. — Moerbeke generally translates συνεχής and συνέχεια by *continuus* and *continuitas* (see our Index or, e.g., the index of the *Meteorologica*). *Constrictionibus* is thus not only strange, but uncharacteristic. Yet, Moerbeke uses *constrictiones* for συνοχαί (II.709°), where it has the sense of 'distress', 'affliction', as are poverty and imprisonment. Is it possible that, in III.1502, he read another word in his Greek model, a specific medical term (συναγχίαις or συνάγχαις), as F. Bossier proposes?[2] Yet, in IV.597° (458), Moerbeke rightly translates the same Greek word with the technical term *quinanchia*.

(++) 1504° (1145) ἀκαταστασίαις V + G (inquietudinibus), *cf.* Procl.] ἀκατασχεσίαις α βγ

(-*) 1516° (1154) ἐστὶν α β] *om.* V γ + G Procl. ('*fort. recte*')

(-?) 1517° (1155) κατὰ τοῦτο (Y: κατ' αὐτὸ Σ βγ) πάλιν τὸ ἐξαίρετον α βγ Heph.] πάλιν ἐξαιρέτως V + G (rursum precipue), *cf.* Procl.

---

[2] Bossier (1994) 129-130.

(-*) 1533° (1166) τὰς ... ὁμιλίας Σ βγ] τὰς (*sc.* συνουσίας) ... ὁμοίως V Y + G (eos qui [*sc.* coitus] ... similiter) + Procl. — The reading of Σ βγ is a superfluous correction. The term ὁμιλία is never used by Ptolemy in this text, he always writes συνουσία. Write τὰς ... ὁμοίως.

(-*) 1534° (1167) γὰρ Σ βγ] δὲ V Y + G (autem) ('*forte recte*')

(++) 1536° (1168) ᾖ Σ + G (sit), *cf.* ἦν V] καὶ Y: *om.* βγ Heph.

(-) 1548° (1176) σαθροὶ V Y Procl.] θρασεῖς βγ + G (proni) Heph., *cf.* θάρσεις Σ. — See II.442°.

(-*) 1555° (1180) σημειώσεως Σ D M γ] δημοσιώσεως S (δημοσίως ἕως V, *cf.* Y Γ) + G (diuulgationem)

(-?) 1566° (1187) εὐπροσκοπώτερον α βγ] εὐπροκοπώτερον V + G (facile profectibilius) (*tacite Boer*), *cf.* Procl. πρακτικώτερον. — Both adjectives are Ptolemaic *hapax legomena*. The variant εὐπρόκοπος is not even mentioned in LSJ. εὐπρόσκοπος occurs twice, here (with the *varia lectio* εὐπρόκοπ-) and in IV.784° (uniformly attested in all mss.), but the signification could not be the same in both places. In LSJ two meanings are distinguished: (A) III.1566°: 'far-seeing, cautious', hence 'increased foresight' (Robb.), 'accroître la perspicacité' (Charvet); (B) IV.784° 'easily taking offense', hence 'easily offended' (Robbins), 'irritée de tout' (Charvet). Signification B is evidently appropriate in IV.784°, where it stands among the debilities of old age (one becomes dispirited, weak, easily offended, hard to please). Moerbeke translates: 'in uita ... debili et facile fatigabili et ad omnia displicentiam habenti'. It is less appropriate, however, in III.1566°. This is probably why Robbins and Charvet chose signification A in that passage. But even signification A is very strange in the context of III.1566°, for it would connect the ignoble effects of Mercury — notoriety, instability in passions, versatility — with the virtues of foresight and perspicacity. Moerbeke translates *facile profectibilius*, which confirms the reading of V: εὐπροκοπώτερον (*cf.* III.1205°/907 ἀπρόκοπος *sine profectu*). The substantive προσκοπή does not occur in Ptolemy. The substantive προκοπή ('progress'), however, is used twice: at IV.804° and at IV.554° where ὠφέλειαι καὶ προκοπαὶ καὶ δωρεαὶ ... (profit, gain, gifts) are mentioned as boons brought in by Mercury. This parallel indicates, we think, that the reading in III.1566° εὐπροκοπώτερον ('easy in making profit') — namely, as an effect of Mercury — as attested by V + G, is authentic. The Arabo-Latin translation seems to confirm it (*Mercurius augmentum adiuvabit*). Noticeable is also the use of the term προκοπτικός in an astrological context by Vettius Valens and Dorotheus. The term εὐπροσκοπώτερον is probably a redactional correction in α βγ.

4. BOOK IV

V (α =YΣ β =DMS γ =ABC)

(++) 3° (3) α΄ προοίμιον V + G (Prohemium ca. p^m) Procl.] *om.* α βγ

(++) 4° (4) β΄ V + G (ca. 2^m) Procl.] α΄ Y D γ: *om.* Σ M S *et sic deinceps*

(++) 13° (13) α΄ προοίμιον V + G (Prohemium capitulum primum) Procl.] *cf. cett.*

(-*) 28° (28) τοὺς τοῦ Σ βγ] τούτου V Y (*Robb.*) + G (hac) Procl.

(-?) 37° (35) πίστεων α βγ] πίστεως V + G (fidem) (*Robb.*). — The plural could be a syntactical adaptation to the next plurals.

(-?) 39° (36) (διὰ) φιλικῶν V, *cf.* Y (φυληκῶν)] (διὰ) φιλίων G° (*cf.* Heph. φιλιῶν) + G (per amicitias): (ἐκ) φιλίας Procl.: (διὰ) φιλῶν Σ βγ. — Both readings are possible.

(++) 40° (36) καὶ[1] V S + G (et[2]) Procl.] ἢ α D M Heph.

(+) 43° (39) ἄνω α M S γ Heph. + G (superioribus) Procl.] ἄλλων V D (γρ. καὶ ἄνω *mg.*)

(+) 48° (43) κτήσεις Σ βγ + G (possessiones), *cf.* Procl.] κτίσεις V Y

(+) 52° (45) λαμβανομένου α βγ + G (accepto) Heph.] λαμβανομένων V (*Boer*)

(+) 73° (63) ἢ Y D γ + G (uel) Procl.] *om.* V: [Σ M S]

(-*) 74° (64) τὰ μὲν φῶτα Σ βγ] τὰ φῶτα V Y + G (luminaria) Heph. Co le Procl. — μὲν seems to be a good syntactical addition by the editor of β, to correspond to δὲ in 75°. Nevertheless, we propose to follow V Y + G Heph. Co le Procl.

(-?) 83° (69) κεκεντρωμένον ᾖ α] κεκεντρωμένον V βγ + G (in angulo) Heph. (*Boer*). — The addition of ᾖ in α may be a conjecture; ᾖ is unnecessary, *cf.* τυγχάνῃ.

(-?) 99° (82) καὶ[3] α βγ Procl.] ἢ V + G (aut)

(-) 124° (100) ἐν α βγ Procl.] καὶ ἐν V + G (et in)

(++) 153° (122) ζωγράφους V + G (pictores) Procl.] στρατιώτας Σ D M γ: *om.* Y S Heph.

(++) 154° (122) παρασχίστας (-σχησ- Y) V Y + G (portionatores) Procl.] περιχύτας Σ βγ Heph., Co le. — This is an interesting case. V Y + G Procl. undoubtedly represent the reading of the common archetype παρασχίστας, a rare term we know through a reference in Diodorus Siculus (*Historiae* 1,91,4), who mentions the role of 'the so-called παρασχίστης' (who 'cuts lengthwise' a corpse) in the embalming process. Ptolemy, writing in Egypt, may have known this 'profession'. The other term, περιχύτης, is no less rare and means 'attendant at bath'. The latter term is found in both Hephaestio and the astrologer Rhetorius, both of whom are quoting from Ptolemy. It is also the reading in a lemma in the commentary and attested by Tib. (*qui in balneis morabitur*). Σ βγ were probably corrected from Heph.

(-) 174° (137) φιλομονομάχους V M S (*Hüb.*), *cf.* Y Heph.] φιλομονάχους D γ + G (amatores monachorum): *om.* Σ

(+) 182° (142) αὐτοῖς Σ M S γ + G (ipsis) Procl.] αὐτὸς V D: αὐτοὺς Y

(+) 189° (148) οἷς ἂν Σ βγ + G (quibuscumque)] οἷς ἐὰν V Y Heph.

(-?) 191° (150) τῶν πράξεων α βγ Procl.] τῆς πράξεως V + G (operationis). — Both readings are possible. Yet we would prefer to write τῆς πράξεως with V + G. The plural might well be a correction with regard to the preceding ποικίλον (*uarietatem*).

(-?) 198° (155) ἢ Σ M S γ Heph.] καὶ Y + G (et) (*Boer*): *om.* V D

(+) 199° (155) δὲ Σ + G (autem) Heph.] τε V Y βγ (*Boer Robb.*)

(+) 203° (159) μάντεις, θύτας α M S + G (diuinatores, immolatores) Heph.] μαντευτὰς V D γ

(§) 213° (167) ὑποτακτικάς Heph. Procl. + G (secundarias operationes)] ὑποπρακτικάς V Y D γ: ὑπὸ τὰς πρακτικάς Σ M S. — Moerbeke translates *secundarias operationes*, taking up *operationes* (πράξεις) from 212° *operationes principales ... secundarias operationes.* — *secundarias* is a free translation of ὑποτακτικάς.

(-*) 225° (171) τῶν ... συμβιώσεων V D γ Heph.] τῆς ... συμβιώσεως Y+ G (conuictu) Procl.: τῆς (τῶν M) ... συμβιβάσεως Σ M S

(-*) 252° (199) συνάπτων V D γ] *om.* α M S + G Heph.

(-?) 262° (209) λέγομεν α βγ, *cf.* Procl.] λέγω μὲν V + G (Dicamus: λέγωμεν)

(+) 266° (211) τοῖς εἰρημένοις Σ βγ + G (dictis) Procl.] τῆς εἰρημένης V Y

(-?) 271° (214) συμβαίνῃ α βγ] συμβαίνει V + G (accidit). — See above, IVB 1.3.3.

(+) 277° (219) ἡδείας Σ β B C + G (delectabiles) Heph.] ἰδίας V Y: [A]

(-*) 293° (228) νομίμους ποιοῦσι βγ] νομίμους V α + G (legitimos)

(+) 308° (237) ἡδείας Σ D M γ + G (delectabiles) Heph.] ἰδίας V Y S

(+) 326° (251) ἐν θηλυκοῖς ... τόποις Σ βγ + G (in femininis ... locis) Heph. Procl.] ἐν θηλυκῶν ... τόποις V Y

V (Y β =DMS γ =ABC)

(-*) 330° (253) καὶ τῆς (*om.* D) κάλπιδος καὶ τῶν ὀπισθίων (τοῦ Λέοντος) V Y D γ] καὶ τῶν ὀπισθίων M S Heph. (*Robb.*) (‘*fort. recte*’) + G (et in natibus et in posterioribus *dupl. lectio Guill.*

(++) 331° (255) κεντρωθέντες V + G (in angulis ... existentes), *cf.* S] -θέντος Y: -θὲν G°: -θέντα D M γ

(-*) 336° (258) αὐλικοὺς V βγ (Co le 157,137) (*Robb.*)] αὐλίσκους Y + G (canales habentes) Co (157,138): *om.* Heph. Procl. — αὐλικός (from αὐλή) means a courtier, someone living at court, but this does not fit well in the series: σπάδοντας αὐλί(σ)κους στείρας ἀτρήτους ‘eunuchs, courtiers (?), sterile women and those without perforation’. The term αὐλίσκος (diminutive from αὐλός) generally indicates a small pipe or tube. In the present context, one should understand ‘small tube’ with a sexual connotation. The term αὐλικός in the lemma in the printed edition of Co might be a mistake of the Renaissance editor, because the commentator explains αὐλίσκους: ὀργασμὸν οὐκ ἔχειν, τὸν αὐλίσκον δὲ μόνον (‘habent non testem, sed fistulam tantum’). In this sense, αὐλίσκος is different from the preceding term: σπάδοντες refers to castrates, to eunuchs in general, while αὐλίσκος denotes a special form of castrates. Moerbeke reads as Y the special term αὐλίσκους and translates *canales habentes*. Write αὐλίσκους.

(++) 352° (270) ὑπερπαθῶς V + G (ualde passionabiliter), *cf.* Procl.] ὑπερ παθ Y: ὑπερπαθῶν β: ὑπερπαθεῖς A C: ὑπερπαθ B + *spat. 3 litt.*

(-?) 394° (304) δὲ M S γ Heph.] δισσὰ V D + G (duplices): δὲ δισσὰ Y (*Boer*), *cf.* διδυμογωνίαν δωροῦνται Procl. (264, -2). — δισσὰ is an anticipation of the δισσὰ in 396°. Was it a copyist’s error in the common archetype ω, corrected later in M S γ?

(++) 396° (305) Ἰχθύσι καὶ Καρκίνῳ καὶ Σκορπίῳ V + G Procl.] *diversimodo* (*sed omnes cum* αἰγοκέροτι) *cett.*, *cf.* Heph.

(-*) 400° (308) (ἢ καὶ ἐν στειρώδεσι) ζῳδίοις βγ Heph.] (...) τόποις V + G (aut etiam in sterilibus locis) (*Robb.*): (...) τόποις ἢ ζῳδίοις Y (‘*fort. recte*’): *simil.* Procl. (265,9-10) — ‘Probably the less usual term *places* (τόποις) is the more original; *signs* (ζῳδίοις) was added as a gloss, and thus came into the text’ (Robbins).

(-?) 409° (315) ὀλίγων α βγ] ὄντων *add.* V + G (existentium)

(-*) 415° (319) ἔσονται βγ] ὄντες V Y + G (existentes) Procl. (*Robb.*)

(-*) 415° (320) ἐὰν δὲ *coni. Hüb.*] ἢ V Y + G (aut) Procl.: εἰ δὲ D M γ (*Boer*): εἰ S Heph. [*sic*] (*Robb. Pingree*). — Write ἤ. The conjecture ἐὰν δὲ at 215° is unnecessary and destroys the opposition ἐὰν μὲν (414°) ... ἐὰν δὲ (417°).

(-?) 448° (345) τριγώνοις εἶεν βγ Heph.] τριγώνοις εἰσὶ Y: τριγώνοις V + G (trigonalibus) + Procl.

(+) 453° (348) σπονδὰς βγ + G (compactiones) Heph.] σπουδὰς V Y

(-) 453° (348) ἀποκαταστάσεις Y M S Heph. (*Robb.* 'reconciliations')] ἀκαταστασίας V D γ + G (inconstantie)

(+) 461° (354) οἱ M S + G (loca), *cf.* οἱ ... τόποι Heph., Procl.] ἡ V Y D γ

(++) 462° (355) καὶ ἔχθρα V Y + G (et inimicitia) Procl.] ἤ τίς *vel* ἥτις β: ἢ ἥτις (ἥτι A) γ

(-*) 464° (356) χρείαν M S Heph. Procl.] χρείας V Y D γ + G (utilitates). — Both readings are possible, but the plural form is better attested.

(++) 469° (360) ἢ τὸ ἔγγιστα V + G (uel propinquissime), *cf.* Procl.] ἔγγιστα Y βγ

(+) 469° (360) ᾖ ζῴδιον γ + G (sit signum), *cf.* Heph.] ἦν ζῴδιον V β (*Boer*): ἡ ζῴδιον Y

(-*) 469° (360) τῇ ἐπαναφορᾷ M S, *cf.* Heph. (*Robb.*)] ἢ ἐπαναφορά V γ + G (aut superillatio) Procl. (*Boer*): ἐπαναφορά Y: ἡ ἐπαναφορά D. — The reading of V + G ἢ ἐπαναφορά was probably the *lectio difficilior* in the archetype ω, which was differently corrected (in the best way by MS). We would follow G + V (γ) for the entire line 469°: ἢ ἐὰν τὸ αὐτὸ ἢ τὸ ἔγγιστα ᾖ ζῴδιον ἢ ἐπαναφορά.

(-*) 484° (372) συστάσεως βγ] συστάσεις V + G (confederationes) Procl. (*Robb.*): συστάσης Y. — Because of the form συστάσεις, Moerbeke also translates the following two terms (γεωργίας and κληρονομίας) as plural accusatives — διὰ συστάσεις ἢ γεωργίας ἢ κληρονομίας *per confederationes aut agriculturas aut hereditates* — whereas they are genitive singulars in the Greek edition, since they follow συστάσεως. The editor of βγ probably intervened in the text because in the next line, 194°, διὰ is construed with a genitive (see also 487°, 490°, 491°). However, this correction is not needed, as we also have constructions with the accusative in the same context (see 486° and 493°). We therefore propose to write, as did Robbins, συστάσεις.

(+) 496° (381) καὶ οὐκ ἐπὶ πολὺ διευθυνούσας Y M + G (et ad multum deficientes [*sic*]) καὶ ἐπὶ πολὺ διευθυνούσας V D S γ

(++) 497° (382) πραγμάτων V AB + G (negotia) Procl.] γραμμάτων Y β C

(-) 506° (388) ἀλλοτριωθέντες Y βγ, *cf.* Procl.] ἀλλοτριωθέντων V + G (alienatis)

(-*) 517° (396) συμφώνως Y β Heph.] συμφωνῶσι V + G (concordes fuerint) ('*fort. recte, recepit Robb.*'): συμφώνους γ. — Write συμφωνῶσι with V + G (*Robb.*).

(-*) 517°-8° (396) τῆς γενέσεως τόποις Y M S] τόποις τῆς γενέσεως V D γ + G (locis natiuitatis) Procl. (*Boer Robb.*)

(-) 518° (396) ἐναντίως V Heph.] ἐναντίους Y βγ + G (contrarias). — Moerbeke is perhaps not significant here, since he could have easily translated the adverb as an adjective.

(++) 518° (397) ποιῶνται V D γ + G (fecerint) Heph.] -οῦνται Y M S

(-*) 524° (403) καὶ αὐτὸς ἢ ... M S γ (*Hüb.*)] ἢ καὶ αὐτὸς ... V + G (aut etiam ... ipse) (*Robb.*): εἴη καὶ αὐτὸς καὶ Y, *cf.* D

(-*) 556° (425) καὶ μάλιστα ἂν βγ] κἂν μάλιστα V Y + G (et si maxime)

(-?) 556° (425-6) διαμηκίσωσιν D M γ Heph.] διαμηκίζουσιν V + G (per longitudinem distant): διαμηκίζωσιν Y, *cf.* S. — See above, IVB 1.3.3.

(+) 559° (428) δυσοδιῶν D γ + G (immeabilia; δυσοδίων *legit uel intell.*)] δυσωδιῶν V Y M S

(-?) 567° (435) ἔτι βγ Procl.] ἐπὶ V Y + G (in). — Good conjecture in βγ. Is ἐπὶ (V Y G) a copyist's mistake in the archetype?

(-*) 569° (436) ἐκ βγ] τουτέστι(ν) V Y + G (et hoc est) (*Robb.*). — Needless correction in βγ.

(-?) 572° (439) ἐκ τῆς τῶν ... ἐπεμβάσεων ποιότητος Y βγ Heph.] ἐκ τῆς <τῶν> τῶν ε' πλανωμένων ἐπεμβάσεων ποιότητος V + G (ex qualitate accessuum quinque planetarum) Procl. (*Boer Robb.*). — One could follow, like Boer et Robbins, V + G Procl, but Heph. confirms the reading without addition

(-*) 600° (460) ἐπισυμπίπτουσιν βγ, *cf.* Procl. (ἐπισυμβαίνοντας)] ἐπισυνάπτουσιν V Y + G (copulantur) (*Robb.*). — Write ἐπισυνάπτουσιν ('are joined') with V Y + G. Both verbs are possible, but ἐπισυνάπτουσιν is the *lectio difficilior*, and ἐπισυμπίπτουσιν could well be an alternative introduced by βγ.

(+) 637° (487) περὶ βγ + G (circa), *cf.* Procl. (πρὸς)] *om.* V Y

(++) 643-4° (492) ἢ τετραπόδοις V + G (aut quadrupedibus) Procl.] τετραπόδοις (ἢ *om.*) Y: *om.* βγ

(-?) 663° (505) καὶ συρμάτων Heph. (*Hüb.*): ἢ συρμάτων βγ (*Boer*)] καὶ συμπτωμάτων V Y + G (et simptomatibus) (*Robb.*), *cf.* Procl. (ἀπὸ συμπτωμάτων καὶ κλασμάτων). In this passage, Ptolemy examines various causes of death, including 'collapses (of buildings)' and 'breaking (of bones)' and συρμάτων/συμπτωμάτων. The rare term σύρμα denotes both 'dragging' and the effects of dragging on the skin ('abrasion'). The more common is συμπτώματα, which can be used for all kinds of accidents. Robbins translates it freely as 'crushing' (perhaps influenced by Procl. κλασμάτων). It is difficult to imagine that a copyist would have inadvertently written the rare form συρμάτων instead of συμπτωμάτων. It is more likely that a substitution would go the other way round, the rare term (συρμάτων) being replaced by a common term (συμπτωμάτων). Therefore, Hübner is right in opting for the rare term, which is confirmed by Hephaestio. Still, this reading is only found in βγ. We therefore have to assume that it is a good conjecture inspired by Heph. The primitive (though erroneous) reading in the archetype was certainly συμπτωμάτων.

(-) 670°-1° (509-10) (τῆς κατὰ τὸ ποῖον κυρίας περὶ τὸν αὐτοῦ τοῦ ἀναιρετικοῦ) τόπου τυχόντα γινομένης M S γ, *cf.* Heph., *cf.* Procl.] (...) τόπον τυχόντα γινομένης V Y D + G (eo quod penes qualitatem dominio facto [κυρίας γινομένης] circa ipsius peremptiui locum existentem [τόπον τυχόντα]).

(-*) 671°-2° (510-1) ἢ καὶ πολλῶν τῶν θανατικῶν συμπτωμάτων ἢ δισσῶν ἤτοι ... βγ] ἢ καὶ διὰ τῶν τῶν θανατικῶν συμπτωμάτων δισσῶν ἤτοι... V Y + G (et hoc aut per simptomata mortis duplicia aut): ἢ καὶ διπλοῖς τῶν θανατικῶν συμπτωμάτων ἢ Heph. (*cf. adn.* Pingree, 189,10). — These two items 670°-2° should be treated together. In the first part, the correction τόπου, inspired by Heph., seems necessary. In the second part, however, one better stays with V Y + G. The text of βγ in the second part is again the result of an editorial intervention: a distinction is made between πολλῶν and δισσῶν, which is not needed here. We propose to follow the main tradition and to write ἢ καὶ διὰ τῶν τῶν θανατικῶν συμπτωμάτων δισσῶν ἤτοι, which reads in translation: 'or also through the fatal occurrences which are twofold, i.e. effected according to quality or quantity'.

(++) 698° (532) παρόμοιον V S + G (consimilitudinem), *cf.* Procl.] παρίμειον Y: προοίμιον D M γ

(++) 699° (533) γεννώμενον V + G (nascitur), *cf.* Procl.] γενόμενον Y βγ (*Robb.*)

(+) 706° (538) προθέμενος βγ + G (proponens)] μὴ προθέμενος V Y

(+) 706° (538-9) δέον βγ + G (quod oportunum erit)] οὓς δέον V Y

(-*) 718° (548) ἐφαρμόζωμεν γ, *simil.* Y β, *cf.* Heph.] ἐφαρμόζειν V + G (adaptare): συνεφαρμόζειν Procl. — See above, IVB 1.3.3.

(-?) 719° (549) ἐπιβολὴ V βγ Procl.] τις ἐπιβολὴ Y + G (quedam adiectio)

(+) 719° (549) ἡ αὐτὴ M S γ + G (eadem) Procl.] αὐτῶν V D: αὐτὴ Y

(+) 724° (552) ἡλικιῶν Y M S + G (etatem) Heph.] ἡλιακῶν V D γ

(+) 738° (561) τοῖς Y S B C + G (hiis) Heph.] τῆς V D M A

(++) 743° (564) καὶ (*ante* μαθημάτων) V Y + G (et[3]) Heph. Procl.] τῶν D M: καὶ τῶν S γ

(+) 750° (570) πόρων Y D γ + G (meatuum) Heph. Procl.] σπόρων V M S

(-?) 750°-1° (570) κατὰ τὴν πλήρωσιν αὐτῶν καὶ ὁρμὴν D γ (*Hüb.*)] κατὰ τὴν πλήρωσιν αὐτῶν ὁρμὴν V Y: et secundum repletionem ipsorum impetum (καὶ κατὰ τὴν πλήρωσιν αὐτῶν ὁρμὴν) G: καὶ τὴν πλήρωσιν αὐτῶν καὶ ὁρμὴν M S. — There was probably a confusion of κατὰ and καὶ in the archetype. Both are necessary for the sentence. The diversity of the readings might stem from an addition of καὶ on top of κατὰ. Perhaps V Y had the non-corrected primitive form, to which a καὶ was later added. The addition of καὶ before ὁρμὴν in DMS (=β) is probably a correction. The version at the basis of Moerbeke's translation, which put the καὶ before κατὰ, seems the most plausible. We propose to write: καὶ κατὰ τὴν πλήρωσιν αὐτῶν ὁρμὴν ἐπὶ τὴν τῶν ἀφροδισίων συνέλευσιν, and to adapt (in this way) the translation of Robbins: 'Venus begins, as is natural, to inspire an activity of the seminal passages, *and at their completion* an impulse towards the embrace of love'.

(-?) 760° (576) ἀνεπιστάτων D Heph.] ἀνεπιπλάστων V Y γ + G (fictis) (*Robb.*): ἀνεπιστάτων πλάστων M S: ἀταξιῶν Procl. — ἀνεπιστάτων is a correction in D from Heph.; the only other place where we find ἀνεπιστάτων + ἁμαρτημάτων is in the TLG. This correction was probably made *supra lineam* in the hyparchetype β, as is clear from the copies M and S, where the correction is misinterpreted as an addition. Hübner opted for the reading of D because it is confirmed by Heph. and seems to fit best in the context. Ptolemy talks of a transition from 'infantile and ἀνεπιστάτων/ἀνεπιπλάστων errors to attention (προσεκτικόν) and sense of shame and honour'. Children easily make mistakes because they are given to play and lack seriousness and attention. The opposition to προσεκτικόν makes clear that ἀνεπιστάτων, here, is the required reading. However, the archetype ω undoubtedly had ἀνεπιπλάστων. Robbins opts for ἀνεπιπλάστων, which he translates as 'ingenuous'.

(++) 770° (584) ἡλικίαν λαχὼν V + G (etatem sortiens), *cf.* Procl.] *inv.* Y βγ

(-) 782° (592) ταχείαις *scr. Hüb.*] καὶ ταχείαις V Y + G (et celeritatibus) Heph. (*ms.* P, Pingree 193,5 *adn.*) (*Robb.*): καὶ -ας γ (*Boer, coni. Pingree*): -ας β. — The addition of καὶ before the adjective is difficult, but it is already attested in Heph. Because of the extra καὶ, Moerbeke translates ταχείαις as a dative substantive form (*cf.* Robbins: 'and speeds'), whereas it is in fact an adjective. This καὶ is found in the whole tradition (except β, but β has -ας), and καὶ ταχείαις probably represents the reading of the archetype ω.

(++) 782°-3° (592) τῆς ἐπὶ τὴν φύσιν ... ἐπιγινομένης V + G (facta in naturam)] ταῖς ἐπὶ τὴν φύσιν ... ἐπιγινομένης Y: τῇ φύσει ... ἐπιγινομένης βγ

(-*) 783° (593) τῷ βίῳ Y βγ] ἐν βίῳ V + G (in uita) Heph. ('*haud scio an recte*'). — Both readings are possible, but see the precedent item: it would be more consistent to follow here also V + G (Heph.).

(+) 792° (599) τῶν ... χρόνων Y M S γ + G (temporibus) Procl.] τὸν ... χρόνον V D

797° (602) *hic desinit* V

Y β (= DMS) γ (= ABC)

(-?) 799° (606) γὰρ Y, *cf.* Procl.] γὰρ ἂν βγ + G (enim utique) (*Boer*)

(+) 800° (606) εἷς βγ + G (unus) Procl.] *om.* Y

(+) 805° (610) ἐν ἀπραγίᾳ βγ + G (sine opere) Procl.] ἐν εὐπραγίαις Y

(-?) 809° (614) συμβαίνοι *Robb. tacite, Hüb.*] συμβαίνει M S + G (accidit), *cf.* Heph.: συμβαίνειν Y D γ (*Boer*)

(-*) 818° (620) ἐν ταῖς ἀφέσεσιν βγ] ταῖς ἀφέσεσιν Y + G (dimissionibus) Heph.

(+) 821° (621) τοὺς σωματικῶς βγ + G (eos qui corporaliter) Procl.] τοὺς σωματικοὺς Y Heph.

(-*) 834° (632-3) τῷ ... συνεγγισμῷ βγ] ἢ κατὰ τὸν ... συνεγγισμῶν Y: κατὰ τὸν ... συνεγγισμὸν (*coni. Robb.*) + G (secundum eam que ... propinquitatem), *cf.* Procl. — We propose to write κατὰ τὸν ... συνεγγισμὸν with *coni. Robb.* + G.

(+) 841° (637) ἕν βγ + G (unum)] *om.* Y Procl.

(+) 849° (643) (τῶν) χρόνων Y G° S + G (temporum) Heph. Procl.] καθολικοὺς μάλιστα Γ D M γ

(-*) 854° (646) ταῖς τοῦ Ἄρεως Y M S] τοῦ Ἄρεως D γ + G (Martis) Heph. (*Robb.*)

(§) 856° (647) ὡς Heph. + G (Tamquam)] καὶ ὡς *codd.*

(-*) 865-6° (654) ἐπ' αὐτοῖς τοῖς τῆς γενέσεως τόποις D (*Boer*)] ἀπ (ἐπ') αὐτοῖς τῆς γενέσεως τοῖς τόποις Y M S: ἐπ' (ἀπ'?) αὐτῆς τῆς γενέσεως τοῖς τόποις Γ G° γ + G (ab [*sic*] ipsa natiuitate locis) (*Robb.*), *cf.* Procl. — Write ἐπ' (*or* ἀπ') αὐτῆς τῆς γενέσεως τοῖς τόποις with Γ G° γ + G.

(-*) 875° (662) φάσεις βγ (*Robb.*)] ἀφέσεις Y + G (dimissiones)

(+) 883° (667) παραίτιοι Y S + G (causatiui) Heph.] αἴτιοι Γ D M γ Procl.

(+) 888° (670) ῥέπῃ βγ + G (tendat)] ῥέπει Γ: *om.* Y Heph.

(+) 888° (671) πολὺ δὲ πλέον (πλεῖον *Hüb.*) Γ M γ + G (multo autem magis) Procl.] καὶ ἐπὶ πολ(λ)ὺ πλέον Y G°, *cf.* S

(+) 897° (677) τοιοῦτος Y + G (talis) (*Robb.*)] τοσοῦτος βγ (*Boer*)

(-*) 898° (678) εἴη βγ] γένοιτο Y + G (fiet)

(-*) 898°-904° τὰ — συγκράσεως *habent* Γ D M^2 Procl.] C2- C11 *alteram conclusionem praebent* Y G° Φ + G

For the *Conclusio* of the text, see IVB 3. 2 and Appendix II.

APPENDIX I
THE ASTROLOGICAL NOTE TRANSLATED BY MOERBEKE

The Oxford manuscript contains on f. 208vb after the end of the text the following addition (see image p. 152):

> Que sequntur in greco exemplari subiniuncta reperi: *quo mense morietur quis in omni natiuitate*?
>
> <A>[1] sole ad horoscopum quere[2] quot intersunt signa in consequentia, et hiis appone dies contingentis mensis secundum partum et subiunctos dies in canonico, et proice de die quidem a sole, de nocte autem a luna, in consequentia dans unicuique signo diem unam; uerbi gratia methir[3] 4 in 5 hora nocturna 5, horoscopo libra, sole in aquario, luna in geminis; a sole ad horoscopum signa 9; appone ipsius mensis dies 4; fiunt 13; hiis adde appositos aquario in nocturna natiuitate dies 102, et fiunt dies 115; hos proice a luna, hoc est a geminis, dans unicuique signo diem unum. Occurrit in sagittario qui est mensis khiac in quo[4] et morietur, etc.

Aries	Dies	90	Nox	90
Taurus	Dies	96	Nox	84
Gemini	Dies	102	Nox	78
Cancer	Dies	108	Nox	72
Leo	Dies	102	Nox	78
Virgo	Dies	96	Nox	84
Libra	Dies	90	Nox	90
Scorpio	Dies	84	Nox	98 (*mg. corr.* 96)
Sagittarius	Dies	78	Nox	102
Capricornus	Dies	72	Nox	108
Aquarius	Dies	78	Nox	102
Piscis	Dies	84	Nox	96

Explicit Expliceat

The text offers a calculation to determine the month in which a person may die, which is valid for every horoscope, whatever the day of birth may be. The table given above indicates the number of days to be used in the calculation for every

[1] *add.* Hübner
[2] quare *Ox*
[3] *an* mechir *scribendum*?
[4] innquo *Ox*

zodiac sign, depending on whether one is born at night or during the day.[5] One must first calculate the number of zodiac signs between the sign of the sun until the position of the ascendant horoscope (inclusive calculation, counting both the first and the last sign). To this number one should add the number of days of the month when one is born. In the table one will find for each zodiac sign the number to be added to this sum, depending on whether one is born at night or during the day. The final sum should reveal the month of death. An example is given to illustrate the calculus. Take somebody born the fourth day in the month Mechir[6] at the fifth hour 5 minutes[7] of the night, with the sign Libra as ascendant, the sun in Aquarius and the moon in Gemini. From the sign of the sun to Libra we have nine units to which the 4 days of the month should be added, which gives us the number 13. To this number one must add the number found in the table ('in canonico') that corresponds to Aquarius at night, namely 102. The total sum we have, now, is 115. These 115 days must then be counted out starting from the sign Gemini (because this was the sign of the moon at the nativity), regarding each zodiac sign as 1 unit. In this way, at the end of the allotted 115 zodiacal 'days', we will reach the sign Sagittarius in the month Khiac (or Choiac: χοίακ).[8] This is the month in which a person born on the fourth day in the month Mechir, at the fifth hour of the night, Libra sun and moon in the given zodiacal places, will die.

It remains obscure how the data in the table have been determined. The calculation only has a practical use, as is the case with many other handy tables (such as those we know from the Egyptian *Oxyrhynchus* papyri edited by Alexander Jones). The fact that references are made to two Egyptian months may provide an argument that this calculation dates from late antiquity. Not only Ptolemy himself, but also Vettius Valens, Hephaestio and other astrological authors, use the Egyptian names of months. A similar but much shorter version of such a calculation is found in the astrological *Hermetica* edited first by Wilhelm Gundel (see VIII 'De mense in quo natus morietur').[9]

---

5 We are greatly indebted to Wolfgang Hübner for helping us to understand this difficult calculation and for various suggestions to improve the text.

6 The month (μεχείρ) in the Egyptian calendar corresponds in the Julian calendar to 27 January-24 February.

7 For a system of division into a fraction of hours, see Bouché-Leclercq (1899) 496-7, with reference to Vettius Valens.

8 The month χοίακ corresponds in the Julian calendar to 27 november–26 december. For the correspondence between this month and the sign Sagittarius see Jones (1999) 4205, a five-day almanac (A.D. 257/258) where recto, l. 7-8 (p. 294-5) Sagittarius is mentioned under the indication of the month Choiac. See also the correspondence in the table of the Byzantine scholar Libanianus (14th century).

9 Gundel (1936) 28; see now Feraboli (1994) 26 (with reference to Val. IV,28; V,4,1-7).

The addition is introduced by the remark: 'I have found what follows in the Greek *exemplar*'. This note cannot come from the copyist of the Oxford manuscript, as he had no access to a 'Greek *exemplar*'. Only Moerbeke, the translator of Ptolemy's work, could have found the note in his Greek manuscript and decided to translate it. Moreover, the Latin terminology corresponds to the vocabulary used by Moerbeke to translate similar expressions in the text of Ptolemy (e.g. *in consequentia, appone, proice, contingentis, subiunctos, occurrit*). The 'etc' at the end was probably not added by the translator, but found in the Greek text (κ.τ.λ). The presence of the note shows again how scrupulous Moerbeke was in his translation, adding even this surprising text at the end.

APPENDIX II
THE CONCLUSION OF THE *APOTELESMATICA*

In the text tradition one finds two different versions of the general conclusion of Ptolemy's *Apotelesmatica*. Only the opening sentence, which summarizes the last chapter (IV.897°-8°), is common to both versions: ὁ μὲν τύπος τῆς τῶν καιρῶν ἐπισκέψεως τοιοῦτός τις ἂν γένοιτο,[10] 'modus quidem igitur temporum considerationis talis aliquis utique fiet.'

The first version of the conclusion, published by Hübner, is attested in some manuscripts of β (D M^2, *deest* S) and in γ and corresponds to the conclusion of the Proclus Paraphrase. S ends after εἴη (898°) without any conclusion; this was also the original state of M, which ends at ἂν (898°), until another hand (M^2) added the conclusion in the first version. We do not know what text was found originally in V, as the treatise in this manuscript ends abruptly at IV.797°. Hübner uses Γ, which is a copy of V, to replace V in its lost part. This copy has the first version of the conclusion (as in βγ). However, this copy was made when V was already mutilated at the end. What we read as the final part in Γ has been added in another hand, and was probably taken from a manuscript belonging to the βγ class, as several variant readings show. One may, therefore, not simply take the last section in Γ as a replacement of the lost section in V (see Robbins, p. xxi). The second version (*altera conclusio* in Hübner) is found in Y and in other manuscripts from class α; it

[10] There is one difference: γένοιτο *versio* II] εἴη *versio* I. Hübner accidently omits in *versio* II τις after τοιοῦτος.

is the version translated by Moerbeke, and it is also found in the Arabic translation (see below).

Hübner considered the first version as the genuine one (see p. XXXIX) and placed it immediately after Ptolemy's text in the same font (IV.898°-904°). The second version is printed in smaller characters as an appendix to Ptolemy's text: *altera operis conclusio* (p. 393, C2-11). In this Hübner follows Boll–Boer, who adopted the first version as the conclusion of Ptolemy's text, and only gave the text of the second version in the apparatus. No reason is given either by Boll or by Hübner for their preference for version I. Robbins, however, defends the opposite view. At the end of the treatise, he gives both versions (p. 458), but he positions version II first and considers it to be the genuine Ptolemaic text. He gives the reasons for his preference in his introduction (p. xx-xxii). As he argues, the first version, which is written in a much simpler language, is obviously a summary of what we have in the second version. It is difficult to imagine the reverse process, namely that someone would have edited version I so as to produce version II: 'I cannot conceive how anyone (except perhaps Ptolemy) could have reversed the process and evolved the tortuous, crabbed Greek of the latter [i.e. version II] from the comparatively simple language of the former [i.e. version I].' Moreover, the tortuous second version displays the stylistic character of Ptolemy's authentic writing. Robbins points to the presence of many favorite Ptolemaic terms.[11] 'In fact practically every word of the passage (...) is to be paralleled in the *Tetrabiblos*, usually many times; to arrange them in so exact an approximation to Ptolemy's usual style would demand a forger of superhuman ingenuity' (p. XXII, n. 1). Nevertheless, Robbins leaves the question open until the full text tradition has been studied.

In our view there is no reason to keep the question open and to doubt the authenticity of version II. Besides the arguments based on style and vocabulary, there are other decisive arguments from the indirect tradition.

(1) The *altera conclusio* was known to Hephaestio in the early 5th century. In his *Apotelesmatica* (II 26,23), after having quoted a large extract from IV, ch. 10, 683°-896°, Hephaestio briefly summarizes what follows in Ptolemy's general conclusion:

> Ταῦτα μὲν οὖν ὁ Πτολεμαῖος φυσικῶς καὶ καθολικῶς ἐπεκτίθεται περὶ τῆς τῶν καιρῶν ἐπισκέψεως, τῶν δὲ κατὰ μέρος ἐπιβολῶν τῆς ποιότητος τοῦ ἀποτελέσματος ἑκάστου ἀφῆκεν ἡμᾶς καταστοχάζεσθαι ἐκ τῶν καθολικῶν (p. 196,29-197,3).

This clearly corresponds to the beginning of conclusion II:

[11] Only one term poses a problem, Robbins says, χρηματείαις; but as Moerbeke's translation shows, this is an error for πραγματείαις (*negotiis*), and this term is frequently used by Ptolemy.

ὁ μὲν οὖν τύπος τῆς τῶν καιρῶν ἐπισκέψεως τοιοῦτός ἂν εἴη κατὰ τὸν ἁρμόζοντα ταῖς φυσικαῖς πραγματείαις[12] τρόπον. Τὰς δὲ κατὰ μέρος ἐπιβολὰς τῆς ποιότητος.

These parallels prove that Hephaestio read the conclusion as transmitted in Y and Moerbeke, i.e. version II.

(2) The Arabic translation undoubtedly renders the second version of the conclusion. As there is no edition of the Arabic translation, we give the two Latin translations (Plato Tiburtinus and Aegidius de Thebaldis) from the first Renaissance editions.

Hac igitur uia secundum modum naturali rei competentem per res tempus prognosticamur. Viarum autem particularium quibus temporalis futuri qualitates prognosticantur, ob hoc quod inde prouenientium multiformis est collectio necnon grauis enarratio, maxime autem in hoc loco enarrationem proponere idoneum iudicauimus, eo quod ab illius libri primordio opus generalis nature stellarum eiusque cum rebus particularibus ut oportet similitudinem secundum speciei rationalis existimationem in commixtione horum explanare proposuimus. Rebus itaque natiuitatum generaliter explicatis hoc in loco huic libro finem imponere non incongruum existimamus. (Tiburtinus)

Hec est uia per quam debemus perquirere res temporum secundum naturam. Tamen uias particulares, per quas inquirere habemus statum accidentium particularium per id quod acciderit sunt narrare multis uicibus et multis manieribus et graues ad doctrinandum, et maxime in hoc loco, quoniam propositum huius libri est ostendere opera particularium generaliter et quomodo cum rebus particularibus concordant secundum estimationem in modo demonstrandi huius rei commixtiones. Et postquam in modo uniuersali perfecimus iudicium natiuitatum finem in hoc loco huic libro conueniens extimamus. (Aegidius)

(3) The first version of the conclusion is identical with the conclusion of the Proclus Paraphrase. As we know, the author of this Paraphrase intends to reformulate Ptolemy's often tortuous phrases into a more accessible argumentation. This is exactly what we find in the reformulation of conclusion II in the Paraphrase. Moreover, the author of the Paraphrase undoubtedly used himself version II in writing his own conclusion. The first sentence of Ptolemy's conclusion is paraphrased as follows: κατὰ μὲν δὴ πραγματείαν σώζουσαν φυσικὴν ἀκολουθίαν οὕτως ἐπισκεφθήσονται οἱ καιροί. τὰ δὲ εἴδη τῶν ἀποτελεσμάτων ... (Procl. 293,-6 sqq.). It is immediately clear that the paraphraser used the second version of the Ptolemaic conclusion as preserved in Y and in Moerbeke: ὁ μὲν οὖν τύπος τῆς τῶν καιρῶν ἐπισκέψεως τοιοῦτός ἂν εἴη κατὰ τὸν ἁρμόζοντα ταῖς φυσικαῖς πραγματείαις

[12] The transmission of conclusion II is problematic. We corrected the text using Moerbeke's translation, which is for this section the most reliable witness. We write φυσικαῖς πραγματείαις on the authority of G *naturalibus negotiis*. See above in IVB, p. 83.

τρόπον. We have a clear indication in what follows that the paraphraser was working with conclusion II. See the terminology τῶν ἀποτελεσμάτων ... κατὰ διέξοδον ... τῶν ἀστέρων ἡ ποιητικὴ δύναμις ... κατὰ τὸ ἀκόλουθον ἐφαρμόζεσθαι δύναται ... εὐστόχως ... τοῦ μαθηματικοῦ.

To conclude: version II is certainly the authentic conclusion of Ptolemy's *Apotelesmatica.* It is attested in Y, in Moerbeke's translation, in the Arabic translation, and indirectly in Heph. and Procl. Moreover, it is composed in a style, and deploys a vocabulary, characteristic of Ptolemy. Version I of the conclusion is only attested in one member of β, namely in D, and in Γ and γ depending on it. Most likely, there was no conclusion in the hyparchetype of class βγ after ὁ μὲν τύπος τῆς τῶν καιρῶν ἐπισκέψεως τοιοῦτός τις ἂν εἴη (as is still the case in M *ante corr.* and S). The copyist-editor of the text in D (or of his exemplar) decided to complement what was lacking at the end with the conclusion taken from the Proclus Paraphrase. Hübner wrongly took this version as the authentic conclusion. We propose to replace this version by Hübner's *altera conclusio.*

EDITORIAL PRINCIPLES

1. TEXT

1.1. Until II.619 (814°), the text is transmitted in two manuscripts, *Ve* und *Ox*; from II.620 (815°) on, only in *Ox*. Therefore, we have to differentiate the principles for the constitution of the text and the composition of the apparatus in the two parts (A: I-II.619; B: II.620-IV). In the first part, we generally prefer the readings of *Ve* to *Ox*, as *Ve* is the more reliable copy. In the second part, we can only rely on *Ox*, which presents a text affected by many copyist errors. Therefore, we often have to intervene in the text with corrections. For a few passages, we could also rely on Bate's testimony.

1.2. *Numbering.* In the external margins of the text, one finds the line numbering starting in each book from 1. Only every fifth line is numbered. In the interior margins, one finds the line numbering of Hübner's edition of the Greek text: only every fifth line is numbered. The sign (ˈ) in the Latin text indicates where every fifth line of the corresponding Greek text begins. As the Latin translation often radically restructures the Greek text, the beginning of a new line can only approximately be indicated. To avoid confusion between the two numbering systems, the Greek numbers are given in a smaller font than the Latin. Within each chapter, between square brackets, one finds the numbering of the paragraphs as found in Hübner: [1], [2], [3],... We follow the chapter numbering Moerbeke adopted from his Greek model. Since the numbering of the chapters in book III, from chapter 4 on, differs from Hübner's system by one unit, from that chapter on we have added Hübner's chapter numbering between square brackets in the chapter titles.

To make it easier to follow an argument with an enumeration (of a series of planets, for instance), we print, following Boll and Hübner, the enumerated terms in a spaced format. See, for instance, I, chapter 9, IV, chapter 4, 9 and 10.

1.3. *Orthography.* In this edition, we adopt a normalised medieval orthography following Moerbeke's own practise, as we know it from the Archimedes-autograph (*Ottob. lat.* 1850).[1] This is somewhat less evident in this edition, since we have only two indirect copies of the autograph in the first part, and in the second part

[1] On these principles of orthography, see the comments of G. Vuillemin-Diem in her introduction to the edition of the *Meteorologica* (Aristoteles Latinus X 2.1), 2008, p. 380-7.

only *Ox*, which, as we can see in the first part by comparison with *Ve*, often has a peculiar spelling. We therefore had to standardize the spelling. We always spell diphthongs *–ae* and *–oe* as *–e*, make no distinction between *–v-* for consonants and *–u-* for vowels (*V* is only used as capital letter). We retain the particular forms of the declension of the pronoun *hic* (*hii, hee, hiis*). We assimilate prefixes and always write *imp-* (never *inp-*), *imm-* (never *imn-*), *tamquam, numquam, eundem, quondam*, and we follow the classical usages in the word endings on *–tio*(*a*) (*um*) / *-cio* (*a*) (*um*) without indicating variants (*c* or *t*) in the apparatus. We keep, however, the 'medieval' spelling of *habundo/-dantia, as*(*con*)*secuntur, oportunus, ydoneus, ydemptitas, autumpnus, condempnatio, contempno, dampnosus, sompnus*. We write always *exti-* (not *exsti-*). We keep the variation in *optineo* (*obt-*), *hyemalis* (*yem-*), *hyems* (*hiems, yems*), *peri(y)odus* , *regyratio/giratio, symptoma, simp-, yris (hyrys), ystoria, hist-*. For geographical, mythological, historical names we keep the variants of the manuscripts. See for instance Ἀχαΐα: *Acaia, Acaya, Achaya*; Ἰδουμαία: *Idumea, Ydumea*; Κυκλάδες: *Ciclades, Kiklades, Kyklades*; Χηλαί: *Chele, Khele*. We respect variations in the spelling of transliterations such as ἐρυσίπελας *erysipela* (*-pila*), βορραπελιωτικός: *borafelioticus, boreapelioticus, boreafelioticus*; νοταπηλιωτικός: *notapelioticus, notafelioticus, notafilioticus*.

1.4. *Corrections*. As both manuscripts do not depend directly on Moerbeke's autograph, but rather on a common exemplar affected by errors (see chapter I, 3), we have often had to make corrections in the transmitted text. These corrections are particularly necessary in part B, where we have only one manuscript and the less reliable of the two. Most errors in part B are due to the negligence of the copyist of *Ox*, but some may come from the common exemplar *VeOx*. As in part A, there are in part B – but much more frequently – incorrect readings of abbreviations, confusions of characters (*o/a e/a qui/que*), and simple omissions. In making our corrections, we compared the Latin text with the Greek tradition; we took into account Moerbeke's translation methodology and terminology in unproblematic passages of this and other of his translations; and we only proposed corrections that could be explained paleographically. For many corrections, we could rely on conjectures made or suggested by Anthonis and Bossier.

1.5. *Omissions*. The translation omits words, or even whole passages, which are uniformly attested in the Greek tradition. This makes it implausible that these omissions were found in the Greek model of the translation. It is possible that Moerbeke omitted himself, by inadvertence, certain passages or words. Most omissions, however, are found in part B where only *Ox* is available, a copy of mediocre quality. They are therefore most probably errors made by the *Ox* copyist or his model, rather than by Moerbeke.

In part A, we complement omitted words between angular brackets, and indicate the editor's intervention in the apparatus. Examples: II.35: in the text: <hoc quidem>; in the Latin apparatus: hoc quidem *add.*] *om. codd.*; II.102 in the text: <regionis>; in the Latin apparatus: regionis *add.*] *om. codd.* There are two longer omissions in part A. In II.504-5, we add what we suppose was in Moerbeke's translation: <celi supra terram, circa sacra et reges et mediam etatem; que autem apud> and mention the editorial intervention in the Latin apparatus: 504-5 celi — apud *add.*] *hom. om. codd.* (on this passage, see above chapter I, 3). In II.542, because it is uncertain whether the omission occurred in the archetype of the Latin translation, or in the Greek tradition, the omitted text is found in the Greek-Latin apparatus: 542 aeris] + ψύχη φοβερά, παγώδη καὶ ὁμιχλώδη καὶ λοιμικά *codd.* (*cf.* βγ). In the Latin apparatus, the possibility of an omission is only indicated. 542 *an post* aeris *quaedam exciderunt.*

In part B, where we have only manuscript *Ox*, we always complement small omissions, adding the missing words or prefixes between angular brackets, without indicating our editorial intervention in the apparatus. In the case of longer omissions, which are mostly due to homoioteleuton in the Latin tradition, we indicate the omission in the text with angular brackets, adding only the corresponding term of the homoioteleuton. We always mention the omitted Greek text in the Greek-Latin apparatus and, when possible, provide in the Latin apparatus a translation of the omitted text in Moerbeke's style. See IV.230-1: in the text: in trigonalibus <- - - in trigonalibus> inuicem habent domos; in the Latin-apparatus: *an post* trigonalibus* *verba exciderunt fere haec*: <signis inuicem habent exaltationes, ad Saturnum autem secundum faciem senioris status, quoniam iterum in trigonalibus>? IV.238: in the text: bene stabiles < - - - instabiles> et nociuos; in the Latin apparatus: *an post* stabiles* *verba exciderunt fere haec*: <facit conuictus, accedente autem Mercurio et proficuos, accedente autem et Marte instabiles> ?

1.6. *Deletions.* There are a number of superfluous additions of *et* in part A in both manuscripts, and in part B in *Ox*, which must come from copyist's errors in their common exemplar (see Chapter I,3). We delete them in the text with square brackets [et] and, in part A, mention the intervention in the apparatus.

2. LATIN APPARATUS

2.1. The apparatus is negative. We first give the adopted reading before square brackets, followed by the variant reading(s). The abbreviation *codd.* refers in part A to the manuscripts *VeOx*. The following orthographical variants are generally not mentioned:

Ve often has simplifications of double consonants (for instance, *Ve* always has *posunt, posit* etc. instead of *possunt, possit,* etc. I.100; I.140; see also *coasumpta* I.146, *asueti* I.95), or duplications of simple consonants (for instance *occassionem* I.126; *defficere* I.131; *defficiat* I.131). — Often also *equinot-* instead of *equinoct-, e.g.* II.57, 493, 498.

Ox reads generally (from book II on) *contra* instead of *circa* (II.82 circa] contra *Ox, et sic deinceps, fere semper*; III.25 circa] contra *Ox, et sic deinceps fere semper*). — *Ox* has also duplications of simple consonants (for instance, *tolleratiuus* instead of *toleratiuus* III.936, 943, III.1165 illegitima] -gitti- *Ox*, 1169 legitimas] -gitti- *Ox*) or inversely III.835 terrestria] terest- *Ox*, 836 corrosiones] coro- *Ox*, 1087 sollertes] soler- *Ox*, III.35 annotabiliter] anota- *Ox*, 270 communicauerit] comuni- *Ox*, 713 commensuratos] comensu- *Ox*, 714 commensurationem] comensu- *Ox*

As *Ox* is in part B the only witness of the text, we mention in the apparatus all its readings, including meaningless orthographic variants, which do not correspond to the adopted text. However, we exclude meaningless variants *ante correctionem* (for instance III.842 incurabiles] -bibes *Ox*[ac]) meaningless connections of separate terms (such as III.964 ad conuictus] adconuictus *Ox*, III.1096 ad multiformius] admultiformius *Ox*, III.1186 ad uulgabilius] aduulgabilius *Ox)* or, inversely, separations of connected terms (such as III.769 defluxum] de fluxum *Ox*).

2.2. *Editorial corrections.* In part A, editorial corrections are always mentioned in the Latin apparatus as *scr.* or *add.* or *del.* In part B, where we have only *Ox*, the numerous editorial corrections are not explicitly mentioned in the Latin apparatus as *scr.* or *add.* or *del.* We always give, however, the adopted (corrected) reading before the angular brackets, and after the brackets the reading as found in *Ox*. Small additions and deletions are indicated only in the text (not in the apparatus) by the use of angular or square brackets.

3. GREEK-LATIN COMPARATIVE APPARATUS

3.1. The translation is primordially compared with the Greek manuscript V and only subsidiarily with the other Greek manuscripts quoted in Hübner's apparatus. But in each critical unit of the apparatus, the readings of all manuscripts quoted by Hübner are explicitly or implicitly ('*cf.*') given. One finds the sigla of the relevant manuscripts on each page at the beginning of the apparatus between parentheses (see below). The abbreviation *codd, cett.* refers always to the manuscripts listed on each page. As we made no new collations of the Greek manuscripts (except for L), all readings are based on Hübner's apparatus.

In cases of important variations between the manuscripts, we indicate the reading adopted by Hübner (who mostly follows Boll/Boer); in special cases also Boll, Boer, Robbins. Where relevant, we also refer to the indirect tradition: Hephaestio, the anonymous commentary, and the Proclus Paraphrase.

(*Hüb.*) = Hübner only or with Boll (respectively in III-IV Boer),with Robbins or with both.

(*Boll*) (*Boer*), (*Robb.*) are explicitly mentioned for important variant readings only.

c, Co, Co le, Procl., Heph. are mostly quoted according to Hübner's indications. We made, however, a new collation of the Proclus Paraphrase. See *Conspectus siglorum* for the editions.

3.2. The apparatus is generally negative (2.1), but also partially positive (2.2)

3.2.1. We indicate in the apparatus *all deviations* of G with respect to V, excluding however deviations due to Moerbeke's translation method. To reduce the apparatus we do not include some meaningless errors of V, which Moerbeke, even if he had seen them in his Greek model, would never have translated or even have corrected himself. Some examples:

I.38 (44°) faciemus: ποιησόμεθα *cett.*] -σώμεθα **V** — 81 (98°) insensibilius: ἀσημότερον *cett.*] ἀσυμ- **V** — 219 (278°) et uniuersaliter (*cett.*)] *iteravit* **V** — II.20 (21°) exposita sint: ἐφοδευέσθω *cett.*] ἐφοδεύσθω **V** — III.207 (263°) addextrantibus in bono: ἐπὶ καλῷ δορυφορήσασι *cett.*] ἐπικαλωδορήσασι **V** — IV.178 (226°) considerandum: σκεπτέον *cett.*] σκοπτέον **V** — 239 (311°) coetaneas: ὁμήλικας **α D M γ** (*Hüb.*)] ὁμίληκας **V S** — 413 (540°) habitationum: οἰκήσεων **β** (*Hüb.*)] οἰκείσεων **V**, *cf. cett.* — 427 (559°) per difficiles nauigationes: διὰ δυσπλοίων *cett.*] διὰ δυπλοίων **V** — 428 (560°) precipitia: κρημνισμῶν **β γ** Heph. (*Hüb.*)] κριμνισμῶν **V**, *cf.* **Y** – 571 (752°) nociuus appetitus: λύσσα *cett.*] λήσσα **V**

Other exceptions are those cases in which one finds two alternative readings in Greek, only one of which is present in V, but both of which may correspond to the same term in Latin. The most frequent cases are variants such as συσχηματισμός/σχηματισμός and the corresponding verbal forms. When the form συσχημα- is attested in all Greek manuscripts, the normal equivalent is *configura-*, but one find also *figura-*. The same holds for σχηματ-: the normal and dominant equivalent is *figura-*, but *config-* is also found. In the Greek-Latin index we give in such cases the reading of V (as it is plausible that G corresponds to it), but the reference is indexed with *.h.* in superscript to indicate that another reading may be found in Hübner's apparatus. For the same reason, we have not mentioned in the apparatus but only in the index such readings as III.491 (652°) occursus: ὑπαντήσεις **α β γ** (*Hüb.*): ἀπαντήσεις **V**; 737 (980°) qui elevantur: ἐπαναφερομένας **α D M γ** (*Hüb.*): ἀναφερομένας **V S** Heph, Co le («*fort. recte*»); 573 (766°) ascendens: ὡροσκόπος **V**: ὡροσκοπῶν *cett.* (*Hüb.*). A similar case arises with the variant readings of οἰκοδεσποτεία/συνοικοδεσποτεία. In I.705 (936°) συνοικοδεσποτεία is uniformally attested in the Greek tradition and translated by Moerbeke as *dominatio*. In two other passages, however, it is uncertain whether *dominatio* stands for οἰκοδεσποτεία or for συνοικοδεσποτεία: I.702 (931°) dominationem: οἰκοδεσποτείαν **α β γ**, *Hüb.*] συνοικο- **V L**, *Boll*; I.739 (982°) dominationem: οἰκοδεσποτείαν **V α**] συνοικο- **L β γ**, *Hüb.* These readings are therefore mentioned in the apparatus.

3.2.2. The apparatus is also partially positive. We mention readings of V, which are confirmed by Moerbeke's translation, when Hübner does not adopt this reading in his edition. The apparatus is also positive for titles of chapters and numbering of chapters: they are always mentioned, even when they correspond to V and to Hübner's edition.

3.3. The text of V is interrupted at IV.602 (797°-8°). For this last section, we compare the translation with Y (in this part the main witness for α), β =DMS and γ =ABC. We also use Γ, which is an apograph of V — but in this section related to β — and, where relevant, the two other witnesses of α, namely G° Φ.

3.4. The authentic version of the conclusion which was translated by Moerbeke, is only preserved in the α family. We compare therefore the translation with Y G° Φ, which belong to that family.

3.5. According to the presence or absence of the different Greek manuscripts, the manuscript evidence is different in different sections of the text. The list of the sigla at the beginning of each page changes in the following way (the place of change in the text is given, if necessary, inside the apparatus). For the sigla see *conspectus siglorum*:

I.1 ff.	V (α =YΣ β =DMS γ =ABC)
I.187 ff. (235°)	V (Y β =DMS γ =ABC)
I.303 ff. (383°)	V (Y β =DMS γ =ABC, *excerpta* L)
I.404 ff. (519°]	V (α =YΣ β =DMS γ =ABC, *excerpta* L)
I.769 ff. (1020°)	V (α =YΣ β =DMS γ =ABC)
I.879 ff. (1140°)	V (α =YΣ β =DMS γ =ABC, *excerpta* L)
II.1 ff.	V (α =YΣ β =DMS γ =ABC)
II.709 ff. (925°)	V (α =YΣ β =DMS γ =ABC, *excerpta* L)
II.808 ff. (1057°)	V (α =YΣ β =DMS γ =ABC)
III.1 ff.	V (α =YΣ β =DMS γ =ABC)
III.919 ff. (1222°)	V (Y β =DMS γ =ABC)
III.984 ff. (1295°)	V (α =YΣ β =DMS γ =ABC)
IV.1 ff.	V (α =YΣ β =DMS γ =ABC)
IV.252 ff. (329°)	V (Y β =DMS γ =ABC)
IV.603 ff. (797°)	Y, β (= DMS), γ (= ABC) (Γ)
IV.678-686 (C2°-11°)	Y G° Φ

3.6. When the Latin translation deviates from V or even from the whole Greek tradition, and this deviation is probably not due to a different reading in Moerbeke's Greek model, but rather to his translation method, we put the reference in the apparatus between parentheses.

3.7. Omissions of passages of the Greek text are always indicated (with the exception of small omissions corrected in the text itself, such as <in>experientia). Omissions that may be due to Moerbeke's translation method (such as the omission of particles) are only found in the indexes, not in the apparatus.

3.8. Additions of terms that most likely come from the translator have been collected in supplement III of the indices (often repetitions of nouns where the Greek article may have this function). Only in cases where an Latin addition could have come from the Greek exemplar – a scholion, for instance – is it mentioned in the apparatus.

3.9. *Bate*. For a few passages we have quotations in Bate (see Chapter III). In the apparatus we only mention readings in Bate that correct errors in *Ox*.

BIBLIOGRAPHY

PTOLEMAEUS, *APOTELESMATICA*: EDITIONS, TRANSLATIONS, COMMENTARY

Camerarius 1535: *Claudii Ptolemaei Pelusiensis libri quatuor compositi Syro fratri.* [...]. *Traductio in linguam Latinam librorum Ptolemaei duum priorum* [...]. Norimbergae: apud Ioannem Petreium.

Camerarius 1541: *Claudii Ptolemaei mathematici operis libri quatuor* [...] *Ioachimo Camerario interprete.* In *Claudii Ptolemaei Pelusiensis Alexandrini Omnia quae extant Opera*, ed. Hieronymus Gemusaeus. Basileae: apud Henricum Petrum, 429–500.

Gogava 1548: *Cl. Ptolemaei Pelusiensis mathematici Operis quadripartiti in latinum sermonem traductio, adjectis libris posterioribus, Antonio Gogava interprete.* Lovanii: apud P. Phalesium ac M. Rotarium.

Melanchthon 1553: *Cl. Ptolemaei De praedictionibus astronomicis cui titulum fecerunt Quadripartitum, Graece et Latine, libri IIII Ph. Melanthone interprete.* Basiliae: apud J. Operinum.

Robbins, F. E. 1940: *Ptolemy, Tetrabiblos.* Cambridge (M)–London (*The Loeb Classical Library*).

Boll, F. – Boer, Æ. 1957: *Claudii Ptolemaei opera quae exstant omnia* III. 1: Ἀποτελεσματικά. Lipsiae 1940, editio correctior 1954 (= 1957).

Hübner, W. 1998: *Claudii Ptolemaei* Ἀποτελεσματικά, *post F. Boll et Æ. Boer secundis curis* ed. Stuttgart–Leipzig: Teubner.

Charvet, P. 2000: *Ptolémée. Le livre unique de l'astrologie. Traduction et thèmes de naissance.* Paris.

Co 1559: Εἰς τὴν τετράβιβλον τοῦ Πτολεμαίου ἐξήγησις ἀνώνυμος. *In Claudii Ptolemaei quadripartitum enarrator ignoti nominis, quem tamen Proclum fuisse quidam existimant.* Basileae 1559, 1-180.

Procl. 1635: *Procli Diadochi paraphrasis in Ptolemaei libros IV de siderum effectionibus a Leone Allatio e Graeco in Latinum conversa.* Lugduni Batavorum: ex officina Elzeviriana.

Aegidius de Thebaldis 1493: *Liber quadripartiti Ptholemei ; Centiloquium [ejusdem]. Centiloquium Hermetis.* Venetiis: per Bonetum Locatellum.

Plato Tiburtinus 1484: *Liber Ptholomei quattuor tractatuum cum centiloquio eiusdem Ptholomei et commento Haly.* Venetiis: per Erhardum Ratdolt de Augusta.

Plato Tiburtinus. 1533. *Cl. Ptolomaei Pheludiensis Alexandrini Quadripartitum.* Basileae: per Johannem Hervagium.

WORKS QUOTED IN THE INTRODUCTION

Acerbi, F. (forthcoming): 'Une topographie du *Vat. Gr.* 1594' in D. Bianconi – F. Ronconi (eds.), *La 'collection philosophique' face à l'histoire.* Leiden.

Anthonis, L. 1984: *Tekstkritisch en lexicologisch onderzoek over een anonieme Latijnse vertaling van Cl. Ptolemaeus' Tetrabiblos,* unpublished master thesis. KU Leuven.

Anthonis, L. 1989: '*Iudicialia ad Syrum.* Une traduction de Guillaume de Moerbeke du *Quadripartitum* de Cl. Ptolémée' in Brams–Vanhamel (1989), 253-255.

Beullens, P. 2014: 'True colours: the medieval Latin translations of *De Coloribus*' in De Leemans (2014), 165-201.

Björnbo, A. A. 1909: 'Die mittelalterliche lateinische Übersetzungen aus dem Griechischen auf dem Gebiete der mathematischen Wissenschaften', *Archiv für die Geschichte der Naturwissenschaften und Technik,* 1, 385-394.

Boll, F. 1894: 'Studien über Claudius Ptolemäus. Ein Beitrag zur Geschichte der griechischen Philosophie und Astrologie', *Jahrbücher für classische Philologie,* Suppl. 21, 51–244.

Boll, F. 1899: 'Beiträge zur Ueberlieferungsgeschichte der griechischen Astrologie und Astronomie ', *Sitzungsberichte der philosophisch-philologischen und der historischen Classe der k. b. Akademie der Wissenschaften zu München,* 1, 77–140.

Bossier, F. 1989: 'Méthode de traduction et problèmes de chronologie' in Brams–Vanhamel (1989), 257-295.

Bossier, F. 1994: 'Le problème des hapax legomena et composés grecs dans les traductions latines du *Tetrabiblos* de Cl. Ptolémée' in Ch. -M. Ternes – M. Mund-Dopchie, *Méthodologie de la traduction: de l'antiquité à la renaissance.* Luxembourg, 111-151.

Bouché-Leclercq, A. 1899: *L'astrologie grecque.* Paris.

Brams, J. – Vanhamel, W. (eds.) 1986: *Guillaume de Moerbeke. Recueil d'études à l'occasion du 700e anniversaire de sa mort.* Leuven.

Caballero-Sánchez, R. 2013a: 'Historia del texto del Comentario anónimo al *Tetrabiblos* de Tolomeo', *MHNH. Revista internacional de investigación sobre magia y astrología antiguas*, 13, 77-198.

Caballero-Sánchez, R. 2013b: 'El Comentario anónimo al *Tetrabiblos* de Tolomeo. Edición crítica y traducción castellana de los escolios metodológicos del libro I (in Ptol. Tetr. 1. 1. 1-1. 3. 1)', *MHNH. Revista internacional de investigación sobre magia y astrología antiguas*, 13, 221-258.

Cordonier, V. – Steel, C. 2011: 'Guillaume de Moerbeke, traducteur du *Liber de Bona Fortuna* et de l' Éthique *à Eudème*' in A. Van Oppenraay (ed.), *The Letter before the Spirit*, Leiden, 415-416.

De Leemans, P. (ed.) 2014: *Translating at the Court. Bartholomew of Messina and Cultural Life at the Court of Manfred, King of Sicily*. Leuven.

Edwards, D. R. 1984: *Ptolemy's* Περὶ ἀναλήμματος. *An annotated Transcription of Moerbeke's Translation*. Dissertation Brown University.

Feraboli, S. 1994: *Hermetis Trismegisti De triginta sex decanis*, (*Hermes Latinus* IV.1 = *Corpus Christianorum, Continuatio Mediaevalis* 144). Turnhout.

Gundel, W. 1936: *Neue astrologische Texte des Hermes Trismegistos*. München.

Haskins, Ch. 1924: *Studies in the History of Mediaeval Science*. Cambridge.

Heiberg, J. L. 1898–1903: *Claudii Ptolemaei opera quae exstant omnia*. Vol. I. *Syntaxis mathematica*. Leipzig: Teubner.

Heiberg, J. L. 1907. *Claudii Ptolemaei opera quae exstant omnia*. Vol. II. *Opera astronomica minora*. Leipzig: Teubner.

Heiberg, J. L. 1910: 'Eine mittelalterliche Übersetzung der Syntaxis des Ptolemaios', *Hermes*, 45, 57-66.

Heiberg, J.L. 1977: *Euclidis Elementa*. Vol. V Pars 1: *Prolegomena Critica, Libri XIV-XV, Scholia in libros I-V*. Leipzig: Teubner.

Heilen, S. 2010: 'Ptolemy's Doctrine of the Terms and Its Reception', in A. Jones (ed.), *Ptolemy in Perspective, Use and Criticism of his Work from Antiquity to the Nineteenth Century*, (*Archimedes* 23). London–New York.

Hernández, F. J. – Linehan, P. 2004: *The Mozarabic Cardinal: the life and times of Gonzalo Pérez Gudiel*. Florence.

Jones, A. 1999: *Astronomical Papyri from Oxyrhynchus*. Philadelphia.

Lammert, F. 1952: *Claudii Ptolemaei opera quae estant omnia*. Vol. III. 2. περὶ κριτηρίου καὶ ἡγεμονικοῦ. *De iudicandi faculate et animi principatu*. Leipzig: Teubner.

Macray, W. D. 1883/ 1999: *Catalogue of the Bodleian Digby Collection* with additional notes by R. W. Hunt (†) and A. G. Watson (1999) (*Bodleian Library quarto catalogues. 9. Digby manuscripts*)

Minio-Paluello, L. 1947: 'Guglielmo di Moerbeke traduttore della *Poetica* di Aristotele', *Rivista di filosofia neo-scolastica* 39,1-17 [reprinted in *Opuscula. The Latin Aristotle*, Amsterdam, 1972, 40-56].

Molland, G. 2006: 'Bredon, Simon (d. 1372)', *Oxford Dictionary of National Biography*, Oxford, online edn. May 2006

Nybakken, O. E. 1959: *Greek and Latin in Scientific Terminology*, Iowa State University Press.

Pingree, D. 1971: 'The astrological School of John Abramius', *Dumbarton Oaks Papers*, 25, 189-215.

Pingree, D. 1973-4: *Hephaestionis Thebani Apotelesmaticorum libri tres*, 2 vols, Leipzig: Teubner.

Poulle, E. 1972: 'Henry Bate of Malines', *Dictionary of Scientific Biography*, VI, 272-275.

Ronconi, F. 2012: 'La collection brisée. La face cachée de la 'collection philosophique': les milieux socioculturels' in P. Odorico (ed.), *La face cachée de la littérature Byzantine. Le texte en tant que message immédiat.* Paris, 137-158.

Snedegar, K. 1999: 'Simon Bredon, a Fourteenth-Century Astronomer and Physician', in L. Nauta – A. Vanderjagt (eds.), *Between Demonstration and Imagination. Essays in the History of Science and Philosophy presented to John D. North.* Leiden, 285–309.

Steel, C. 1986: 'Guillaume de Moerbeke et saint Thomas' in Brams–Vanhamel (1986) 57-82.

Talbot, C.H. 1962-3: 'Simon Bredon (c. 1300-1372): physician, mathematician and astronomer', *British Journal for the History of Science*, 1, 19-30.

Thorndike, L. – Kibre P. 1963: *A Catalogue of Incipits of Mediaeval Scientific Writings in Latin.* Revised and augmented edition. Cambridge, Mass.

Thorndike, L. 1958: 'Notes upon Some Medieval Latin Astronomical, Astrological and Mathematical Manuscripts at the Vatican, Part II', *Isis*, 49, 34-49.

Torrell, J.-P. 2005: *Saint Thomas Aquinas: the person and his work.* Volume 1. Washington.

Vuillemin-Diem, G. 2014: 'Revision der *Translatio Bartholomaei* oder Neuübersetzung? Zu dem Fragment von *De coloribus* des Wilhelm von Moerbeke', in De Leemans (2014) 203-247.

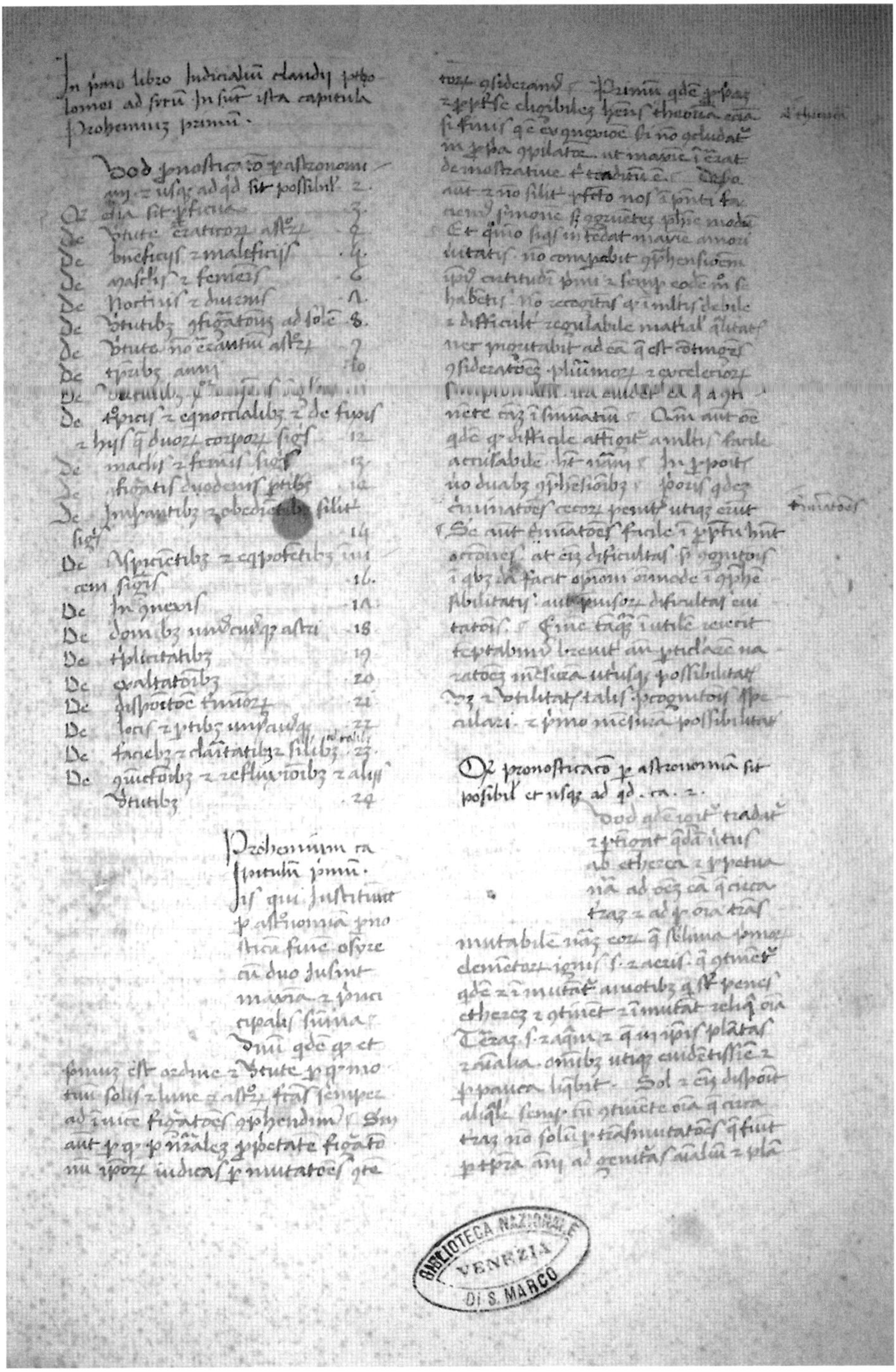

Venezia, Biblioteca Nazionale Marciana, Lat. xiv*, 242 (= 4295), fol. 1r. Reproduced by permission of the library.*

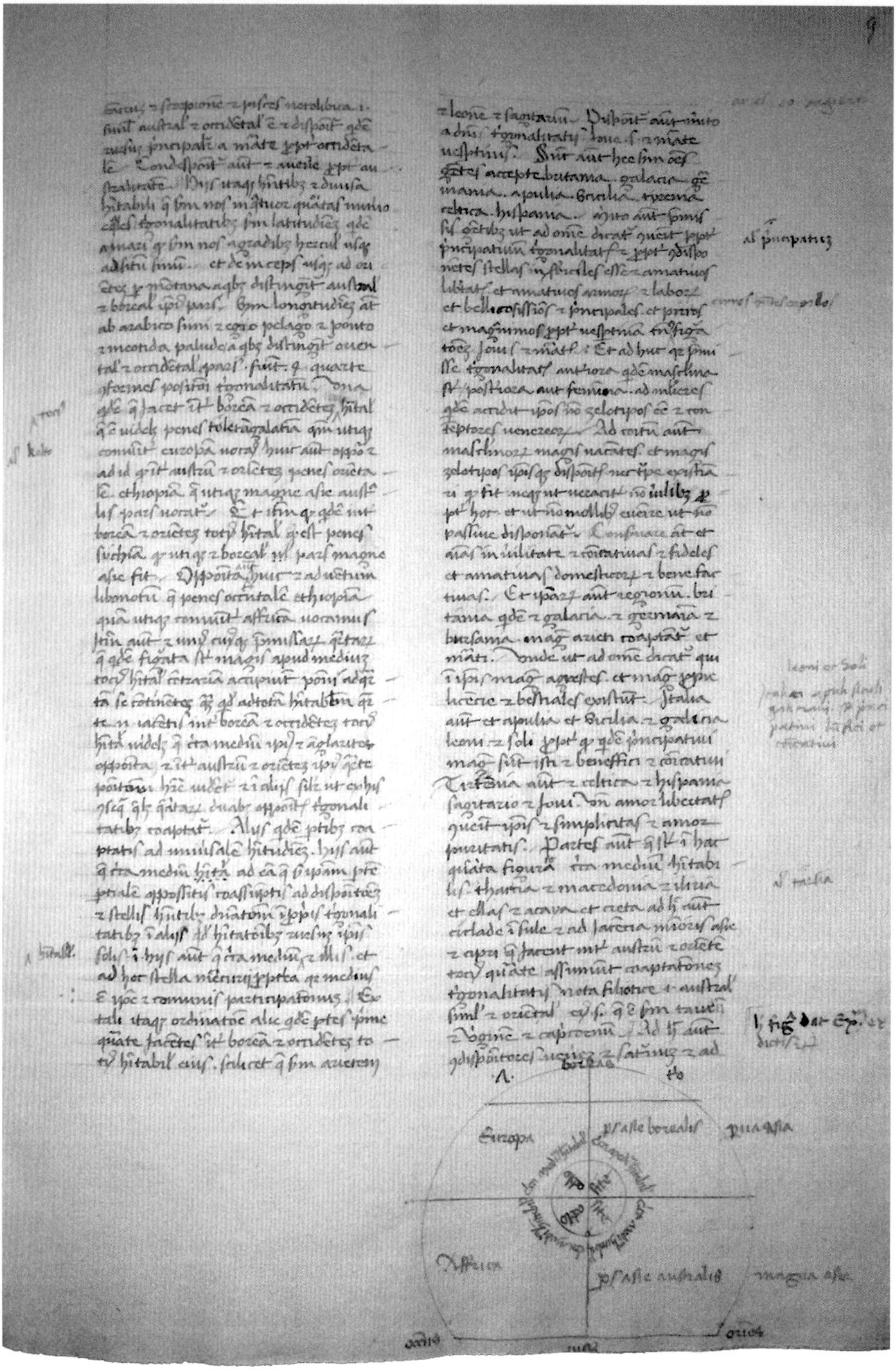

Venezia, Biblioteca Nazionale Marciana, Lat. XIV*, 242 (= 4295), fol. 9r. Reproduced by permission of the library.*

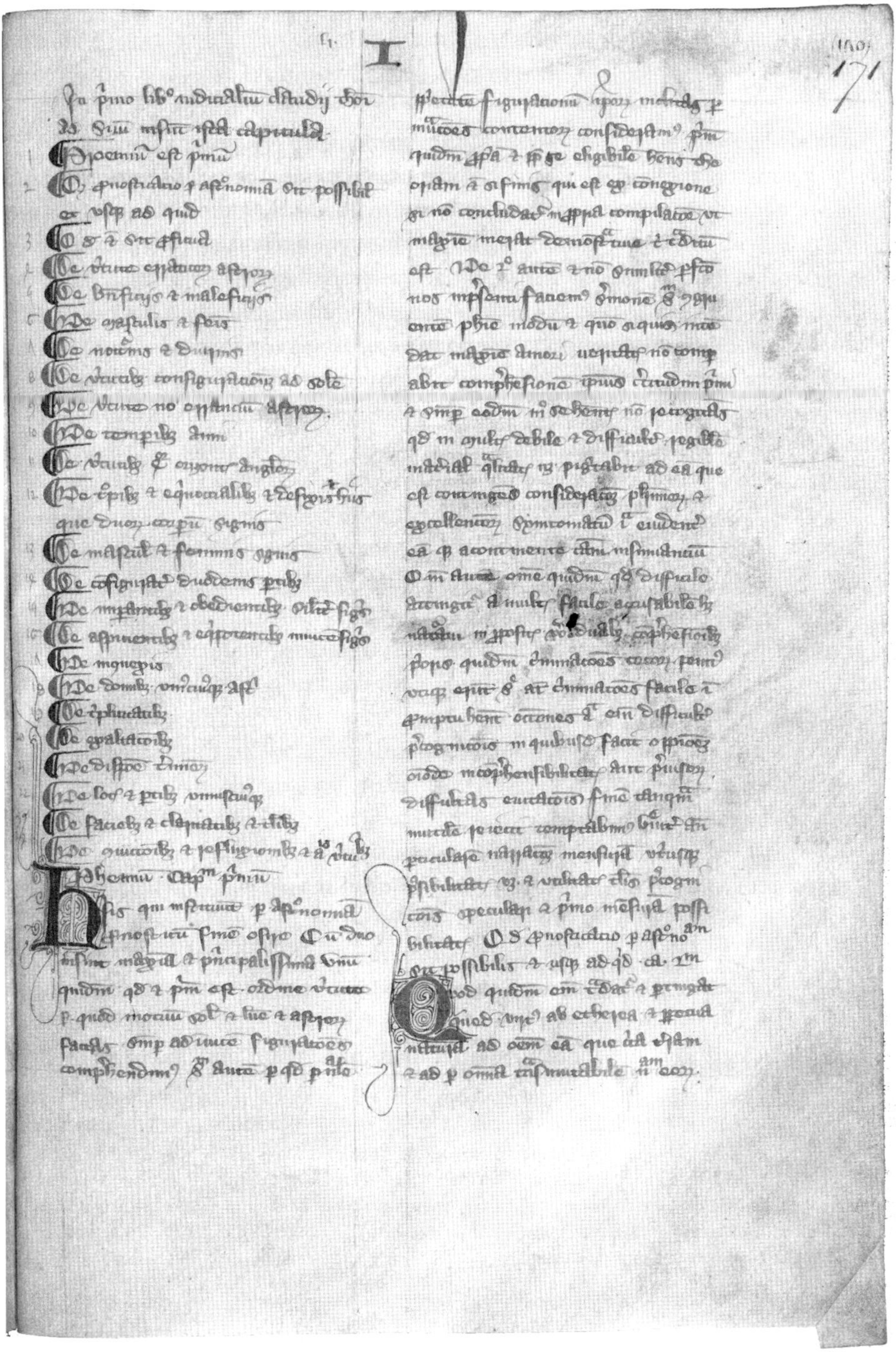

Oxford, Bodleian Library, Digby 179, fol. 171r.
Reproduced by permission of the library.

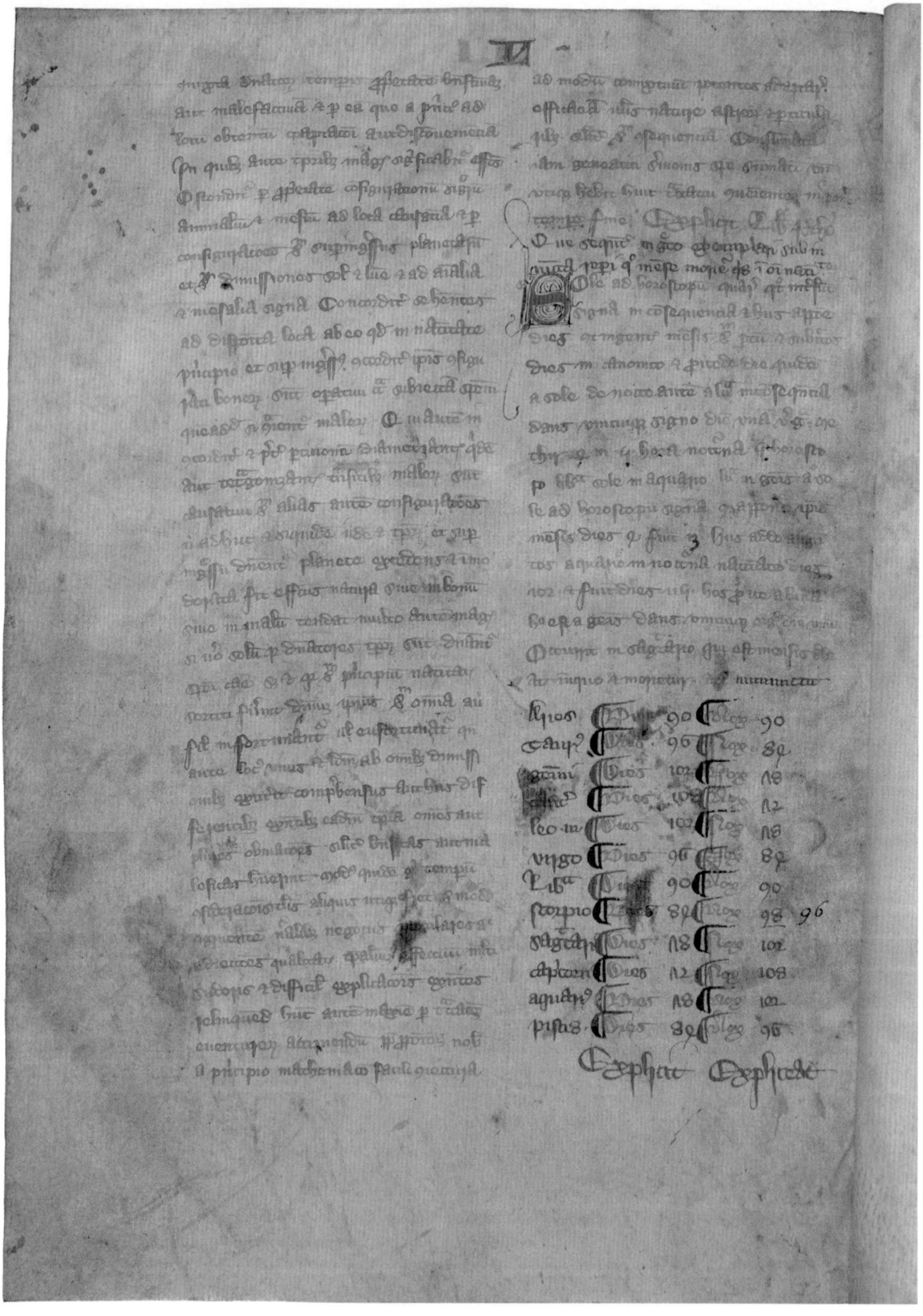

Oxford, Bodleian Library, Digby 179, fol. 208v. Reproduced by permission of the library.

CLAUDII PTOLEMAEI LIBER IUDICIALIUM

TRANSLATIO GUILLELMI DE MORBEKA

CONSPECTUS SIGLORUM

I. TEXTUS

Numeri marginibus exterioribus textui latino ascripti = numeri linearum nostrae editionis.

Numeri marginibus interioribus textui latino ascripti = numeri linearum editionis graecae Hübnerianae (cf. infra).

Numeri in textu latino uncis quadratis inclusi = numeri paragraphorum, quos W. Hübner, secundum F. Boll et AE. Boer, in marginibus editionis graecae ad capitula diuidenda addidit.

Numeri capitulorum translationis quos Guillelmus in suo exemplari graeco invenit: in libris I, II, IV cum numeris capitulorum editionis Hübnerianae concordant, in libro III, inde a capitulo 4, uno numero ab editione Hübneriana differunt.

II. ADNOTATIO LATINA

Ve = cod. Venetus, Bibl. S. Marci, Lat. XIV,242 (= 4295), ff.1r-12v: saec. XIV

Continet librum I et partem libri II, usque ad II 9, 619: « particularia. Observare».

Ox = cod. Oxoniensis, Bibl. Bodl., ms. Digby 179, ff. 171r-208v: saec. XIV

Continet libros I-IV, exceptis duabus particulis quae deficiunt in libro I (lacuna) : I.84–121 et I.187-222.

Prima pars adnotationis (I - II 9, 619). Omnes lectiones codicum *Ve* et *Ox* quae a textu edito (semper ante uncum quadratum) differunt, post uncum quadratum adhibemus (ad raras exceptiones, vide supra p. 139-140).

Ve^{ac}, Ox^{ac}; Ve^{pc}, Ox^{pc} = *Ve, Ox ante/post correctionem prima, ut videtur, manu effectam.*
$Ve^{i.t.}$, $Ox^{i.t.}$; Ve^{mg}, Ox^{mg}; Ve^{ss}, Ox^{ss} = codicum *Ve, Ox manus prima in textu; in margine; superscripsit.*
Ox^{1}, Ox^{2} = *Ox ante / post correctionem altera, ut videtur, manu effectam.*

Secunda pars adnotationis (II 9, 620 - IV). Translatio Guillelmi in hac parte uno solo codice (*Ox*), multis mendis scriptoris depravato, asservata est. Lectiones igitur codicis *Ox* saepe in textu edito corrigendae erant. In adnotatione omnes lectiones codicis *Ox* quae a textu edito differunt, post uncum quadratum adhibemus. Ante uncum quadratum semper lectionem textus editi indicamus, sed, ut apparatus simplicior et brevior sit, sine mentione «scr.» vel «coni.». Vide supra. p. 139-140.

Rarissime citantur lectiones «Quadripartiti» ab Henrico Bate laudatae:

Bate, *Liber Servi Dei* = Henricus Bate, *Liber Servi Dei de Mechlinia super inquisitione et verificatione nativitatis* (ed. E. Van de Vyver, pro manuscripto). Citatur in Lib. III.874 876.

Bate, *Speculum*: Henricus Bate, *Speculum divinorum et quorundam naturalium*, Pars XIX,19. Citatur in Lib. III.852 853 1132 1133 1140 1147; in Lib. IV.157 162 167.

III. ADNOTATIO GRAECO-LATINA

Lectiones codicum graecorum (excepto codice **L**) ex editione W. Hübner (vide infra) citantur; excerpta in **L** de novo contulimus. Lectiones quas uncis rotundis inclusimus, modo transferendi potius quam variis lectionibus graecis attribuendae sunt. Quantum ad principia citationum in adnotatione graeco-latina, vide supra p. 140-143.

Codices citati in editione Hübneriana:

V = Vaticanus gr. 1038, saec. XIII
Continet I 1-IV 10, abrumpitur post IV.797° («πρὸς τὰ»).

raro citantur cognati:

L = Laurentianus gr. 28,34, saec. XI
Continet excerpta e capitulis I 3(finis), 4, 5, 8-12, 14, 18-20, 21 (ultima pars), 22-24, II 12-13.

Γ = Parisinus gr. 2509, saec. XV
Apographum codicis **V**, sed textum a IV.797°, ubi **V** deficit, ex alio codice classis β supplevit

y = Vindobonensis phil. gr. 115, saec. XIII
ff. 10r-16r, capitula II 10-14 (imperfectum) e quadripartito excerpta continet.

stirpis α

Y = Parisinus gr. 2425, saec. XV

Σ = Norimbergensis gr. Cent. V app. 8, saec. XVI
Continet I—IV 5, usque ad 327° ; deficit I. 235°-509° *et* III.1222°-1292°

raro citantur:

G° = Oxoniensis Laudianus gr. 50, saec. XVI in. (in editione Hüb. = **G**)

Φ = Parisinus suppl. gr. 597, saec. XVI

stirpis β

D = Laurentianus gr. 28,20, saec. XIV

M = Marcianus gr. 314, saec. XV in.

S = Monacensis gr. 419, saec. XIV

raro citatur:

Ψ = Parisinus gr. 2363, saec. XV

stirpis γ
A = Angelicus gr. 29, anno 1388
B = Vaticanus gr. 208, saec. XIV (inter 1370-1390)
C = Laurentianus gr. 28,16, anno 1382

Codices in nostra editione citati:

	I 1-986	II 1-869	III 1-1188	IV 1-601	 601-678	 678-685 *conclusio*
V	+	+	+	+	–	–
L (excerpta)	cap. 4, 5, 8-12, 14, 18-20, 21 (a paragr. 22), 22-24	cap. 12-13	–		–	
Y	+	+	+	+	+	+
Σ	1-181 399-968	+	1-919 984-1188	1-252	–	–
β	+	+	+	+	+	–
γ	+	+	+	+	+	–
G°	–	–	–	–	raro	+
Φ	–	–	–	–	–	+
Γ	–	–	–	–	raro	–

Commentaria et editiones :

c = editio Camerarii prior a. 1535.
m = editio Camerarii altera (Melanchthone adiutore) anno 1553.
Co = commentarius anonymus anno 1559 editus.
Εἰς τὴν τετράβιβλον τοῦ Πτολεμαίου ἐξήγησις ἀνώνυμος / In Claudii Ptolemaei quadripartitum enarrator ignoti nominis, quem tamen Proclum fuisse quidam existimant [...], Basileae 1559, 1-180.
Co le / Co^le = lemma huius commentarii.
Heph. = Hephaestionis Thebani Apotelesmaticorum libri tres, ed. D. Pingree, Lipsiae 1973-1974.
Procl. = Procli quae dicitur paraphrasis = Procli Diadochi paraphrasis in Ptolemaei libros IV de siderum effectionibus a Leone Allatio e Graeco [cod. Vat. gr. 1453, saec. X; cf. CCAG V 1 (1904), p. 82. n. 14] in Latinum conversa, Lugduni Batavorum 1635.
Co, Heph., Procl.: partim ex editione W. Hübner, partim ex documentis ipsis citantur.

Boll, Boer = Claudii Ptolemaei opera quae exstant omnia III.1: Ἀποτελεσματικά, ed F. Boll (lib. I-II) et Æ. Boer (lib. III-IV) Lipsiae 1940, editio correctior 1954 (= 1957).

Robb. = Ptolemy, Tetrabiblos, ed., trad. F. E. Robbins, London 1940, (The Loeb Classical Library, [8]2001).

Hüb. = Claudii Ptolemaei opera quae exstant omnia III.1: Ἀποτελεσματικά, post F. Boll et Æ. Boer secundis curis ed. W. Hübner, Stuttgart-Leipzig 1998 (quantum ad codices et libros impressos, vide p. XI-XXV, LXXIV-V).

Charvet = Le livre unique de l'astrologie. Le Tétrabible de Ptolémée. Astrologie universelle et thémes individuels. Présenté, traduit et commenté par Pascal Charvet, Nil éditions, Paris 2000.

IV. IN UTRAQUE ADNOTATIONE

G = Guillelmus (-i) de Morbeka

codd. (*cett.*) = in prima parte adnotationis latinae ad codices *Ve* et *Ox* respicit; in adnotatione graeco-latina ad codices respicit qui ineunte adnotatione in singulis paginis recensentur

«aut — mensuratum» = *a verbo* aut *usque ad verbum* mensuratum

«aut ... mensuratum» = *verbum* aut *et verbum* mensuratum

«[**Y**], [**B**], .. [*Ve*], [*Ox*] » = testimonium codicis **Y**, **B**, ... *Ve*, *Ox*, deficit

* stellula in utraque adnotatione ad alteram adnotationem refert lectorem

abbr.	=	abbreviavit, -atio	*lac.*	=	lacuna
add.	=	addidimus, -idit, -endum	*lat.*	=	latinus, -a
adn.	=	adnotatio, -ionem	*loc.*	=	locus, -um
coni.	=	coniecit, -erunt, -iectio	*mg.*	=	margine
corr.	=	correxit, correctio *etc.*	*mut.*	=	mutavit, -erunt
del.	=	delevimus, -it	*obsc.*	=	obscurus, -re *etc.*
dub.	=	dubitanter	*om.*	=	omisit, omiserunt
excid.	=	excidit, -dunt	*praem.*	=	praemisit, -erunt
exp.	=	expunxit	*redd.*	=	reddidit
fort.	=	fortasse	*sc.*	=	scilicet
gen.	=	genitivus, -um	*scr.*	=	scripsimus, -it, -erunt
gr.	=	graecus, -a	*scrib.*	=	scribendum
hab.	=	habet, -ent	*sent.*	=	sententia, -iae, -iarum
hom.	=	homoeoteleuton, -o	*s.l.*	=	super lineam
insequ.	=	insequens, -tem	*spat.*	=	spatium, -io
intell.	=	intellexit	*ss.*	=	superscripsit
interp.	=	interpunxit, -ximus	*suppl.*	=	supplevit
inv.	=	invertit, -erunt	*tit.*	=	titulus, -um
it., iterat.	=	iterauit, iteratio, -ne	*tr.*	=	transposuit, -erunt
l. v.	=	lectio varia, -nes variae	*trad.*	=	traditio, -ne

<LIBER I>

In primo libro Iudicialium Claudii Ptholomei ad Syrum insunt ista capitula

---

V (**α**=**YΣ** **β**=**DMS** **γ**=**ABC**) 1 *tit. om.* (**V**)] Κλαυδίου Πτολεμαίου τῶν πρὸς Σύρον ἀποτελεσματικῶν βιβλίον α′ **D** (*Hüb.*): *diversimode cett.* 2-3 In primo — capitula: Τάδε ἔνεστιν ἐν τῷ α′ βιβλίῳ κεφάλεια Κλαυδίου Πτολεμαίου τῶν πρὸς Σύρον ἀποτελεσματικῶν **V S** (*Hüb.*)] *aliter* **Y γ**: *om. cett.* 4 Prohemium primum: α′ προοίμιον **V** (*Hüb.*)] προοίμιον **S**: *om. cett.* 5 2: β′ **V** (*Hüb.*)] α′ **D γ**: *om. cett.* 6 3: γ′ **V** (*Hüb.*)] β′ **D γ**: α′ **Y**: *om. cett.*: *et sic pro numeris deinceps* 8 beneficis: ἀγαθοποιῶν *cett.*] ἀγαθῶν **V** 11 uirtutibus] τῆς δυνάμεως *codd.* 14 uirtutibus] τῆς δυνάμεως *codd.* (orizontis] *om. codd., fort. add.* **G**) 17 configuratis: συσχηματιζομένων *cett.* (*cf.* **Y**)] συσχατιζ- **V** 18 imperantibus: προστασσόντων *cett.*] προτ- **V** similiter: ὁμοίως (*coni. Boll Hüb.*)] ὁμοίων *codd.* (19 signis] *om. codd., fort. add.* **G**) 23 exaltationibus: ὑψωμάτων *cett.*] ἰδιωμάτων **V**

---

2 Ptholomei] th'oi *Ox* Syrum] sirum *Ox* 4 Prohemium primum] prohemium est primum *Ox* 5 sit possibilis] *post* quid *Ve* 8 beneficis … maleficis *scr.*] -ficiis … -ficiis *codd., cf. infra* 334 9 masculinis] masculis *Ox, abbr. Ve* femininis *scr.*] feminis *Ox*: femineis *Ve* 15 tropicis] -pibus *Ox* 16 masculinis … femininis *scr.*] masculis … feminis *perscr. Ox, abbr. Ve* (*cf. infra, cap.* 6) 26 talibus] similibus *Ve*[i.t]: al' talibus *Ve*[ss]

Prohemium
Capitulum primum

[1] Hiis que instituunt per astronomiam pronosticum finem, o Syre, cum duo insint maxima et principalissima: unum quidem quod [et] primum est ordine et uirtute, per quod | motuum solis et lune et astrorum factas semper ad inuicem figurationes comprehendimus, secundum autem, per quod per naturalem proprietatem figurationum ipsorum inditas permutationes contentorum consideramus.

Primum quidem, propriam et propter se eligibilem habens theoriam, etiam si finis qui est ex connexione secundi non concludatur, in propria compilatione ut maxime inerat demonstratiue tibi traditum est.

[2] De secundo autem et non similiter perfecto nos in presenti faciemus sermonem secundum | congruentem philosophie modum, et quomodo, si quis intendat maxime amori ueritatis, non comparabit comprehensionem ipsius certitudini primi et semper eodem modo se habentis, non recogitans quod in multis debile et difficulter regibile materialis qualitatis, neque | pigritabit ad eam que est contingens considerationem plurimorum et excellentiorum simptomatum, ita euidenter eam que a continente causam insinuantium.

[3] Quoniam autem omne quidem quod difficile attingitur a multis facile accusabilem habet naturam, in propositis uero duabus comprehensionibus | prioris quidem criminationes cecorum penitus utique erunt, secunde autem criminationes facile in promptu habent occasiones (aut enim difficultas precognitionis in quibusdam facit opinionem omnimode incomprehensibilitatis, aut preuisorum difficultas euitationis finem tamquam inutilem reiecit): temptabimus breuiter ante | particularem narrationem mensuram utriusque, possibilitatis uidelicet et utilitatis, talis precognitionis speculari, et primo mensuram possibilitatis.

V (**α**=**ΥΣ** **β**=**DMS** **γ**=**ABC**) 28-9 Prohemium capitulum primum: α´ προοίμιον **V** (*Hüb.*)] προοίμιον **β γ**: *om.* **α** 32 ad inuicem] + τε καὶ τὴν γῆν *codd.* 36 finis (*cett.*)] τέλους **V** 43 excellentiorum: ὁλοσχερεστέρων **Y**, Heph., *cf.* Procl. (*Hüb.*)] ὁλοσχερῶν *cett.* 48 facile in promptu: εὐπροφασίστους *cett.* (*Hüb.*)] ἀπρο- **V S A** 51 ante: πρὸ *cett.*] πρὸς **V** utriusque: ἑκατέρου *cett.*] ἐκ κέντρου **V** 54 Capitulum 2ᵐ: β´ (*ante titulum*) **V** (*Hüb.*)] α´ **D γ**: *om. cett.*

---

30 que *scr.*] qui *codd.* (*cf. infra* 156, 256, 372, 526) Syre] Sire *Ox* 31 principalissima] principalis summa *Ve* et[2] *del.*] *hab. codd.* et[3]] *om. Ox* 34 inditas] inclitas *Ox* 35 theoriam] al' theoricam *add.* *Ve*[mg] 41 *ante* Non (*sic*) *interp. Ve* 42 regibile] regulabile *Ve* neque *scr. cum Ox*] nec *Ve* 47 criminationes] t'minationes (?) *add.* *Ve*[mg] 48 difficultas] -culter *Ox* 54 Capitulum 2[m]] .ca. 2[m] *codd.* ; *et sic deinceps pro omnibus capitulis*

Quod pronosticatio per astronomiam sit possibilis et usque ad quid Capitulum 2[m]

[1] Quod quidem igitur tradatur et pertingat quedam uirtus ab etherea et perpetua natura ad omnem eam que circa terram et ad per omnia transmutabilem naturam eorum que sub luna primorum elementorum, ignis scilicet et aeris, que continentur quidem et immutantur a motibus qui sunt penes etherem, et continent | et immutant reliqua omnia, terram scilicet et aquam et que in ipsis plantas et animalia, omnibus utique euidentissime et per pauca liquebit.

[2] Sol et enim disponit aliqualiter semper cum continente omnia que circa terram, non solum per transmutationes que fiunt per tempora anni ad geni-turas | animalium et plantarum fructificationes et fusiones aquarum et corporum transmutationes, sed etiam per periodos in singulis diebus, calefaciens quidem et humectans et exsiccans et infrigidans ordinate et consequenter consimilibus [et] figurationibus factis ad locum qui est secundum uerticem capitum nostrorum.

[3] Luna uero, tamquam propinquissima terre, exhibet super eam plurimum defluxum, compatientibus sibi et commutatis plurimis et inanimatis et animatis, et fluuiis quidem augmentantibus et minuentibus inundationes suas cum lumine ipsius, maribus autem conuertentibus ad | orientes et occidentes proprios impetus, et plantis et animalibus, uel totis uel secundum partes aliquas, completis cum ipsa et condiminutis.

[4] Astrorum etiam non errantium et errantium accessus plurimas faciunt diuersitates continentis, estuosas et uentosas et | nebulosas, a quibus et ea que sunt circa terram conformiter disponuntur.

[5] Iam autem et que ad inuicem ipsorum configurationes, concurrentibus aliqualiter et commixtis uirtutibus, plurimas et uarias transmutationes efficiunt. Optinente quidem solis uirtute ad uniuersalem | impressionem qualitatis, cooperantibus autem uel impedientibus secundum aliquid reliquis, et luna quidem euidentius et frequentius ut in coniunctionibus, dichotomiis et pleniluniis, astris autem tardius et insensibilius, ut in apparitionibus et occultationibus et adnuitionibus.

V (**α**=**ΥΣ** **β**=**DMS** **γ**=**ABC**) 66 locum qui est] τὸν **V S** («*scilicet* τόπον» *Hüb.*): τῶν, τῷ, τὸ, τῆς *cett.* 67 exhibet (**α**, *Robb.*)] + ἡμῖν **V β γ** (*Hüb.*) 70 ipsius (αὐτῆς **α S γ**, *Hüb.*)] αὐτοῖς **V M D** autem: δέ *cett.*] *om.* **V** (73 diuersitates] ἐπισημασίας *codd.*; *fort. ad sensum* **G**) (77 uirtutibus] διαδόσεων *codd.*; *fort. ad sensum* **G**) 78 impressionem: τεταγμένον *legit* **G**, *tacite Robb.*] -μένης *codd. sec. Boll et Hüb.*

---

55 igitur] enim *Ox* 59 terram] ignem *Ox* 64 periodos] peraydos *Ox* 65 et *del.*] *hab. codd.* 68 commutatis] immutatis *Ve, an recte*? et animatis] *om. Ve*[i.t.], *suppl. Ve*[mg] 69 et[1]] *om. Ox* 71 uel] *om. Ox* 72 cum] tamen *Ox* 73 etiam] et *Ox* 77 commixtis] -motis *Ox* 80 dichotomiis] dit'omis (*cum corr.*) *Ve* 81 insensibilius] -bilis *Ve*

[6] Quod autem hiis ita consideratis, non solum que iam concreta sunt necessarium aliqualiter disponi ab horum motu, sed et spermatum principia et fructificationes informari et regulari ad propriam continentis qualitatem, omnibus utique uidebitur | consequens esse.

[7] Qui uero magis obseruatiui agricolarum et pastorum ab accidentibus uentis penes coitus et seminum proiectiones coniecturant qualitatem euenturorum. Et uniuersaliter grossiora quidem et que per notabiliores configurationes | solis et lune et astrorum significantur, et ab hiis qui non naturaliter sed obseruanter solum intendunt, [et] ut ad omne prenosticari uidemus.

[8] Que quidem ex maiori uirtute et simpliciori ordine etiam a penitus idiotis, magis autem et a quibusdam | irrationalium animalium, ut puta differentias annuales temporaneitatum et uentorum; horum enim ut in totum sol causa est. Que autem minus sic se habent, ab hiis qui iam assueti sunt circa obseruationes penes necessarium, sicut a nautis particulares tempestatum et spirituum presignificationes | queque fiunt penes periodalius a configurationibus lune uel etiam non errantium astrorum ad solem.

[9] At tamen quia neque ipsorum horum tempora et loca ab <in>experientia certitudinaliter possunt comprehendere nec errantium astrorum accessus, | que et ipsa per plurimum conferunt, sepe ipsis falli euenit.

[10] Quid igitur prohibet eum qui certitudinaliter nouerit omnium astrorum et solis et lune motus, ita ut nullius figurationum nec tempus nec locus lateat ipsum, receperitque ex longe precedente | continua ystoria ipsorum ut ad omne dicatur, quamuis non eas que secundum ipsum quod subicitur sed uirtute effectiuas (puta ueluti naturam solis quia calefacit et lune quia humectat, et in reliquis similiter), et sufficiens fuerit ad hec talia existentia, naturaliter simul et bene coniecturabiliter, deprehendere eam que ex | commixtione omnium qualitatis proprietatem, ut possit quidem in unoquoque datorum temporum ex habitudine tunc apparentium proprietates continentis dicere (ut puta quia calidus aut humidus erit) ?

V (**α**=**YΣ** **β**=**DMS** **γ**=**ABC**) 85 continentis] τότε *praem. codd.* (100 accessus] περιόδους *codd.*) 103 tempus ... locus] *inv. codd.* 104 ipsorum: αὐτῶν **β** **Y** (*Hüb.*)] αὐτὰ **V** **Σ**: αὐτὰς **γ** 105 quamuis] τὰς φύσεις *praem.* **β** **γ** Procl. (*Hüb.*): φύσει *praem.* **V** **Σ**: φήση *praem.* **Y** 106 naturam] τὴν *codd.*; *fort. suo arbitrio* **G**, *cf. adn. praec.* (107 hec talia] τοιαῦτα *codd.*: *an inter* ταῦτα *et* τοιαῦτα *haesit.* **G** ?) 108 eam que ... proprietatem: ... τὸ ἴδιον **Σ** **β** **γ** (*Hüb.*)] ... ἴδιον (*om.* τὸ) **V** **Y** commixtione: συγκράσεως **V** **Σ** **β** **γ**] -κρίσεως **Y** (*Hüb.*)

---

83 que] quem *Ox* 84 et fructificationes — partium theoria (121)] *deficit* (*lacuna 32 uersuum*) *Ox* 90 significantur] *e corr. Ve*: [*Ox*] 91 et *del.*] *habet Ve*: [*Ox*] 99 <in>experientia] experientia *Ve*: [*Ox*] 103 nullius *scr.*] nulla *Ve*: [*Ox*] 104 precedente *scr.*] -tis *Ve*: [*Ox*] ystoria *scr.*] -ias nos *Ve*: *an* nos *post* precedente *scrib.* ?: [*Ox*] 107 hec talia *Ve*] *an* hec *delend.* ?: [*Ox*] 110 calidus ... humidus *Ve*] *an* calidius ... humidius *scrib.* ? [*Ox*]

[11] Et possit etiam penes unumquemque hominum | uniuersalem qualitatem proprie complexionis conspicere ab eo quod secundum constitutionem continente (puta quod secundum corpus quidem talis et secundum animam [et] talis), et euentus per tempora, propterea quod tale continens tali complexioni commensuratum aut etiam aptum sit ad bonam habitudinem, | tale incommensuratum et procliue ad malignationem ?

[12] Ideoque quod quidem possibile talis comprehensionis per hec et similia est conspicere. Quod autem facile occasionabiliter quidem, non oportune tamen que ad <im>possibile habet criminationem, sic utique deprehendemus. Primo quidem enim | peccata eorum qui non certitudinaliter examinant opus, que sunt multa ueluti in magna et multarum partium theoria, et uerum dicentibus faciunt opinionem quod a casu, non recte. Quod enim tale non scientie, sed utentium est impotentia.

[13] Deinde et plurimi lucri gratia | artem alteram a suo nomine credentes, ydiotas quidem decipiunt, insinuantes se predicere multa etiam eorum que nullam naturam habent ut precognoscantur, studiosioribus autem per hoc dant occasionem ex equo despiciendi pronosticationem naturam habentium ut predicantur. Neque hoc | oportune. Neque enim philosophiam destruendum, quia aliqui fingentium ipsam nequam uidentur.

[14] Sed tamen euidens est quod, et si perscrutabiliter aliquis ut possibile est maxime et familiariter ad mathemata accedat, contingit ipsum sepe deficere, propter nullum quidem dictorum, sed propter | ipsam rei naturam et propter infirmitatem ad magnitudinem pronuntiationis.

[15] Vniuersaliter enim, cum hoc quod theoriam que est circa qualitatem materie omnem accidit esse uerisimilatiuam et non assertiuam, et maxime collectam ex multis dissimilibus, adhuc antiquis planetarum | configurationibus, <a> quibus tamquam eodem modo se habentibus nunc adaptamus obseruatos a prioribus effectus illorum, quasi similes quidem enim possunt esse magis aut minus

---

V (**α**=**YΣ** **β**=**DMS** **γ**=**ABC**) 114 complexioni: συγκράσει **V Σ β γ**] -κρίσει **Y** (*Hüb.*) 122 quod: τοῦ **Y** (*Hüb.*)] τούτου **V** *cett.* (*Robb.*) 137 (tamquam eodem modo] ὡσαύτως *codd.*) (obseruatos: παρατετηρημένας **α**, *Hüb.*] παρατηρημένας **V**: τετηρημ- **β**) (138 illorum] ἐπ' ἐκείνων *codd.*)

---

113 talis *scr.*] tali *Ve*: [*Ox*] et² *del.*] *habet Ve*: [*Ox*] 114 propterea *scr.*] propria *Ve*: [*Ox*] 115 aut] autem *Ve*: [*Ox*] aptum *scr.*] aptom (?, *abbr.*) *Ve*: [*Ox*] tale incommensuratum et *scr.*] talia incommensurantur *Ve*: [*Ox*] 116 malignationem *scr.*] malistionem *Ve*ᵃᶜ: maliationem *Ve*ᵖᶜ (*cf.* III.1113): [*Ox*] 119 <im>possibile] possibile *Ve*: [*Ox*] 121 theoria *scr.*] theorica, *sed* -ca *e corr. Ve* (*cf. supra* 35, *infra* 134): [*Ox*] uerum *Ox*] + et *Ve*: *abhinc Ox denuo adest* faciunt] facit *Ox* 122 opinionem] -ni *Ve* tale] tule *Ox* 126 precognoscantur] prenoscantur *Ox* 127 ex equo] ex quo *Ox* 131 mathemata] methe- *Ve* 132 nullum quidem] *inv. Ve* 134 theoriam] theorica *mg. add. Ve* 135 omnem *scr.*] omni (*ex* omnem *ut videtur*) *Ve*: omnium *Ox* uerisimilatiuam] -simulatiuam *Ve* collectam *scr.*] -iuam *codd.* 136 a *add.*] *om. codd.* 138 effectus] affectus *Ox* similes] similies *Ve* quidem] *om. Ox*

et hee post longa tempora. [2.16] In nullo autem dissimiles nequaquam, conrestitutione omnium que in celo ' cum respectu ad terram examinate non concessa, aut non totaliter aut non secundum tempus sensibile homini (nisi aliquis uane opinetur circa notitiam et comprehensionem incomprehensibilium aut circa uite hominis tempus mensuratum), ita ut propter hoc predictiones ' aliquando fallant dissimilibus existentibus suppositis exemplis.

[17] Circa considerationem quidem igitur simptomatum que fiunt secundum continens, in hoc solo utique erit difficultas, nulla hic causa coassumpta motui celestium.

[18] Circa geneaticas autem rationes et uniuersaliter circa rationes ' commixtionis singulariter uniuscuiusque, non paruas neque contingentes est uidere concausas proprietatis constitutorum. Spermatum enim differentie plurimum possunt ad proprietatem generis, quoniam quidem continente et orizonte supposito ' eodem, optinet spermatum unumquodque in figurationem proprii formabilis uniuersaliter, puta hominis et equi et aliorum; loca etiam generationis non modicas faciunt circa ea que constituuntur uarietates.

[19] Et enim spermatibus secundum genus suppositis ' eisdem, puta humanis, et continentis consistentia eadem, preter differentiam regionum qui generantur multum differunt et corporibus et animis. Ad hec autem alimenta et consuetudines, omnibus premissis suppositis indifferentibus, conferunt aliquid ad ' deductiones particulares uitarum. Quorum unumquodque si non conferatur cum causis continentis, et si quam maxime plurimam istud habeat uirtutem (eo quod continens illis ipsis quod talia sint fiat concausa, huic autem illa nequaquam), multum errorem ' possint facere hiis qui in talibus opinantur posse cognoscere a solo altorum motu omnia, etiam que non ex toto sunt in ipso.

V (**α**=**YΣ** **β**=**DMS** **γ**=**ABC**) (140-1 non concessa — sensibile homini *hic* **G**] ἢ μηδ' ὅλως — ἀπαρτιζομένης *ante* ὡς (ita, 143) *codd.*: *fort. ordinem sententiarum mut.* **G**) (142 notitiam et comprehensionem] *inv. codd.*) 142-3 aut — mensuratum] *om. codd.* 145 simptomatum (**Y M S**, *Hüb.*)] συναπτομάτων **V D γ**: *om.* **Σ** 146 nulla (*cett., Hüb.*)] καὶ *praem.* **V** (*Boll*) 148 geneaticas rationes: γενεθλιαλογίας *cett.* (*Hüb.*)] -λογικὰς **V**, *cf.* **Σ** commixtionis: συγκράσεως **γ**, *an legit vel coni.* **G** ?] -κρίσεως **V Σ β** (*Hüb.*) 150 concausas] + καὶ αὐτὰ γινόμενα *codd.* 152 optinet (κατακρατεῖ *coni. Boll Robb. Hüb.*)] -κρατεῖν *codd.* (*cf.* **Y**) 156 qui generantur: οἱ γενόμενοι **V Σ** (*Robb.*), *cf.* **Y**] οἱ γεννώμενοι **β γ** (*Boll Hüb.*) 159 uitarum: βίων *cett.*] νέων **V** (160 continentis] ἀπὸ τοῦ περιέχοντος *codd.*)

---

139 hee *scr.*] hec *codd.* dissimiles] -milies *Ve* conrestitutione *scr.*] -nem *Ve*: cum restitutione *Ox* 140 examinate] *fort.* certitudinaliter examinate *scrib.* (*cf. supra* 100, 102, 120) 142 opinetur] optinetur *Ve*[i.t.]: al' opinetur *Ve*[mg] 148 geneaticas] genetiticas *Ox* 149 non] neque *Ve* 150 concausas] al' eciam causas *add.* *Ve*[mg] 156 preter] propter *Ve* qui *scr.*] al' qui que *Ve*: que *Ox* 157 animis *scr.*] -malibus *codd.* hec] *potius* hoc (*abbr.*) *Ox* 161 autem] aut *Ve* 163 altorum] astrorum *Ve*

[20] Hiis autem ita se habentibus, conueniens utique erit neque quoniam fallere quandoque contingit talem pronosticationem | totum ipsius perimere, sicut neque gubernatiuam reprobamus propterea quod sepe deficiat. Sed ut in tam magnis et diuinis pronuntiationibus acceptare et amabile existimare quod possibile est, neque rursum omnia nobis humane et coniecturabiliter requirere ab ipsa, sed | coaduocare et in quibus non est in ipsa ipsum totum dirigere; et sicut de medicis, cum requirunt aliquem, de ipsa egritudine et de proprietate infirmi, non querelamus interrogantes, ita et hic genera et regiones et alimenta aut etiam aliqua eorum que iam acciderunt | supponentes non subsannandum.

Quod etiam sit proficua
Capitulum 3^{m}

[1] Quo quidem igitur modo possibilis sit pronosticatio per astronomiam et quod pertingat utique usque ad sola illa symptomata que eueniunt secundum ipsum continens et ad ea que | consecuntur hominibus a tali causa (hec autem erunt utique circa eas que ex principio ydoneitates potentiarum et operationum corporis et anime, et circa ipsorum per tempora passiones, multasque et modicas durabilitates, adhuc autem et quecumque extrinsecorum propriam et naturalem habent ad | prima connectionem, ut ad corpus quidem possessio et conuictus, ad animam autem honor et dignitas, et horum per tempora fortunas), fere ut in summa factum est nobis palam.

[2] Reliquum autem utique erit propositorum eam que secundum utile considerationem breuiter facere | prius sumentibus, quomodo et ad quem finem aspicientes uirtutem ipsius utilitatis suscipiamus. Si quidem enim ad bona anime, quid

---

V (**α**=**YΣ** **β**=**DMS** **γ**=**ABC**) 168 omnia (**Σ M S γ**, *Hüb.*)] μὴ *praem.* **V Y D** 169 in ipsa*: ἐν αὐτῇ **β**, *cf. Hüb.*] ἐπ' αὐτὴν **V α**: [**γ**] 170 ipsa: αὐτῆς *cett.* (*Hüb.*)] αὐτοῦ **V** 171 querelamus: μεμφόμεθα **Y β γ**] -ψόμεθα **V Σ** (*Hüb.*) interrogantes: λέγοντες **V Y β γ**] λέγουσιν **Σ** (*Robb. Hüb.*) 174 Capitulum 3^{m}: γ′ (*ante tit.*) **V** (*Hüb.*)] *aliter vel om. cett.* (176 illa symptomata que eueniunt] τῶν ... συμπτωμάτων *codd., fort. dupl. lectio* **G**) 180 autem (*cett.*)] *om.* **V** 181 possessio: κτῆσις **Y Σ S γ** (*Hüb.*)] κτίσις **V D M** (*corr.* **D**^{2}) conuictus: ἡ συμβίωσις **Σ M S γ** (*Hüb.*)] εἰ συμβίωσις **V Y D**: ***abhinc usque ad*** 396 ***deficit* Σ**

---

164 ita] *om. Ox* 169 ipsa^{1,2} *scr.*] ipso *codd.* 171 aut] ante *Ox* 172 subsannandum (-sanan- *Ve*)] seruandum *Ox* 173 sit proficua *scr.*] fit proficua sit *codd.*: *fort.* sit *corr. pro* fit *erat* 176 symptomata] simpt- *Ox* 178 ydoneitates *scr.*] ydonietates *Ve*: ydometates *Ox* 179 anime] -o *Ox* 181 prima *scr.*] primam *Ve Ox*^{mg}: p^{u}nam *Ox*$^{i.t.}$ 182 autem *scr.*] aut *codd.* 184 que] quo *Ve*

utique erit magis conferens ad bonam operationem et gaudium et uniuersaliter boni placentiam tali cognitione studiosius, secundum quam humanorum | et diuinorum efficimur conspectores?

[3] Si autem ad ea que corporis, omnibus utique magis talis comprehensio cognoscet conueniens et oportunum ydoneitati que secundum unamquamque complexionem. Si<n> autem ad diuitias uel gloriam uel talia, licebit ulterius et de tota philosophia | idem hoc dicere, nullius enim talium est quantum in se ipsa acquisitiua. Sed neque illam propter hoc utique neque hanc iuste contempnemus, relinquentes id quod ad maius conferens est.

[4] Totaliter autem utique exquerentibus dicent utique qui inutilitatem | comprehensionis querelantur ad nichil principalissimorum aspicientes, sed ad ipsum hoc solum, quod eorum que omnino penitus futurorum precognitio superflua sit. Et hoc simpliciter ualde et non bene diiudicatum.

[5] Primo quidem enim oportet considerare, quod etiam in hiis que ex necessitate euentura sunt quod quidem | inexpectatum turbationes dementantes et gaudia extasim facientia maxime natum est facere, precognoscere autem assuescit et preparat animam meditatione absentium ut presentium et facit cum pace et tranquillitate singula euenientium recipere.

[6] Deinde quod neque sic | omnia oportet putare hominibus assequi ab ea que desuper causa tamquam ex principio ab aliquo insolubili et diuino precepto secundum unum quodcumque lege statuta et ex necessitate euentura, nulla alia simpliciter causa potente contraagere. Sed oportet tamquam quidem tempore | motus celestium secundum fatum diuinum et intransponibile a seculo perseuerante, ea autem terrestrium alteratione secundum fatum naturale et transponibile, primas causas desuper suscipiente secundum accidens et secundum assequentiam.

[7] Et tamquam hiis quidem propter uniuersaliores circonstantias | hominibus accidentibus et non ex propria quacumque et naturali idoneitate (ut quando secundum magnas et difficile effugibiles immutationes, adustionis uel pestilen-

V (Y **β**=**DMS** **γ**=**ABC**) 187-8 magis conferens ... studiosius] συμφορώτερον (*ante* πρὸς εὐπραγίαν) **V Y** (*Hüb.*): σπουδαιότερον (*post* προγνώσεως **γ**, *post* γνώσεως **M S**) **M S γ**: *om.* **D**: *fort. ambas lectiones legit* **G** 188 cognitione: γνώσεως **M S**] προγνώσεως **V Y D γ** (*Hüb.*) 192 complexionem: σύγκρασιν *ut* **γ** *legit vel coni.* **G** (*Robb.*)] -κρισιν **V Y β** (*Hüb.*) talia (**β γ**)] + συνεργεῖ **V Y** (*Hüb.*) 195 maius: τὸ μεῖζον **β γ**] τὰ μείζω **V Y** (*Hüb.*) 196 dicent (φαῖεν ?)] φανεῖεν *codd.* 205 omnia: ἅπαντα **β γ** Heph. (*Hüb.*).] ἕκατα **V**: ἕκαστα **Y** (*Robb.*) 207 unum quodcumque: ἓν ἕκαστον **β**] ἕνα ἕκαστον **V Y γ** (*Hüb.*) 208 oportet] *om. codd.*: *an add.* **G** ? tempore motus] τῆς ... κινήσεως *codd.*: *an suo ingenio* **G** ? (213 et[2]] *om. codd.*) 214 immutationes] τοῦ περιέχοντος *praem. codd.* adustionis: ἐκ πυρώσεως **β γ**: ἐκπυρώσεως *legit* **G**] ἐκ -εων **V Y** (*Hüb.*)

187 et uniuersaliter — contrapassionem inferentibus (222)] *deficit* (*lacuna 13 uersuum*) *Ox* 192 Sin *scr.*] Si *Ve*: [*Ox*] 198 omnino *scr.* (παντῇ)] omnia (*abbr.*) *Ve*: [*Ox*] 202 assuescit *scr.*] assuessant (*abbr.*) *Ve*: [*Ox*] 213 et[2] *sic Ve, cf.* 956: [*Ox*] 214 immutationes] *an* continentis *ante* immutationes *add.* ?: [*Ox*]

tiarum uel diluuiorum, in multitudinibus corrumpantur), contractiore causa cadente semper sub maiori et | fortiori. Hiis autem que secundum quamcumque naturalem proprietatem, propter paruas et contingentes continentis contrapassiones.

[8] Hiis enim ita receptis manifestum quia et uniuersaliter et particulariter quorumcumque quidem accidentium prima | causa inuincibilis utique et maior omni contraagente fuerit, omnino penitus euenire necessarium. Symptomata autem que non ita se habent, horum que quidem sortiuntur contrapassionem inferentibus facile auertibilia fiunt; que autem non bene habundabunt curantibus et ipsa primas naturas assecuntur, propter ignorantiam tamen | et non adhuc propter uirtutis necessitatem.

[9] Idem autem uidebit utique aliquis accidere simpliciter et in omnibus habentibus naturalia principia. Et enim lapidum et plantarum et animalium, adhuc autem et uulnerum et passionum et infirmitatum, hec quidem ex necessitate facere aliquid nata sunt, hec autem si nichil | contrariorum in contrarium agat.

[10] Ita igitur oportet et que euenient hominibus predicere accedentes naturaliter ad talem pronosticationem et non secundum uanas opiniones; tamquam hiis quidem, quia multa et magna existunt factiua, ineuitabilibus existentibus, hiis autem | suscipientibus immutationes propter contrarium. Quemadmodum et medicorum, quicumque possint per signa cognoscere passiones, pronosticant et que omnino periment et que capiunt auxilium.

[11] In hiis autem que possunt uitari ita audiendum est geneaticum sermonem, uerbi gratia, quod tali complexioni secundum talem continentis | proprietatem, mutatis suppositis commensurationibus in plus uel minus, talis passio consequetur. Sicut et medicum quidem, quia istud ulcus ampliationem aut putrefactionem efficit; metallici autem sermonis gratia, quod magnetes ferrum trahit.

[12] Sicut enim horum utrumque, | quantum ex se quidem propter ignorantiam eorum que contraagent, omnino penitus assequetur uirtutem prime nature, neque autem ulcus ampliationem aut putrefactionem operabitur si cura in con-

---

V (**Y β**=**DMS γ**=**ABC**) (216 quamcumque (ἑκάστην ?)] ἑκάστου **M S γ**: ἑνὸς ἑκάστου **V Y**, *Hüb.*) (220-1 utique ... fuerit: ἄν τε ἐστι?] τέ ἐστι ... *codd.*) 221 omnino: καὶ πάντῃ **β**] ταῦτα καὶ πάντῃ **V Y γ** (*Hüb.*) Symptomata: συμπτώματα (*post* ἔχει) **β**] *om.* **V Y γ** (*Hüb.*) 222 que[1]: ἃ **β γ**] ὅσα **V Y** (*Hüb.*) non (*cett.*)] *om.* **V** (223 curantibus: *an* τῶν θεραπευόντων *legit vel* curantibus *add.* **G** ?, *cf. infra* 248] *om. codd.*) 226 uidebit: ἴδοι **β γ** (*Hüb.*)] ἴδη **V**: εἴδη **Y** (accidere] συμβεβηκὸς *codd.*) 230 oportet] + νομίζειν *codd.* accedentes naturaliter ...: τοὺς φυσικῶς ... προσερχομένους ?] τοὺς φυσικοὺς ... προσερχομένους *codd.*

---

219 particulariter *scr.*] participaliter *Ve*: [*Ox*] 222 inferentibus] inferentibus *Ve*: [*Ox*] 223 facile auertibilia] *abhinc Ox denuo adest* 230 predicere] -dicte *Ox* 231 uanas] nonas *Ox* 234 pronosticant] -cantur *Ox* 239 istud *scr. cum Ve*] i[d] *Ox* 240 efficit] efficiet *Ve* magnetes (μαγνῆτις) *scr. cum Ox*] magnes *Ve* 242 assequetur] -entur *Ox*

trarium adhibeatur, neque magnetes ferrum trahet confricato ipso allio; et hec ipsa autem | prohibentia naturaliter et secundum fatum contrapatiuntur. Ita et in illis ignorata quidem euentura hominibus aut cognita quidem, non sortientia autem contrariis agentibus, omnino penitus consequentur ordinem prime nature. Precognita autem et bene habundantia curaturis, | naturaliter rursum secundum fatum, aut non fient penitus aut moderatiora constituentur.

[13] Totaliter autem tali uirtute eadem existente in hiis que in totalitate quadam uidentur et in hiis que particulariter, mirabitur aliquis propter quam causam de uniuersalibus | credunt omnes, et possibilitati pronosticationis et utilitati precauendi.

[14] Temporaneitates enim et non erraticorum significationes et lune figurationes plurimi precognoscere confitentur et multam prouidentiam faciunt solliciti de cautela ipsorum, | semper ad estatem quidem que possunt infrigidari, ad hiemem autem calentium, et uniuersaliter preparantes naturas ipsorum ad temperantiam; et adhuc ad securitatem temporum et regiminum obseruantes stellarum fixarum significationes, | ad principia uero coituum et plantationum configurationes lune secundum impletionem luminis, et nullus aliquatenus talia despicit neque tamquam impossibilia neque tamquam inutilia.

[15] In particularibus autem ex reliquis proprietatibus complexionis (puta ex maiori et | minori hieme et estu et ex proprietate complexionis secundum unumquemque) neque pronosticari adhuc possibile existimant aliqui neque multa posse uitari; quamuis manifestum sit, quod ad uniuersales estus si forte preinfrigidauerimus nos ipsos, minus estuabimus, et potest | simile operari et ad ea que singulariter hanc complexionem augent in intemperantiam calidi.

[16] Sed causa quidem talis peccati est difficultas et inconsuetudo pronosticationis particularium, quod quidem et in aliis fere omnibus incredulitatem facit; et ea que penes primas naturas efficiuntur ineuitabiliter, non collata ut ad omne

---

V (**Y β**=**DMS γ**=**ABC**) 248 curaturis: θεραπευσόντων **β γ**] -πευόντων **V Y** (*Hüb.*) (250 quadam] *om. codd.*) 251 (quam] + δήποτε *codd.*) de (**V**)] + μὲν *cett.* (*Hüb.*) (252 precauendi: πρὸς τῷ προφυλάσσεσθαι **V D γ**, *cf. cett.*] πρὸς τὸ φ., *Hüb.*) (260 nullus: οὐδεὶς **Y γ**] οὐδὲ εἷς **V β**, *Hüb.*) 262 ex[1] (*cett.*, *Hüb.*)] καὶ ἐκ **V** (*Robb.*) (263 hieme ... estu] χειμόνων ... καυμάτων *codd.*) (266 et[1]] καὶ ὅτι **γ**: *om. cett.*, *Hüb.*) 267 intemperantiam: ἀμετρίαν **Y M S γ** (*Hüb.*)] μετρίαν **V D** 268 inconsuetudo: ἄηθες *cett.* (*Hüb.*)] ἀληθὲς **V** 270 et ea que — efficiuntur ineuitabiliter *hic* **G**] *post* διαθέσεως (dispositionis 271/346°) *codd.*: *fort. ord. sententiarum mutavit* **G** (*cf. supra* 140) et ea que penes primas naturas efficiuntur ineuitabiliter: καὶ τὰ περὶ τὰς πρώτας φύσεις ἀνεμποδίστως ἀποτελούμενα *reddit* **G**] καὶ περὶ τὰς πρώτας φύσεις ἀνεμποδίστως ἀποτελουμένης (*sc.* διαθέσεως) **V** (*Robb. Hüb.*): τὰς πρώτας φύσεις ἀνεμποδίστως ἀποτελουμένας **Y β γ** 270-1 non collata ... uirtute ...] καὶ τὸ μὴ συναπτομένης ... τῆς ... δυνάμεως *codd.*

---

244 adhibeatur] -eant (*abbr.*) *Ox* magnetes *scr. cum Ox*] magnes *Ve* trahet] trahit *Ve* 246 in] *om. Ve* quidem] quam *Ve* 247 autem *scr.*] aut *codd.* 249 constituentur] -stitientur *Ox* 256 que *scr.*] qui *Ox*: quod *Ve* 257 hiemem] yemalem *Ox* autem] aut *Ve* 258 adhuc] ad hec *Ve* 260 impletionem] -plectionem *Ve* 263 hieme] yeme *Ox* 264 adhuc] ad hec (*abbr.*) *Ve* aliqui] aliquos (*e corr.* ?) *Ox* 269 quod] *om. Ox* 270 ineuitabiliter] ineuitabil' *Ox*

dicatur | uirtute contrapatiente pronostice propter raritatem ita perfecte dispositionis, opinionem fecerunt de omnibus simpliciter euenturis tamquam de immutabilibus et ineuitabilibus.

[17] Estimo autem sicut de ipsa pronosticatione, et si non ex toto sit sine errore, tamen quod possibile est de ipsa maximo studio dignum uidetur, eodem modo et de euitatione, et si non est omnium remediatiua, tamen in quibusdam, quamuis pauca quamuis parua sint, diligenda est et | acceptanda et non lucrum quodcumque existimare conuenit.

[18] Hiis autem, ut uidetur, consentientes tamquam ita se habentibus Egiptii, qui maxime talem artis uirtutem produxerunt, coniunxerunt ubique medicinalem cum pronosticatione per astronomiam. Non enim umquam utique instituissent in contrarium regimina aliqua | et preseruatoria contra dispositiones continentis superuenientes aut presentes uniuersales et particulares, si qua ipsis immutabilitatis et ineuitabilitatis euenturorum inesset opinio.

[19] Nunc autem et quod potest contraagere secundum inferiores naturas in secundo loco | sermonis qui est secundum fatum ponentes, copulauerunt cum pronosticationis uirtute eam que est secundum utilitatem et profectum, per uocatas ab ipsis medicinales compilationes, quatinus per astronomiam quidem suppositarum complexionum qualitates scire accidat et | simptomata euentura propter continens et proprias ipsorum causas (tamquam sine horum notitia et auxilia secundum plurimum deficere debeant, uelut non omnibus corporibus aut passionibus eisdem commensuratis existentibus), per medicinalem autem a compatientibus aut contrapatientibus propriis unicuique | preseruationes futurarum passionum et iam presentium curas indeficientes, ut contingit maxime, facientes perficiant.

[20] Sed hec quidem nobis usque ad tanta summatim figuraliter predicta sint. Faciemus autem iam sermonem | secundum introductorium modum, incipientes de proprietate uniuscuiusque celestium quantum ad ipsam effectiuam uirtutem, sequendo aptas obseruationes factas ab antiquis secundum naturalem modum et primas uirtutes errantium siderum, solis scilicet et lune.

V (**Y** **β**=**DMS** **γ**=**ABC**) 282 preseruatoria] φυλακτήρια (φυκτηρία **V**) καὶ θεραπείας *codd.* (continentis] ἐκ τοῦ περιέχοντος *codd.*) 289 complexionum: συγκράσεων **Y** **γ**, *fort.* **G**] -κρίσεων **β** (*Hüb.*): -κρίσεως **V** accidat: συμβαίνῃ *cett.*] -νει **V** (293 propriis: οἰκείοις ?] -ως *codd.*)

---

275 tamen] tantum (*abbr.*) *Ox* uidetur] uidet *Ve* 276 remediatiua] -ditiua *Ox* 284 inesset] inde esset *Ox* 287 profectum] proficium (*ut patet*) *Ve* 288 uocatas] notatas *Ox* compilationes] -nis *Ve* 294 et] *om.* *Ve* 296 hec] hoc *Ox*

De uirtute errantium astrorum
Capitulum 4^m^

[1] S o l deprehensus est efficaciam substantie habere in calefaciendo et remisse exsiccando. Hec autem maxime plus aliis facile sensibilia nobis fiunt propter magnitudinem ipsius | et propter euidentiam transmutationum secundum tempora, quoniam quanto magis utique nobis appropinquat quantum ad locum qui est secundum uerticem, tanto magis nos ita disponit.

[2] L u n a autem plurimum quidem uirtutis habet in humectando propter uicinitatem ad terram uidelicet et propter | humidorum exalationem, et disponit ita a regione corpora digerens et putrefaciens plurima; communicat autem debiliter et cum eo quod est calefacere propter illuminationes a sole.

[3] Astrum autem S a t u r n i plus qualitatis habet | in infrigando et in exsiccando debiliter secundum infrigidationem, propterea quod plurimum, ut uidetur, distat simul a solis caliditate et ab exalatione circa terram humidorum. Constant autem uirtutes in hoc et in reliquis per obseruationes in configurationibus ad solem et lunam, | quoniam hii quidem isto modo, hii autem illo, continentis consistentiam uidentur in maius aut minus immutare.

[4] M a r s autem maxime habet naturam exsiccandi et urendi conuenienter igneitati coloris et | propinquitati ad solem, supposita sibi solari spera.

[4.5] I u p i t e r autem temperatam habet efficaciam uirtutis, motu ipsius facto intermedio infrigidationis secundum Saturnum et ustionis secundum Martem; | calefacit enim simul et humectat et, quia magis est calefactiuus a suppositis speris, fit fecundorum uentorum factiuus.

[6] V e n u s autem eorundem quidem est secundum temperantiam factiua, sed secundum contrarium: calefacit quidem | enim debiliter propter propinquitatem ad solem, maxime autem humectat sicut luna, propter magnitudinem proprii luminis attrahens exalationem ab humidis continentibus terram.

[7] M e r c u r i u s autem ut ad omne ex equo aliquando quidem exsiccatiuus | deprehenditur et humidorum consumptiuus, propterea quod numquam multum distat secundum longitudinem a caliditate solis; aliquando autem rursum

---

V (Y β=DMS γ=ABC) ***abhinc excerpta codicis* L *adsunt*** 302 Capitulum 4^m^: δ′ **V** (*Hüb.*)] γ′**D γ**: β′ **Y**: *om.* **M S**, *et sic deinceps pro omnibus numeris capitulorum* 307 tanto: τοσούτῳ (*coni. Hüb.*)] τότε **V L**: *om. cett.* 309 exalationem: ἀναθυμίασιν *cett.*] ἀθυμίασιν **V** 315 in² (*cett.*)] *om.* **V L** (*Hüb.*) 326 luna] + καὶ αὐτός (*ante* καθάπερ **γ**) *codd.* (*Hüb.*) 329 consumptiuus: ἀναπωτικὸς **β γ** (*Hüb.*)] ἀεὶ ποτικὸς **V**: ἀνειπωτικὸς **L**, *cf.* **Y**

---

303 est] *om.* *Ve*^ac^ calefaciendo *scr.*] tale faciendo *codd.* 304 Hec] hoc *Ox* facile] facere *Ve* 307 ita] ipsos *Ox* 309 a] *ras.* *Ve*^i.t.^: ex *Ve*^ss^ 312 in] *om.* *Ox* 313 propterea] propter *Ve*^i.t.^: ea *suppl.* *Ve*^ss^ quod] quia *Ve* 314 Constant] -at *Ve* 315 obseruationes] + et *Ox* 316 continentis] -tes *Ve* 324 eorundem] *ex* eodem *Ve* 328 ex equo] ex quo *Ox*

humectatiuus, propterea quod superponitur spere lune propinquissime terre. Celeres autem facit in ambobus permutationes, uentosus factus tamquam a | ueloci motu circa ipsum solem.

De beneficis et maleficis
Capitulum 5[m]

[1] Hiis ita se habentibus, quoniam quatuor passionum due quidem sunt fecunde et factiue, calidum uidelicet et humidum (per hec enim omnia congregantur et augentur), | due autem corruptiue et passiue, siccum scilicet et frigidum (per que omnia disgregantur et pereunt), duos quidem planetarum, Iouem scilicet et Venerem et adhuc lunam, tamquam beneficos antiqui acceperunt, propter temperantiam et quia | plus habent in calido et humido.

[2] Saturnum autem et Martem maleficos tamquam contrarie nature factiuos, hunc quidem propter excessum infrigidationis, hunc autem propter excessum exsiccationis; solem autem et Mercurium propter communitatem naturarum potentes ambo, et magis simul uertibiles, | cum quibuscumque aliorum socientur.

De masculinis et femininis
Capitulum 6[m]

[1] Rursum quoniam prima genera naturarum duo sunt, masculinum scilicet et femininum, propositarumque uirtutum que humide substantie maxime feminina existit (plus enim accidit | uniuersaliter hec pars omnibus femininis, alia autem magis masculinis), merito lunam quidem et Venerem femininos nobis tradiderunt quia plurimum habent in humido, solem autem et Saturnum et Iouem et Martem propter siccitatem masculinos. | Mercurium autem communem ambobus generibus, secundum quod ex equo factiuus est humide substantie et sicce.

V (**Y β=DMS γ=ABC**, *excerpta* **L** *ad* 345) 336 passionum] χυμάτων **V L β γ** (*Hüb.*): σωμάτων **Y**; *cf.* Χύματα ἐκάλεσι τὰ στοιχεία Co (20,23) 338 et[2]: τε καὶ **L**: καὶ **α β γ**] *om.* **V** siccum … frigidum (**V L D**, *Robb.*)] *inv.* **Y M S γ** (*Hüb.*) 339 omnia (**V L M S**)] + πάλιν **Y D γ** (*Hüb.*) 342 maleficos: κακοποιοὺς **β γ**] *om.* **V L Y** (*Hüb.*) 345 ambo: ἀμφότερα **V L**] ὡς *praem.* **Y β γ** (*Hüb.*) 353 propter siccitatem (διὰ τὴν ξηρότητα ?)] *om. codd.*

---

331 humectatiuus] -ctiuus *Ox* spere] + et *Ox* 333 circa] contra *Ox* 334 beneficis … maleficis] beneficiis … malfic[s] *Ox* 337 calidum] caliditas (?, *abbr.*) *Ox* congregantur (συγκρίνεται) *scr.*] concreantur *codd.* 339 pereunt] perdunt *Ve* 342 maleficos] caleficos *Ox* 346 masculinis (*perscr. Ve*)] *ut* masculis *abbr. Ox* femininis *scr.*] *ut* feminis *abbr. codd.* 349 femininum (*perscr. Ve*)] *ut* feminum *abbr. Ox* feminina *scr.*] *ut* femina *abbr. codd.*; *et sic deinceps*

[2] Aiunt autem astra masculina et feminina fieri et secundum figurationes ad solem: matutinos quidem enim existentes et precedentes masculinos fieri, uespertinos autem et sequentes | femininos. Et adhuc penes figurationes ad orizontem: in locis quidem enim ab oriente usque ad medium celi uel ab occidente usque ad oppositum medii celi sub terra tamquam orientales masculinos fieri, in reliquis uero duabus quartis tamquam occidentales femininos esse.

De diurnis et nocturnis
Capitulum 7m

[1] Similiter autem quoniam tempus facientium manifestissime distantie due hee existunt, que uidelicet diei masculina facta magis (propter calorem qui est in ipso magis et actiuam uirtutem), et que noctis feminina facta (propterea quod | in ipsa est humiditas et pausatio), nocturnos quidem consequenter tradiderunt lunam et Venerem, diurnos autem solem et Iouem, communicantem autem secundum hec Mercurium et in matutina quidem figuratione diurnum, in uespertina autem nocturnum.

[2] Distribuerunt | autem utrique diuisionum et duos corruptiue nature, non tamen secundum easdem nature causas, sed secundum contrarias. Hiis quidem enim que bone complexionis copulata similia maiorem ipsorum profectum faciunt, hiis autem que corruptiue dissimilia admixta | dissoluunt multum de malitia ipsorum. Hinc Saturnum quidem infrigidatiuum existentem calido diei aptauerunt, Martem autem siccum existentem humido noctis; ita enim uterque a complexione temperamentum adipiscens conueniens fit diuisioni bonam temperantiam precedenti.

V (Y β=DMS γ=ABC) (358 locis] σχηματισμοῖς *codd.*) 359 orientales (ἀπηλιωτικοὺς ?)] -κοῖς *codd.* 360 occidentales: λιβυκοὺς **γ**] -κοῖς **V Y β** (*Hüb.*) 363 facientium: τῶν ποιούντων *cett.*] ποιοῦντα **V** 365 facta (**V M S**)] + μᾶλλον **Y D γ** (*Hüb.*) 368 hec: ταῦτα *legit* **G** (*Boll*)] ταὐτὰ *codd.* (*Robb. Hüb.*) (370 non] + ἔτι *codd.*) 377 precedenti (προεχούσης ?)] παρεχούσης **V γ** (*Hüb.*), *cf.* **Y**: περι- **β**; ***abhinc* L *denuo adest***

---

355 astra] + et *Ox* et[2]] *om. Ox* 357 orizontem] orientem *Ox* 363 autem] *om. Ox* due] scilicet dies et nox *supra lin. add. Ve, an nota Guillelmi ?* 367 solem ... Iouem] *tr. Ox* 368 hec] hoc *Ox* 370 diuisionum] -ni *Ve* 372 que bone *scr.*] qui bene *codd.* profectum] defectum *Ox* 373 corruptiue] corrupte *Ve* 375 aptauerunt] non *praem. Ve*

De uirtute configurationum ad solem
Capitulum 8^m^

[1] Iam tamen et penes configurationes ad solem, luna et tres errantium maius | et minus accipiunt in propriis suis uirtutibus. Luna enim secundum incrementum quidem ab ortu usque ad primam dichotomiam magis est factiua humiditatis, secundum illud autem quod est a prima dichotomia usque ad plenilunium caliditatis, secundum illud uero quod a plenilunio usque ad secundam | dichotomiam siccitatis, secundum autem illud quod est a secunda dichotomia usque ad occultationem frigiditatis.

[2] Errantes uero matutini quidem ab ortu usque ad primam stationem magis sunt humectatiui, a prima autem statione usque ad acronictum magis calefactiui, ab | acronicto autem usque ad secundam stationem magis exsiccatiui, a secunda autem statione usque ad occultationem magis infrigidatiui.

[3] Palam autem est quod ad inuicem commixti multas ualde differentias qualitatum efficiunt in continens nos, optinente quidem ut sepius | propria uniuscuiusque uirtute, mutata autem secundum quantum a uirtute configuratorum.

De stellarum fixarum uirtute
Capitulum 9^m^

[1] Cum autem sit consequens stellarum fixarum naturas secundum propriam ipsarum efficaciam percurrere, exponemus | obseruatas in ipsis proprietates secundum similitudinem ad naturas planetarum manifestationem facientes. Et primo de hiis que habent figuras circa ipsum zodiacum.

[2] A r i e t i s igitur <que> quidem in capite efficaciam | similiter habent contemperatam uirtuti Martis et Saturni; que autem in ore, uirtuti Mercurii et debiliter Saturni; que autem in posteriori pede, uirtuti Martis; que autem in cauda, uirtuti Veneris.

V (**Y** **β**=**DMS** **γ**=**ABC**, *excerpta* **L**)   383-384 secundum illud — caliditatis (*cett.*)] *om.* **V**   387 (uero] τε ... καὶ)   matutini quidem ab: *an* ἑῷοι μὲν ἀπὸ *legit* **G** ?] ἑῷοι ἀπὸ μὲν **V L**: ἑω μένω ἀπὸ μὲν **Y**: ἑῷοι μόνον ἀπὸ μὲν **β γ** (*Hüb.*)   388-389 magis — acronicto (*cett.*)] *om.* **V**   389 secunda: τοῦ δευτέρου *cett.*] κέντρου **V**   390 occultationem (κρύψεως *leg.* **G**)] δύσεως *cett.*: *lac.* **V**   396 Cum autem sit consequens: ἑξῆς δὲ ὄντος **V L Y** (*Robb.*)] ἑξῆς ὄντος ἀκολούθου **β γ** (*Boll*): ἑξῆς δέοντος *Hüb. secund. Kroll*   399 (zodiacum: ζῳδιακὸν **Σ**]: τὸν διὰ μέσων κύκλον *cett.*, *Hüb.*; ***abhinc* Σ *denuo adest***)

---

380 ad solem] *iter. Ox*   382 magis est factiua] i (*pro* .i. = id est) augmenta plenilunii *Ve*, *an nota ad* incrementum ?   383 dichotom-] dithom- (*vel* dichom-) *Ve*, *etiam* 384 *et* 385   388 acronictum *scr.*] acronixtum *Ox*: cronyctum *Ve*   389 acronicto] acronyco *Ve*   393 uirtute] *om. Ox*   400 que *add.*] *om. codd.*   401 debiliter] debilitater *Ox*   403 uirtuti] -tis *Ox*

[3] Stellarum autem que sunt in Tauro que quidem in abscisione | similem habent complexionem Veneri et debiliter Saturno; que autem in Pliade lune et Marti; earum autem que sunt in capite, clara quidem Yadis et subrubea (uocata autem Lampauras) Marti; relique autem Saturno et debiliter | Mercurio; que autem in summitatibus cornuum Marti.

[4] Earum autem que in Geminis que quidem sunt in pedibus participant qualitatem similem Mercurio et remisse Veneri, que autem circa crura clare, Saturno; duarum autem clararum que sunt in capitibus | que quidem in precedente Mercurio (uocatur autem Apollinis), que autem in sequenti Marti (uocatur autem Eraclei).

[5] Earum autem que in Cancro due quidem que sunt in oculis eiusdem actus sunt effectiue cum Mercurio | et remisse cum Marte; que autem in ungulis, Saturno et Mercurio; que autem in pectore nebulosa rotunditas (uocata autem Presepe), Marti et lune; due autem que sunt ex utraque parte illius (uocate autem Asini), Marti et soli.

[6] Earum autem que circa Leonem due quidem que sunt in capite faciunt simile Saturno et Marti remisse; tres autem que sunt in collo, Saturno et debiliter Mercurio; que autem in corde clara (uocata Basilicus), Marti et Ioui; | que autem in lumbis et que in cauda clara, Saturno et Veneri; que autem in cruribus, Veneri et debiliter Mercurio.

[7] Virginis autem stellarum que quidem sunt in capite et que in extremitate ale australis efficaciam habent similem | Mercurio et debiliter Marti; relique autem ipsius ale clare et que sunt circa cinctoria, Mercurio et debiliter Veneri; que autem in boreali ala clara (uocata autem Preuindemiator), Saturno et Mercurio; uocata autem Spica, | Veneri et debiliter Marti; que autem in summitatibus pedum et in <- - ->, Mercurio et debiliter Marti.

[8] Vngularum autem Scorpionis que quidem in extremitatibus ipsis disponunt consimiliter Ioui et Mercurio; que | autem in mediis Saturno et debiliter Marti.

V (**α**=**YΣ** **β**=**DMS** **γ**=**ABC**, *excerpta* **L**)   (409, 414 Earum] τῶν ... ἀστέρων *codd.*)   414 oculis: ὀφθαλμῶν **V L α D**[1] Procl. (*Robb.*)] ποδῶν **β** (-**D**[1]) **γ** Heph. (*Hüb.*)   420 Marti: Ἄρεως *cett.* (*Hüb.*)] Διός **V**   429 <- - ->] τῷ σύρματι **V Y β γ** (*Hüb.*): τῷ σύρματι τοῦ ἱματίου **L Σ**   436 Veneri: τῆς Ἀφροδίτης **α L** Heph. (*Hüb.*)] τῆς Ἑρμῆς **V**: τοῦ Ἑρμοῦ **β γ**

---

404 quidem] autem *Ox*   406 subrubea] -rubem *Ox*   407 Lampauras] lanp- *Ox*   411 clararum] stellarum *Ox*   416 autem[2]] que *Ox*   419 autem] uero *Ve*   421 Basilicus *scr. cum Ve*] Basilius *Ox*: *an scrib.* Basiliscus ?   426 debiliter] -litati *Ox*   427 autem[2]] ut (*exp.*) aut *Ox*   Preuindemiator] Preuendimiator *Ve*   429 <- - ->] *spat.* ± 4 *litt. Ox*: libra *Ve*; *fort.* σύρματι *graece scripsit* **G** *vel spatium reliquit*; libra *fort. lectio marginalis* (**G** ?) *ad* Vngularum Scorpionis (430) *erat quam Ve hic perperam inseruit*   430 disponunt *scr.*] -untur *codd.*

[9] Earum autem que sunt in corpore Scorpionis que quidem in fronte clare idem faciunt cum Marte et cum Saturno debiliter; tres uero que sunt in corpore, quarum media subrubea et clarior (uocatur autem Anthares), | Marti et debiliter Ioui; que autem sunt in spondilibus, Saturno et debiliter Veneri; que autem sunt in aculeo, Mercurio et Marti; sequens autem nebulosa rotunditas, Marti et lune.

[10] Earum autem que sunt in Sagittario que quidem sunt in cuspide sagitte similem habent efficaciam Marti et lune; que autem circa arcum et capturam manus, Ioui et Marti; que autem in facie conuolutio, soli et Marti; que autem sunt in uexillis | et in dorso, Ioui et debiliter Mercurio; que autem sunt in pedibus, Ioui et Saturno; quod autem in cauda quadrilaterum, Veneri et debiliter Saturno.

[11] Capricorni autem stellarum que quidem sunt in | cornibus eodem modo operantur cum Venere et debiliter cum Marte; que autem in ore, cum Saturno et debiliter cum Venere; que autem sunt in pedibus et uentre, Marti et Mercurio; que autem sunt in cauda, Saturno et Ioui.

[12] Earum autem que sunt circa Aquarium ille quidem que sunt in humeris disponunt consimiliter Saturno et Mercurio cum illis que sunt in sinistra manu et in fronte; que autem sunt in cruribus, magis quidem Mercurio, minus autem Saturno; ille autem que sunt in fusione aque, Saturno | et debiliter Ioui.

[13] Stellarum autem circa Pisces que quidem sunt in capite australioris piscis efficiunt idem cum Mercurio et cum Saturno debiliter; que autem sunt in corpore, Ioui et Mercurio; que autem sunt in cauda et in australi | lino, Saturno et debiliter Mercurio; que autem sunt in corpore et in spina borealis piscis, Ioui et debiliter Veneri; que autem in boreali lino, Saturno et Ioui; clara autem que est in ligamine, Veneri et debiliter Mercurio.

[14-19] Earum autem que sunt in borealioribus figurationibus a zodiaco, que quidem sunt circa Vrsam minorem clare similem habent qualitatem cum Saturno et debiliter cum Venere.

Que autem sunt circa Vrsam maiorem, Marti. Que autem sunt | sub cauda ipsius in conuolutione <- - ->, lune et Veneri.

V (**α**=**YΣ** **β**=**DMS** **γ**=**ABC**, *excerpta* **L**)   437 sequens: ἑπομένη **L M S γ** Heph. (*Hüb.*)] ἐπιμένη **V**: λεγομένη **α**: *om.* **D**   438 in[1]: ἐπὶ **V L**] περὶ *cett.* (*Hüb.*)   438-9 que[2] ... habent: οἱ ... ἔχουσιν **V L** (*Boll*)] ὁ ... ἔχει **β γ** Heph. (*Hüb.*), *cf.* **Y**   (441 uexillis *sic, fort. suo arbitrio* **G**] ἐφαπτίσι **V L B γ**, *Hüb.*: πτέρυξι **G**[o] **Σ**, *Boll*, *cf.* **Y**: πτέρυξιν ἤτοι ἐφαπτίσι Heph.)   449 fronte: μετώπῳ **V L** Procl.] ἱματίῳ (χύματι **Σ**[1]) *cett.*, Heph. (*Hüb.*)   455 lino: λίνου **L M S γ** (*Hüb.*), *cf.* λίνῳ **α** Heph.] αἴνου **V D**   456 Mercurio (**α M S**)] + Περὶ τῶν βορειοτέρων τοῦ ζῳδιακοῦ (*Tit.*) **V L D** Heph. (*Hüb.*), *cf.* **γ**   459 Veneri] τῷ τε τοῦ Ἄρεως *codd.*   460 Que autem sunt[2]: οἱ δὲ **L D γ** (*Hüb.*)] ἡ δὲ **V Y**, *cf.* **M S**   461 <- - ->] τοῦ Πλοκάμου *codd.*

---

436 spondilibus] spodi- *Ve*   438 Sagittario *scr.*] -riis *codd.*   442 quod] que *Ve*   447 que sunt[1]] *om.* *Ve*   454-5 borealis ... boreali] borialis ... boriali *Ox*   et debiliter Veneri] *om.* *Ox*   461 <- - ->] *spat. vac.* ±12 *litt. Ox*: *om. Ve*; *an litt. graec. vel spat. vac. in autographo* ?

Que autem sunt in Dracone clare, Saturno et Marti.

Ille autem que sunt Cefeos, Saturno et Ioui.

Que autem sunt circa Bootem, Mercurio et Saturno. Clara autem et subrufa uocata Arcturus, Marti et Ioui.

Que autem sunt in Boreali Corona, Veneri et Mercurio.

Que autem sunt penes illam que est In genibus, Mercurio.

Que autem sunt in Lira, Veneri et Mercurio.

Et ille que sunt in Aue eodem modo.

Que uero sunt secundum Cassiepeiam, Saturno et Veneri.

Que autem sunt penes Perseum, Ioui et Saturno; que autem est in captura gladii conuolutio, Marti et Mercurio.

Clare autem que sunt in Auriga, Marti et Mercurio.

Que autem sunt penes Serpentarium, Saturno et debiliter Veneri.

Que autem sunt penes Serpentem ipsius, Saturno et Marti.

Que autem sunt secundum Sagittam, Marti et debiliter | Veneri.

Que autem sunt circa Aquilam, Marti et Ioui.

Que autem sunt in Delfino, Saturno et Marti.

Clare autem que sunt penes Equum, Marti et | Mercurio.

Que autem sunt in Andromeda, Veneri.

Stelle autem Trianguli, Mercurio.

[20-24] Earum autem que sunt in a u s t r a l i o r i b u s a zodiaco figurationibus que | quidem est in ore Australis Piscis clara similem habet operationem Veneri et Mercurio.

Ille autem que sunt penes Cete, Saturno.

Earum autem que sunt circa Orionem que quidem sunt in humeris Marti et Mercurio, relique autem clare | Ioui et Saturno.

Earum uero que sunt in Flumine ultima quidem et clara Ioui, relique autem Saturno.

V (**α**=**ΥΣ** **β**=**DMS** **γ**=**ABC**, *excerpta* **L**) 476 Veneri: τῆς Ἀφροδίτης *cett.* (*Hüb.*)] τῆς Ἑρμῆς **V** 478 Saturno: τῷ τε τοῦ Κρόνου *cett.* (*Hüb.*)] τῷ περὶ τοῦ Κρόνου **V** 481 Mercurio (**α M S**)] + Περὶ τῶν νοτιωτέρων τοῦ ζῳδιακοῦ (*Tit.*) **V L D** (*Hüb.*), *cf.* **γ** 483 Veneri: τῆς Ἀφροδίτης *cett.*] τῆς Ἑρμῆς (*sic*) **V** (485 Cete] τὸ Κῆτος *codd.*)

---

462 Dracone] drachone *Ve* 463 sunt] + realibus (*ex* borealibus ?) *Ve* Cefeos *scr.*] cefex *codd.* 464 Bootem *scr.*] boetem *codd.* 467 genibus] generibus (?) *Ve* 469 Aue] auem (?) *Ve*[i.t.], .s. aprima *add.* *Ve*[ss] 470 Cassiepeiam *scr.*] cassepiam *Ve*: cassiopeiam *Ox* 474 sunt] *om.* *Ox* 479 Clare] clara *Ox* 480 Andromeda] -mede (*an recte* ?) *Ox* 481 autem] aut *Ox* 482 australioribus *scr.*] austrariolibus *Ve*: astralioribus *Ox* 483 in ore] more *Ox* 486 Orionem] orizonata *Ve*

Que autem sunt in Lepore, Saturno et Mercurio.

Earum uero que sunt circa Canem alie quidem Veneri, clara autem que est circa os Ioui et debiliter Marti.

Que autem est in Precane clara, Mercurio et debiliter Marti.

Clare autem que sunt penes Ydrum, Saturno et Veneri.

Que autem sunt in Cratere, Veneri et debiliter Mercurio.

Que uero sunt circa Coruum, Marti et | Saturno.

Clare autem que sunt Nauis, Saturno et Ioui.

Earum autem que sunt circa Centaurum que quidem sunt in humano corpore, Veneri et Mercurio; | clare autem que sunt in equo, Veneri et Ioui.

Que uero sunt circa Bestiam clare, Saturno et debiliter Marti.

Que autem est in Australi Corona clara, Saturno et Mercurio.

Virtutes quidem igitur stellarum secundum se ipsas talem sortite sunt ab antiquis obseruationem.

De temporibus anni
Capitulum 10^m

[1] Temporibus autem anni quatuor existentibus, uere scilicet et estate et autumpno et hieme. Ver quidem habet plus in humido propter diffusionem frigoris precedentis incipiente caliditate. Estas autem habet plus in calido propter | propinquitatem solis ad summitates capitum nostrorum. Autumpnus autem plus in sicco propter consumptionem humidorum ex precedenti estu. Hyems autem plus habet in frigido, quia sol tunc plurimum distat a loco qui est penes uerticem capitum nostrorum.

[2] Propter quod quidem et zodiaci circuli nullo existente | principio per naturam in quantum circuli, duodecimam partem incipientem ab equinoctio uernali, Arietem, aliorum principium supponunt, uelut animati animalis zodiaci humidam aeris habundantiam primitiuam facientes et consequenter reliqua tem-

---

V (**α**=**YΣ** **β**=**DMS** **γ**=**ABC**, *excerpta* **L**) 490 Saturno: Κρόνου *cett.*] Κριὸς **V** 500 Marti] + Οἱ δὲ ἐν τῷ Θυμιατηρίῳ τῷ τε τῆς Ἀφροδίτης καὶ ἠρέμα τῷ τοῦ Ἑρμοῦ (Κρόνου **Σ**) *codd.* 501 Que … est … clara: ὁ … λαμπρὸς **V L β γ**] οἱ … λαμπροὶ **α** Heph. (*Hüb.*) 504-5 *Tit. hic habet* **G** *ut* **V L D γ** (*Robb.*)] *ante* αἱ … δυνάμεις (uirtutes 502/673°) **α M S** (*Hüb.*) (509 ad summitates capitum nostrorum] πρὸς τὸν κατὰ κορυφὴν ἡμῶν τόπον *codd., cf.* **Σ**) 511 tunc (**β γ**)] *om.* **V L α** (*Hüb.*) 515 aliorum] τῶν ὅλων *codd.* 516 aeris: ἀέρος **V L Y D γ**] ἔαρος M S (*Hüb.*) primitiuam: προκαταρκτικὴν **L α M S** (*Hüb.*)] -αρτικὴν **V**, *cf. cett.*

---

491 circa[2]] contra *Ox* 495 Cratere *scr.*] cartere *codd.* 496 Coruum] eorum *Ox*[1] 500 clare] + autem *Ve* 507 hieme] yeme *Ox* propter] per *Ox* 508 autem] uero (*e corr.*) *Ve* 510 Hyems] hiens *Ox* 514 quantum] *om. Ve* 515 animalis] *obscure Ve*

pora, propterea quod omnium animalium | prime quidem etates plurimum habent humide substantie consimiliter aeri, tenere adhuc et delicate. Secunde autem que sunt usque ad statum plus habent in calido consimiliter estati. Tertie autem et iam in declinatione et principio detrimenti plus iam | habent in sicco consimiliter autumpno. Vltime autem et que tendunt ad dissolutionem, plus habent in frigido, sicut et hyems.

De uirtute <quatuor> angulorum
Capitulum 11^m

[1] Similiter autem et quatuor orizontis locorum et angulorum, | a quibus et uenti per omnes partes flantes principia habent. Qui quidem apud orientes ipse plus habet in sicco [et], propterea quod, sole facto penes ipsum, que in nocte humefacta sunt tunc primo incipiunt exsiccari; uenti ab ipso flantes, quos | communiter subsolanos uocamus, sine humore sunt et exsiccatiui.

[2] Locus autem qui apud meridiem ipse est calidissimus propter igneitatem que fit a sole in meridie, et quia quantum ad declinationem nostre habitationis hoc magis uergit ad meridiem; | uenti flantes ab ipso, quos communiter austros uocamus, calidi et repletiui.

[3] Locus autem qui est apud occidentem ipse est humidus, quia cum fit apud ipsum sol, que in die subtiliata sunt tunc primo incipiunt humectari; uentique ab ipso delati, quos communiter | zephiros uocamus, recentes sunt et humectatiui.

[4] Locus autem qui est uersus Vrsas ipse est frigidissimus, et quia nostra habitatio decliuis est et plus distat a sole celum mediante qui est caliditatis causa, quasi in opposito solis in medio celi; ideo | uenti flantes ab ipso, uocati autem communiter boree, frigidi sunt et compactiui.

V (**α**=**YΣ** **β**=**DMS** **γ**=**ABC**, *excerpta* **L**) 517 propterea quod: διὰ τὸ *cett.*] διατάττουσι **V** 518 humide substantie: ὑγρᾶς οὐσίας **L α M S** (*Robb.*)] ὑγρασίας **V D γ** (*Hüb.*) aeri*] τῷ ἔαρι *codd.* tenere] + οὖσας *codd.* 520 iam] + καὶ αὐτὰς *codd.* (531-2 a sole in meridie ... hoc] τῶν τοῦ ἡλίου μεσουρανήσεων ... ταύτας *codd.*) 533 calidi] + εἰσι *codd.* 538 et plus distat a sole celum mediante qui est caliditatis causa] τὰς τῆς θερμότητος αἰτίας τοῦ ἡλίου μεσουρανήσεις (μεσουρανοῦντος **γ** **Σ**^mg, *cf.* **G**) πλεῖον αὐτοῦ (*cf. om.* **G**) διεστάναι **V L β** (*Hüb.*), *cf.* **α**: *totam sententiam suo ingenio redd.* **G**

---

518 aeri*] *an* ueri *scrib.* ? que] *om. Ve* 522 hyems] yems *Ox* 523 quatuor *add.*] *om. codd.* angulorum] + orizontis *Ve* 526 Qui *scr.*] que *Ve*: quod *Ox* 527 habet] -ent *Ox* et *del.*] *hab. codd.* 530 meridiem] medium *Ox* 531 declinationem] declarationem *Ve* 532 hoc (*sc.* sol in meridie)] h° *Ox*: h· *Ve* quos] quo *Ox* 533 calidi] -da *Ox* 534 cum fit] non sit *Ox* 536 humectatiui] humectaui *Ox*^i.t., ti *suppl.* *Ox*^ss 538 celum] celi *Ve* quasi] *mutil. Ve* 539 flantes] perflantes (*an recte* ?) *Ox* communiter] *om. Ox* 540 compactiui] comactiui *Ve*

[5] Vtilis autem est horum distinctio ad posse discernere per omnem modum
complexiones semper; facile enim intelligibile est quod penes tales consisten-
725 tias ˈ uel temporum uel etatum uel angulorum uertitur aliqualiter efficacia uir-
tutis astrorum, et in congruis sibi consistentiis puriorem habent qualitatem et
operationem fortiorem (puta in calidioribus calefactiui per naturam et in humidis
732 humectatiui), ˈ et in aliis autem eodem modo secundum proportionem qualitatis
contemperate per mixturam.

736 [12.1] Hiis autem ita preexpositis consequens utique erit connectere traditas
naturales proprietates ipsarum duodenarum partium zodiaci. Totaliores quidem
enim complexiones secundum unumquodque ipsorum proportionaliter se habent
740 ad ˈ tempora que fiunt secundum ipsa; consistunt autem [et] quedam ipsorum
proprietates et a conuenientia ad solem et lunam et stellas, quas in sequentibus
745 disseremus, cum preinstituerimus ˈ uirtutes obseruatas ipsarum duodenarum
partium secundum se et ad inuicem sine permixtione.

735 De tropicis et equinoctialibus et fixis ˈ et bicorporeis signis Capitulum 12^m^

746 [2] Prime quidem igitur sunt differentie uocatorum tropicorum et equinoctia-
lium et fixorum et bicorporeorum. Duo quidem enim sunt tropica; illud scilicet
750 primum ab estiuali uersione triginta partium, Cancer, et primum ab ˈ hyemali
uersione, quod secundum Capricornum. [3] Hec autem ab accidente acceperunt
denominationem; uertitur enim in principiis ipsorum existens sol, rediens in con-
traria motuum secundum latitudinem, et apud Cancrum quidem estatem faciens,
755 apud Capricornum autem yemem. ˈ Duo autem uocantur equinoctialia; primum
uidelicet a uernali equinoctio triginta partium, scilicet Aries, et quod ab autump-
nali equinoctio, quod scilicet Chelarum. Nominata sunt autem et hec rursum

---

V (**α**=**YΣ** **β**=**DMS** **γ**=**ABC**, *excerpta* **L**) (544 sibi] *om. codd.*) 545 calidioribus] θερμασίαις **V**: θερμαῖς *cett.* (*Hüb.*), *an etiam* **G** ? 546 humectatiui (**γ**)] + ἐν δε ταῖς ἐναντίαις — οἱ ὑγραντικοί *cett.* (*Hüb.*) 548 Hiis autem — permixtione (554) *hic habet* **G** *ut* **V L**] *post tit.* (556) *cett.* (*Hüb.*) 551 ipsa: αὐτὰ **L Σ β γ** (*Hüb.*)] αὐτὸ **V**, *cf.* **Y** 553 ipsarum] + μόνον **V L**: + μόνων *cett.* (*Hüb.*) 559 estiuali — ab (*cett.*)] *om.* **V** (Cancer] τὸ τοῦ Καρκίνου *codd.*) 564 triginta partium (τριακοντάμοιρον ?)] δωδεκατημόριον *codd.*

---

545 operationem] opositionem *Ve*^i.t.^, al' operationem *Ve*^mg^ calidioribus *sic codd.* calefactiui] -factum *Ox* 546 humectatiui] humectum (*abbr.*) *Ox* 547 mixturam] misturam *Ox* 549 ipsarum] -orum *Ox* 550 ipsorum (*sc.* signorum) *codd.* 551 et *del.*] *hab. codd.* quedam] quod *Ox* 555 bicorporeis] -iis *Ve* 559 estiuali *scr.*] estuali *codd.* triginta] 30 *Ve* hyemali] yemali *Ox* 561 sol] *om. Ox* rediens] radiens *Ox* 563 autem] aut *Ox* 564 Aries] aeris *Ox* 565 quod *scr. cum Ve*] *om. Ox* Chelarum *scr.*] telarum *Ox*: celorum (-o- *in ras.*) *Ve*^i.t.^, al' celarum *Ve*^mg^ hec rursum] hoc sursum *Ox*

ab accidente, quoniam in principiis ipsorum existens sol equales facit ubique noctes | diebus.

[4] Reliquarum autem <octo> duodenarum partium quatuor quidem uocantur fixa, quatuor autem bicorporea. Et fixa quidem sunt que secuntur tropica et equinoctialia, scilicet Taurus, Leo, Scorpius, Aquarius, quoniam | temporum incipientium in illis humiditates et caliditates et siccitates et frigiditates, sole existente in illis, magis et firmius nos occupant, non quod consistentie que fiunt per naturam sint tunc magis impermixte, sed quia nos iam morati sumus in ipsis et propter hoc quod | sensibilius est uirtutis percipimus.

[5] Bicorporea autem sunt que fixa consequuntur, Gemini, Virgo, Sagittarius, Pisces, quia sunt in medio fixorum et tropicorum et quasi communicant penes fines et principia naturali | proprietate duarum consistentiarum.

De masculinis et femininis signis
Capitulum 13^m^

[1] Iterum autem eodem modo sex quidem duodenarum partium attribuerunt nature masculine et diurne, sex autem feminine et nocturne. Et ordo quidem datus est ipsis ponendo unum post unum, quia | connexa sunt et prope semper dies nocti et femininum masculino.

[2] Principio autem accepto ab Ariete propter causas quas diximus, et eodem modo masculino principante et primoritatem habente, quoniam factiuum semper prius est uirtute passiuo, | Arietis quidem signum et Libre masculina putauerunt et diurna, quoniam equinoctialis circulus descriptus per ipsa primam et fortissimam lationem totorum efficit. Que autem sunt ipsis deinceps consequenter secundum ordinem unius post unum, ut diximus.

[3] Vtuntur autem aliqui ordine masculinorum et femininorum et ab oriente signo, quod itaque uocant horoscopum, principium masculini facientes. Sicut enim principium tropicorum a signo lunari accipiunt quidam, quia ipsa aliis citius | mouetur, ita et principium masculinorum ab horoscopizante, quia est magis orientale.

V (**α**=**YΣ** **β**=**DMS** **γ**=**ABC**, *excerpta* **L** *ad* 577) 569 equinoctialia (*cett.*)] + ὡς μέσοις τοῖς τε τροπικοῖς καὶ (*sequente lac. 7 litt.*) **V**; *cf.* **γ** (571-2 in illis*] ἐν τούτοις *codd.*; *an perperam* **G** ?) 574 uirtutis: ἰσχύος **L Y M**] ἰχθύος **V**: ἰσχύος αὐτῶν **Σ D S γ** (*Hüb.*) 576 tropicorum (**V L β** Procl., *cf. Boll*)] + καὶ ἰσημερινῶν **α γ** (*Hüb.*) 585 primoritatem habente: πρωτεύοντος *cett.*] -εύον **V** 586 et (**V**)] + ἔτι *cett.* (*Hüb.*) 588 totorum (**α M S**, *Hüb.*)] *om.* **V D γ**

---

568 octo *add.*] *om. codd.* 571-2 in illis*] nullis *Ox*: *an* in hiis *scrib.* ? 577 et] *om. Ve* 584 Principio] primo *Ox* diximus — prius est (586)] *om. Ve* 585 primoritatem *scr.*] primo raritatem *Ox*: [*Ve*] 588 efficit] -citur *Ox* consequenter *scr.*] cñr *Ve*: cñs *Ox* 589 unius] ipsius *Ox* 591-3 masculini — principium] *om. Ve*^i.t.^, *suppl. Ve*^mg^

[4] Et hii quidem utentes rursum similiter ordinem unius post unum, hii autem penes totas quartas diuidentes et matutina quidem existimantes et masculina quod est ab ascendente usque ad id quod est in medio celi, et | secundum oppositionem quod est ab occidente usque ad oppositum medii celi sub terra, uespertina autem et feminina relique due quarte.

[5] Et alias quasdam appellationes duodecim signis adaptauerunt a figuris que sunt penes ipsa (dico autem puta | quadrupedia et campestria et rectiua et multe dispersionis et talia) quas, quia ex se causam et manifestationem habent, superfluum existimamus dinumerare, cum qualitas que ex talibus occurrit figurationibus possit preexponi in quibuscumque iudiciis oportuna esse uideatur.

De signis duodecim que ad inuicem configurantur
Capitulum 14^{m}

[1] Conueniunt autem inuicem partium zodiaci primo que configurantur. Hec autem sunt quecumque oppositionem habent, continentia duos rectos angulos et sex signa | et gradus 180; et quecumque trigonalem habent situm, continentia unum rectum angulum et tertiam et quatuor signa et gradus 120; et quecumque dicuntur tetragonizare, continentia unum rectum angulum et tria signa et gradus 90; et adhuc quecumque exagonum faciunt situm, continentia | duas partes anguli recti et signa duo et gradus 60.

[2] Propter quam autem causam hee sole distantiarum accepte sunt, ex hiis utique addiscemus. Eius enim que est secundum oppositionem manifesta est ratio ex se ipsa, quoniam super unam rectam faciunt coobuiationes. Acceptis autem | duabus maximis et per consonantiam partibus et superpartientibus, partibus quidem (medietate scilicet et tertia) per comparationem ad oppositionem duorum rectorum: quod quidem in duo eam que quadrati fecit, quod autem in tria eam que exagoni et eam que trigoni.

V (**α**=**YΣ** **β**=**DMS** **γ**=**ABC**, *excerpta* **L** *ab* 605) 598 oppositum medii celi sub terra: *an* τοῦ ὑπὸ γῆν ἀντιμεσουρανοῦντος *leg.* **G** ?] τοῦ ὑπὸ γῆν μεσουρανοῦντος **V β γ** (*Hüb.*): τοῦ ἀντιμεσουρανοῦντος **α** 603 cum qualitas: τῆς ... ποιότητος **α** (*Hüb.*)] τοῖς ... ποιότητος **V**; *cf.* **β γ** (605 ad inuicem] *om. codd.*) 608 duos rectos angulos: δύο ὀρθὰς γωνίας **L α** (*Hüb.*)] ὀρθὰς ἴσας γωνίας **V**, *cf. cett.* 610 tertiam: τρίτην **V**] -ον **L Y** (*Hüb.*), *cf. cett.* 611 angulum (**M S γ**)] *om.* **V L α D** (*Hüb.*); *fort. suppl.* **G** 612 anguli: γωνίας **V L** (*Boll*)] μίας **α β γ** (*Hüb.*) 613 60 (*cett.*)] ζʹ **V** 618 oppositionem ...: τὸ ... διάμετρον **L β**: τὴν ... διάμετρον **α** (*Hüb.*)] τὸ ... (*om.* διάμετρον) **V** (*Boll*) 619 tria (*cett.*)] ς **V**

---

596 quidem] quidam *Ox* 601 campestria *scr.*] -postria *codd.* 603 existimamus] extim- *Ve* qualitas] quantitas *Ox* 604 oportuna] -num *Ve* 607 configurantur] cum figurantur *Ve* 608 oppositionem] opinionem *Ox* 609 habent] *om. Ox* 617 consonantiam] -ia *Ox* 618 medietate] mediante (?) *Ox* 619 quod[2]] que *Ve*

[3] Superpartientibus autem (emiolii scilicet et supertripartientis) per comparationem ad quadraturam unius recti anguli, | intermedium acceptum: emiolium quidem rursum fecit eam que quadrati ad eam que exagoni, supertripartiens autem eam que trigoni ad eam que quadrati.

Harum tamen configurationum triangulares quidem et | sextiles consone uocantur, quia ex eiusdem generis signis constituuntur, scilicet ex omnibus femininis uel ex omnibus masculinis. Inconsone autem que ex opposito et que ex quadrato, quia secundum oppositum omogeneorum situm accipiunt.

De imperantibus et obedientibus similiter signis
Capitulum 15^{m}

[1] Eodem autem modo imperantia et obedientia dicuntur signa penes equalem distantiam ab eodem et quocumque equinoctialium punctorum figurata, quia in | equalibus temporibus eleuantur et descendunt et quia in equalibus sunt parallelis.

[2] Horum autem que quidem in estiuo semicirculo imperantia uocantur, que autem in hyemali obtemperantia, propterea quod apud illa existens sol diem facit maiorem nocte, apud | ista autem minorem.

De aspicientibus et equipotentibus inuicem signis
Capitulum 16^{m}

[1] Iterum autem equaliter posse aiunt inuicem partes quasdam, que equaliter distant ab eodem et quocumque tropicorum punctorum, | quia apud utrumque ipsorum sol existens dies diebus et noctes noctibus et distantias propriorum temporum in equali tempore perficit. Hec autem et respicere inuicem dicuntur propter predicta et quia utrumque ipsorum ex eisdem partibus | orizontis oritur et in easdem occumbit.

V (**α**=**YΣ** **β**=**DMS** **γ**=**ABC**, *excerpta* **L** *ad* 628) 622 acceptum: λαμβανόμενον **L** **γ**] -νομεν **V**: -νωμεν **Y**: -νομένων **β** (*Hüb.*): -νομένου **Σ** (*Robb.*) 627 que ex opposito et que ex quadrato: οἵ τε διάμετροι καὶ οἱ (αἱ **L**, *om.* **M S**) τετράγωνοι **L** **β** **γ**] οἵ τε τρίγωνοι καὶ οἱ τετράγωνοι **V**: οἱ τετράγωνοι καὶ οἱ κατὰ διάμετρον **Σ** (*Hüb.*); *cf.* **Y** 629 similiter: ὁμοίως **V** (*Hüb.*)] *om.* **α** **β** **γ** Heph. («fort. recte») (632 et] ἢ καὶ *codd.*) 633 et descendunt: τε καὶ καταφέρεσθαι **β** **γ** Procl. (*Hüb.*)] *om.* **V** **α** 636 illa: ἐκείνων **β** **γ** (*Hüb.*)] -νον **V** **Y**: -νο **Σ** 640 (aiunt: φασιν *cett.*] φησὶν **V**) quasdam: τά τινα **V**] *om. cett.* (*Hüb.*)

---

621 superpartientibus] -tientis *Ve* emiolii] emielii *Ve* 622 emiolium *scr.*] -iam *codd.* 623 rursum] sursum *Ox* 625 configurationum] figurationum *Ve* sextiles] sestiles *Ox* 626 constituuntur] -tuunt *Ox* 635-6 uocantur — obtemperantia] *it.* *Ox*^{ac} 636 apud] potentia (*abbr.*) *Ox* 637 apud] a potentia (*abbr.*) *Ox* 638 inuicem] in lucem *Ox* 640 equaliter] equali (*abbr.*) *Ox* 645 easdem] eadem *Ox*

De inconnexis
Capitulum 17m

[1] Inconnexa autem et abalienata uocantur signa quecumque nullam habent ad inuicem rationem prenumeratarum conuenientiarum. Hec autem sunt que neque existunt imperantium | uel obedientium neque aspicientium aut equipotentium, adhuc autem et expositarum quatuor figurationum (diametralis scilicet et trigonalis et quadrature et exagonalis) penitus sine participatione deprehendimus que post unum aut post quinque fiunt | signa.

[2] Quoniam que quidem post unum, aduersa sunt sicut secus inuicem et, cum sint ipsa duo, unius continent [et] angulum, que autem post quinque in inequalia diuidunt totum circulum, aliis figurationibus facientibus diuisionem perimetri in equalia.

De domibus uniuscuiusque planete
Capitulum 18m

[1] Coaptantur autem et planete zodiaci partibus secundum uocatas domos et trigonalitates et exaltationes et terminos et talia. Et quod quidem domorum talem habet naturam.

[2] Quoniam enim XII[cim] animalium borealiora | et plus aliis appropinquantia loco qui est penes uerticem capitum nostrorum caliditatisque et estus propter hoc factiua existentia sunt Cancer et Leo, ideo duo hec maximis et principalissimis, hoc est duobus luminaribus, assignauerunt domos.

[3] Leonem quidem masculinum | existentem s o l i, Cancrum autem femininum existentem l u n e. Et consequenter semicirculum quidem qui est a Leone usque ad Capricornum solarem supposuerunt, illum autem qui est ab Aquario usque ad Cancrum lunarem, quatinus in utroque semicirculorum unum animalium secundum unumquemque 5 planetarum conuenienter | assignetur (hoc quidem ad solem, hoc autem ad lunam figuratum), consequenter speris motuum ipsorum et proprietatibus naturarum.

V (**α**=**YΣ** **β**=**DMS** **γ**=**ABC**, *excerpta* **L** *ab* 658) 649 rationem(**V**)]+ἁπλῶς*cett.*(*Hüb.*) 651 autem et: δὲ καὶ **Σ**: τε καὶ **V** (*Hüb.*)] καὶ *cett.* 652 deprehendimus: καταλαμβανόμεθα **V**: -νομεν **Y**] -νόμενα **Σ** (*Hüb.*), *cf.* **β γ** que: τὰ **V**] καὶ *cett.* (*Hüb.*) 663 borealiora: βορειότερα **V L D M γ**] -τατα **α S** (*Hüb.*) 665 (ideo] *om. codd.*) 667 existentem (*cett.*)] *om.* **V L** 669 ab Aquario: ἀπὸ Ὑδροχόου **L α β** (*Hüb.*)] τοῦ Διδύμου **V**, *cf.* **γ** (670 unum animalium] ἓν ζῴδιον *cett.* (*Hüb.*): ἐν τῷ ζῴδιον **V**)

---

648 autem] sunt *Ox* 655 et *del.*] *hab. codd.* in] *om. Ox* 660 planete] plante *Ox* 663 borealiora] boreariola *Ve* 665 et Leo] *tr. Ox* 667 autem *scr.*] aut *Ox*: *om. Ve* 668 qui] quod *Ox* 669 supposuerunt] super (?) posuerunt *Ve* 672 autem] quidem *Ox* figuratum] -tus *Ox*

[4] Saturno quidem, infrigidatiuo existenti magis secundum naturam secundum contrarietatem calidi et habenti | zonam supremam et elongatam a luminibus, data sunt animalia diametralia Cancri et Leonis, Aquarius scilicet et Capricornus, cum eo quod etiam hec signa frigida et hyemalia existunt et adhuc quod configuratio diametralium inconsona est ad beneficia.

[5] Ioui autem, qui est temperatus et sub Saturni spera, data sunt duo que secuntur premissa, uentosa existentia et fecunda, Sagittarius et Pisces, secundum triangularem ad lumina distantiam, que est consone et benefice configurationis.

[6] Deinceps autem Marti exsiccatiuo | magis existenti secundum naturam et sub Ioue habenti speram, signa habita illis rursum data sunt similem habentia naturam, Aries scilicet et Scorpius, conformiter corruptiue et inconsone qualitati, habentia situm quadrature ad luminaria.

[7] Veneri autem temperate existenti et sub sole, habita duo animalia data sunt fecundissima existentia, Khele uidelicet et Taurus, seruantia consonantiam sextilis distantie, et quia non plus duobus signis hec stella ad utraque secundum plurimum distat a | sole.

[8] Tandem autem Mercurio, numquam plus uno signo facienti distantiam a sole ex utraque parte et existenti sub aliis, magis autem prope aliqualiter ambobus luminaribus, reliqua duo signa et continua domibus illorum data sunt, | Gemini scilicet et Virgo.

V (**α**=**ΥΣ** **β**=**DMS** **γ**=**ABC**, *excerpta* **L**)   674 quidem (**β**)] + διὰ **V**: + δὴ **L**: + οὖν **γ** (*Hüb.*); *cf. cett.*   675 zonam: ζώνην *cett.*] ζωὴν **V**   676 diametralia: τὰ διάμετρα *cett.*] *om.* **V**   676-7 Aquarius ... Capricornus: Ὑδροχόος ... Αἰγόκερως **L M**] *inv. cett.* (*Hüb.*): Ζυγὸς ... Αἰγόκερως **V**   677 cum eo quod etiam hec: μετὰ τοῦ καὶ ταῦτα *cett.*] μετὰ ταῦτα **V**   678 quod configuratio diametralium: τοῦ τὸν τῶν διαμέτρων συσχηματισμὸν] τοῦτον κατὰ διάμετρον συσχηματισμὸν **V**: τῶν (+ τῶν **Υ**) διαμέτρων συσχηματισμῶν **L Υ**: τὸ τοῦ κατὰ διάμετρον συσχηματισμοῦ **γ**: τοῦ τὸν κατὰ διάμετρον συσχηματισμὸν *Hüb.*; *cf.* **β**   682 existenti: ὄντι *cett.*] ὅτι **V**   683 signa: δωδεκατημόρια *cett.*] -μορίων **V**   (685 situm] διάστασιν *codd.*)   686 Veneri: τῷ δὲ τῆς Ἀφρωδίτης **L Υ D M γ** (*Hüb.*)] ... τῆς Ἑρμῆς **V**; *cf.* **S**   sub sole (ὑπὸ τὴν τοῦ ἡλίου ?)] ὑπὸ τὴν τοῦ Ἄρεως *codd.* (*cf.* **Υ Σ M**)   687 et Taurus: καὶ ὁ Ταῦρος *cett.*] εὖ **V**   688-9 secundum plurimum (*cett.*, *Hüb.*)] *om.* **V L**   692 signa: δωδεκατημόρια *cett.*] -μορίοις **V S**

---

675 elongatam] -tum *Ox*   luminibus] *abbr. codd.*: *an* -aribus *scrib.* ?   677 hyemalia *scr.*] hyemalea *Ve*: yemalia *Ox*   adhuc *scr.*] ad h' *Ve*: ad hoc *Ox*   678 configuratio] con- Ox^{ss} 680 secuntur *scr. cum Ve*] consequuntur *Ox*, *an recte* ?   681 lumina ($Ve^{i.t.}$)] unam *Ox*: luminarium Ve^{mg}: *an* -aria *scrib.* ?   benefice] -ficie *Ve*   683-4 et sub Ioue — naturam] *hom. om. Ve*   684 Aries] aer[i]es *Ve*   conformiter *scr.*] -firmiter *Ve*: -formis *Ox*   686 animalia] alia *Ox*   687 fecundissima] fecon- *Ve*   691 existenti] -te *Ox*

De triplicitatibus
Capitulum 19[m]

[1] Que autem est ad trigonalitates conuenientia talis quedam existit. Quoniam enim trigonum equilaterum figura est maxime consona sibi ipsi et zodiacus a tribus circulis | terminatur, equinoctiali scilicet et duobus tropicis, ideo diuiduntur 12 ipsius partes in 4 trigona equilatera. P r i m u m quidem quod est per Arietem et Leonem et Sagittarium, ex tribus masculinis animalibus compositum et domos habens solis et Iouis et Martis. | Datum est Ioui et soli, extra partitionem solarem existente Marte. Accipiunt autem ipsius primam dominationem de die quidem sol, de nocte autem Jupiter. Et est Aries quidem magis apud equinoctialem circulum, Leo autem apud estiualem, Sagittarius autem apud hyemalem.

[2] Fit autem hec triplicitas principaliter borealis propter dominationem Iouis, quoniam iste fecundus est et spirituosus, conformiter uentis qui sunt ab Vrsa. Propter domum autem Martis accipit quandam admixtionem occidentalis uenti et fit borealibica, quia | Mars talium uentorum est factiuus, propter partitionem autem lune et regione occasuum feminina facta.

[3] S e c u n d u m autem trigonum, quod est per Taurum et Virginem et Capricornum, compositum ex tribus femininis, | consequenter datum est lune et Veneri, dominante sibi nocte quidem luna, de die autem Venere. Et est Taurus quidem apud estiualem circulum magis, Virgo autem apud equinoctialem, Capricornus autem apud hyemalem.

[4] Fit autem hoc trigonum | principaliter quidem australe propter dominationem Veneris, quoniam hec stella similium spirituum est propter calidum uirtutis et uaporatiuum effectiua; assumens autem mixturam subsolani (quia domus

---

V (**α**=**YΣ** **β**=**DMS** **γ**=**ABC**, *excerpta* **L**)   697 trigonum] + καὶ *codd.*   (sibi ipsi: ἑαυτῷ **L Σ β γ**] αὐτῷ **V Y:** αὑτῷ *coni. Hüb.*)   698 tribus (*cett.*)] *om.* **V**   (ideo] δὲ *post* διαιρεῖται *codd.*, *del. Hüb.*)   700 ex (*cett.*)] + γὰρ **V L**   701 Martis: Ἄρεως **L β** (*Hüb.*)] ἄρρεν **V**; *cf. cett.*   Ioui et soli: τῷ Διὶ καὶ ἡλίῳ **L D γ** (*Hüb.*)] δι (*sequente lacuna*) **V**; *cf. cett.*   extra: παρὰ *cett.*] *om.* **V**   702 Marte (*cett.*)] *om.* (*spatio relicto*) **V**   (dominationem: οἰκοδεσποτείαν **α β γ**, *Hüb.*, *an etiam* **G** ?] συνοικο- **V L**, Boll; *sed cf.* 705/936°)   705 principaliter (**V L β**)] + μὲν **α γ** (*Hüb.*)   708 fit: γίνεται **V L**] συνίσταται **α β γ** (*Hüb.*)   712 sibi: αὐτῷ **V**] αὐτοῦ **L Y β γ** (*Hüb.*): αὐτῶν **Σ**   Venere (*cett.*)] τοῦ τῆς Ἑρμῆς **V**   Taurus (*cett.*)] Κριὸς **V**   713 Virgo — equinoctialem (*cett.*)] *hom. om.* **V**

---

698 et] a *Ox*   699 12 *scr.*] ēi *Ve*: ii. (*vel* n.) *Ox*   700 Sagittarium *scr.*] sagita- *Ve*: sagirta- *Ox*   701 solis] solum *Ox*   703 quidem[1]] quod *Ox*   704 Sagittarius] sagita- *Ve*: sagirta- *Ox*   autem[2]] nunc *Ox*   hyemalem] yem- *Ox*   705 hec] hoc *Ox*   706 iste] ipse *Ve*   fecundus] fertilius *Ve*   spirituosus *scr.*] spumosus *codd.*   uentis] uenteris *Ox*   707 admixtionem] admirtio- *Ox*   708 borealibica] -libixa *Ve*   partitionem] participationem *Ox*   709 regione … feminina (femina *Ox*) facta *sic codd.*   occasuum] -sum *Ve*   711 consequenter] *spat. Ox*   714 hyemalem] yem- *Ox*

Saturni in ipso existit, scilicet Capricornus) consistit | et ipsum notapelioticum secundum oppositionem primi, quoniam est Saturnus talium uentorum factiuus, conformis et ipse orienti propter partitionem ad solem.

[5] Tertium autem trigonum, quod est per Geminos et Libram | et Aquarium, ex tribus masculinis signis compositum, et ad Martem quidem nullam rationem habens, ad Saturnum autem et Mercurium propter domos; hiis attributum est rursum dominantibus, de die quidem Saturno propter partitionem, de nocte autem | Mercurio.

[6] Et est Geminorum quidem signum apud estiualem, Khelarum autem apud equinoctialem, Aquarii uero apud hyemalem. Exstat autem et hoc trigonum principaliter quidem subsolanum propter Saturnum, secundum mixturam autem boreapelioticum, propter | partitionem Iouis ad diurnalem rationem coaptari ei que Saturni.

[7] Quartum autem trigonum, quod est per Cancrum et Scorpionem et Pisces, relictum est quidem reliquo existenti, scilicet Marti, et rationem habenti ad ipsum propter domum Martis, scilicet | Scorpionem; condominantur autem ipsi propter partitionem et propter feminitatem animalium nocte quidem luna, de die autem Venus. Et est Cancer quidem apud estiualem circulum, Scorpius autem apud hyemalem magis, Pisces autem apud equinoctialem.

[8] Et hoc autem trigonum | exstat principaliter quidem occidentale propter dominationem Martis et lune, secundum mixturam autem notolibicum propter dominationem Veneris.

De exaltationibus
Capitulum 20^m

[1] Vocate autem planetarum exaltationes rationem habent | talem. Quoniam enim sol in Ariete quidem existens facit transitum in semicirculum altum et borealem, in Khelis autem in humile et australe, merito Arietem quidem tamquam ex-

V (**α**=**ΥΣ** **β**=**DMS** **γ**=**ABC**, *excerpta* **L**) 718 notapelioticum (*cett.*)] ὄντα ἀφηλιωτικὸν **V** 721 et ... et (*cett.*)] *om.* **V** Aquarium (*cett.*)] Σκορπίου **V** 727 Aquarii (*cett.*)] Σκορπίου **V** 729 diurnalem: ἡμερινὸν *cett.*] ἰσημερινὸν **V** ei que: τῇ *cett.*] τὴν **V**: τῷ **Σ** 732 quidem: μὲν **V L** (*Robb.*)] μόνῳ *cett.* (*Hüb.*) (733 Martis] *om. codd.*) 735 Venus (*cett.*)] ὁ τῆς Ἑρμῆς **V** (739 dominationem: οἰκοδεσποτείαν **V α**, *Robb.*] συνοικο- **L β γ**, *Hüb.*)

718 scilicet] si *Ox* notapelioticum] -lieticum *Ve* 726 estiualem] estualem *Ox* 727 equinoctialem] eque- *Ox* Exstat] existat *Ox* 728 mixturam] mixtam *Ox* 729 partitionem] partionem *Ox* Iouis] + et *Ox* ei que] eique *Ox* 733 Scorpionem] -ni *Ox* 734 partitionem] -cipationem *Ox* 736 hyemalem *scr.*] hymalem *Ve*: yem- *Ox* 737 exstat] extat *Ox* 738 notolibicum *Ve*[i.t.]] notobilicum *Ox*: notabilicum *Ve*[mg] 743 transitum] -tura *Ox* 744 exaltationem] exacta- *Ox*

altationem reposuerunt ipsi, penes quem incipit diei magnitudo et | calefactiuum sue nature augeri, Khelas autem tamquam humiliationem propter contraria.

[2] Saturnus autem rursum, ut ad solem habeat diametralem situm quemadmodum in domibus, Libram quidem opposite uelut exaltationem accepit, Arietem autem uelut humiliationem. Vbi enim calidum augetur, | minuitur frigidum, et ubi frigidum augetur, minuitur calidum.

[3] Rursum quoniam in exaltatione solis, Ariete uidelicet, pertransiens luna primam apparitionem facit et principium augmenti luminis et quasi exaltationis in primo animali proprii trigoni, | Tauro uidelicet, hoc quidem ipsius exaltatio uocatum est, diametrale autem, Scorpius scilicet, humiliatio.

[4] Post hec autem Jupiter quidem borealium et fecundorum spirituum effectiuus existens, in Cancro maxime borealissimus factus, augetur rursum et adimplet propriam | uirtutem; unde hoc signum exaltationem ipsius fecerunt, Capricornum autem humiliationem.

[5] Mars autem natura estuosus existens et magis in Capricorno, quia australissimus est, estuosissimus factus, et ipse exaltationem quidem accepit merito secundum oppositionem exaltationi | Iouis, Capricornum, humiliationem autem Cancrum.

[6] Rursus Venus humectatiua existens per naturam et magis in Piscibus, in quibus magis notatur aeris humidi principium, et ipsa augens propriam uirtutem, exaltationem quidem habet in Piscibus, humiliationem autem in | Virgine.

[7] Mercurius autem contrarium magis subsiccus existens merito secundum oppositum in Virgine quidem, secundum quam siccum autumpnale prenotatur, et ipse quasi exaltatur, apud Pisces autem quasi humiliatur.

V (**α**=**YΣ** **β**=**DMS** **γ**=**ABC**, *excerpta* **L**) 748 Libram: ζυγὸν *cett.*] *om.* (*lac.*) **V** 749 (Vbi enim calidum augetur) minuitur frigidum: μειοῦται τὸ ψυχρόν] (...) ἐκμειοῦται ἐκεῖ τὸ ψ. **V**: (...) μειοῦται ἐκεῖ τὸ ψ. **α M S** (*Hüb.*): (...) ἐκεῖ μειοῦται τὸ ψ. **L** (*Boll*): (...) μειοῦται τὸ ψ. ἐκεῖ **D γ**: *fort.* ἐκεῖ *correctio ulterior erat, quam exemplar Guillelmi nondum habuit; cf. adn. insequ.* 750 ubi frigidum augetur, minuitur calidum: ὅπου τὸ ψυχρὸν αὔξεται μειοῦται τὸ θερμόν] ὅπου τὸ ψυχρὸν αὔξεται, ἐκμειοῦται (μηοῦτε **L**) ἐκεῖ τὸ θ. **V L** (*cf. Hüb.*): ὅπου ἐκεῖνο μειοῦται, τὸ θερμὸν αὔξεται **Y**: ὅπου ἐκεῖνο μειοῦται, τὸ ψυχρὸν αὔξεται **β**, *cf. cett.* 752 quasi: ὥσπερ ἂν εἰ **Y** (*cf. Boll Hüb.*): ὡσπερεὶ **Σ β γ** (*Robb.*)] ὥσπερ ἂν εἴη **V**, *cf.* **L** 753 Tauro (*cett.*)] *om.* (*lac.*) **V** 757 signum (*cett.*)] *om.* (*lac.*) **V** 761 Capricornum: τὸν Αἰγόκερων **L α**] τῶ (*sic*) τοῦ Ἀιγοκέρου **V**: *om. cett.* 763 Venus (**V L D γ**)] μὲν (μέντοι **Σ**) *praem.* **α M S** (*Hüb.*) 764 magis: μᾶλλον **V L βγ**] *om.* **α** (*Hüb.*) aeris] ἔαρος *codd.* 767 autumpnale (μετοπωρινὸν ?)] μετόπωρον *codd.* ***inter 768 et 878 excerpta cod. L deficiunt***

---

745 diei] dierum (?) *Ve* 748 Arietem] ariete *Ox* 751 Ariete] in *s.l. praem.* (*an recte* ?) *Ve* 753 hoc] hec *Ox* 755 effectiuus] affect- *Ox* 756 maxime] -mo *Ox* 760 et *scr.*] est *codd.* 763 Rursus *codd.*] *fort.* rursum *scrib.* 765 habet] h' *Ve* 767 oppositum] -tam *Ve*

De terminorum dispositione
Capitulum 21^m

[1] De terminis duo modi feruntur maxime, et unus quidem est Egiptiorum, alius autem Caldeorum (qui ad trigonorum dominationes). E g i p t i a c u s quidem igitur qui est terminorum communiter dictorum, non omnimode saluat consequentiam neque | ordinis neque quantitatis per singula.

[2] Primo quidem enim in ordine, hac quidem dominis domorum primos dederunt, hac autem dominis trigonorum, aliquando autem dominis exaltationum. Quoniam exempli gratia, si domos sequebantur, propter quid Saturno, si forte, primo dederunt | in Libra et non Veneri ? Et propter quid in Ariete Ioui et non Marti ?

[3] Si uero trigona, propter quid Mercurio dederunt in Capricorno et non Veneri ? Si autem exaltationes, propter quid Marti in Cancro et non Ioui ? Siue autem | habentibus plura horum, propter quid in Aquario Mercurio dederunt, habenti trigonum solum, et non Saturno ? Huius enim et domus est et trigonum.

[4] Aut propter quid omnino in Capricorno Mercurio primo dederunt, nullam rationem dominationis habenti ad animal ? Et in reliquo autem | ordine eandem contrarietatem inueniet utique aliquis.

[5] Secundo autem et quantitas terminorum nullam consequentiam habere uidetur. Numerus enim qui colligitur ex omnibus secundum unum quodcumque astrum, ad quem aiunt ipsorum cronica in summam redigi, nullam conuenientem | habet rationem neque acceptabilem. Si etiam huic secundum collectionem numero crediderimus tamquam e regione ab Egiptiis confesso, multipliciter quidem et aliter quantitate permutata penes signum, numerus utique collectus inuenietur.

V (**α**=**YΣ** **β**=**DMS** **γ**=**ABC**) 771 terminis] + δὲ *codd.* Egiptiorum (**V β γ**, Procl.)] + ὁ πρὸς τὰς — κυρίας **α** (*Hüb.*) 774 per singula: καθ' ἕκαστον *cett.*] καθ' ἑκάτερα **V** 775 enim (**α**, *Robb.*)] *om.* **V β γ** (*Hüb.*) (777 si¹: εἰ **D**: εἴ γε **M S**, *Robb.*] εἴτε **V**, *Hüb.*: *cf.* **α γ**; *fort. hic et infra* (*cf.* 780, 781) εἴτε *sua sponte redd.* **G**) 777-778 Saturno si forte (*cett.*, *cf.* **γ**)] *om.* (*lac.*) **V** 778 non (*cett.*)] *om.* (*lac.*) **V** Veneri (*cett.*)] ὁ τῆς Ἑρμῆς **V** 779 non (*cett.*)] οὕτω **V** (780 Si uero] εἴτε **V Y**, *Hüb.*: εἰ δὲ **Σ γ**) (781 Si autem] εἴτε καὶ **V α β**, *Hüb.*: *cf.* εἰ δέ γε καὶ **γ**) (Siue autem] εἴτε **V α**, *Hüb.*: εἰ δὲ **β γ**) 783 habenti (*cett.*)] ἔχουσι **V** 785 autem ordine: δὲ τάξεως **V** *cett.*] διατάξεως **G**°(*Hüb.*), -αιως **Y** (786 contrarietatem] ἀναλογίαν *codd.*) 790-1 secundum collectionem: κατὰ συναγωγὴν *cett.*] κατὰ -γῆς **V** 792 numerus ὁ ἀριθμὸς **V β**] ὁ αὐτὸς ἀριθμὸς α *Hüb.*

---

771 quidem] quidemque *Ve* 772 qui] quod *Ox* 774 singula] singulam *Ox* 775 hac] hos hac *Ve*: *an* hac *correctio pro* hos *erat* ? primos [*sc.* terminos] *codd.* 776 hac] hos hac *Ve* 777 quid] quod *Ox* 778 primo (πρώτῳ) *scr.*] primos *codd.* quid] quod *Ox* 781 Siue] sine *Ox* 782 plura *perscr. Ox*] pl'a *Ve* 786 contrarietatem* *sic codd.* 788 quodcumque] quodque *Ox* 789 summam *scr. cum Ve*] summa *Ox* 790 Si] sed *Ox* 791 crediderimus *scr.*] reddid- *Ox*: credidimus *Ve* 792 permutata *scr.*] -tate *codd.*

[6] Et quod quidam conantur persuadere et ratiocinari de ipsis, ascensionali ratione scilicet in omni climate, quod scilicet tempora que colliguntur aliqualiter secundum unumquodque astrum ad eandem conducant quantitatem, falsum est.

[7] Primo quidem enim secuntur communem traditionem, et ad regulares | excessus ascensionum constitute neque ad modicum propinque ueritati, secundum quam in parallelo quod est per regionem Egipti, Virginis quidem et Libre signum in 38 partibus utrumque et adhuc tertia uolunt eleuari, Leo autem et | Scorpio utrumque in 35, cum sit demonstratum per lineas quod hec quidem adhuc pluribus quam 35 temporibus sursumferuntur, Virgo autem et Libra paucioribus.

[8] Deinde et qui tale conati sunt instituere non adhuc uidentur assecuti neque sic | quantitatem terminorum que fertur apud plurimos, et in multis compulsi mentiri et alicubi partibus partium uti gratia saluandi propositum suum, cum ipsi, ut diximus, non habeant ueram intentionem.

Que tamen feruntur apud multos propter | traditionem hactenus dignam fide, hoc supponuntur modo.

V (**α**=**YΣ** **β**=**DMS** **γ**=**ABC**) 794 (et[2]: καὶ **Σ D γ**, *Boll*] δὲ καὶ **V Y**, *cf.* **M S**: τε καὶ *coni. Hüb.*) 794-5 (ascensionali ratione* scilicet in omni climate, quod] ὅτι κατὰ παντὸς κλίματος ἀναφορικὸν λόγον *codd.*) 798 constitute: συνισταμένῃ *cett.*] συνιστάμενα **V** neque: μηδὲ **Σ β γ**] μὴ **V Y** (*Hüb.*) propinque (*modus dativus ut* constitute): *fort.* ἐγγιζούσῃ (*coni. Boll*) *legit* **G**] ἐγγιζούσης **V M S γ** (*Hüb.*), *cf.* **D**: ἐγγὺς οὔσῃ **α** (*Robb.*) 799 regionem] κάτω *praem. codd.* 801 adhuc: ἔτι **V**] ἐν *cett.* (*Hüb.*) 802 paucioribus: ἐλάττοσι **V β**] ἐν *praem.* **α γ** (*Hüb.*) 805 uti: χρήσασθαι **M S** (*Hüb.*)] ἐχρήσασθαι **V**: ἐχρήσαντο **Σ D γ**: [**Y**] 807 hactenus: ἐπάνωθεν **α M S**, *Hüb. Robb. vel* ἔτι ἄνωθεν **V D γ**, *Boll* ?

---

794 ratiocinari *scr.*] ratio terminari *Ve*: rationari *Ox* 794-5 ascensionali ratione scilicet *scr.* (*cf.* κατά *cum acc.*: *per abl.*)] ascensionalis (+ scilicet *sic Ox*) ratio scilicet *codd.* 798 constitute *codd.*: *fort. dativum* συνισταμένῃ (*qui cum dativo praecedente* πραγματείᾳ = traditionem *concordat*) *perperam per dativum* (constitute) *loco accusativo redd.* **G**, *cf.* propinque* 799 parallelo *scr.*] paralello *codd.* 800-1 38 *et* 35 *in numeris* (*ut in graeco*) *Ve*] *in litteris Ox* 801 per lineas quod hec] persias et hoc *Ox* 802 *post* paucioribus *tabula* (*cf. infra* 811-835) *inserta est in Ve, verbis praecedentibus*: Hec est tabula terminorum planetarum (*fort. explicatio translatoris, cf. subscriptionem* 835-836) 803 assecuti] -cati *Ox* 804 fertur] feruntur *Ve* in] *om. Ve* 805 uti] utique *Ve* 807 propter] quod *Ox* 808 hoc] hac *Ve*

[9-10]

Termini secundum Egiptios 1072

Aries	Iouis 6	Veneris 6	Mercurii 8	Martis 5	Saturni 5
Taurus	Veneris 8	Mercurii 6	Iouis 8	Saturni 5	Martis 3
Gemini	Mercurii 6	Iouis 6	Veneris 5	Martis 7	Saturni 6
Cancer	Martis 7	Veneris 6	Mercurii 6	Iouis 7	Saturni 4
Leo	Iouis 6	Veneris 5	Saturni 7	Mercurii 6	Martis 6
Virgo	Mercurii 7	Veneris 10	Iouis 4	Martis 7	Saturni 2
Libra	Saturni 6	Mercurii 8	Iouis 7	Veneris 7	Martis 2
Scorpio	Martis 7	Veneris 4	Mercurii 8	Iouis 5	Saturni 6
Sagittarius	Iouis 12	Veneris 5	Mercurii 4	Saturni 5	Martis 4
Capricornus	Mercurii 7	Iouis 7	Veneris 8	Saturni 4	Martis 4
Aquarius	Mercurii 7	Veneris 6	Iouis 7	Martis 5	Saturni 5
Pisces	Veneris 12	Iouis 4	Mercurii 3	Martis 9	Saturni 2

In hac tabula tractatur de dispositione terminorum planetarum seu de diuisione.

V (**α**=**YΣ** **β**=**DMS** **γ**=**ABC**) 810 Termini secundum Egiptios (**V** **α** **D** **M**, *Hüb.*)] *om.* **S**, *cf.* **γ** 811-34 *tabulam habent* **Σ** **Y**[2] **β**(-**S**) **γ** (*non solum singulos numeros terminorum planetarum habent, sed etiam summas numerorum addunt, quae semper numerum* λ′ [= 30] *dant et quae deficiunt* (*an recte* ?) *in tabula latina*)] *tabulam om.* **V** (*spatio relicto*) **Y**[1] **S** 813 *de Tauro cum Mercurio*: 6 (ς′ *cett.*)] η′ **α**: [**V Y**[1] **S**] 817 *de Cancro cum Venere*: 6 (ς′ *cett.*)] ζ′ **γ**: [**V Y**[1] **S**] 829 *de Capricorno cum Venere*: 8 (η′ *cett.*)] ζ′ **α**: [**V Y**[1] **S**] *de Capricorno cum Marte*: 4 (δ′ *cett.*)] ε′ **α**: [**V Y**[1] **S**] 835-6 In — diuisione] *om. codd.*: *fort. subscriptio ipsius Guillelmi*

810 *inscriptio ad tabulam* «Termini secundum Egyptios» *habet hic ut in graeco** *Ox*, *cf. adnot. insequ.* 811-836 *Tabulam ipsam et subscriptionem hic inserui, habet post* «paucioribus» (802) *Ve*] *om.*, *sed post* «In secundo autem trigono» (848) *spatium* 16 *versuum habet Ox*

[11] Colligitur autem uniuscuiusque ipsorum numerus ita: Saturni quidem 57, Iouis 79, Martis 66, Veneris 82, Mercurii 76; fiunt 360.

[12] Caldaicus autem modus quandam quidem simplicem consequentiam et quantitatis ordinem habet et magis probabilem, non sic autem sufficientem ad | trigonorum dominationes, ita ut possit aliquis de facili scire ipsos etiam sine tabulari descriptione.

[13] In primo quidem enim trigono, Ariete, Leone et Sagittario, eandem apud ipsos habenti diuisionem penes signum, primo quidem accipit | trigoni dominus, Jupiter, deinde consequenter dominus consequentis trigoni, dico autem Venerem, postea consequenter domini Geminorum, Saturnus scilicet et Mercurius, ultimus autem reliqui trigoni dominus, scilicet Mars.

[14] In secundo autem trigono, Tauro et Virgine et Capricorno, rursum | eandem habenti penes signum diuisionem, Venus quidem primo, deinde iterum Saturnus et Mercurius, post hec autem Mars, ultimo autem Jupiter. Fere autem et in reliquis duobus trigonis ordo conspicitur.

[15] Duorum autem dominorum eiusdem trigoni, | dico autem Saturni et Mercurii, prioritatem quidem ordinis secundum conueniens de die quidem Saturnus accipit, nocte autem Mercurius, et quantitas autem per singula simpliciter quedam existit.

[16] Vt enim secundum subiectionem ordinis priorum et quantitas terminorum uniuscuiusque | uno gradu deficiat a premissa, primo quidem semper dederunt gradus 8, secundo autem 7, tertio autem 6, quarto autem 5, quinto autem 4, completis sic 30 gradibus per signum.

V (**α**=**YΣ** **β**=**DMS** **γ**=**ABC**) 837 quidem: μὲν **Y**] μὲν ἔτη **V D M**: μοῖραι **γ** (*Hüb.*) 838 fiunt (**V D**)] γίνεται ὅμου γ (*Hüb.*), *cf. cett.* (839-40 consequentiam — ordinem *hic* **G**] *post* δεσποτείας (dominationes 841/1090°) *codd.*: *ordinem sententiarum fort. mut.* **G**) 840 autem: δὲ *cett.*] *om.* **V** 841 ita ut: ὥστε Procl. (66,11)] ὥστε μέντοι *codd.*: *ad hunc locum cf. Robb.* (*p.* 99, *n.* 2), *sensum totius textus ut Procl. intell.* **G** (scire] ἐπιβάλλειν **M S**: ἐπιβαλεῖν **α**: διαβάλλειν **V D γ** (*Hüb.*); *fort.* ἐπιβάλλειν *ad sensum per* scire *redd.* **G**, *cf.* «comprendre» Charvet, «understand» Robbins) ipsos: αὐτοὺς *coni. Boll Hüb.*] αὐτοῖς **V β**, *cf. cett.* 844 primo: πρῶτον **V**] πρῶτος *cett.* (*Hüb.*) 845 consequentis: ἐφεξῆς *cett.*] *om.* **V** Venerem (*cett.*)] ὁ τῆς Ἑρμῆς **V** 846 domini: οἱ **β**] ὁ **V Y** Procl. (*Hüb.*); *cf.* **Σ γ** 848 Tauro (*cett.*)] *om.* **V** Capricorno (*cett.*)] Παρθένῳ (*iter.*) **V** (849-50 primo ... ultimo] πρῶτος ... τελευταῖος *codd.*) 851 ordo (**V**)] + ἥδε *cett.* (*Hüb.*) 853 prioritatem: τὸ πρωτεῖον *cett.*] τὸ πρῶτον **V** quidem[1]] *om. codd.* (854-5 simpliciter ... existit] ἁπλῆ ... οὖσα τυγχάνει *codd.*) 858 5 (*cett.*)] μοίρας πέντε **V** quinto ... 4: πέμπτῳ ... δ′ **S** (*Hüb.*)] πέμπτῳ ... μοίρας δ′ **V**: τελευταίῳ ... δ′ *cett.* 858-9 completis — signum (*cett.*)] *om.* **V**

---

838 79] 59 *Ox* 841 aliquis] alias *Ox* 844 diuisionem *scr.*] -nes *codd.* 848 trigono] + *spatium* 16 *versuum loco tabulae hab. Ox* 849 Saturnus] + scilicet *Ve* 854 nocte] noctem *Ox* simpliciter] -plices *Ox* 856 subiectionem] subicionem *Ox* 859 sic *scr.*] sit *codd.*

[17] Colliguntur autem et ex hoc Saturni quidem gradus de die | quidem 78, de nocte autem 66, Iouis autem 72, Martis uero 69, Veneris autem 75, Mercurii uero de die quidem 66, de nocte autem 78, et fit totum simul 360.

[18] Hiis uero terminis qui sunt secundum Egiptiacum modum existunt magis fide digni, ut diximus, et propter | collectionem ipsorum apud Egiptios scriptores tamquam utilem tabulari descriptione dignam habitam, et quia consonant ipsis ut in omnibus gradus terminorum, qui ab ipsis ponuntur in exemplaribus natiuitatibus.

[19] Ipsis tamen conscriptoribus hiis nusquam coordinationem ipsorum | neque numerum manifestantibus, suspectum utique merito fiet et facile criminabile quod est inconfessum circa ordinem.

[20] Iam uero incidimus nos cuidam e x e m p l a r i a n t i q u o et multum corroso, continenti naturalem et consonam rationem ordinis et quantitatis ipsorum, | et cum hoc predicte circa natiuitates graduum descriptiones et numerus collectorum terminorum consona inueniebantur antiquorum descriptioni.

[21] Verba autem libri ualde quidem prolixa erant et cum multa quadam ostensione, non salua autem propter corrosionem libri, et uix nobis poterat | intentionem in summa ostendere et exprimere, illos coadiuuante et ipsa terminorum descriptione, que salua erat magis aliqualiter, quia stabat apud finem libri.

[22] Habet autem insinuatio totius ipsorum intentionis modum | istum. De ordine quidem enim qui per singula 12 signa, accipiuntur et exaltationes et trigonalitates et domus. Vniuersaliter quidem stella que duas harum habet dominationes in eodem signo preordinatur, etiam si malefica sit.

[23] Vbi autem non hoc accidit, | malefici quidem semper ultimi ponuntur, primi autem domini exaltationis, deinde domini trigonalitatis, postea domini domus consequenter ei qui deinceps ordini signorum. Rursum duas consequenter habentibus dominationes preordinatis ante unam dominationem habentem in eodem signo.

V (**α**=**ΥΣ** **β**=**DMS** **γ**=**ABC**) 860 Colliguntur: συνάγονται **α** **M** **S** (*Robb.*)] συνάγεται **V** *et cett.* (*Hüb.*) 861 75 (*cett.*)] μοίρας *praem.* **V** **β** 862 66 (*cett.*)] ξε′ **V**, *cf.* ξη′ **Σ** (et] *om. codd.*) totum simul (τὸ πᾶν ὅμου ?)] τὸ πᾶν **V** **D** **γ** (*Hüb.*): ὅμου **M** **S**: μοῖραι **α** 869 criminabile] + *αὐτῶν codd.* 873 predicte circa natiuitates graduum descriptiones: τῶν γενέσεων προειρημένας μοιρογραφίας **V**] τῶν προειρημένων γενέσεων μοιρογραφίας **β** (*Hüb.*): *diversimode cett.* 875 multa: πολλῆς **V** (*Boll*)] περισσῆς *cett.* (*Hüb.*) 876 non salua: ἀδιάσωστον **V** **α**] ἀδιάγνωστον **β** **γ** (*Hüb.*) intentionem] αὐτὴν τῆν ... προαίρεσιν *codd.* 877 ostendere et exprimere: ὑποδεῖξαι καὶ τυπῶσαι **V** (*Boll*)] ὑποτυπῶσαι *cett.* (*Robb. Hüb.*) (illos] καὶ ταῦτα *codd.*) ***abhinc excerpta codicis* L *denuo adsunt*** 879 istum (*cett., Hüb.*)] + (*tit.*) περὶ ὁρίων ὡς (**V**: ὁ **D** **γ**) Πτολεμαῖος **V** **D** **γ**, *cf.* **L** 881 quidem] + *γὰρ codd.* 882 etiam si: κἂν *cett.*] καὶ **V** (885 duas] ἀνὰ δύο *codd.*) 886 preordinatis: προτασσομένων **V** **β**] *-νου* **γ** (*Boll Hüb.*): *-νοι* **α** (*Robb.*): *-νος* **L**

860-1 de nocte] *iter. Ox* 869 criminabile] erumnabile *Ox* 870 est] *Ve*[ss] 871 uero] numero *Ox* 872 continenti (*abbr. Ox*)] constituti *Ve* 872 et[1]] *om. Ve* 877 summa] -am *Ox* coadiuuante] -tem *Ox* 880 qui *scr.*] quoniam ($\overline{\text{qm}}$) *codd.* 881 domus] domos *Ox* 882-3 male-[1,2]] mali- *Ox* 883 semper] super *Ox*

1150 [24] Cancer tamen et Leo, | domus cum sint solis et lune, quia non dantur luminaribus termini, attribuuntur maleficis, Cancer quidem Marti, Leo autem Saturno, et quia in ordine quantum ad alia signa postponuntur; in hiis ordo ipsis proprius conseruatur.

1155 [25] De quantitate autem terminorum, nullo inuento secundum duas rationes domino aut in eodem signo aut in hiis que deinceps usque ad quadraturam, beneficis quidem, hoc est Ioui et Veneri, unicuique dant gradus 7, maleficis autem,
1160 hoc est | Saturno et Marti, unicuique gradus 5, Mercurio autem communi existenti gradus 6 ad complementum 30 graduum.

[26] Quoniam autem habent aliqui semper duas rationes (Venus enim sola fit
1165 dominatrix trigonalitatis secundum Taurum, luna | ad gradus terminorum non assumpta), addunt quidem unicuique sic habentium siue in eodem signo, siue in consequentibus usque ad quadraturam, gradum unum, quibus et apponuntur puncta.

[27] Auferuntur autem que apponuntur duplicibus a reliquis et singularibus,
1170 ut ad plurimum autem | a Saturno, deinde et a Ioue propter tarditatem motus ipsorum.

1186 [30] Fiunt autem et horum ex collectione Saturni gradus 57, Iouis 79, Martis 66, Veneris 82, Mercurii 76.

V (**α**=**ΥΣ** **β**=**DMS** **γ**=**ABC**, *excerpta* **L**) 890 et (*om. codd.*) quia in ordine — postponuntur *hic* **G**] *post* κακοποιοῖς (maleficis 889) *codd.*; *fort. ordinem sententiarum mut. et* et *praem.* **G** (quantum ad alia signa] *om. codd.*; *fort. add.* **G**) (892 rationes] τρόπους *codd.*; *fort. sic* **G**) (893 in eodem signo] ἐν αὐτῷ τῷ ζῳδίῳ *codd.*; *fort. sic* **G**, *cf.* 899) 898 Taurum: Ταῦρον **L** (*Robb.*)] *om. in textu* **V**[1]: Καρκίνον καὶ Αἰγόκερων **α**: Καρκίνον **β γ**: Ταῦρον καὶ <Παρθένον καὶ> Αἰγόκερων *Hüb.*, *cf. mg.* **V**[2]: Ταῦρον καὶ Κριόν Procl. 899 addunt: προσδίδο(ν)ται *cett.* (*Hüb.*)] -δίδοντες **V L** 903 ipsorum] + ἔστι δὲ καὶ ἡ τούτων τῶν ὁρίων ἔκθεσις τοιαύτη· ὅρια κατὰ Πτολεμαῖον **V L α D M** (*Hüb.*), *cf.* **γ** *et* **S**: *deinde tabulam praebent* **L α D M γ** (Procl.) *quam om.* **V**(*spatio relicto*) **S** 905 autem] *om.* **V β γ**: [**L α**] 906 82 Mercurii (**β γ**)] *om.* **V**: [**L α**] 76] + γίνονται τξ′ **V S**: + ὁμοῦ τξ′ **D M γ** (*Hüb.*): [**L α**]

---

888 Cancer] Cancri *Ox* 890 ordine — hiis] *it. Ox* 890-1 ipsis proprius] in ipsis propriis *Ox* 893 domino] duo *Ox* aut[1] *scr.*] autem *codd.* beneficis] -ficiis *Ox* 898 luna *scr.*] bona *codd.* 899 quidem] quidam *Ox* sic] sicut *Ox* habentium] hictium (?) *Ox* 903 deinde] *om. Ox* 905 79] 59 *Ve*

De locis et partibus uniuscuiusque
Capitulum 22m

[1] Diuiserunt autem aliqui et adhuc in subtiliores diuisiones dominationes, nominantes loca et partes; et locum quidem supponentes signi duodecimam (hoc est gradus duos et dimidium) et dantes ipsius dominationem sequentibus signis. Alii autem penes alios | quosdam ordines irrationales, partes quidem singulas rursum dantes a principio unicuique astrorum conformiter ordini Caldaicorum terminorum.

[2] Hec quidem igitur, neque naturalem neque probabilem habentia rationem sed uane opinionis, preteribimus. Illud autem quod est dignum attentione, non pretermittemus: quia et | duodenarum partium et terminorum principia a tropicis et equinoctialibus punctis rationabile est facere, et auctoribus hoc aliqualiter insinuantibus, et maxime quia naturas et uirtutes et conuenientias ipsorum uidemus ex preostensis a | tropicis et equinoctialibus principiis et non ab alio aliquo habentes causam.

[3] Aliis enim principiis suppositis, aut cogemur non adhuc uti naturis ipsorum quantum ad effectus, aut, si utimur, intermittere, pretergressis et alienatis | zodiaci distantiis immittentibus uirtutes ipsis.

De faciebus et claritatibus et talibus
Capitulum 23m

[1] Conuenientie quidem igitur astrorum et 12 signorum fere utique erunt tot. Dicuntur autem et proprias habere facies, | quando unusquisque ipsorum eandem conseruat configurationem ad solem uel lunam, quam quidem et domus ipsius ad illorum domos, puta quando Venus uerbi gratia sextilem distantiam habuerit ad luminaria, sed ad solem quidem uespertina existens, ad lunam autem matutina, | conformiter hiis que a principio domibus.

V (**α**=**YΣ** **β**=**DMS** **γ**=**ABC**, *excerpta* **L**) (909 subtiliores] τούτων *praem. codd.*, *fort. non redd.* **G**) 911 gradus: μοίρας *cett.*] μόρια **V** 915 Hec: ταῦτα **α** (*Hüb.*); *cf.* τὰ **β γ**] *τοῦτο **V** quidem igitur neque naturalem neque probabilem: μὲν οὖν οὔτε φυσικὸν οὔτε πιθανὸν *legit* **G**] οὔτε πιθ. οὔτε φυσ. Co le: μὲν οὖν (οὖν *om.* **β γ**) πιθ. καὶ οὐ φυσ. *codd.*: μὲν οὐ πιθ. καὶ οὐ φυσ. *coni. Hüb.*, *cf. Boll* 916 quod: τὸ **β γ**] *om.* **V L α** (*Hüb.*) 922 uti: συγχρῆσθαι **L α** (*Hüb.*), *cf.* **β γ**] -χωρῆσθαι **V** (923 alienatis: ἀλλοτριωθέντων **Σ β γ**] ἀπαλλο- **V L Y**, *Hüb.*)

---

909 adhuc] ad h' *Ve* 910 nominantes] -nationes *Ox* 913 Caldaicorum] Caldayc- *Ve* 915 Hec *scr. cum Ox*] *Hoc *Ve* 916 attentione] -em *Ox* 918 rationabile] -nale *Ve* auctoribus] -rabilius *Ve* 919 insinuantibus] insumantibus *Ox* 921 causam] + in principiis *Ve* 925 et[1]] *om. Ox* 927 erunt] erint *Ve* 928 ipsorum] eorum *Ox* 931 ad[2]] *om. Ve* ad lunam autem] aut luna aut *Ox*

[2] Dicuntur autem in claritatibus propriis esse et thronis et talibus, quando secundum duos aut plures preexpositorum modorum conuenientes fuerint locis in quibus deprehenduntur, tunc maxime uirtute ipsorum ad effectum augmentata | propter similitudinem et cooperationem proprietatis conformis signorum.

[3] Gaudere autem aiunt ipsos, quando etiam si non ad ipsos fuerit conuenientia continentium signorum, sed tamen ad eos qui sunt earundem partitionum; a remotis quidem magis sic | facta compassione, communicante autem tamen secundum talem modum similitudine. Quemadmodum quando in alienis et in locis contrarie partitionis deprehenduntur, multum dissoluitur de propria ipsorum uirtute, cum efficiat quandam mixtam naturam ex ea que secundum | dissimilitudinem continentium signorum complexione.

De coniunctionibus et refluxionibus et aliis uirtutibus Capitulum 24^m

[1] Et uniuersaliter autem coniungi quidem dicuntur precedentibus qui secuntur, refluere autem precedentes sequentibus, | quamdiu utique non magna fuerit distantia intermedia ipsorum. Accipitur autem tale siue accidat corporaliter siue etiam secundum aliquam traditarum figurationum.

[2] Nisi quod ad eas quidem que per ipsa corpora coniunctiones et refluxiones latitudines ipsorum obseruare oportunum, | ad hoc quod suscipiantur soli secus inuicem transitus ad easdem partes transeuntis per media animalia. Ad eas autem que per figurationes superfluum est quod tale, radiis omnibus semper delatis ad eadem, hoc est ad centrum terre, et similiter undecumque coincidentibus.

V (**α**=**YΣ** **β**=**DMS** **γ**=**ABC**, *excerpta* **L**) 933 Dicuntur autem in claritatibus propriis ... talibus] λαμπήναις δ' ἐν ἰδίαις ... τοιούτοις **V Y**, *cf.* **L**: καὶ ἐν λαμπ. ἰδ. λέγονται ... τοιούτοις **Σ**: λαμπ. δ' ἐν ἰδ. ... τοιούτοις οἱ ἀστέρες λέγονται **β**, *simil.* **γ**: ἐν ἰδ. δὲ λαμπ. ... τοιούτοις λέγονται Heph.: λαμπ. δ' ἐν ἰδ. λέγονται ... τοιούτοις *Hüb.* 936 conformis: ὁμοφύλου **L** (*Hüb.*): -φυοῦς **α β γ**] -φίλου **V** signorum] περιεχόντων *praem. codd.* 937 Gaudere: χαίρειν *cett.*] φέρειν **V** conuenientia: συνοικείωσις *cett.*] -σεις **V** (938 a remotis] ἐκ μακροῦ *codd.*) 942 quandam] ἄλλην *praem. codd.* 942-943 ex ea que ... complexione (ἐκ τῆς ... κράσεως ?)] τῆς ... κράσεως *codd.* 946 Et uniuersaliter autem: καὶ καθόλου δὲ **α**] καθόλου **V β γ** (*Hüb.*), + δὲ **L** 946-947 coniungi ... precedentibus qui secuntur, refluere ... precedentes sequentibus: *an* συνάπτειν ... τοῖς προηγουμένοις οἱ ἑπόμενοι, ἀπερρυηκέναι ... οἱ προηγούμενοι τῶν ἑπομένων *legit* **G** ? ; *cf.* Co (50,19-20)] συνάπτειν ... τοῖς ἑπομένοις οἱ προηγούμενοι, ἀπερρυηκέναι ... οἱ ἑπόμενοι τῶν προηγουμένων *codd.* 951-952 secus inuicem transitus (*an* ἀλλήλων παρόδους *legit* **G** ?)] εὑρισκομένας παρόδους **V L Σ D γ** (*Hüb.*); *cf.* **Y M S** (952 transeuntis per media animalia] τοῦ διὰ μέσων *codd.*) 953 est: ἐστι **α γ** (*Hüb.*)] ἔτι **V β**: ἔτη **L**

933 thronis] inthronis *Ox* 936 proprietatis] -tas *Ve* 937 ad] ab *Ve* 938 sunt] fiunt *Ox* 940 similitudine *scr.*] -ni *codd.* 942 cum] est (?) *Ox* que] quod *Ox* 943 complexione] -plect- *Ve* 946 coniungi quidem *scr.*] *inv.* *Ve*: coniungi *Ox* 954 undecumque *scr.*] unumcumque *codd.*

[3] Ex hiis itaque omnibus facile uisibile est quod qualitatem quidem uniuscuiusque astrorum considerandum ex propria et naturali ipsorum proprietate et adhuc ex proprietate continentium signorum, uel etiam ex proprietate figurationum ad solem et ad angulos secundum modum expositum nobis de hiis ˈ ipsis. Virtutem autem primo quidem ex eo quod ipsi sunt orientales et adicientes super proprios motus (tunc enim maxime sunt fortes) aut ex eo quod sunt ablatiui a proprio motu et occasiui (tunc enim debiliorem habent effectum).

[4] Deinde et ex eo quod qualiter habent ad orizontem, attingentes ˈ quidem enim ad medium celi uel accedentes ad medium celi maxime sunt uirtuosi; secundo autem quando in ipso orizonte fuerint uel accedunt ad ipsum, et magis quando in orientali, minus autem quando sub terra fuerint in medio celi aut aliter configurentur orienti loco. Non sic autem ˈ se habentes impotentes penitus existunt.

Finis libri primi

V (**α**=**YΣ** **β**=**DMS** **γ**=**ABC**, *excerpta* **L**) (956 et[1]] *om. codd.*; *cf. supra* 213) 957 figurationum: σχηματισμῶν **V L Σ**] συσχημα- *cett.* (*Hüb.*) 958 angulos: τὰς γωνίας **L α M S** (*Hüb*)] τῶν γωνίων **V D γ** 960-1 ablatiui ... et occasiui: ἀφαιρετικοὺς καὶ δυτικούς **V L**] *inv.* **α β γ** (*Hüb.*) 966 Non (*cett.*)] *om.* **V** impotentes: ἀδύναμοι **α S** Procl. (*Hüb.*)] ἀδύνατοι **V L D M γ** 968 Finis libri primi: τέλος τοῦ α΄ βιβλίου **D γ**] + Κλαυδίου Πτολεμαίου **V**; *cf. cett.*

955 est *scr. cum Ve*] *om. Ox* 957 continentium] -tis *Ox* 959 adicientes] *in* adiecientes *mut. Ve* 960 maxime sunt] *tr. Ox* 962 habent ad orizontem] in orizonte *Ox* 963 accedentes] accid- *Ox* 964 fuerint] fuerunt *Ox* 965 in orientali — autem] *it. Ox* aliter] + aut (*exp.*) *Ox* 966 configurentur] -antur *Ox* se] *om. Ox* impotentes] -ponentes *Ox*

<LIBER II>

Hec sunt capitula libri secundi

Diuisio totius intentionis	ca. 1^{m}
De proprietatibus penes omnia climata	ca. 2^{m}
De coaptatione regionum ad trigonalitates et ad stellas	ca. 3^{m}
Expositio regionum que referuntur ad unumquodque signorum	ca. 4^{m}
Accessus in effectus particulares eclipsium	ca. 5^{m}
De consideratione dispositarum regionum	ca. 6^{m}
De tempore effectuum	ca. 7^{m}
De genere dispositorum	ca. 8^{m}
De qualitate effectus ipsius	ca. 9^{m}
De coloribus eclipsium et de cometis et talibus	ca. 10^{m}
De initio anni	ca. 11^{m}
De particulari signorum natura ad consistentias	ca. 12^{m}
De consideratione particulari consistentiarum	ca. 13^{m}
De metheororum significatione	ca. 14^{m}

Diuisio totius intentionis
Capitulum primum

[1] Principaliora quidem preexpositorum nobis tabulariter ad | considerationem particularium prenuntiationum ut summatim usque ad tanta nobis exposita sint. Copulabimus autem iam secundum rationem consequentie particularia negotii de incidentibus in possibilitatem talis prenuntiationis, in operando adherentes ubique | preinductioni secundum naturalem modum.

V (α =**YΣ** **β** =**DMS** **γ** =**ABC**) 2 Hec sunt capitula libri secundi: τάδε ἔνεστιν ἐν τῷ β´ βιβλίῳ (*coni. Boll Hüb.*)] *κεφάλαια τοῦ δευτέρου βιβλίου **V D**, *cf.* **M**: τάδε ἔνεστιν ἐν τῷ β´ βιβλίῳ τῶν πρὸς Σύρον συμπερασματικῶν **γ**; *cf. cett.* 3 ca. 1^{m} *etc.*: *capitum numeros praebent* **V Y D γ** (*Hüb.*)] *om. cett.* 12 eclipsium*: ἐκλείψεων **Y M**, *cf. tit. cap.* 10^{m} (638)] ἐκλείψεως **V** *et cett.* (*Hüb.*) 19 quidem: μὲν **V γ**] μὲν δὴ **α** (*Hüb.*): μὲν οὖν **β** nobis (**V D γ**)] *om. cett.* (*Hüb.*) 21-22 particularia negotii de incidentibus ... in operando: τὰ καθ' ἕκαστα τῆς πραγματείας (!) τῶν ... ἐμπιπτόντων πραγματείας **V** (*fort. primo loco verbum* πραγματείας *e glossa pro* τῆς *in archetypo codd. inmissum*)] καθ' ἕκαστα τῶν ... ἐμπιπτόντων τῆς πραγματείας **Y**: τὰς καθ' ἕκαστον τῶν (τῆς **D**) ... ἐμπιπτόντων πραγματείας **Σ β γ**: τὰ καθ' ἕκαστα τῆς τῶν ... ἐμπιπτόντων πραγματείας *coni. Hüb.*

---

2 Hec sunt capitula *scr. cum Ox*] *Capitula *Ve, an recte?* 3 Diuisio] Qussio *Ox* ca. 1^{m} *Ve*] ca. 2. (*sic*) *Ox* 4 ca. 2^{m} *Ve*] ca. 2. *Ox*; *et sic deinceps (sed cf. infra* 13) 5 ad[1]] et *Ox* 6 Expositio] De expositione *Ve* signorum] *om. Ox* 7 effectus] + et *Ox* eclipsium *scr.*] eclipsī *Ve*: eclips' *Ox* 10 dispositorum *scr.*] -o[2]- *e corr. Ve*: -arum *Ox* 12 eclipsium] eclips' *Ox* 13-16 *ordinem capitulorum inv.* (13^{m}-11^{m}-14^{m}-12^{m}) *Ox* 14 consistentias *scr.*] constantias *codd.* 19 quidem] + finis libri primi. Hec sunt capitula libri secundi (*cf. supra* 2) *add. Ox* pre-] primi *Ox*[ac], prius *Ox*[pc] 23 preinductioni *scr.*] -ductis *codd.*

[2] In duas igitur maximas et principalissimas partes diuisa pronosticatione per astronomiam, et primo quidem existente et generaliori illo quod secundum omnes gentes et regiones uel ciuitates accipitur (quod uocatur uniuersale), secundo autem | et magis proprio illo quod secundum unumquemque hominum (quod uocatur geneaticum), conuenire existimamus de uniuersali primo facere sermonem. Quoniam itaque hec quidem secundum maiores et fortiores causas uerti | nata sunt magis quam ea que particulariter efficiuntur, [3] cadentibusque | semper debilioribus naturis sub potentioribus et particularibus sub uniuersalibus, penitus necessarium utique erit uolentibus speculationem facere de uno quocumque multo prius de totalioribus prepertractasse.

[4] Et ipsius autem uniuersalis speculationis hoc quidem | secundum regiones totas accipitur, hoc autem secundum ciuitates, et adhuc <hoc quidem> secundum maiores et magis periodales effectus, puta bellorum aut pestilentiarum aut famium aut terremotuum aut diluuiorum et talium, hoc autem quantum ad minores et temporaliores, quales sunt que temporum anni et que secundum magis et minus | alteratio, quantum ad remissiones aut intentiones hyemis et ardoris et uentorum, fertilitatesque et sterilitates et talia.

[5] Vtrumque autem horum precedit merito quod penes regiones totas et quod secundum maiores effectus, propter causam eandem cum predicta. Ad horum autem | considerationem maxime coacceptis duobus hiis (conuenientia uidelicet signorum zodiaci et adhuc stellarum ad quelibet climata et hiis que fiunt in propriis partibus per tempora significationibus, [6] secundum coniugationes quidem solis et lune eclipticis, | secundum planetarum autem transitus hiis que circa ortus et stationes), proexponemus dictorum affectuum naturalem rationem, simul notantes summatim consideratas in omnibus gentibus ut ad omne dicatur corporales et morales proprietates, non alienas existentes | a naturali condicione conuenientium, scilicet stellarum et signorum.

V (**α** =**YΣ** **β** =**DMS** **γ** =**ABC**) 27 magis proprio (ἰδικωτέρου *redd.* **G**)] εἰδικωτέρου *codd.* quod[2]: ὅπερ **V**] ὃ καὶ αὐτὸ **α β γ** (34 quidem] + πάλιν *codd.*) 35 totas: ὅλας *cett.*] ὅλαι **V** 36 aut pestilentiarum aut famium (ἢ λοιμῶν ἢ λιμῶν)] ἢ λοιμῶν ἢ λειμῶν **S γ**: ἢ λιμῶν ἢ λοιμῶν **V M** (*Hüb.*): *l.v. cett.* 38-9 et que ... alteratio: καὶ ἡ ... ἀλλοίωσις **V** (*Boll*)] ... ἀλλοιώσεις **α D γ** (*Hüb.*), *cf.* **M S** 40 fertilitatesque: εὐφορίας τε **V**] ἢ *praem.* **α** (*Hüb.*): καὶ εὐφορίας **β γ** 45 per: κατὰ *cett.*] ἐπὶ **V** 47 ortus: ἀνατολὰς *cett.*] *om.* (*lac.*) **V**

---

27 et] *om. Ve* hominum]-em *Ve* 29 hec] hoc *Ox* 32 uno quocumque] unocumque *Ox* 33 totalioribus] -litatibus *Ox* prepertractasse *scr. cum Ox*] pretractasse *Ve* 35 adhuc] ad hoc *Ox* hoc quidem *add.*] *om. codd.* 36 magis] maius *Ox* 37 minores] maiores *Ve* 38 magis] maius *Ox* 41 quod] *om. Ve* 46 secundum] in *praem. Ve* circa] contra *Ox* 47 affectuum *scr.*] effectuum *codd.* 49 alienas] alienos *Ox* 50 scilicet] sibi *Ve*

De proprietatibus secundum omnia climata Capitulum 2^m

[1] Proprietatum igitur gentium has quidem penes totos parallelos et totos angulos diuidi accidit, ab | habitudine ipsorum ea que ad zodiacum circulum et ad solem.

[2] Ea enim que secundum nos habitabili in una quartarum borealium existente, qui quidem sub australioribus parallelis (dico autem eos qui ab equinoctiali usque ad estiualem tropicum), secundum uerticem accipientes solem | et adusti, sunt nigri corporibus et capillis crispi et asperi et figuris contracti et magnitudinibus tenues et natura calidi, et moribus ut ad omne dicatur agrestes existunt propter continuitatem habitationum sub estu, quos utique uocamus communiter | Ethiopes. [3] Et non solum ipsos uidemus ita se habentes, sed etiam consistentiam aeris ipsos continentem et alia animalia et plantas que sunt apud ipsos uidemus insinuantia adustionem.

[4] Qui autem sub borealioribus parallelis (dico autem eos qui | sub Vrsis habentes locum qui secundum uerticem), multum a zodiaco et calore solis distantes, frigidi quidem sunt. Propter hoc autem quia participant habundantius substantia humida, existente maxime uegetatiua et a nullo exalata calido, albique | sunt colore, dependentesque capillis et corporibus magni et bene referti in magnitudine, et frigidi secundum naturam; agrestes autem et ipsi moribus propter continuitatem habitationis in frigore. [5] Concomitatur autem hec et hyemps aeris ipsos continentis et plantarum magnitudo et | ferocitas animalium. Vocamus autem hos ut ad omne dicatur Scytas.

[6] Qui autem medii sunt inter estiualem tropicum et Vrsas, neque sole facto apud ipsos secundum uerticem neque multum distante secundum equinoctiales transitus, bonam contemperantiam | aeris acceperunt, et ipsa quidem differente, sed non ualde magnam immutationem estuum ad frigora accipiente. [7] Hinc coloribus sunt medii et magnitudinibus mediocres et naturis bene temperati et habitationibus continui et moribus domestici | existunt.

V (**α** =**YΣ** **β** =**DMS** **γ** =**ABC**) 67 quia participant habundantius: δαψιλεστέρως μεταλαμβάνοντες **Σ G**° **γ**] δαψιλεστέρας μεταλαμβάνοντες **V M S**, *Hüb.*, *cf.* **Y** 68 exalata: *fort.* ἀναπνουμένης *legit vel coni.* **G**, *cf.* ἐξικμαζομένης Procl.] ἀναπουμένης **V**: ἀναπινομένης **Σ** (*Hüb.*), *cf. cett.* 69 referti: εὐτραφεῖς *cett.*] εὐστραφεῖς **V** (71 habitationis] οἰκήσεων *codd.*)

51-2 De proprietatibus — 2^m] Diuisio totius intentionis ca. primum (*sic*) *Ox* 53 totos] totas *Ox* 54 ipsorum] -arum *Ve* 57 qui[1]] que *Ox* australioribus] austraiolibus *Ve* eos] *corr. ex* eas (*sic*) *Ve* qui[2]] quod *Ox* 62 ita se] *tr. Ox* 63 continentem] -es *Ox*: al' continentes *mg. add. Ve* 67 autem] aut *Ox* 69 colore] calore *Ox* capillis *scr.*] capilli *codd.* 71 habitationis] *an* -num* *scrib.*? hec] hoc *Ox* 72 plantarum *scr.*] planetarum *codd.* ferocitas] societas *Ox* 73 Scytas *scr. cum Ve*] seytas *Ox* 76 contemperantiam] con- *exp. Ox* 77 ualde magnam] *tr. Ox* immutationem] mutationem *Ox* 78 et[3]] + ad *Ve*

[8] Horum autem ut dicatur ad omne qui ad austrum sollertiores et artificia-
liores magis et circa diuinorum historiam copiosiores, quia appropinquat zodiaco
locus qui secundum uerticem et errantium circa zodiacum astrorum. Quibus
congrue habent et ipsi anime motus obseruatiuos ' et perscrutatiuos et percuncta- 105
tiuos uocatorum proprie mathematum.

[9] Vniuersaliter rursum qui quidem apud orientem magis sunt masculini et
robusti animabus et omnia elucidantes. Quoniam quidem orientem utique quis
merito solaris nature existimet et partem illam ' diurnam et masculinam et dex- 110
tram, secundum quod etiam in animalibus uidemus dextras partes magis ydonei-
tatem habentes ad ualitudinem et robur.

[10] Qui autem ad occasum feminini magis sunt et teneriores magis secundum
animas et multa occultantes, quoniam rursum pars ista lunaris existit, ' luna 115
semper primos ortus et post coniunctionem apparentias ab occasu faciente.
Propter hoc itaque nocturnum uidetur esse clima et femininum et sinistrum, op-
positum orientali.

[3.1] Iam autem et in singulis hiis totarum partium quedam singulares proprie- 121
tates morum et legitimorum naturaliter assecute sunt. Sicut enim in consistentiis
continentis et in ipsis ut ad omne dicatur debitis caloribus aut ' frigoribus aut tem- 125
periebus, secundum partem singularizant loca et regiones alique in magis et minus
aut propter positionis ordinem, aut eleuationem aut demissionem, aut propter
uicinantiam. Adhuc autem sicut equestres quidam sunt magis propter campe-
stritatem regionis et nautici propter maris uicinitatem et domestici propter ' apti- 130
tudinem <regionis>, [3.2] ita et ex conuenientia naturali particularium climatum
ad stellas et signa, singularis modi inueniet utique quis naturas apud singulos; et
ipsas autem, ut ad omne dicatur, non ut secundum unumquemque, semper inexi-
stentes. Ideoque necessarium ' summatim superuenire super quodcumque fuerit 135
oportunum ad particulares considerationes.

V (**α** =**ΥΣ** **β** =**DMS** **γ** =**ABC**) 81 appropinquat] + αὐτῶν **V α** (*Hüb.*): + αὐτοῖς **β γ**: *fort. non redd.* **G** (82 zodiacum] αὐτὸν *codd.*) 82-83 Quibus congrue ... et (*cett.*)] *om.* (*lacuna*) **V** 83 obseruatiuos: εὐεπιβόλους **Σ D γ** (*Hüb.*), *cf.* **M S**] -βούλους **V Υ** 85 Vniuersaliter (**V**)] + δὲ *cett.* (*Hüb.*) 86 Quoniam quidem orientem (*cett.*)] *om.* (*lac.*) **V** 88 etiam: καὶ *cett.* (*Hüb.*)] *om.* **V** 90 magis² (**V**)] *om. cett.* (*Hüb.*) 93 clima et femininum: κλίμα καὶ θηλυκὸν *coni. Boll*] κλίμα θηλυκὸν **V** (*Robb.*): καὶ θηλυκὸν **α β γ** (*Hüb.*) 97 ipsis: αὐτοῖς *cett.*] τοῖς **V** (*Hüb.*) 99 eleuationem: ὕψος **V α**] διὰ ὕψος **β γ** (*Hüb.*) 101 aptitudinem (εὐφυΐαν ?)] εὐθηνίαν *codd.*

---

80 artificialiores] -lores *Ve* 81 circa] contra *Ox* 82 circa] contra *Ox*, *et sic deinceps, fere semper* 83 obseruatiuos] -tionis *Ox*: *an etiam* **G** ?, *cf. indicem ad* ψυχικός, σινωτικά, παθητικός perscrutatiuos] -tionis (*sic*) *Ox* percunctatiuos] -tionis (*sic*) *Ox* 87 existimet] eistimet *Ve* 90 teneriores] temeriores *Ox* 93 clima] *obscure* (dñm) *Ox* 99 demissionem] dimiss- *Ox* 100 Adhuc] ad hoc *Ox* 102 regionis *add.*] *om. codd.* ita et] *tr. Ox* 103 signa] significat (*abbr.*) *Ox* inueniet] -nit *Ve* 104 unumquemque] -quodque *Ox* 105 summatim] sinuatim *Ox* superuenire] *om. Ox* quodcumque *scr.*] qdcum *Ve*: quodque *Ox*: dictum *mg. add. Ve* fuerit] *om. Ve*

De conuenientia regionum ad trigonalitates et ad stellas Capitulum 3[m]

[3] Quatuor itaque trigonalibus figuris in zodiaco consideratis, ut ostensum est nobis per priora, quod trigonalitas quidem que est secundum Arietem, Leonem et Sagittarium | borealibica (id est simul aquilonaris et occidentalis) est, et disponitur quidem principaliter a Ioue propter borealitatem, condisponitur autem et a Marte propter occidentalitatem. [4] Trigonalitas autem que secundum Taurum et Virginem et Capricornum notafiliotica (id est simul australis et subsolana) est, et disponitur quidem rursum | principaliter a Venere propter australitatem, condisponitur autem et a Saturno propter subsolanitatem. Illa uero trigonalitas que est secundum Geminos et Khelas et Aquarium boreafeliotica est (id est simul borealis et subsolana), et disponitur quidem principaliter a Saturno propter subsolanitatem, | condisponitur autem et a Ioue propter borealitatem. [5] Trigonalitas autem que secundum Cancrum et Scorpionem et Pisces notolibica (id est simul australis et occidentalis) est, et disponitur quidem rursum principaliter a Marte propter occidentalitatem, condisponitur autem et a Venere propter australitatem.

[6] Hiis | itaque ita se habentibus et diuisa habitabili que secundum nos in q u a t u o r q u a r t a s numero equales trigonalitatibus (secundum latitudinem quidem a mari quod secundum nos a Gadibus Herculis usque ad Issicum sinum et deinceps usque ad orientem per montana, a quibus distinguitur | australis et borealis ipsius pars, [7] secundum longitudinem autem ab Arabico sinu et Egeo pelago et Ponto et Meotida palude, a quibus distinguitur orientalis et occidentalis pars), fiunt quatuor quarte conformes positioni trigonalitatum.

V n a quidem | que iacet inter boream et occidentem totius habitabilis, que est uidelicet penes Keltogalatiam, que utique communiter Europam uocamus. [8] Huic autem opposita et ad id quod inter austrum et orientem penes orientalem Ethiopiam, que utique Magne Asie australis pars uocatur. Et iterum quod quidem

---

V (*α* =**ΥΣ** *β* =**DMS** *γ* =**ABC**) 107 *Tit. hic ut* **V** *β* *γ* Procl. (*Robb.*) **G**] *ante* Iam autem (95/121°) *α* (*Hüb.*) 110 nobis (*cett.*)] *om.* **V** (111 id est — occidentalis: *explicatio translatoris, cf.* 114, 117, 120, 177 *et al.*) 113 Taurum: Ταῦρον *cett.*] Ζυγὸν **V** (*corr.* **V**[2]) (116 et: καὶ **M S** *γ*] *om.** **V** *α* **D**, *Hüb.*) 117 Khelas: τὰς Χηλὰς *cett.*] τὰ (*ex* τοῦ) Κριοῦ **V** (122 et: καὶ *γ*] *om. cett.*, *Hüb.*) 126 a quibus distinguitur (*cett.*)] *om.* **V**

---

110 trigonalitas] -tatis *Ve*[ac] 111 borealibica] bora- *Ve*: boria- *Ox* 112 condisponitur] Cumdis- *Ve* 113 Trigonalitas] -tatis *Ve*[ac] 114 subsolana] -laria *Ve* 116 condisponitur] Cum disponitur *Ve* et] *om.** *Ox* 117 Khelas *Ve*[mg]] chelas *Ve*[i.t.]: Khelos *Ox* boreafeliotica] borafe- *Ve* 118 principaliter] *om. Ox* 120 autem] *om. Ox* 122 occidentalitatem] -talem *Ve* condisponitur] cum disp- *Ox* 123 itaque ita se *scr.*] itaque *Ve*: ita se *Ox* secundum nos] seruos *Ox*[ac] 125 Gadibus] gradibus *Ve* ad Issicum] ad situm (*ex* adisitum *corr.*) *Ve* 126 montana] mundana *Ox* 127 sinu] simi *Ve* Egeo] egio *Ve* 128 a quibus] *om. Ox* 130 totius] *mg. Ve* habitabilis] -talis *Ve* 131 Keltogalatiam] toletumgalatiam *Ve*[i.t.], kelto *Ve*[mg] que] quoniam *Ve* 132 Huic] Hiis *Ox* austrum] astrum *Ox*

inter boream et orientem | totius habitabilis quod est penes Scithiam, quod utique et borealis pars Magne Asie fit. Opposita autem huic et ad uentum libonotum, que penes occidentalem Ethiopiam, quam utique communiter Affricam uocamus.

[9] Iterum autem et uniuscuiusque premissarum quartarum | que quidem figurata sunt magis apud medium totius habitabilis contrariam accipiunt positionem ad quartam se continentem quam quidem illa ad totam habitabilem. [10] Quarte enim iacentis inter boream et occidentem totius habitabilis, uidelicet Europe, que circa medium ipsius et | angulariter opposita [et] inter austrum et orientem ipsius quarte positionem habere uidentur; et in aliis similiter, ut ex hiis consequitur: quelibet quartarum duabus oppositis trigonalitatibus coaptatur.

[11] Aliis quidem partibus coaptatis ad uniuersalem habitudinem, hiis autem que circa | medium ad eam que secundum ipsam partem partialem oppositam; coassumptis ad dispositionem et stellis habentibus dominationem in propriis trigonalitatibus: in aliis quidem habitationibus rursum ipsis solis, in hiis autem que circa medium habitabilis et illis, et adhuc stella | Mercurii, propterea quia medius est ipse et communis partitionum.

[12] Ex tali itaque ordinatione alie quidem partes prime quarte, iacentes inter boream et occidentem totius habitabilis, | eius scilicet que secundum Arietem et Leonem et Sagittarium. Disponitur autem merito a dominis trigonalitatis, Ioue scilicet et Marte, uespertinis.

V (**α** =**YΣ** **β** =**DMS** **γ** =**ABC**) 134 totius — penes (*cett.*)] *om.* (*spat.*) **V** (135 et[1]] + αὐτὸ *codd.*) 137 quartarum: τεταρτημορίων *cett.*] -μορίου **V** 139 quam quidem: ἥνπερ **Y β γ** (*Boll*), *cf.* **Σ**] ὥσπερ **V** (*Robb. Hüb.*) 142 quarte: τεταρτημορίου *cett.*] τεσσάρων μοιρῶν **V** consequitur] *om. codd.*: *an* ἀκολουθεῖ *legit vel* consequitur *suo arbitrio suppl.* **G** ?, *cf.* φανερὸν Procl. (88,3) 143 coaptatur: συνοικειοῦται **V** Procl.] -οῦσθαι *cett.* (*Hüb.*) 144 habitudinem] πρόσνευσιν **V α** (*Hüb.*): *om.* **β γ**: *an* σχέσιν *legit uel potius* πρόσνευσιν *hic ad sensum redd.* **G** ? 145 secundum ipsam partem partialem (κατ' αὐτὸ τὸ μέρος μερικὸν)] κατ' αὐτὸ τὸ μερικὸν **V Y**: κατ' αὐτὸ τὸ μέρος **Σ** (*Hüb.*): κατὰ μέρος **β γ** 148 adhuc: ἔτι *cett.*] ἔστι **V** 149 ipse: αὐτὸν **Σ β γ** (*Hüb.*)] αὐτῶν **V**: *om.* **Y** 150 quarte] + λέγω δὲ τοῦ κατὰ τὴν Εὐρώπην *codd.* 151 habitabilis] + συνοικειοῦται τὸ βορρολιβικὸν τρίγωνον **V**: + συνοικειοῦται μὲν τῷ βορρολιβυκῷ τριγώνῳ *cett.* (*Hüb.*) eius scilicet que (*sc.* quarte ?) secundum Arietem] τῷ (*sc.* τριγώνῳ) κατὰ Κριὸν *codd.*: *an* τοῦ *pro* τῷ *legit vel coni., praemissa verba non habens* (*cf. adnot. praec.*) **G** ?

---

134 habitabilis *scr.*] habitalis *Ve*: *om. Ox* Scithiam] suchiam *Ve* 135 borealis] uel *add. et del. Ve* autem] *mg. Ve* et ad uentum] admentum *Ox* 138 positionem] -ni *Ve* 139 illa] *om. Ve* 140 Europe] *om. Ve* 141 et[2] *del.*] *hab. codd.* 142 et in aliis] *etiam ante* uidentur *Ox* 144 uniuersalem] utilem *Ox* 145 medium] habitabilis *add. et exp. Ve* partem partialem *sic codd., cf. adn. gr.-lat.* oppositam *scr.*] -tis *codd.* 148 habitabilis] *mg. Ve* adhuc] ad hec *Ve* 149 partitionum] participationum *Ve* 150 Ex] Et *Ox* tali] ab isto loco deficit commentum usque ad tale signum o-ff-o *add. mg. Ox* (*alia manu* ?; *sed istud signum non invenitur*)

[13] Sunt autem hee secundum omnes gentes accepte: Britannia, Galatia, Germania, Apulia, Sicilia, Tyrrenia, | Celtica, Hispania. Merito autem premissis gentibus ut ad omne dicatur conuenit, propter principatiuum trigonalitatis et propter condisponentes stellas, insubicibiles esse et amatiuos libertatis et amatiuos armorum et laborum et bellicosissimos et principales et puros et magnanimos.

[14] Propter uespertinam tamen configurationem Iouis et Martis et adhuc quia premisse trigonalitatis anteriora quidem masculina sunt, posteriora autem feminina, ad mulieres quidem accidit ipsos non zelotipos esse et contemptores uenereorum, ad | coitum autem masculinorum magis uacantes et magis zelotipos, ipsisque dispositis neque turpe existimari quod fit neque ut ueraciter non uirilibus propter hoc et non ut mollibus euenire, ut non passiue disponantur, conseruare autem et animas in uirilitate et communicatiuas et fideles | et amatiuas domesticorum et benefactiuas.

[15] Et ipsarum autem harum regionum Britannia quidem et Galatia et Germania et Barsania magis Arieti coaptantur et Marti; unde ut ad omne dicatur qui in ipsis magis agrestes et magis proprie licentie et bestiales existunt. [16] Italia autem et Apulia | et Sicilia et Gallia, Leoni et soli; propter quod quidem principatiui magis sunt isti et benefici et communicatiui. Tyrenia autem et Celtica et Hispania, Sagittario et Ioui; unde amor libertatis conuenit ipsis et simplicitas et amor puritatis.

[17] Partes autem que sunt in hac quarta figurate circa medium habitabilis, Thracia et Macedonia et Illiria et Ellas et Acaya et Creta, adhuc autem Ciclades insule et adiacentia Minoris Asie et Cipri (que iacent inter austrum et orientem totius quarte) | assumunt coaptationem trigonalitatis notafiliotice (id est australis simul et orientalis), eius scilicet que est secundum Taurum et Virginem et

---

V (**α** =**ΥΣ** **β** =**DMS** **γ** =**ABC**) 154 Germania (**V** Procl. Heph. I 1,6, *cf. Robb. ad loc.*)] + Βασταρνία Ἰταλία Γαλλία **α** **β** **γ** (*Hüb.*) 157 condisponentes: συνοικοδεσποτήσαντας **α** **β** (*Hüb.*), *cf.* **γ**] οἰκοδεσποτοῦντας **V** 161 quidem: μὲν *cett.* (*Hüb.*)] τε **V** (164 non ut*] *om. codd., fort. add.* **G**) 168 Barsania: βαρσανία **V**] βασταρνία *cett.* (*Hüb.*): *om.* Heph. 170 Apulia et Sicilia et Gallia (*coni. Hüb.*)] σικελία καὶ γαλλία **V** **D**: γαλλία καὶ ἀπουλία καὶ σικελία **γ** (*cf.* Heph.): γαλλία καὶ σικελία **α** **M** **S** 172 Sagittario: Τοξότῃ *cett.*] τοῦ Ἄρεως **V** (conuenit] *om. codd.*) 174 que sunt: μὲν ὄντα *cett.* (*Hüb.*)] μένοντα **V** 176 insule: νῆσοι **β** **γ**] *om.* **V** **α** (*Hüb.*), *cf. infra* 358 176 Cipri: Κύπρου **V** Procl.] Κύπρος *cett.* (*Hüb.*) 178 secundum Taurum: τοῦ κατὰ τὸν Ταῦρον *cett.* (*Hüb.*)] τοῦ τε καὶ κατὰ τὸν Ζυγὸν **V**

---

155 Sicilia] Scicilia *Ve* 156 principatiuum *Ve*[i.t.]] principatuum *Ox*: al' principatum *Ve*[mg] 157 insubicibiles] -ciles *Ve* amatiuos[1]] ambitionis *Ox* amatiuos[2]] amationis *Ox* 158 puros *Ox*[pc]] piros *Ve Ox*[ac] 159 configurationem] figurationem *Ve*[i.t], con *Ve*[ss] 160 autem] aut *Ve* 162 masculinorum] -lorum *Ox* 163 uirilibus] in uilibus *Ox* 164 non ut] *inv. Ve* 165 communicatiuas] consumatiuas *Ox* 167 quidem] autem *Ox* 168 Barsania] boirsania (?) *cum corr. Ve* ad] *om. Ox* 170 Gallia] *ex* Gallea *corr. Ox:* Galicia (i *primum e corr.*) *Ve* 171 benefici] -ficii *Ox* Tyrenia] Tirrennia (?) *cum corr. Ve* 175 Thracia *scr.*] Tharcia *Ve*[i.t.] *Ox*: al' tarsia *Ve*[mg] Illiria] Hilliria *Ox* Ellas] *ex* Elleas *corr. Ox* Acaya] akaia (?) *Ox* 175-6 Ciclades insule *scr.*] ciclade insule *Ve*: insule ciclades *Ox*

Capricornum, [18] adhuc autem condispositores Venerem et Saturnum et adhuc Mercurium. Vnde habitantes regiones has commixte figurationis | magis dicti sunt et mixti corporibus et animabus; principatiui quidem et ipsi existentes et generosi et insubicibiles propter Martem; [19] amatoresque libertatis et sue legis et democratici et legislatiui propter Iouem; amatores autem musice et amatores discipline | et amatores exercitii propter Mercurium, et mundi in conuersatione propter Venerem; communicatiuique et amatores extraneorum et amatores iustitie et amatores litterarum et in sermonibus operosissimi propter Mercurium; misteriorum autem maxime perfectiui propter Veneris uespertinam figurationem.

[20] Iterum autem | et harum ipsarum pars, qui quidem circa Kyklades et adiacentia Asie Minoris et Cipri, Tauro et Veneri magis coaptantur; unde ut in pluribus delicati sunt et mundi et corporis curam facientes. [21] Qui autem circa Elladem et Acaiam | et Cretam, Virgini et Mercurio; propter quod logici magis existunt et amatores discipline et circa ea que sunt anime primo studentes quam circa ea que corporis. Qui uero circa Macedoniam et Traciam et Illiridem, Capricorno et Saturno; propter quod amatores quidem pecuniarum, non | ita autem domestici neque communicatiui legibus.

[22] S e c u n d e autem q u a r t e que est penes australem partem Magne Asiae, alie quidem partes continentes Indicam, Arianem, Gedrusiam, Parthiam, Mediam, Persidem, Babiloniam, Mesopotamiam, Assiriam et positionem | habentes inter austrum et orientem totius habitabilis, merito et ipse coaptantur quidem notafeliotice trigonalitati, Tauro scilicet et Virgini et Capricorno.

V (**α** =**YΣ** **β** =**DMS** **γ** =**ABC**) 179 condispositores: συνοικοδεσπότας **α β γ** (*Robb.*)] -οῦντας **V** (*Hüb.*) 180 has] ἐκείνας *cett.* (*Hüb.*): *om.* **V Y**; *fort. verbum* ἐκείνας *non legit* (*cf.* **V Y**), *sed* has *suo arbitrio add.* **G** 184 propter Mercurium] *om. codd.* (188 et harum ipsarum pars] κατὰ μέρος καὶ τούτων αὐτῶν *codd.*, *cf.* **α**) 189 Cipri: κύπρου **V Y** Procl. (*Robb.*)] Κύπρον *cett.* (*Hüb.*) Tauro (*cett.*)] *om.* (*spat.*) **V** (pluribus] πλεῖστον **V D γ**, *Hüb.*: πολὺ **α M S**, *Robb.*) (192 primo ... quam: πρῶτα τῶν *legit vel coni.* **G**] πρῶτα **V Y M S**: πρὸ τὰ **D**: πρὸ **Σ γ** (*Boll Robb.*): πρὸ τῶν *coni. Hüb.*) 193 circa: περὶ *cett.*] ἐπὶ **V** et Illiridem: καὶ ἰλλυρίδα **α** (*Hüb.*)] καὶ ἰλλυρία **β γ**: *om.* **V** 193-194 Capricorno — pecuniarum (*cett.*)] *om.* **V** 195 legibus: νόμοις **V α** (*Robb.*)] νομίμοις **β γ** (*Hüb.*) 197 Gedrusiam: γεδρουσίαν **β γ** (*Hüb.*), *cf.* α] τε δρουσίαν V 200 Tauro (*cett.*)] *om.* (*spat. relicto*) **V**

---

179 adhuc[1]] Ad hec *Ve*: hec figura dat exemplum (+ ex dictis et cetera ?) *mg. add. Ve*; *ad figuram in fine fol. 9 refert, cf. appendicem* adhuc[2]] + et ad (ad *exp.*) *Ox* 180 figurationis] + et *Ox* 182 insubicibiles] -ciles *Ve* -que] *om. Ox* 186 operosissimi] composissimi *Ve* 187 figurationem] config- *Ve* 188 Kyklades] alliades *Ve* 189 Minoris et *scr.*] et minoris et *Ve*: et minoris *Ox* Cipri] cirpi *Ve* 191 Acaiam] acayam *Ve* 193 Traciam] Traiciam *Ox* et[2]] *om. Ox* 197 Indicam] indicia *Ox* Arianem *scr.*] Ariaenē *Ve*: Ariani (+ *spat. 3 litt.*) est *Ox* Gedrusiam] +edrusiam (*sic*) *Ve* Parthiam] partitariam *Ox* 198 Assiriam] asyria *Ox*

[23] Condisponuntur autem a Venere et Saturno in matutinis figurationibus; propter quod et naturas eorum qui in ipsis | utique assequentes quis inueniet hiis que efficiuntur a sic dominantibus. Colunt enim Venerem ipsam Isim nominantes, Saturnum autem Mithram Solis et prediuinant multi futura. [24] Immolanturque apud ipsos genitalia membra propter configurationem premissarum | stellarum spermaticarum existentium per naturam. Adhuc autem calidi et coitiui et impetuosi ad uenerea sunt, saltatiui et tripudiatores et ornatus amatores propter Venerem, simplicis autem diete propter Saturnum.

[25] In manifesto autem faciunt et non latenter coitus cum | mulieribus propter orientalitatem figurationis, coitus autem ad masculos execrantur. Propter hec autem et plurimis ipsorum accidit ex matribus ipsorum natos procreare et adorationes pectore facere, propter matutinos ortus et quia principale cordis conuenienter habet ad | solarem uirtutem.

[26] Sunt autem ut ad omne dicatur quantum ad alia quidem que circa stolas et ornatus et totaliter ad corporales habitudines delicati et feminini propter Venerem. Quantum ad animas et electiones magne sapientie et generosi et bellicosi, propterea quia Saturnus congrue se habet | ad figuram ortuum.

[27] Secundum partem iterum Tauro quidem et Veneri magis coaptantur Parthia et Media et Persia; unde qui ibi stolis floretis utuntur, operientes se totos excepto pectore, et omnino sunt splendide diete et | mundi. [28] Virgini autem et Mercurio que circa Babiloniam et Mesopotamiam et Assiriam; propter quod et apud illos qui ibi morantur accidit esse precipuam mathematicam et obseruationem quinque astrorum. Capricorno autem et Saturno que circa Indiam et Arianem et | Gedrusiam; unde et proprietas habitantium regionem est incompositio et immunditia et bestialitas.

V (**α** =**YΣ** **β** =**DMS** **γ** =**ABC**) 201 Condisponuntur (συνοικοδεσποτεῖται ?)] *οἰκοδεσποτεῖται **β γ** (*Hüb.*), *cf.* **Y Σ**: συνοικειοῦται **V** a: ὑπὸ *cett.*] τῷ **V** 202 assequentes (ἀκολούθους ?)] ἀκολούθως *codd.* 202-3 hiis que efficiuntur: τοῖς ... ἀποτελουμένοις **V**] τοῖς ... -μένας **Σ β γ** (*Hüb.*) *cf.* **Y** (204 Solis: Ἡλίου ?] Ἥλιον **V** *et plur.*, *Hüb.*) 206 spermaticarum existentium: σπερματικῶν ὄντων (*sc.* ἀστέρων) *legit vel coni.* **G**] σπερματικόν ὄντα (*sc.* τὰ γεννητικὰ μόρια) *codd.* 209-10 mulieribus — ad] *om.* (*spat.*) **V** 211 et plurimis ipsorum (*cett.*)] *om.* (*spat.*) **V** matribus ipsorum: μητέρων αὐτῶν **β γ**] ἀμητέρων **V**: μητέρων **α** (*Hüb.*) 214 ad alia quidem que: τἄλλα μὲν τὰς **V**] τὰ μὲν ἄλλα **Y β γ** (*Hüb.*), *cf.* **Σ** 216 et[1] (*cett.*)] + κατὰ **V** 217 ortuum (*cett.*)] *om.* (*spat.*) **V** 221 Babiloniam: βαβυλονίαν *cett.* (*Hüb.*)] βαβυλῶνα **V** Procl. (*Robb.*) 224 Gedrusiam: γεδρουσίαν **β γ** (*Hüb.*)] γεδροσίαν **V**: γεδρωσίαν **α** (proprietas ... est incompositio] τὸ ... ἄμορφον *codd.*, *cf.* **Σ**) regionem: τὰς χώρας **V Y β**] τὰς χώρας ἐκείνας **Σ γ** (*Hüb.*)

---

201 Condisponuntur] *Disponuntur *Ox* (*recte*?) in] et *Ox* 202 quod] *om. Ve* qui *scr.*] que *codd.* 203 Isim] ysim *Ox* 204 Mithram] mithiam *Ox* futura] figura fitura *Ox* 206 coitiui] coiatiui *e corr. Ve* 207 uenerea sunt (τυγχάνουσιν) *scr.*] uenerea *Ve*: ueñ est *Ox* 209 In manifesto] inmanifesto *Ve* 210 execrantur *sic codd.* hec] hoc *Ox* 212 pectore] pector *Ve* propter] *om. Ve* cordis] corde *Ox* 214 quidem] quedam *Ox* stolas] scolas *Ve* 219 floretis *scr.*] flortis *Ve*: ploratis (f —> p) *Ox* 221 Mesopotamiam *scr.*] -potanian *Ve*: -potamicam (?) *Ox* 222 qui ibi] *om. Ox* mathematicam] mathecam *Ox* 224 Arianem] auria- *Ox* regionem] -num *Ve*

[29] Relique autem huius quarte partes circa medium figurate totius habitabilis, Ydumea, Profunda Syria, Iudea, Fenicia, Caldea, Orchenia, Arabia felix, ¦ positionem habentes inter boream et occidentem totius quarte, accipiunt iterum coaptationem trigonalitatis borealibice, Arietis scilicet, Leonis et Sagittarii, habent autem condispositores Iouem et Martem et adhuc Mercurium. [30] Propter quod isti magis aliis sunt negotiatiui ¦ et magis commutatiui et cum formidine despectatiui et insidiatiui et dolose anime et totaliter diuersa simulantes propter configurationem premissorum astrorum.

[31] Et horum autem rursum qui quidem circa Profundam Syriam et Ydumeam et Iudeam Arieti et ¦ Marti magis coaptantur; propter quod quidem ut ad omne dicatur audaces sunt et sine deo et insidiatiui. Fenices autem et Caldei et Orchenii, Leoni et soli; propter quod quidem simpliciores sunt et humani et amatores astrologie et maxime omnium colentes solem.

[32] Qui autem circa Arabiam ¦ Felicem, Sagittario et Ioui; unde consequenter appellationem regionis felicitas consecuta est et multitudo aromatum et hominum bona armonizatio ad conuersationes liberas et commutationes et negotiationes.

[33] T e r t i e uero q u a r t e que est penes borealem partem Magne Asie alie quidem partes, continentes Yrcaniam, Armeniam, Matianem, Bactrianem, [et] Caspiriam, Sericam, Sauromaticam, Oxianem, Sugdianem <et> iacentes inter boream et orientem totius habitabilis, coaptantur quidem ¦ trigonalitati boreafiliotice (id est aquilonari simul et orientali), Geminis scilicet, Libre et Aquario, disponuntur autem merito a Saturno et Ioue in figurationibus orientalibus.

[34] Propter quod quidem qui istas habent regiones colunt quidem Iouem et solem. Ditissimi autem sunt et multi auri, circa ¦ dietas autem mundi et bone conuersationis, sapientes autem circa diuina et magi et secundum mores iusti et liberales et quantum ad animas magni et generosi, odientes malum et federatissimi et supermorientes prompte pro domesticis gratia boni et sancti. Ad uenereo-

V (*α* =**ΥΣ** *β* =**DMS** *γ* =**ABC**) (226 huius] τοῦ *codd.*) 227 positionem: τὴν θέσιν **α β γ** (*Hüb.*)] καὶ κατὰ τὴν θέσιν **V** (229 scilicet ... et] *om. codd.*) 229-230 habent autem: ἔχει δὲ **β γ** Procl. (*Hüb.*)] ἔτι δὲ **V**; *cf.* **α** 231 magis commutatiui: συναλλακτικώτεροι **V** (*sec. Boll.*, *cf. adn. Hüb.*)] συναλλακτικοί, πανουργότεροι δὲ **α** (*cf.* Co[le] *Boll Hüb.*): συναλλακτικοί, πανοῦργοι τε **β γ** 234 circa Profundam Syriam: περὶ τὴν Κοίλην Συρίαν **α β γ** (*Hüb.*)] κατὰ τὴν κοιλίαν Συρία (*sic*) **V** 239 Sagittario: Τοξότῃ **α β γ**] ἄρει **V** 241 conuersationes: διαγωγὰς **α β γ** (*Hüb.*)] διὰ τὰς **V** 244 Sericam: Σηρικὴν *cett.*] συρικὴν **V** 251 odientes malum: μισοπόνηροι **α β γ** Procl. (*Hüb.*)] καὶ πόνηροι **V** 252 domesticis: οἰκείων **β γ** Procl. (*Hüb.*)] οἰκειωτάτων **V α**

227 Syria] Siria *Ve* Fenicia *scr.*] fenix *codd.* 230 adhuc] ad hec *Ve* 231 et[1]] *om. Ox* 234 quidem] -dam *Ox* circa] propter *Ox* Profundam] -dum *Ve* Syriam] siriam *Ve* 236 Caldei] calidi *Ve* Orchenii] Orthenii *codd.* 237 Leoni] Ioui *Ve* 238 et] *om. Ve* omnium] -nes *Ox* 239 consequenter] conuenienter *Ox* 240 felicitas] *an* fertilitas *scrib.* ? 241 ad conuersationes] aduersationes *Ox* 243 Yrcaniam] Yrtan- *codd.* et *del.*] *hab. codd.* 244 Oxianem] Oxsan- *Ve* et *add.*] *om. codd.* 246 aquilonari] -lionari *Ox*, (?) *Ve* 248 qui] quod *Ox* 248-9 Iouem et solem] *inv. Ox* 249 Ditissimi] Dilectissimi *Ox*

rumque usus reuerentes | et puri, in uestitu pretiosi, donatiuique et alta sapientes; que quidem ut ad omne dicatur orientalis configuratio Saturni et Iouis operatur.

[35] Et harum autem gentium rursum que quidem circa Hyrcaniam et Armeniam et Matianem magis coaptantur Geminis | et Mercurio; propter quod quidem sunt facilioris motus magis et subdoli.

[36] Que autem circa Bactrianem et Caspiriam et Sericam, Libre et Veneri; unde qui optinent regiones ditissimi sunt et amatores musicorum et magis splendide diete. Que autem circa Sauromaticam et | Oxianem et Sugdianem, Aquario et Saturno; propter quod et hee gentes magis immansuete et austere et bestiales.

[37] Relique autem huius quarte et iacentes circa medium totius habitabilis, Bithinia et Frigia, Colchica, | Syria et Commagena, Cappadocia, Lidia, Licia, Cilicia, Pamphilia, situm habentes ad libonotum ipsius quarte, accipiunt coaptationem quarte austrolibice, Cancri scilicet, Scorpionis et Piscium, et condispositores Martem et Venerem et adhuc | Mercurium.

[38] Propter quod quidem qui circa regiones has colunt ut in pluribus Venerem tamquam matrem deorum, uariis et congruis nominibus appellantes, et Martem tamquam Adonidem aut aliter qualiter iterum nominantes et misteria quedam cum lamentis attribuentes ipsis. Circumuoluti | sunt et dolose anime et laboriosi et maligni, et stipendiariis militiis et rapinis et captiuationibus se ingerentes et seruituti adicientes se ipsos et bellicis exterminationibus incidentes.

[39] Propter Martem et Venerem et orientalem configurationem (quia in | trigonali quidem Veneris animali, scilicet Capricorno, Mars exaltatur, in trigonali autem animali Martis, scilicet Piscibus, Venus exaltatur) — propter quod mulieres accidit omnem beniuolentiam ad uiros ostendere, affectuosasque existentes et domus curam habentes et laboriosas | et ministratiuas et totaliter laboriosas et subiectas.

V (**α** =**YΣ** **β** =**DMS** **γ** =**ABC**) 254 ut: ὡς *cett.*] *om.* **V** orientalis: ἀνατολικὸς **β γ**] -λιῶν **V**: -λικῶν *coni. Robb. Hüb.* ; *cf.* **Y** *et* **Σ** 257 subdoli: ὑποπόνηρα *cett.*] οἱ πόνηροι **V** 258 Sericam: Σηρικὴν *cett.*] συρικὴν **V** 259 magis (*cett.*)] *om.* **V** 263 (et ... et] *om. codd.*, *cf.* **V**) Colchica — Cappadocia (*cett.*)] *om.* **V** Colchica: κολχική **α** (*Hüb.*)] κολχίς **β γ**: [**V**] Licia: λυκία **V** Procl.] *om. cett.* (*Hüb.*) (265 et[1]: καὶ **α**] *om. cett.*) 270 Circumuoluti (*an* περίκυκλοι *legit* **G** ?)] περίκακοι δὲ *codd.* (271 stipendiariis militiis ... se ingerentes: ἐν μισθοφόροις στρατείαις ... γινόμενοι **α β** (*Hüb.*), *cf.* **γ**] μισθοφορικαῖς στρατιαῖς ... γινόμενοι **V**) 273 (Propter] + δὲ *codd.*) et[2] (*an* καὶ *legit*?)] κατὰ *codd.* 275 propter quod: διὸ **V**] διὰ τοῦτο **α** (*Hüb.*): *om. cett.*

---

254 Iouis] Ioui *Ve* 255 Hyrcaniam] Ircaniam *Ve* 256 Matianem] marianem *Ve* Geminis] gemis *Ve* 257 facilioris] -res *Ve* 259 ditissimi] diriss- *Ox* 263 Bithinia] -thuria *Ox*, (?) *Ve* Colchica] -chicha *Ox* Syria] Siria *Ve* 264 libonotum] -narum *Ve* 264-5 accipiunt — quarte] *hom. om. Ox* 265 austrolibice *scr. cum Ve*[mg]] -libie *Ve*[i.t.]: -lobice *Ox* 268 matrem] martem *Ve* 269 aut] autem *Ve* 270 Circumuoluti] -te *Ox* 271 stipendiariis] -ris *Ve* 276 beniuolentiam] bene- *Ox*

[40] Et horum autem rursum qui quidem circa Bithiniam et Frigiam et Colchicam coaptantur magis Cancro et lune. Propter quod quidem ut ad omne dicatur uiri quidem sunt reuerentes et subditi, mulierum autem plurime propter | lune orientalem et masculinizatam figuram presunt uiris et principatiue et bellice sunt, quemadmodum Amazones, fugientes quidem uirorum concubitus, amatrices autem armorum existentes et masculinizantes feminina omnia a pueritia abscisione dextrarum mamillarum propter militaria exercitia et | denudatione harum partium quando instruunt, ad ostensionem nature que deposuerit feminitatem.

[41] Qui autem circa Syriam et Commagenam et Cappadociam, Scorpioni et Marti; propter quod multum apud ipsos accidit proprietas audacie et malignitas et insidiositas et laboriositas. Qui | autem circa Lidiam et Ciliciam et Pamphiliam, Piscibus et Ioui; unde isti magis sunt multarum possessionum et acquisitiui et communicatiui et liberales et fideles circa commutationes.

[42] Relique autem quarte, que est penes uocatam communiter | Libiam, alia quidem continentia Numediam, Carchedoniam, Africam, Fazaniam, Nassamonitem, Garamanticam et Mauritaniam, Getuliam, Metagonitidem et situm habentia ad liuonotum totius habitabilis, coaptantur quidem trigonalitati notolibice, scilicet Cancro, Scorpioni et Piscibus, | disponuntur autem merito a Marte et Venere secundum figurationem uespertinam.

[43] Propter quod accidit plurimis ipsorum propter dictam stellarum conuenientiam gubernari sub uiro et muliere qui sunt eiusdem matris frater et soror, uiro quidem | presidente uiris, muliere autem mulieribus, conseruata tali successione. Calidique sunt ualde et precipites ad mulierum coitus, ita ut nuptie fiant per raptus, et in multis locis reges primo coeunt cum hiis que desponsantur, apud aliquos autem et | communes esse mulieres omnium.

[44] Sunt autem amatores pulcritudinis et ornatus muliebres circumcingunt propter Venerem, uiriles tamen sunt quantum ad animas et subdoli et magici, seductores autem et fallaces et proiectores periculi propter Martem. Horum autem

---

V (**α** =**ΥΣ** **β** =**DMS** **γ** =**ABC**) 280 coaptantur: συνοικειοῦνται *cett.*] -οῦται **V** 282 et[3] (*cett.*)] *om.* **V** 283 fugientes: φεύγουσαι **Σ** Procl. (*Hüb.*): φύγουσαι **V**] φεύγουσι(ν) *cett.* 285 denudatione: ἀπογυμνώσει **V**] ἀπογυμνοῦσαι **α** (*Hüb.*): -νοῦσι δὲ **M S γ**, *cf.* **D** 285-6 harum partium: τούτων τῶν μερῶν **V**] ταῦτα τὰ μέρη **α β γ** (*Hüb.*) 296 Cancro: Καρκίνῳ **β**] Καρκίνου **α** (*Hüb.*): Κρώνου **V**: *om.* **γ** Scorpioni: Σκορπίῳ **β**] Σκορπίου **V α γ** (*Hüb.*) (et[1]: καὶ **α γ**] *om.* **V β**, *Hüb.*) Piscibus: ἰχθύσιν **β**] ἰχθύιν **V**: ἰχθύων **α γ** (*Hüb.*) (299 qui sunt] δυεῖν *codd.*; *ad sensum* **G**, *cf. Robb. ad loc.*: «who are») 301 ita ut: ὥστε **β γ**] ὡς **α** (*Hüb.*): *om.* **V** (302 primo: πρῶτα **α γ**] πρώτους **V M S**, *Hüb.*) 305 magici: μαγευτικοί **α M S γ** (*Hüb.*)] γαμευτικοί **V D**

---

282 presunt] possunt *Ve* 284 autem] *om.* *Ox* a pueritia] prouincia *Ve* 286 deposuerit] despos- *Ve* 289 Lidiam *Ve*[i.t.]] al' libiam *Ve*[mg]: Lydiam *Ox* 290 multarum *scr.*] -orum *codd.* 295 liuonotum *sic codd.* notolibice *scr.*] nota- *codd.* 299 qui *scr.*] que *codd.* sunt] sint *Ve* 302 reges] *om.* *Ox* coeunt] coheunt *Ox* desponsantur] dis- *Ox*

iterum qui | quidem circa Numediam et Carchidonem et Africam coaptantur magis Cancro et lune; propter quod quidem isti communicatiui et acquisitiui existunt et in habundantia omni permanentes.

[45] Qui autem circa Metagonitidem et Mauritaniam et Getuliam, Scorpioni et | Marti; unde isti bestialiores sunt et pugnacissimi et carnifagi et ualde proiectores periculi et despectores eius quod est uiuere, ut neque ab inuicem separentur. Qui autem circa Fazaniam et Nassamonitidem et Garamanticam, Piscibus et Ioui; propter quod liberales et | simplices moribus et amatores operis et bene sententiantes, mundique et insubicibiles sunt ut ad omne dicatur, et Iouem tamquam Ammonem religione honorant.

[46] Relique autem huius quarte partes et figurate circa medium totius habitabilis, Cyrenaica, Marmarica, | Egiptus, Thebaida, Oasis, Trogloditica, Arabia, Azania, Media Ethiopia, site ad boream et orientem totius quarte, accipiunt coaptationem trigonalitatis boreafeliotice, scilicet Geminorum, Libre, Aquarii, et condispositores propter hoc Saturnum et | Iouem et adhuc Mercurium.

[47] Vnde qui secundum ipsas regiones fere participant quinque planetarum dominatione uespertina deorum amatores quidem facti sunt et timidi ad deos et deo supplices et amatores planctus, et mortuos terra occultant et sepeliunt propter uespertinam | figurationem; omnimodis autem consuetudinibus et legitimis et deorum omnimodorum religionibus utentes; et in subiectionibus quidem humiles et timidi, in presulatibus autem boni animi et magna sapientes.

[48] Multarum autem mulierum et multi uiri et precipites et sororibus coaptati, | et uiri multi spermatis, mulieres autem facile concipientes, conformiter regionis fecunditati. Multi autem masculorum audaces et feminini quantum ad animas, quidam autem et genitalia membra contempnentes propter configurationem maleficorum siderum cum Venere uespertina.

V (**α** =**YΣ** **β** =**DMS** **γ** =**ABC**) 307 Carchidonem: Καρχηδόνα **V** Procl. (*Robb.*)] Καρχηδο(ω)νίαν *cett.* (*Hüb.*) 310 circa: περὶ *cett.*] ἐπὶ **V** 317 Relique autem (*cett.*)] + περὶ τὸ μέσου **V** (huius] τοῦ *codd.*) 319 boream et orientem: βορραπηλιώτην *cett.* (*Hüb.*)] ἀπηλιώτην **V** Procl. 323 dominatione: οἰκοδεσποτείας **Σ M S γ** Procl. (*Hüb.*)] οἰκοδεσπότας **V Y D** 327 timidi] + καὶ μικρολόγοι καὶ ὑπομονητικοί *codd.* (328 Multarum ... mulierum] πολυγύναιοι ... *codd.*) 330 audaces: θρασεῖς **β γ**] σαθροί **V α** (*Hüb.*)

---

309 omni] omnium (*abbr.*) *Ve* 310 Getuliam] + et *Ox* 314 liberales et simplices] simplicitates et liberales *Ox* 316 Ammonem] amno- *Ve* religione *scr.*] -nes *Ve*: -nem *Ox* 317 et] *om. Ox* 318 Cyrenaica *scr.*] Cureniaca *codd.* 321 adhuc] ad hoc *Ox* 323 dominatione] + per *Ox* 324 supplices] simplices *Ve* planctus] plantus *Ve* propter] *ss. Ox* 325 autem] aut *Ve* 326 in] *om. Ox* 327 timidi] timili *Ox* autem] *it. post* boni *et exp. Ox* 331 genitalia] gentalia *Ox* contempnentes] continientes (?) *Ve* 332 maleficorum] -ciorum *Ox*

[49] Et horum autem <qui> quidem circa Cyrenaicam et Marmaricam et maxime qui circa inferiorem regionem Egipti magis coaptantur Geminis et Mercurio; propter quod quidem isti sunt meditatiui et intelligentes et facile applicabiles ad omnia et maxime circa | inuentionem sapientie et diuinorum, magici etiam et occultorum misteriorum obseruatiui et omnino sufficientes circa mathemata.

[50] Qui autem circa Thebaydem et Oasim et Trogloditicam, Libre et Veneri; unde et ipsi calidiores sunt secundum naturas et moti et in habundantiis habentes | conuersationes. Qui uero circa Arabiam et Azaniam et Mediam Ethiopiam, Aquario et Saturno; propter quod et isti sunt carnifagi et piscifagi et pascuales sunt, agrestem et bestialem uitam uiuentes.

[4.1] Coaptationes quidem igitur stellarum et duodecim signorum ad partiales gentes et proprietates ipsarum ut in pluribus summatim sint nobis dicta figuraliter hoc modo.

Expositio regionum que referuntur ad unumquodque signorum Capitulum 4^{m}

Exponemus autem propter | facilitatem usus secundum nudam descriptionem singulas gentes coaptatas ad unumquodque signorum, conformiter hiis que preaccepta sunt de ipsis, secundum hunc modum.

V (*α* =**YΣ** *β* =**DMS** *γ* =**ABC**) 333 Cyrenaicam: Κυρηναϊκὴν *cett.*] Κραϊκὴν **V** 340 circa: περὶ *cett.*] ἐπὶ **V** 343-345 Coaptationes — modo *hic (ante tit.) ut* **V**] *post tit.* **D** (*Hüb.*); *cf.* **M** *γ* Co

---

333 qui *add.*] *om. codd.* quidem] *ss. Ox* Cyrenaicam] cyinaicam *Ve* 334 Egipti] -tii *Ox* 336 maxime] + qui contra (*sic*) — intelligentes (334-335) *it. et uacat ss. Ox* magici *scr.*] magicis *Ve*: magis *Ox* 337 misteriorum] + et *Ve* 338 Thebaydem] tebay- *Ox* et Oasim] *ss. Ox* Trogloditicam] troglodicam *Ve* 340 conuersationes] conseruationes *Ox* 341 pascuales] pasculiles *Ox* 348 Capitulum] capitulu. *Ve* 349-51 Exponemus — modum *hic hab. Ox*] *post tabulam* (*ante* Regiones omnes ...) *hab. Ve*

[2] **Arietis**	**Tauri**	**Geminorum**	**Cancri**
Britannia	Parthia	Hircania	Numedia
Galatia	Media	Armenia	Karchedonia
Germania	Persia	Matiane	Affrica
Barsania			
circa medium	*circa medium*	*circa medium*	*circa medium*
Profunda Siria	Kiklades	Cirenaica	Bithinia
Palestina	Ciprus	Marmarica	Frigia
Idumea	Parua Asia	Egipti inferioris <regio>	Kolchica
Iudea			
[3] **Leonis**	**Virginis**	**Libre**	**Scorpionis**
Italia	Mesopotamia	Bactriane	Methagonites
Gallia	Babilonia	Caspiria	Mauritania
Sicilia	Assiria	Serica	Getulia
Apulia			
circa medium	*circa medium*	*circa medium*	*circa medium*
Fenicia	Ellas	Thebais	Siria
Caldea	Achaya	Oasis	Commagena
Orchenia	Creta	Trogloditica	Capadocia
[4] **Sagitarii**	**Capricorni**	**Aquarii**	**Piscium**
Tirenia	India	Sauromatica	Farzania
Celtica	Ariane	Oxiane	Nassamonitica
Hispania	Gedrusia	Sugdiane	Garamantica
circa medium	*circa medium*	*circa medium*	*circa medium*
Arabia felix	Tracia	Arabia	Lidia
	Macedonia	Azania	Cilicia
	Illiria	Media Ethiopia	Pamphilia

V (**α** =**ΥΣ** **β** =**DMS** **γ** =**ABC**) 352-378 *tabulam habet* **G** *ut* **V M S** (*Hüb.*)] *ordinem pervertit* **D**, *planetas add.* **γ**: *om.* **α** 355 Matiane (Ματιανή *Hüb.*)] ματτι- **α**: μαντι- **β γ**: γαμπανή **V** 356 Barsania (βαρσανία ?)] *om.* **V**: Βασταρνία *cett.*, *cf. supra* 168 358 Kiklades: Κυκλάδες **V** Heph. («*fort. recte*»)] + νῆσοι *cett.* (*Hüb.*); *cf. supra* 175-6 359-360 Palestina *super* Idumea (Heph. I 1, 6)] Παλαιστίνη *infra* Ἰδουμαία **V β γ** (*Hüb.*): *om.* **α**; *cf. Boll ad loc.* 360 Parua Asia: μίκρα Ἀσία **V** Heph.] παράλια Μικρᾶς Ἀσίας *cett.* (*Hüb.*) 365 Serica (*cett.*)] συρική **V** 373 Nassamonitica: νασαμωνιτική **V M**] Νασαμωνῖτις *cett.* (*Hüb.*)

352 Arietis] Aries *Ox*^{ac} Britannia] -ania *Ox.* 353 Parthia] -tica *Ox*^{ac} Numedia] Sinimedia *Ve* 355 Matiane] Martiane *Ox* 358 Profunda Siria] *ss. Ox*: *om. Ve* Kiklades *scr.*] -kaldes *codd.* Cirenaica] -naika *Ve* Bithinia] *obscure in textu*, bichinia *ss. Ox* 359 Ciprus] -pris *Ox* 360 regio *add.*] *om. codd.* 363 Mesopotamia] -tania *Ve* 364 Caspiria] Kaspira *Ve* 365 Assiria] -iria *Ox*$^{i.t.}$: -uria *Ox*^{ss} 369 Achaya] Acaia *Ox* Commagena] Oromag- *Ve* 372 Tirenia] Sire- *Ve*: Tirre- (?) *Ox* Sauromatica] *e corr. Ox* 373 Nassamonitica] nassamoniatica *Ve*: Nassominitica (*exp.* -ca *et ss.* de) *Ox* 374 Hispania] Hys-*Ve* 378 Pamphilia] *Ox*^{ss}, -filia *Ox*$^{i.t.}$

[5] Regiones omnes sunt 72. Expositis autem hiis, rationabile et illa huic parti adicere: quia et non errantium | stellarum singule coaptantur regionibus, quibuscumque uidentur facientes compassionem et zodiaci partes, in quibus habent non erratice prosneuses (id est adnuitiones) in circulis descriptis per polos ipsius.

[6] Et quod in principalibus ciuitatibus illa loca | zodiaci maxime compassionem efficiunt, in quibus in fundationibus ipsarum a principio sol et luna transitum faciebant, quemadmodum in hominis generatione, et angulorum maxime ascendens. In quibus autem tempus fundationis non inuenitur ascendens, in quibus | euenit medium celi in natiuitatibus pro tempore principantium aut regnantium.

Accessus in particulares effectus eclipsium
Capitulum 5^{m}

[1] Hiis autem ita preconsideratis, consequens utique erit de cetero effectuum accessus summatim percurrere, | et primo eorum qui accipiuntur penes totas condiciones regionum aut ciuitatum. Erit autem modus considerationis talis. Prima quidem igitur et fortissima talium euenientium causa fit penes eclipticas coniugationes solis et lune et configurationes | stellarum in ipsis.

[2] Effectus autem ipsius hoc quidem est locale, secundum quod oportet pronosticari quibus regionibus aut ciuitatibus partiales eclipses aut etiam erraticarum per tempora mansorie stationes, hee autem sunt Saturnus, Iupiter et Mars quando steterint; faciunt | enim tunc significationes. [3] Hoc autem aliquid temporale, secundum quod tempus significationum et quantum extendi potest

V (**α** =**YΣ** **β** =**DMS** **γ** =**ABC**) 379 Regiones omnes sunt 72: γίνονται χῶραι οβʹ **V M** Procl. (*Hüb.*)] *om. cett.* (*Boll Robb.*) 381 (uidentur — compassionem] *ante* καὶ ὅτι [Et quod 383] *codd.*; *fort. tr.* **G**) in quibus: καθ' ὧν Heph. (*Hüb.*)] καθὸν **V**: καθ' ὃ **Y**: μεθ' ὧν **Σ β γ** 382 prosneuses (*cett.*)] ἐποχὰς *praem.* **V** (in circulis descriptis ...] ἐπὶ τοῦ ... γραφομένου κύκλου *codd.*) 383 quod: ὅτι *cett.*] ὅτε **V** illa: ἐκεῖνοι **β γ** (*Hüb.*), *cf.* **α**] ἐκείνων **V** (zodiaci *ut* **V**] + κύκλου *cett.*) 384 in quibus: καθ' ὧν **Y β γ** (*Hüb.*)] καθ' ὅν **V Σ** 385 (quemadmodum — generatione *hic*] *post* αὐτῶν [ipsarum 384] *codd.*; *fort. tr.* **G**) (hominis] *om. codd.*; *fort. add.* **G**) 386 fundationis: κτίσεων *cett.*] κτήσεων **V** in quibus^{2}: καθ' ὧν **β γ** (*Hüb.*)] καθ' ἣν **V**: *om.* **α** 393 Erit: ἔσται *cett.*] ἐπεὶ **V** (395 configurationes] παρόδους *codd.*) 396 quidem: μὲν **β γ** (*Hüb.*)] μέντοι **V**: μέν τι **α** 399 aliquid (τι ?)] ἐστι *codd.*

379 72] septuaginta due *Ox* 380 adicere *scr.*] addicere *Ve*: dicere *Ox* 382 prosneuses *scr.*] prosnentes *Ve*: pronenses *Ox* adnuitiones *scr.*] adnritiones (?) *Ve*: admissiones *Ox* 385 hominis] omnis *Ve* 386 ascendens^{2} *scr.*] accendens *Ox*: accendendum *Ve* 391 effectuum] -tiuum *Ox* 392 qui] que *Ox* 394 et] *om. Ox* euenientium] + in *Ox* 396 est] *om. Ox* 397 erraticarum] + et *Ox* 398 mansorie *scr.*] -sione *codd.* 399-401 Hoc — pronosticari] *mg. Ve* 400 significationum *scr.*] -torum *codd.*

oportebit pronosticari. Hoc autem g e n e r i s , secundum quod oportebit accipere circa qualia genera euenient accidentia. Vltimo autem quod s p e c i e i , secundum quod ǀ qualitatem ipsius effectus considerabimus.

De consideratione dispositarum regionum Capitulum 6^{m}

[1] De primo quidem igitur et locali talem faciemus tractatum. In hiis enim que fiunt eclipticis coniugationibus ǀ solis et lune (et maxime circa magis sensibiles) considerabimus locum zodiaci eclipticum et regiones coaptatas trigonalitati ipsius loci, et similiter que ciuitatum habeant compassionem ad signum eclipsis, uel ex ascendente et loco luminarium in constructione ciuitatis, uel ex ǀ mediatione celi tunc regnantium.

[2] In quibuscumque autem regionibus uel ciuitatibus inuenerimus premissam coaptationem, circa omnes quidem ut ad omne dicatur existimandum fore aliquod accidens, maxime autem circa eas que rationem habent ad ipsum signum eclipsis, ǀ et in quibuscumque ipsarum apparet eclipsis super terram existens.

De tempore effectuum Capitulum 7^{m}

[1] Secundum autem et temporale capitulum, secundum quod tempora significationum et quantitatem durationis ǀ conuenit discernere, considerabimus modo tali. Hiis enim que in eodem tempore fiunt eclipsibus solaribus et lunaribus non secundum omnem habitationem in eisdem horis temporalibus factis, neque quantum ad magnitudines obtenebrationum aut quantum ad tempus durationum secundum equalitatem ǀ ubique acceptis, primo quidem secundum eclipticam

---

V (**α** =**YΣ** **β** =**DMS** **γ** =**ABC**) 406 De: περὶ **V β γ**] *om.* **α** (*Hüb.*) 408 trigonalitati: τῷ ... τριγώνῳ **γ** (*Boll*)] τῶν ... τριγώνων **V β** Procl. (*Robb. Hüb.*), *cf. cett.* (409 habeant — eclipsis *hic habet* **G**] *in fine sententiae* [411] *codd.*; *fort. tr.* **G**) (412 regionibus: *cett.*] ἢ *praem.* **V D**) 413-4 aliquod accidens: τι σύμπτωμα *tacite Robb.*] τὸ σύμπτωμα *codd.* 415 apparet*: φαίνεται **G**° **Σ**] φένεται **Y**: ἐφαίνετο **V β γ** Heph. Procl. (*Hüb.*) 420 eclipsibus solaribus et lunaribus *hic*] *codd. post* ἀποτελουμένων (factis 421) *sic*: τῶν τε ἡλιακῶν δηλαδὴ καὶ τῶν σεληνιακῶν **Σ γ**: τῶν τε ἡλιακῶν τῶν αὐτῶν **V Y** Procl. (*Robb. Hüb.*): τῶν τε ἡλιακῶν καὶ τῶν σεληνιακῶν τῶν αὐτῶν **β**: *totum locum ut glossema damnavit Boll*

---

403 quod] *om. Ox* 404 dispositarum *scr.*] -torum *codd.* 408 eclipticum et] eclipticionum esse *Ve* 412 autem] tunc (*abbr.*) *Ve* 413 existimandum] exsti- *Ox* 415 ipsarum] ipsum *Ve* apparet *scr. cum Ox*] appareat *Ve*: *an* *apparebat *scrib.* ? 419 discernere] discerne *Ox* 420 enim] autem *Ox* que] *exp. et* qui *ss. Ox* 421 temporalibus] -oribus *Ox* 423 equalitatem] qualitatem *Ox*

horam in unaquaque habitationum rationem habentium ascendens et angulos, sicut in natiuitate, disponemus.

[2] Deinde et ad quot horas equinoctiales in unaquaque habitatione protenditur obtenebratio eclipsis. Hiis enim exquisitis | quotcumque equinoctiales horas inuenerimus, in solari quidem eclipsi ad tot annos durare quod efficitur existimabimus, in lunari autem ad tot menses. Principiis tamen et maioritate totius consideratis ex | habitudine loci ecliptici ad angulos.

[3] Apud orientalem enim orizontem locus accidens inchoationem simptomatis in primis quatuor mensibus a tempore eclipsis significat, et totales excessus circa primam tertiam partem temporis, quod est ad totam durationem; apud angulum autem | medii celi, in secundis quatuor mensibus et in media tertia parte; apud occidentalem autem orizontem, in tertiis quatuor mensibus et in ultima tertia parte.

[4] Particularibus autem remissionibus et intentionibus a medio coniugationibus, quando in locis causam efficientibus | uel in locis configuratis eis euenerint, similiter consideratis, et ab aliis ad hec accedentibus, quando scilicet astra factiua effectus ortus uel occasus uel stationes uel in extremo noctis apparentiam fecerint, configurata fuerint signis causam habentibus; quoniam orientia | quidem uel stationaria intentiones faciunt simptomatum, occumbentia autem et sub radiis existentia aut extremitate noctis facientia precessiones (id est retrogradationes) remissionem effectuum faciunt.

V (**α** =**ΥΣ** **β** =**DMS** **γ** =**ABC**) 424 ascendens] καὶ τὸ τοῦ πόλου ἔξαρμα *codd.*; *fort. sic redd.* **G** et angulos: τά τε κέντρα **α**: καὶ τὰ κέντρα **M S**] κέντρα **V D γ** Procl. (*Hüb.*) 429 tamen: μέντοι **V α** (*Robb.*)] *om.* **β γ**: μὲν *coni. Hüb.* (et maioritate totius] καὶ τῶν ὁλοσχερεστέρων ἐπιτάσεων *codd.*, *cf.* **γ**) (433-4 apud angulum ... medii celi] πρὸς ... τῷ μεσουρανήματι) 439 (similiter consideratis] *om. codd.*; *cf.* «we deduce from» *add. Robb.*, «nous pourrons déduire» *add. Charvet*) ad hec accedentibus* (παροδευόντων ?)] παρόδων **V** Procl. (*Hüb.*), *cf.* **Υ**: παρανατελλόντων **Σ D γ**: *om.* **M S** 441 signis: δωδεκατημορίοις *cett.*] -μόριον **V**

---

427 enim] *om. Ox* 428 existimabimus] exti- *Ve* 429 lunari] luminari *Ox* 433 quod] que *Ox* 439 ad hec accedentibus* *scr.*] ad hoc accidentibus *codd.* scilicet] *ss. Ve* factiua] tractiua *Ox* 441 fuerint] fuerit *Ox*

De genere [autem] dispositorum
Capitulum 8[m]

[1] Tertio autem existente capitulo eo quod generis, per quod oportet accipere | circa qualia genera eueniet simptoma. Accipitur et hoc per proprietatem signorum et formam, in quibuscumque extiterint eclipsium loca et ille stellarum, erraticarum et fixarum, que habent dominium | signi eclipsis et eius quod est penes angulum qui est ante eclipsim.

[2] Accipitur autem horum dominium in erraticis quidem stellis sic. Qui enim plures habet rationes ad ambo exposita loca, scilicet ad locum eclipsis et ad locum | anguli ipsum sequentis, secundum propinquissimas et apparentes coniunctiones aut defluxus et configurationes rationem habentes et adhuc secundum dominium domorum et trigonalitatum et exaltationum aut etiam terminorum, ille accipiet solus gubernationem.

[3] Si autem non idem inueniatur | dominus eclipsis et anguli, duorum illum assumendum, qui ad utrumque locum plures habet coaptationes, ut premissum est. Si autem ambo easdem habeant rationes, preferimus dominum eclipsis. Si autem plures inueniantur secundum utrumque similes, eum qui fuerit magis in cuspide aut in usu fortiorem aut | eum qui est magis dominus partitionis preferimus in gubernationem.

[4] In fixis autem coassumemus illam que cum loco ecliptico in usu fortior fuerit primam fulgidarum in ascendente angulo, secundum apparentes figurationes determinatas nobis in prima compilatione nouem | modorum, et eam que fuerit in apparente dispositione secundum horam eclipticam uel simul oriens uel celum medians cum eo qui secundum sequentia angulo loci eclipsis.

V (**α** =**YΣ** **β** =**DMS** **γ** =**ABC**) 451 qui est ante: τὸ πρὸ **G°** (*Hüb.*)] τὸ πρῶ **Y**: τοῦ πρὸ **β γ**: *om.* **V Σ** (453 plures: πλείους **α**] πλείστους **V β γ**, *Hüb.*; *cf. supra* 189) 454 et apparentes (**V α**, *Robb.*)] φαινομένας **β γ** (*Hüb.*) 458 duorum: δύο **V** Procl.] ἀλλὰ δύο **β γ** Heph. (*Hüb.*): δύο δὲ **α** illum ... qui habet: τὸν ... ἔχοντα **β γ** (*cf.* Heph.)] τοὺς ... ἔχοντας **V Y** Procl. (*Hüb.*), *cf.* **Σ** 460 Si autem ambo easdem habeant rationes] *om. codd.*; *an suo sponte vel e graeco add.* **G** ? 461 utrumque: ἑκάτερον *cett.*] ἕτερον **V** 462 in usu fortiorem: χρηματιστικώτερον *cett.*] σχημα- (ρ *super* χ *add.*) **V** (eum qui est magis dominus] μᾶλλον ὄντα *codd.*; dominus *fort. add.* **G**) 464 cum loco ecliptico: πρὸς τῷ ἐκλειπτικῷ τόπῳ **M**: τῷ ἐκλειπτικῷ τόπῳ **D S γ**] αὐτῷ τῷ ἐκλειπτικῷ τόπῳ **Σ** (*Hüb.*), *cf.* **Y**: αὐτῷ τῷ ἐκλειπτικῷ χρόνῳ **V** Procl. 466 eam: τὸν **β γ** (*Hüb.*)] τῶν **V α** 468 cum eo qui secundum sequentia: τῷ κατὰ τὰ ἑπόμενα *cett.*] κατὰ τὰ ἐπικείμενα **V**

---

445 autem *del.*] *hab. codd.* 448 per] propter *Ox* 449 extiterint *sic codd.* 452 Accipitur] -piatur *Ox* 453 eclipsis] -im *Ve* 456 aut] autem *Ve* 462 fortiorem] fortior est *Ox* 464 illam] *om. Ox* 465 fulgidarum] -orum *Ox* 466 nobis] uobis *Ox*

[5] Consideratis autem ita hiis que ad causam simptomatis | coassumuntur astris, consperculabimur signorum formationes in quibus fuerint eclipsis et astra dominium habentia, tamquam ab horum proprietate et qualitas dispositorum [et] generum ut ad omne dicatur accipienda sit.

[6] Signa enim que sunt humane forme, | circa circulum qui est per medium animalium et que penes stellas fixas, circa hominum genus faciunt quod efficitur. Aliorum autem terrestrium quadrupedia quidem circa similia brutorum animalium, reptilia autem circa serpentes et talia. [8.7] Et iterum | feralia quidem <circa> indomita animalia et nociua humani generis. Domestica autem circa utilia et mansueta et cooperatiua in bonum, proportionaliter hiis que per singula formis, ut puta equorum uel bouum uel ouium et talium.

[8] Adhuc autem terrestrium que quidem apud | Vrsas magis circa repentinos terre motus, que autem circa meridiem circa inexpectatos ex aere fluxus. Rursum autem principalia quidem loca existentia in alatorum formis (ut in Virgine et Sagittario, in Aue, Aquila et talibus) circa uolatilia, et maxime circa ea que ueniunt | in alimentum hominum simptoma faciunt, in natatilibus autem circa aquatilia et pisces.

[9] Et horum in marinis quidem, puta in Capricorno, Cancro, Delfine, circa marina et adhuc circa deductiones uasorum marinorum; in fluuialibus autem, uelut in Aquario et Piscibus, circa fluuialia | et fontalia; circa Nauem autem circa ambo genera.

[10] Eodem autem modo in tropicis aut equinoctialibus existentia communiter quidem circa consistentias aeris et circa conuenientia singulis ipsorum tempora efficiunt significationes, singulariter | autem et circa terre nascentia. Existentia enim circa uernale equinoctium, circa germina fructuum arborum, puta uinee, ficulnee, et congerminantium. Secundum estiualem autem uersionem, circa fructuum collectiones et repositiones, in Egipto autem singulariter | circa

---

V (**α** =**YΣ** **β** =**DMS** **γ** =**ABC**) 470 astra: ἀστέρες **α β γ** (*Hüb.*)] τῶν ἀστέρων **V** (*Boll*) (481 fluxus: ῥύσεις *cett.*] χύσεις **V**) 482 ut: οἷον *cett.*] *om.* **V** (et] *om. codd.*) 484 natatilibus: νηκτοῖς **Σ β γ** (*Hüb.*)] νυκτοῖς **V Y** 486 Capricorno (Αἰγοκέρωτι *cett.*: Αἰγοκέρωτος **V**) Cancro] *inv. codd.* (492 terre] ἐκ τῆς γῆς *codd.*)

---

469 coassumuntur *scr.*] -metur *Ve*: quo assumuntur *Ox* 470 formationes] conformationes *Ox* fuerint] fuerit *Ox* 471 et *del.*] *hab. codd.* 473 circulum qui est per] *om. Ve* 476 circa² *add.*] *om. codd.* 478 in *scr. cum Ve*] et *Ox*: *an scrib.* ad ? proportionaliter] -nalia *Ox* 481 meridiem *scr.*] medium *codd.* 482 autem] *om. Ox* 483 Sagittario] sagita- *Ve* Aquila] aquilia *Ox* 484 hominum] -ni (?) *Ox* natatilibus *scr.*] natalibus *codd.* 486 Capricorno] -nio (*cum corr.*) *Ve* Delfine *scr. cum Ve*] delphine (*an recte* ?) *Ox* 487 et] *om. Ve* uasorum] uascorum *Ox* marinorum] marinariorum *Ox* autem] aut *Ve* 490 aut] autem *Ve* 491 conuenientia] -ter *Ox* 492 autem] *om. Ox* 493 fructuum] fluctuum *Ox* 494 ficulnee *scr.*] ficulnie *Ve*: ficulne *Ox* congerminantium] cum germinantium *Ve*: *an* cumgerminantium *scrib.* ? 495 fructuum] + et *Ox*

inundationem Nili. [11] Penes autumpnale autem equinoctium, circa sementem et herbalia et talia. Secundum hyemalem autem uersionem, circa olera et circa genera auium uel piscium illo tempore apparentia. Adhuc autem equinoctialia quidem sacris | et religionibus circa deos significationes insinuant; tropica autem circa transmutationes aerum et ciuilium consuetudinum; fixa autem circa fundamenta et edificia; bicorporea autem circa homines et reges.

[12] Similiter autem et que apud orientem magis | habent positionem in tempore eclipsis, circa fructus et iuuenilem etatem et circa fundamenta quod futurum est significant; que autem apud medium <celi super terram, circa sacra et reges et mediam etatem; que autem apud> occasum, circa legitimorum transmutationes | et senilem etatem et circa mortuos.

[13] Circa quotam autem partem subiecti generis dispositio eueniet, et magnitudo obtenebrationis eclipsium insinuat et habitudines astrorum causam imprimentium ad eclipticum locum. Vespertina | enim figurata ad solares eclipses, matutina uero ad lunares, super minus ut ad omne dicatur disponunt; diametrizantia autem, circa dimidium; matutina autem figurata ad solares uel uespertina ad lunares, super plurimum.

De qualitate ipsius effectus
Capitulum 9[m]

[1] Quartum capitulum de ipsa qualitate effectus, hoc est utrum bonorum uel contrariorum sit effectiuus et quantorum in utroque secundum | proprietatem speciei. Hoc autem deprehenditur ab effectiua <natura> stellarum dominantium super principalia loca, et a commixtione ad inuicem et ad loca in quibus utique fuerint existentes.

[2] Sol quidem enim et luna ordinatores et quasi presides | sunt aliorum, cum sint ipsi causa eius quod secundum actum totius et dominationis astrorum, et

---

V (**α** =**YΣ** **β** =**DMS** **γ** =**ABC**) 500 consuetudinum: ἐθισμῶν **β γ** Heph. cod. A (*Hüb.*)] ἐθισμένων **V** (*cf. Robb.*); *cf.* **α** 507 Circa: περὶ **V** (*Boll*)] καὶ περὶ **α β γ** (*Hüb.*) quotam autem: πόστον δὲ **Σ β γ** Heph. (*Hüb.*)] πόσῶν δὲ **V**: μέσον **Y** eueniet: ἐπελεύσεται **α β γ** Heph. (*Hüb.*)] ἐλεύσεται **V** 511 circa dimidium: περὶ τὸ ἥμισυ **β γ** Heph.: ἐπὶ τὸ ἥμισυ **α** (*Hüb.*)] πρὸς τῷ ἡμίσει **V** 515 Quartum (**α**)] + δὲ **V β γ** (*Hüb.*) 516 quantorum: *an* ποσῶν *legit vel potius sic* ποταπόν (-πῶν ?) *redd.* **G** ?] ποταπόν (ποδαπόν) *codd.* 520 sunt (*cett.*)] *om.* **V**

---

496 Nili] *ex* vili *Ox*: + al' nauili *Ve*[mg] autem] *om. Ox* sementem] *dub. Ox* 498 auium] aequm (?) *Ve*[i.t.]: al' auium *Ve*[mg] 499 circa] quibus *Ox* 501 edif-] hedif- *Ve* 502 in] *om. Ox* 503 fundamenta] fonda- *Ve* 504-5 celi — apud *add.*] *hom. om. codd.* 507 autem] *ss. Ve* et] *om. Ve* 508 imprimentium] -tum *Ox* 509 enim] *om. Ve* 511 autem[1]] aut *Ve* 516 effectiuus *scr.*] effectus *codd.* 517 ab] ad *Ve* natura *add.*] *om. codd.* 518 a commixtione] ad -nem *Ox* 520 enim] + est *Ox*

adhuc fortitudinis et debilitatis dominantium. Speculatio autem commiscitiua accipientium dominium ostendit effectuum qualitatem.

[3] Incipimus autem a proprietate effectiua per singulos planetarum, illud communiter preexponentes, gratia admonitionis capitularis: quod quando uniuersaliter dicimus aliquod quinque astrorum, complexionem et efficaciam similis nature existimandum, siue ipsa in propria fuerit | dignitate siue fixarum aliqua uel zodiaci locorum aliquis secundum congruam sibi complexionem consideretur, [4] quemadmodum utique si naturarum et qualitatum ipsarum et non astrorum extiterint appellationes; et quod in commixtionibus rursum non solum | commixtionem planetarum ad inuicem oportet considerare, sed etiam commixtionem ad ea que participant eandem naturam, siue sint stelle fixe siue loca zodiaci, et hoc secundum assignatas ipsorum coaptationes ad planetas.

[5] S a t u r n u s quidem igitur solus dominationem | accipiens, uniuersaliter quidem causa est corruptionis per infrigidationem. Singulariter enim circa homines quidem facto simptomate, egritudines longas et ptises et consumptiones et humidorum molestationes et reumatismas et quartanarias molestationes et cordis refugationes et defectus et constrictiones, luctus | et timores et mortes, maxime etate prouectorum, efficit.

[6] Brutorum autem animalium circa utilia, ut ad omne dicatur, raritatem et eorum que fuerunt corruptiones corporales et egritudinales, a quibus et homines utentes condispositi corrumpantur. Circa consistentiam autem aeris | grauitates aeris et nubositates et caligines, adhuc nebularum multitudinem non bonarum sed corruptiuarum, a quibus et ledentia humanam naturam reptilium congregantur.

[7] Circa flumina autem uel maria communiter quidem hiemes | et caruanarum nauigia et nauigationis difficultates et piscium indigentiam et corruptionem; singulariter autem in mari quidem retractationes aquarum et refluxus, in fluminibus autem excrescentias et malignitatem fluuialium aquarum.

V (**α** =**YΣ** **β** =**DMS** **γ** =**ABC**) 522 adhuc fortitudinis ...: ἔτι τῆς ... ἰσχύος *cett.*] εἴ τις ... ἰσχύος **V** 524 Incipimus*] ἀρξό(ώ)μεθα *codd.* per singulos: καθ' ἕκαστον **V β γ**] καθ' ἕνα (ἕν **Σ**) ἕκαστον **Y Σ** (*Hüb.*) 527 ipsa: αὐτὸς *cett.*] αὐτὸ **V** 528 consideretur: θεωρῆται *cett.*] θεωρητέον **V** (532 et hoc] *om. codd.*: *fort. add.* **G**) (537 et cordis refugationes] φυγαδείας τε *codd.*) 542 condispositi: συνδιατιθέμενοι *cett.*] διατιθέμενοι **V** aeris] + ψύχη φοβερά, παγώδη καὶ ὁμιχλώδη καὶ λοιμικά *codd.* (*sed cf.* **β γ**) (546 nauigia: *fort.* ναυά(ή)για *transscr.* **G**] ναυάγια *codd.*, *cf. indic.*) 547 fluminibus — malignitatem (*cett.*)] *hom. om.* **V**

---

522 commiscitiua] cominstitiua *Ve* 524 Incipimus *codd.*] *an* -piemus *scrib.?*, *sed cf.* III.185 per] uel per *Ox* 525 uniuersaliter] utiliter *Ox* 526 aliquod] ad *Ox* complexionem] -plectio- *Ve* 528 complexionem] -plectio- *Ve* 530 extiterint *scr.*] existerit *codd.*, *cf.* III.171 531 oportet] debet *Ox* 533 ad planetas] *Ox*[mg]: planetarum (*exp.*) *Ox*[i.t.] 534 uniuersaliter] naturaliter *Ox* 535 enim] *an* autem *scrib.* ? 536 egritudines] -nis *Ox* 537 humidorum] humarum *Ox* 538 refugationes] *dub.* *Ve*[i.t.]: al' sufocationes *Ve*[mg] 540 ut] *om.* *Ox* 542 condispositi] cum dispositi *Ox* *an post* aeris* *quaedam exciderunt* ? 544 congregantur] -atur *Ox* 545 flumina] fluuia *Ox* hiemes] hemes *Ve* et] *om.* *Ve* 546 nauigia *sic codd.*, *hic et alibi* nauigationis] + et *Ox*

[8] Circa terre autem fructus defectum et raritatem et perditionem, maxime
725 eorum que producuntur ad necessarias | oportunitates, ab eruca uel locusta uel a
kathaclismis uel inundatione aquarum uel a grandine uel a talibus, ut pertingat ad
pestilentiam et talem hominum perditionem.

730 [9] I u p i t e r autem solus dominationem sortiens, uniuersaliter quidem | aug-
mentationis est factiuus. Singulariter autem circa homines eueniente effectu,
glorias perficit et annos bonos et habundantias et consistentias pacificas et ydo-
neorum incrementa, bonasque habitudines corporis et anime, adhuc autem bene-
735 ficia et dona a regibus et | illorum ipsorum augmentationes et magnitudines et
magnanimitates, uniuersaliterque felicitatis est causa.

[10] Circa bruta autem animalia eorum quidem que ad humanum usum boni-
tatem et magnam copiam facit, eorum autem que ad contrarium corruptionem et
741 interitum. Bene temperatam autem aeris consistentiam | terre nascentium efficit,
caruanarumque bonas nauigationes et fluminum moderatas excrescentias et fruc-
tuum felicitatem et quecumque hiis similia.

745 [11] M a r s autem solus gubernationem accipiens, uniuersaliter | quidem est
causa corruptionis per exsiccationem. Singulariter autem circa homines quidem
eueniente simptomate, bella efficit et seditiones intestinas et captiuitates et ser-
uitutes et insurrectiones et iras presidum et per talia mortes repentinas; adhuc
750 autem egritudines | calidas et tertianas molestias et sanguinum eductiones et
acutas per uiolentiam mortes, maxime eorum qui perfecte etatis; similiter autem
et uiolentias et iniurias et illegalitates, incensionesque et homicidia et rapinas et
depredationes.

755 [12] Circa aeris autem consistentiam estus et uentos calidos, | pestilentiales
et contabefactiuos, et fulminum dimissiones et incensionum et carentias pluuie.
Circa maria autem caruanarum repentina nauigia per uentos inordinatos uel per
tempestates uel per talia, fluminum autem defectus aquarum et exsiccationes fon-
tium et corruptionem aquarum [et] potabilium.

V (**α** =**ΥΣ** **β** =**DMS** **γ** =**ABC**) (551 inundatione aquarum] ὑδάτων ὀμβρίων ἐπιφορᾶς *codd.*) 552 pestilentiam: λοιμοῦ **α**] λιμοῦ **V** **β** **γ** Procl. (*Hüb.*) 553 sortiens: λαχῶν **V**] λαβῶν *cett.* (*Hüb.*) (559 bonitatem] δαψίλειαν *codd.*; *sic suo ingenio redd.* **G**, *cf. indicem*) 561 consistentiam*] + καὶ ὑγιεινὴν καὶ πνευματώδη καὶ ὑγρὰν καὶ θρεπτικὴν *codd.* 565 circa: περὶ *cett.*] τε **V** 569 acutas: ὀξείας *cett.*] ὀρείας **V** 574 maria: θάλασσας **V**] θάλασσαν *cett.* (*sed cf.* **α**) (*Hüb.*)

---

549 raritatem] cari- *Ox* maxime] *om. Ox* 550 ab] ad *Ox* 550-1 a kathaclismis] *coniuncte Ox*: ab atha dismis *Ve* 551 inundatione *scr.*] in undatione *Ve:* in nudatione *Ox* 553 uniuersaliter] soliter (?) *Ox* 556 bonasque] bonos- *Ox* 559 animalia] *ss. Ox* 560 ad] *om. Ve* 561 interitum] -tus *Ox* *an post* consistentiam* *quaedam exciderunt* ? 562 caruanarumque] carnauarum- *Ox* bonas] bonos *Ox* fluminum] fluuium *Ox* excrescentias] existentias (?) *Ve* 564 autem] est *Ox*^{ac} 570 incensionesque] intentio- *Ox* homicidia et] *om. Ox*$^{i.t.}$: homicia *Ox*^{ss} 572 uentos] -us *Ox* calidos] + et *Ox* 573 contabefactiuos *scr.*] -satiuos *Ve*: -saturus *Ox* 574 per uentos] peruentes *Ox* 576 et^{2} *del.*] *hab. codd.*

[13] Circa ea autem que super terram ydonea | ad usum humanum, brutorumque animalium et ex terra nascentium raritatem et corruptionem fructuum que fit ex inflammationibus caumatis, aut per brucum aut per concussionem uentorum aut ex combustione in apotecis.

[14] Venus autem sola domina existens accidentis, uniuersaliter quidem consimilia Ioui efficit cum quadam ueneritate. Singulariter autem circa homines quidem glorias et honores et alacritates et bonos status, bonasque nuptias et multitudinem filiorum et complacentias ad omnem | coaptationem et possessionum augmentationes et dietas mundas et decentes et ad uenerabilia preciosas. [15] Adhuc autem corporales euexias et ad presides coaptationes et principum luxurias.

Circa aerem autem uentorum temperantias et humectationum et uegetatiuissimorum | consistentias, aeris bonitatem et puritatem et fecundarum aquarum delicatas exuberantias; caruanarumque faciles nauigationes et eufortunia et lucra et fluminum plenas inundationes. In animalibus autem boni usus et terre fructibus maxime habundantiam et fertilitatem et profectum efficit.

[16] Mercurius autem dominationem accipiens, uniuersaliter quidem, ac si esset commixtus unicuique aliorum, coaptatur illorum naturis. Singulariter autem est omnium magis commotiuus et in humanis quidem effectibus acutus et operosissimus et ad intentum bene industrius; | predatiuorum processuum et temptatiuorum et latrociniorum et inuasionum, adhuc autem difficilis nauigationis factiuus est in configurationibus ad maleficos; egritudinumque siccarum causa et efimerinarum significationum et ptisis cum tussi et resolutione. [17] Effectiuus est et eorum que accidunt circa sacerdotalem sermonem | et circa deorum religiones et regales processus, et permutationis consuetudinalium aut legitimorum per tempora, conuenientes complexioni ad unumquodque astrorum semper.

V (**α** =**YΣ** **β** =**DMS** **γ** =**ABC**) 579 inflammationibus: καταφλέξεων **V**] καταφλέξεως *cett.* (*Hüb.*) 588 Circa aerem autem uentorum] περὶ δὲ τὰ τοῦ ἀέρος πνεύματα (-μάτων **α M S**) *codd.*: *an sua sponte redd.* **G** ? 591 In: ἐν **α**] ἔτι **V β γ** (*Hüb.*) 593-4 ac si esset: ὡς ἂν ᾖ (*coni. Boll Robb. Hüb.*)] ὡς ἀνεὶ **V**: ὁσανεὶ **Y**: ᾧ ἂν ᾖ **Σ β γ** 597 temptatiuorum et latrociniorum (πειρατικῶν καὶ κλοπῶν)] *inv. codd.* 598 configurationibus: συσχηματισμοῖς **β γ** (*Hüb.*): σχηματισμοῖς **α**] σχηματισμοὺς **V** (maleficos: κακοποιοὺς *cett., Hüb.*] -οῦντας **V**) (599 et ptisis cum tussi et resolutione] καὶ βηχικῶν καὶ ἀναφορικῶν καὶ φθίσεων *codd.*) 602 conuenientes: οἰκείας **V M S**] -ως **α D γ** (*Hüb.*)

---

577 -que] *om. Ox* 579 uentorum] -rem *Ox* 582 ueneritate *sic codd.* 585 mundas] et mendas *Ox* 584 coaptationem et possessionum] *it. Ox* 586 principum] -ium *Ve* luxurias *scr.*] -iarum (?, *e corr.*) *Ve*: -iosas *Ox* 588 uegetatiuissimorum *scr. cum Ve*] uegetissimorum *Ox* 592 profectum] perfectum *Ve* 596 intentum] incentium *Ox* 598 -que] *om. Ox* 599 efimerinarum *scr.*] -orum *codd.* significationum *scr.*] -torum *codd.*; *cf.* 400 ptisis] ytisis *Ox* Effectiuus est et] effectiuus est. Et (*sic*) *Ve* 600 sacerdotalem] + et *Ox* deorum religiones] *inv. Ox* 601 permutationis] -nes *Ox* legitimorum] *cum corr. Ox* 602 complexioni] -plectio- *Ve*

[18] Per comparationem autem ad continens magis siccus existens et facile mobilis propter propinquitatem ad solem | et propter uelocitatem reuolutionis, uentorum inordinatorum et acutorum et facile mutabilium maxime motiuus existit, tonitruorumque merito et incensionum et hyatuum et terremotuum et coruscationum effectiuus est. Et corruptionis que circa utilia animalium et plantarum euenit per hec aliquando factiuus; | aquarumque et fluminum in occasu quidem priuatiuus, in ortu autem adimpletiuus.

[19] Singulariter quidem igitur propriam naturam optinens unusquisque talia efficit. Commixtus autem alius alii penes configurationes et animalium diuersitates | et secundum habitudines ad solem, proportionaliterque in effectibus commixtionem accipiens, proprietatem uariam existentem circa effectum, mixtam ex communicatis naturis, operatur.

[20] Cum autem sit infinitum et impossibile secundum unamquamque commixtionem proprium effectum describere | et omnes simpliciter secundum quemcumque modum figurationes pertractare, tam multipliciter consideratas: merito utique delinquatur species talis sollicitudini et studio mathematici circa iudicia particularia.

[21] Obseruare autem oportet et qualiter se habent astra habentia dominium effectus in coaptatione ad ipsas regiones et ciuitates quibus simptoma futurum significatur. Benefica quidem enim existentia astra et coaptata dispositis et non superata ab hiis que sunt | contrarie [aut] partitionis, adhuc magis efficiunt commodum quod est secundum suam naturam; sicut non coaptata uel superata ab oppositis minus prosunt.

V (**α** =**ΥΣ** **β** =**DMS** **γ** =**ABC**) 604 reuolutionis: ἀνακυκλήσεως **β** (*Hüb.*), *cf.* **α**] ἀνακλήσεως **V** **γ** 609 ortu (*cett.*)] *om.* (*spat. relicto*) **V** 611 alii: ἄλλῳ *cett.*] ἄλῳ (*sic*) **V** 612 habitudines (*an* σχέσεις *legit vel coni.* **G** ?)] φάσεις *codd.* 616 describere: ὑπομνηματίζειν **α** **γ** (*Hüb.*)] -ζόμενον **V D S**, *cf.* **M** 617 consideratas: νοουμένους **D** **γ** (*Hüb.*), *cf.* **M S**] -μενος **V**, *cf.* **α** 618 sollicitudini (τῇ ... ἐπιβολῇ *coni. Boll*)] ἐπὶ τῇ ... ἐπιβολῇ *codd.* (*Hüb.*) 621 in coaptatione: συνοικειώσεως *cett.* (*Hüb.*)] οἰκειώσεως **V** (*Robb.*) quibus: αἷς **S** Heph. (*Hüb.*)] αἱ **V**: ἐν αἷς *cett.* (futurum] *om. codd.*)

---

603 autem] aut (*abbr.*) *Ox* 605 in-] etiam (*abbr.*) *Ox* 606 hyatuum] -tiuum *Ox* 607 effectiuus] -tius *Ox* 608 hec] hoc *Ox* aliquando] *ut* animalium *abbr. Ox* -que] *om. Ox* 609 in ortu] mortus *Ox* 610 unusquisque] unusque *Ox* talia] *it. Ox* 611 animalium] + .i. zodiaci *ss. Ox, an nota* **G**? 612 habitudines] consuetudines (*an alia lectio* **G** ?) *Ox* -que] que *Ox* 614 communicatis *scr. cum Ox*] coniugatis (*an alia lectio* **G**?) *Ve* 615 secundum] *om. Ox* 619 particularia. Obseruare] *a* particu- (*finis fol.* 12) *textus codicis Ve perditus est* (*verba* «particularia obseruare» *ut «Reklame» in fine folii addita sunt*)
620 *abhinc codex Ox solus adest* 622 Benefica] -cia *Ox* non] enim *Ox* 625 minus] unus *Ox*

[22] Nociue autem complexionis existentia illa que dominium effectus habent, si quidem ǀ fuerint coaptata dispositis uel superata fuerint ab hiis que sunt contrarie partitionis, minus ledunt; si autem neque gubernationem habeant regionum neque superentur ab hiis que se habent domestice ad istos, ualidiorem ex complexione corruptionem inferunt.

[23] Attamen, ut ad omne dicatur, illi hominum coincidunt uniuersalibus passionibus quicumque utique secundum proprias natiuitates principalissima loca, dico autem loca luminarium uel loca angulorum, habuerint eadem hiis que efficiunt causam uniuersalium simptomatum, ǀ hoc est locis eclipticis uel etiam diametraliter oppositis. Horum autem <in>certissime et ineuitabiles maxime si sunt per gradum ydemptitates aut diametrales oppositiones locorum eclipticorum ad alterum luminarium.

De coloribus eclipsium et cometarum et talium
Capitulum 10ᵐ

[1] Obseruandum autem ad uniuersales effectus et colores eclipsium, uel ipsorum luminarium uel consistentiarum que fiunt circa ipsa, puta uirgarum uel halo uel talium. Nigri quidem <enim uel> uirides apparentes ǀ fiunt significatiui eorum que dicta sunt de natura Saturni, albi autem eorum que in Ioue, rubei autem eorum que de Marte, flaui autem eorum que de Venere, uarii autem eorum que de Mercurio.

[2] Et si quidem in totis corporibus luminarium aut in totis locis circa ipsa ǀ proprietas coloris facta uideatur, circa plurimas partes regionum erit euentus; si autem ex parte quacumque, circa illam partem solam ad quam nuitio proprietatis facta fuerit.

[3] Obseruandum autem adhuc et apparitiones cometarum consistentes aut penes ǀ ecliptica tempora aut quandocumque ad uniuersales effectus, puta uo-

---

V (**α** =**YΣ** **β** =**DMS** **γ** =**ABC**) 634 uniuersalium: καθολικῶν **Σ β γ** (*Hüb.*)] κολικῶν **V Y** (diametraliter oppositis] τοῖς τούτων διαμέτροις *codd.*) (636 per gradum ydemptitates: αἱ μοιρικαὶ θίξεις *coni. Boll Hüb.*: αἱ μοιρικαὶ θήξεις **V β γ**] αἱ μοιρικαὶ καθέξεις **α**, *Robb.*) (640 eclipsium] περὶ τὰς ἐκλείψεις **V β γ**, *cf.* **α**) 642 halo: ἁλώνων **β γ**: ἅλων **Σ**[pc] (*Hüb.*)] ἄλλων **V Σ**[ac], *cf.* **Y** 644 flaui — Venere (*cett.*)] *om.* **V** 648 nuitio: πρόσνευσις *cett.*] προσνεύσεις **V** 650 aut (*cett.*)] *om.* **V**

---

627 fuerint[1]] fiunt *Ox* superata] separata *Ox* 628 minus] unus *Ox* 630 corruptionem] *an* corruptiuum *scrib.* ? 634 uniuersalium] naturalium *Ox* 635 si sunt] *an* existunt *scrib.*? 636 ydemptitates] -tis (*abbr.*) *Ox* 637 alterum] ultimum *ante corr. Ox* 639 10ᵐ] 20 *Ox* 643 que dicta] *it. Ox* 644 flaui] flauii *Ox* 648 nuitio] nuntio *Ox* 651 uocatarum] -orum *Ox*

catarum trabium uel tubarum uel pithiarum et talium, tamquam effectiuas quidem per naturas proprietatum Martis et Saturni, et bellorum et estuum aut commotiuarum | seditionum et hiis accidentium. Significantes autem per partes quidem zodiaci, in quibus consistentie ipsorum apparuerint, et per adnuitiones secundum figuras come loca quibus inferunt simptomata.

[4] Per ipsas autem quasi | formas consistentie speciem effectus et genus circa quod passio eueniet, per tempus autem mansionis durationem simptomatum; per habitudinem autem ad solem inchoationem, quoniam quidem matutine quidem ad multum apparentes celerius insinuant, uespertine autem | tardant<ia>.

De innouatione anni
Capitulum 11^{m}

[1] Ostensa autem uia circa uniuersalia condicionum regionum et ciuitatum, reliquum utique erit et de particularioribus pertractare; dico autem | de hiis que annuatim circa tempora anni efficiuntur, ad quam considerationem et de uocata neomenia anni congruum utique erit preaccipere. Quod quidem igitur conuenit esse principium hoc restitutionis solis circa unamquamque girationem, palam est ex se, et a uirtute et a | nominatione.

[2] Quod autem principium ponet utique aliquis, in circulo quidem ipso solo simpliciter neque utique intelligetur, in zodiaco autem sola utique merito principia accipiet puncta que ab equinoctiali et tropicis determinantur, hoc est duo equinoctialia et duo tropica. | Hic tamen aliquis iam dubitabit quo de quatuor principali utetur; secundum simplicem quidem igitur et circularem naturam nichil ipsorum est tamquam in uno principio principale.

[3] Vsi sunt autem qui de hiis scripserunt unum aliquod supponentes, differenter unoquoque de quatuor tamquam principio, | inducti per aliquas congruas

---

V (**α** =**YΣ** **β** =**DMS** **γ** =**ABC**) 652 pithiarum*: πιθείων **V** (*cf.* Procl. 131,14: πιθείας): πιθίων **Y**] πίθων **Σ β γ** Heph. (*Hüb.*) 653 (Martis] ἐπὶ *praem. codd., fort. non redd.* **G**) Saturni (*fort.* Κρόνου *legit* **G**)] Ἑρμοῦ *codd.* (estuum: καυσώνων **Σ**] καυσωδῶν **V β γ**, *Hüb.*: *om.* **Y**) 654 seditionum (*an* στάσεων *legit vel coni.* **G** ?)] καταστημάτων *codd.* (655 apparuerint: ἂν ... φαίνωνται **Σ**, *Hüb.*] ἂν ... -νονται **V Y D γ**; *cf. cett.*) 657 per ipsas ... formas: διὰ ... τῶν αὐτῶν ... μορφώσεων ?] διὰ ... τῶν αὐτῆς ... μορφώσεων *codd.* quasi: ὥσπερ **α**: ὡσπερεὶ Heph. (*cod.* A) (*Hüb.*)] ὡς περὶ **V** Heph. (*cod.* B): περὶ **β γ** (660 tardant<ia>*] βράδιον [*sc.* συμπτώματα] *codd.*) 663 condicionum: περιστάσεων *cett.*] παραστάσεων **V** (669 ipso solo] αὐτὸ μόνον *codd.*) 672 iam: ἤδη **Σ γ** (*Hüb.*)] εἴδη **V Y**, *cf. cett.* 673 principali (προηγουμένῳ)] ὡς προηγουμένῳ **α** (*Hüb.*): προηγουμένων **V D**: -μενον **M S**: -μένως **γ** (Procl.) utetur (**α**)] + ἂν *cett.* (*Hüb., sed «**α** fort. recte»*) 676 principio: ἀρχῇ (*Boll Hüb.*)] ἀρχὴν *codd.*

---

652 pithiarum*] pithicorum *Ox* 653 quidem] -dam *Ox* 658 durationem] -nes *Ox* 660 tardant<ia>*] tardant *bis scr. Ox* 665 de uocata] denotata *Ox* 667 hoc] h[i] *Ox* 669 aliquis] al'iis *Ox* 670 utique[1]] itaque *Ox* 672 Hic] hoc *Ox* 673 secundum] in *Ox*

rationes <et> naturales compassiones. Et enim habet singulare aliquid partium
harum unaqueque, a quo utique principium et nouus annus merito iudicetur.
Vernale quidem equinoctium propterea quod primo tunc incipit fieri dies maior
nocte et quia ' humidi temporis est, hanc autem naturam, tamquam etiam priorem 895
dicimus, incipientibus generationibus plurimam inesse.

[4] Tropicum autem estiualem propterea quod penes ipsum maxima dies ef-
ficitur, apud Egiptios et Nili inundationem et Canis stelle ortum ' deprehendi. Au- 900
tumpnale autem equinoctium quia facta est iam fructuum omnium collectio, et
tunc ab alio principio sementem futurorum seminant. Yemalem autem tropicum
propterea quod tunc primo incipit magnitudo diei incrementum ' accipere post 905
diminutionem.

[5] Conuenientius autem mihi uidetur et magis naturale ad annuales considera-
tiones quatuor principiis uti, obseruando propinquissime sibi prius factas solis et
lune coniugationes coniunctionales aut pleniluniares, et harum maxime rursum
eclipticas, ut a ' principio quidem quod est penes Arietem uer, quale sit, consi- 910
deremus, ab eo autem quod est penes Cancrum estatem, ab eo uero quod est penes
Chelas autumpnum, ab eo autem quod est penes Capricornum yemem.

[6] Vniuersales quidem enim temporum qualitates et consistentias ' sol facit, 915
secundum quas et penitus inexperti mathematicorum habent pronosticationem
futuri. Adhuc autem et signorum proprias ad hoc significationes uentorum et to-
taliorum naturarum assumendum.

[7] Alterationes autem in magis aut minus secundum tempora uniuer-
saliter ' quidem rursum coniugationes que fiunt circa predicta puncta et pla- 920
netarum ad ipsas configurationes ostendunt, particulariter quidem et que per
singula XII signa coniunctiones et plenilunia et astrorum concursus, quam utique
mensualem speculationem aliquis appellabit.

[12.1] Cum autem ad hoc debeant preexponi particulares secundum animalia 927
ad anni consistentias naturales proprietates et adhuc secundum unumquodque
astrum, coaptationem quidem ' planetarum et similis complexionis fixarum ad 930

V (**α** =**YΣ** **β** =**DMS** **γ** =**ABC**) 678 utique: ἂν *cett.*] ἐν **V**: *om.* **Σ** (679, 686 primo: πρῶτον **V Y**] πρώτως *cett.*, *Hüb.*) 680-1 priorem (προτέραν ?) dicimus] πρότερον ἔφαμεν *codd.* 691 a principio quidem quod est penes Arietem: ἀπὸ μὲν τῆς περὶ Κριὸν ἀρχῆς **γ**] ἀπὸ μὲν τῆς ἐν τῇ (τῷ **β**) περὶ Κριὸν ἀρχῆς **V β** (*Robb.*): *cf.* **α** *et coni. Hüb.* 696 significationes uentorum: παρασημασίας ἀνέμων **V α** (*Robb.*)] περὶ σημασίας ἀ. **D γ**: περὶ σημασίας περὶ ἀ. **M S**: παρασημασίας περὶ ἀ. (*Hüb.*)

---

678 unaqueque a quo] unaquaque aqua *Ox* 680 hanc autem] *inv.* *Ox* 683 inundationem] in nudationem *Ox* 684 equinoctium] eque- *Ox* 692-3 ab eo uero — Chelas (Clelas ?)] *mg.* *Ox* 694 temporum] teperorum (*abbr.*) *Ox* 695 inexperti] -te *Ox* 695 mathematicorum] *an* mathematorum *scrib.* ? 701 plenilunia] plenissima *Ox* 702 mensualem] -salem *Ox* 703 animalia] aliam (*abbr.*) *Ox*

aerem et uentos et adhuc conuenientiam totorum duodecim signorum ad uentos et tempora singula declarauimus in prioribus. Reliquum autem utique erit et de particulari animalium | natura dicere.

De particulari animalium natura ad consistentias Capitulum 12^m

[2] A r i e t i s quidem igitur signum uniuersaliter quidem est propter equinoctialem significationem tonitruosum et grandinosum. Particulariter autem, in magis aut minus <a> proprietate stellarum fixarum que sunt penes ipsum, priora quidem | ipsius pluuiosa et uentosa, media autem bene temperata, sequentia autem estuosa et pestilentialia. Borealia autem estuosa et corruptiua, australia autem cristallosa et frigida.

[3] T a u r i autem signum uniuersaliter quidem est significatiuum ambarum complexionum et subcalidum. | Particulariter autem precedentia quidem ipsius, et maxime que penes Pliades, commotionalia et uentosa et caliginosa, media autem humectatiua et frigida, sequentia autem et que penes Yadem ignea et tonitruosa et coruscationis factiua. Borealia autem bene temperata, australia autem motiua et | inordinata.

[4] G e m i n o r u m autem signum uniuersaliter quidem est temperantie factiuum. Particulariter autem precedentia quidem ipsius humectantia et corruptiua, media autem temperata, sequentia autem mixta et inordinata. Borealia autem spirituosa et terremotus factiua, australia autem sicca et | estuosa.

Quod autem C a n c r i est signum uniuersaliter quidem est serenum et calidum. Particulariter autem precedentia quidem ipsius et que penes Presepe suffocatiua, media autem temperata, sequentia autem spirituosa. Borealia autem et australia ignea et | estuosa.

V (**α** =**YΣ** **β** =**DMS** **γ** =**ABC**, *excerpta* **L**) 709 *Tit. hic praebet* **G** ut **V β** (*cf.* **S**) **γ** Procl.] *post* προσαγορεύοι (appellabit 702/924°) **α**, *Hüb.*: ***abhinc usque ad* II. 808 *excerpta codicis* L *denuo adsunt*** 717 significatiuum: ἐπισημαντικὸν *cett.*] -κὰ **V** 719 Pliades: τὰς πλειάδας (πλοιά- **Y**) **α**] τὴν πλειάδα **V L β γ** Heph. (*Hüb.*) 720 tonitruosa: κεραυνώδη **V L D γ**] κεραυνῶν **α M S** Heph. (*Hüb.*) 728 suffocatiua (**V L** Procl.)] + καὶ σεισμοποιὰ καὶ ἀχλυώδη **α β γ** (*Hüb.*)

---

708 natura] -am *Ox* 709 particulari animalium] particularium *Ox* consistentias] -iam *Ox* 714 uentosa] -tuosa *Ox* 715 et pestilentialia — estuosa] *mg. Ox* 719 Pliades] peliades *Ox* 720 humectatiua] -ectiua *Ox* 721 coruscationis] *an* -num *scrib.* ? 723 signum] sigiram (*abbr.*) *Ox* 726 spirituosa] speciosa *Ox* sicca] *mg. Ox* 727 serenum] -nosum *Ox* 728 quidem] quid *Ox*

[5] Quod autem Leonis uniuersaliter quidem <est> spirituosum et suffocatiuum. Particulariter autem precedentia quidem ipsius suffocatoria et pestilentia, media autem temperata, sequentia autem uaporatiua et corruptiua. Borealia autem motiua et ignea, australia autem humectatiua.

Quod autem | Virginis uniuersaliter quidem <est> humectatiuum et tonitruosum. Secundum partem autem precedentia quidem ipsius calidiora et corruptiua, media autem temperata, sequentia autem aquosa. Borealia autem spirituosa, australia autem temperata.

[6] Quod autem Chelarum uniuersaliter quidem | <est> uersatiuum et permutatiuum. Secundum partem autem precedentia quidem ipsius et media sunt temperata, sequentia autem aquosa. Borealia autem spirituosa, australia autem uaporosa et pestilentialia.

Quod autem Scorpionis uniuersaliter quidem est tonitruosum et igneificum. Secundum partem autem | precedentia quidem ipsius nebulosa, media autem temperata, sequentia autem terremotus factiua. Borealia uero <estuosa, australia autem uaporosa>.

[7] Quod uero Sagittarii uniuersaliter quidem est spirituosum. Secundum partem autem precedentia quidem ipsius humectatiua, media autem temperata, sequentia autem | igneifica. Borealia autem spirituosa, australia humectatiua et permutabilia.

Quod autem Capricorni uniuersaliter quidem est humectatiuum. Secundum partem autem precedentia quidem ipsius estuosa et corruptiua, media autem temperata, sequentia autem inundationum motiua. Borealia uero | et australia humectatiua et corruptiua.

[8] Quod autem Aquarii uniuersaliter quidem est frigidum et aquosum; secundum partem autem precedentia quidem ipsius humectatiua, media autem temperata, sequentia autem spirituosa. Borealia autem estuosa, australia autem nebulosa.

Quod autem est | Piscium uniuersaliter quidem est frigidum et spirituosum. Secundum partem autem precedentia quidem ipsius temperata, media autem humectatiua, sequentia autem estuosa. Borealia autem spirituosa, australia autem aquosa.

V (**α** =**YΣ** **β** =**DMS** **γ** =**ABC**, *excerpta* **L**) 731 spirituosum* (πνευματῶδες ?)] καυματῶδες *codd.* 739 uersatiuum: τρεπτικὸν **α β γ** Heph. (*Hüb.*)] τροπικὸν **V L**

---

731 spirituosum* (*sic*)] *an perperam pro* estuosum ? 732 pestilentia] *an* -ialia *scrib.*? 739 uersatiuum] uersinum *Ox* 740 precedentia] -tem *Ox* 743 igneificum (*vide infra* 749)] igneiacum (fi—> a) *Ox* 753 inundationum] in nudatione (-atione *abbr.*) *Ox*

De particulari | consideratione consistentiarum Cap. 13^{m}

[1] Hiis autem ita expositis qui secundum partem accessus significationum continent modum hunc. V n u s quidem enim est qui magis in totalitate consideratur ad quartas, secundum quem obseruare, ut diximus, oportebit eas que fiunt proxime | ante tropica <et> equinoctialia puncta coniunctiones aut plenilunia et secundum gradum uel coniunctionalem uel pleniluniarem in unoquoque inquisitorum climatum angulos ut in natiuitate disponere.

[2] Deinde dominatores accipere coniunctionalis aut pleniluniaris | loci et anguli sequentis ipsum, secundum modum ostensum in prioribus de eclipsibus, et ita uniuersale quidem considerare ex proprietate quartarum, quod autem magis aut minus intentionum et remissionum ex | natura dominantium pertractantes, qualisque qualitatis sint et qualium consistentiarum motiui.

[3] S e c u n d u s autem est supergressus qui mensis, secundum quem oportebit <- - -> illud | solum obseruantes quod, coniunctione quidem eueniente propinquissime adiacenti tropico aut equinoctiali puncto, utemur etiam coniunctionibus usque ad sequentem quartam, plenilunio autem pleniluniis.

[4] Considerare <autem> similiter angulos et dominatores amborum locorum et maxime | propinquissimas apparitiones, contactus et defluctus errantium astrorum, proprietatesque ipsorum et locorum, et qualium uentorum sint motiua ipsa et partes signorum in quibus utique existunt, adhuc autem et cui uento [et] latitudo lune adnuit secundum obliquationem | a zodiaco: quatinus ex omnibus hiis secundum dominationem consistentias ut in omnibus mensibus et uentos precognoscamus.

[5] T e r t i u m autem est adhuc magis particulares significationes remissionum et intentionum obseruare. Consideratur autem et | hoc per particulares solis et lune configurationes, non solum per coniunctiones coniunctionales aut pleniluniares, sed etiam per eas que secundum dichotomias, inchoante ut ad omne di-

---

V (**α** =**YΣ** **β** =**DMS** **γ** =**ABC**, *excerpta* **L**) 769 gradum: μοῖραν **L y β** Procl. (*Hüb.*)] *om.* **V Γ α**, *cf.* **γ** (772 ostensum] + ἡμῖν *codd.*) 774 aut: ἤ **L A B**] καὶ **V α β C** (*Hüb.*) dominantium (**V L D γ**)] + ἀστέρων **α M S** (*Hüb.*) pertractantes: διαλαμβάνοντας **L α** (*Hüb.*): -τες **β γ**] διαβάνοντας **V** 776 qui mensis: ἡ μηνιαία *cett.*] ἡμῖν ἰδία **V** 777 <- - ->] τὰς καθ' ἕκαστον — ἐπισκοπεῖν *codd.* (1013°-4°); *an in traditione lat. excid.* ? 787 remissionum: ἀνέσεων *cett.*] ἀνανέσεων **V** (789 coniunctiones coniunctionales] συνοδικῶν *codd.*, *cf. supra* 690)

---

765 partem accessus] proaccessus *Ox* 766 continent] *ex* -nentia *Ox* 771 coniunctionalis] -les *Ox* pleniluniaris] -lunares *Ox* 775 qualisque] qualis que *Ox* 777 <- - ->] *an verba fere haec* <secundum unumquodque signum prius factas coniunctiones aut plenilunias secundum eundem modum considerare> *exciderunt* ? quod (... utemur)] *an* ut (ἵνα) *scrib.* ? 781 contactus] con-t^{a}ct^{us} (contractus ?) *Ox* 783 ipsa] -am *Ox* 784 lune] *it. Ox* 785 dominationem] diuinas *Ox* mensibus] *an* mensium *scrib.* ? 787 significationes] + significationum *Ox* 789 configurationes] -nis *Ox* coniunctiones (-iuct-)] *an* coniugationes *scrib.* (*cf.* 690)?

catur effectus alteratione ante tres dies, aliquando autem et post tres | occursus ad solem equilibrati; et per id quod penes singulas talium stationum aut etiam aliarum, puta trinarum aut sextilium, ad planetas configurationum.

[6] Conformiter enim harum nature et alt<er>ationis proprietas deprehenditur, consone naturalibus coaptationibus inspectorum | astrorum et signorum ad continens et ad uentos.

[7] Ipsarum autem harum particularium qualitatum cotidiane intentiones efficiuntur, maxime quidem quando fixarum fulgentiores et efficaciores apparitiones, matutinas | aut uespertinas, orientales aut occidentales, faciunt ad solem. Vertunt enim ut in pluribus particulares consistentias ad suas naturas, nichil autem minus et quando alicui angulorum luminaria superferuntur.

[8] Ad tales enim ipsorum habitudines que per horam fiunt | remissiones et intentiones consistentiarum permutantur, sicut ad eas que lune effusiones maris et refluxus. Et uentorum uersiones maxime circa tales luminarium angulationes efficiuntur, ad quoscumque uentos latitudo lune | deprehensa fuerit faciens adnuitiones.

[9] Vbique tamen oportet attendere quomodo principali quidem uniuersaliter et prima supposita causa, sequente autem ea que particularium euentuum; et quod confirmatur maxime et fortificatur efficacia, quando | stelle que dominantur uniuersalibus naturis particularibus fuerint configurate.

De ea que ex metheoris significatione
Capitulum 14^m^

[1] Oportune autem erunt utique ad pronosticationes particularium significationum et obseruationes signorum que fiunt circa | solem et lunam et astra. Solem quidem igitur obseruandum ad consistentias quidem diei orientem, ad nocturnas autem occumbentem, ad durantes autem secundum <configurationes> ad lunam, tamquam unaquaque configuratione presignificante consistentiam usque ad sequentem | configurationem ut ad omne dicatur.

V (**α** =**ΥΣ** **β** =**DMS** **γ** =**ABC**, *excerpta* **L**) (791 effectus] κατὰ τὴν ἐπισημασίαν *codd.*) 792-3 per id quod ... configurationum: διὰ τὸ ... συσχηματισμῶν] διὰ τοῦ ... συσχηματισμῶν (συσχημῶν **V**) **V L**: διὰ τοῦ ... -μοῦ **α β γ** (*Hüb.*) (799 orientales: ἀνατολικὰς **α M S**, *Hüb.*] ἀνατολὰς **V L D γ**) 800 ut in: ὡς ἐπὶ *cett.*] κατὰ **V** 801 alicui: τινὶ **α M S** (*Hüb.*)] τι **V L D γ** 803 effusiones: ἀμπώτεις **β A B** (*Hüb.*)] ἀμπώτις (*sic*) **V**: -της **L**, *cf. cett.* (maris] *om. codd.*) 805 uentos] + ἐπὶ τὰ αὐτὰ **V L α** (*Hüb.*): + ἐν αὐταῖς **β γ** 808 causa: αἰτίας *cett.*] αἰτίως **V**: ***abhinc excerpta codicis L denuo desunt*** 815 igitur (*cett.*)] *om.* **V** quidem²: μὲν **α M S** Heph. (*Hüb.*)] *om.* **V D γ**

---

792 equilibrati] que librati *Ox* 793 aliarum] -orum *Ox* 794 harum] horum *Ox* 797 harum] et horarum *Ox* particularium] -rum *Ox* 803 sicut] *ex* ut *corr. Ox* 805 faciens] *ss. Ox* 808 euentuum] euentiunt *Ox*

[2] Mundus quidem enim et non adumbratus et bene status et sine nube oriens aut occumbens, serene consistentie est significatiuus. Varium autem circulum habens aut rubeum uel radios rubeos emittens uel ad exteriora uel ad se ipsum in profundum, uel | eas que dicuntur nubes parelie habens ex una parte, uel figuras nubium ruphas et quasi longos radios protendens, uentorum ualidorum est significatiuus et talium ad quos angulos utique predicta signa fuerint.

[3] Niger aut subuiridis oriens aut occumbens, cum nubositate uel halo | habens circa se unam uel etiam duas, aut secundum unam partem aut etiam ex ambabus partibus nubes parelias <et radios> subuirides aut nigros, tempestatum et pluuiarum est significatiuus.

[4] Lunam etiam obseruandum <in transitu> ante tres uel post tres dies a coniunctionibus et pleniluniis et | dichotomiis. Subtilis quidem enim et munda apparens et nichil habens circa se, serene consistentie <est> significatiua. [5] Subtilis autem uel rubicunda et totum circuitum eius quod non illuminatur habens patentem et submotum, uentorum est significatiua ad quamcumque partem maxime fecerit adnuitionem. | Nigra autem uel uiridis uel rubicunda et spissa considerata, tempestatum et imbrium est significatiua.

[6] Obseruandum autem et areas que fiunt circa ipsam. Si quidem enim fuerit una et ipsa munda et quiete marcescens, serenam consistentiam significat; si autem due uel etiam | tres fuerint, tempestates significant; rubee quidem existentes et quasi disrupte, eas que per ualidos uentos; caliginose autem et grosse, eas que per nubositates; uirides autem uel nigre et rupte, eas que per ambo; et quanto utique plures fuerint, tanto maiores.

V (**α** =**YΣ** **β** =**DMS** **γ** =**ABC**) 821 uel[3]] + ὡς *codd.* 822 profundum: κοιλούμενον **V β**, *cf.* **γ**] κυκλουμένας **y** (*Robb. Hüb.*); *diversimode cett.* que dicuntur: λεγόμενα **α** Heph.] καλούμενα **V β γ** Procl. (*Hüb.*) 823 et: καὶ **α M S** (*Hüb.*)] *om.* **V D γ** 825 halo: ἅλως **Σ D S y** (*Hüb.*), *cf.* **Y**] ἄλλως **V M** Heph. 826 unam uel etiam duas aut (*an glossa graeca* ?)] *om. codd.* 833 submotum (**α** Heph., «*fort. recte*» *Hüb.*)] ὅλον *praem. cett.* (*Hüb.*) 834 uel uiridis uel rubicunda: ἢ χλωρὰ ἢ ὠχρὰ **S**] ἢ ὠχρὰ ἢ χλωρὰ **D M**: ἢ ὠχρὰ **α γ** Heph.: ἢ χλωρὰ (χλόρα **V**) *Boll Robb. Hüb.* 839 disrupte: ἐκρηγνύμεναι (Heph., *Hüb.*): περιρρηγνύμεναι **β**] ῥυγνύμ- **V**: ῥηγνύμ- **γ** (*Boll*), *cf.* **Y Σ** 840 uirides: ὑπόχλωροι **α** Procl.] ὕπωχροι **V D S**, *cf.* **M** (*Hüb.*)

---

822 habens] *it.* *Ox* 825 Niger] nigri *Ox* subuiridis] -di *Ox* uel] + sub (*expunxit*) *Ox* 827 parelias] paralias *vel* peralias *Ox* 828 significatiuus] -catim (*abbr.*) *Ox* 830 dichotomiis] dicotoniis *Ox* 832 uel] *an* et *scrib.*? rubicunda] rubei- *Ox* 833 quamcumque] quac- *Ox* 834 adnuitionem] admistionem *Ox* rubicunda] -do *Ox* 837 munda] monda *Ox* 838 tres] + significant *Ox* 839 eas que[1,2]] easque *Ox* caliginose] -gniose *Ox* 840 nubositates (*cf.* 543, 825)] multositates *Ox*

[7] Et aree autem consistentes circa stellas erraticas | et circa fulgentes fixarum 1100
insinuant conuenientia coloribus suis et appellationibus eorum que intus deprehenduntur.

[8] Et fixarum autem que ad aliquam multitudinem uicine sunt, obseruandum
colores et magnitudines. Fulgentiores | enim et maiores uise ultra consuetas fan- 1105
tasias ad quamcumque partem fuerint, uentos a proprio loco significant.

[9] Quin immo etiam nebulosarum conuolutionum singulariter, puta Presepis
et talium, si serenitate existente consistentie obscure | et quasi non apparentes uel 1110
ingrossate uideantur, copie aquarum sunt significatiue; pure autem et quasi saltantes continue, ualidorum spirituum sunt significatiue.

[10] Et eorum autem que fiunt per tempora in excelsis, cometarum quidem 1116
conuolutiones semper siccitates et uentes significant, et tanto maiores, quanto
utique ex pluribus partibus et ad multum consistentia fuerit. | Discursus autem et 1120
iaculationes astrorum, si quidem ab uno fiant angulo, uentum ab illo significant; si autem a contrariis, instabilitatem uentorum; si autem a quatuor, omnimodas tempestates usque ad tonitrua et coruscationes et talia.

[11] Eodem autem modo et nubes uelleribus | lanarum existentes similes, fiunt 1125
aliquando presignificatiue hyemalis temporis. Yrides autem per tempora consistentes, yemem quidem ex serenitate, serenitatem autem ex yeme presignificant. Vt ad omne dicatur colligendo, uniuersaliter fantasie aeris que fiunt proprii coloris,
similia significant hiis que efficiuntur | a propriis simptomatibus <secundum> 1130
preostensa per priora.

[12] Speculatio quidem igitur uniuersalium, et eorum que magis in totalitate considerantur et eorum que in particulari, usque ad tanta nobis capitulariter
insinuata sit. Eius autem que est secundum | geneatici sermonis speciem prono- 1135
sticationis tractatus in consequentibus secundum conuenientem consequentiam aggrediemur.

Explicit liber secundus

V (**α** =**YΣ** **β** =**DMS** **γ** =**ABC**) 842 aree: ἅλως **Σ β γ** Heph. (*cod.* A) (*Hüb.*)] ἄλλως **V Y** Heph. (*cod.* L P) 843 appellationibus (φάσεσιν ?)] φύσεσιν *codd.* 845 aliquam: τι **α M S** Heph. (*Hüb.*)] *om.* **V D B C**, *cf.* **A** 850 copie: φορᾶς (*cett.*)] *lac.* **V** 850-1 pure — spirituum (*cett.*)] *lac.* **V** (851 sunt significatiue] *om. codd.*) 852 Et (*cett.*)] *om.* **V** 853 semper: πάντοτε **α M S** Heph.] ὡς ἐπίπαν **V D γ** (*Hüb.*) 856 instabilitatem: ἀκαταστασίαν *cett.*] εὐκατα- **V** omnimodas: παντοίους *cett.*] παντοίως **V** 860 serenitatem autem: εὐδίαν δὲ **β γ** Heph. (*Hüb.*)] εὐδίας δὲ **V**: καὶ εὐδίαν **α** 861 uniuersaliter fantasie (*cett.*)] *lac.* **V** 862 a propriis simptomatibus (*cett.*)] *lac.* **V** 866 geneatici sermonis: γενεθλιαλογικὸν **Y γ**] γενεθλιακὸν **V Σ β** Heph. 868 Explicit liber secundus: Τέλος τοῦ β′ βιβλίου **D γ**] Τέλος τοῦ β′ βιβλίου Πτολεμαίου **V**: *om. cett.*

---

845 que] que quo *Ox* uicine] uicime *vel* intime *Ox* 846 uise] in se *Ox* 848 Presepis] presepē *Ox* 859 presignificatiue] presignatōē *Ox* Yrides] Hyrydes *Ox* 860 presignificant] p'signāt *Ox* 865 considerantur] -atur *Ox* 866 sit] sunt *Ox* geneatici] generatio *Ox*

<LIBER III>

Hec sunt capitula libri tertii

V (**α** =**ΥΣ** **β** =**DMS** **γ** =**ABC**) 1 *tit. om.* (**V S**)] ΒΙΒΛΙΟΝ Γ, *Hüb., cf. cett.* 2 Hec sunt capitula libri tertii: Τάδε ἔνεστιν ἐν τῷ γ´ βιβλίῳ **V** (*Hüb.*), *cf.* **D γ**] τὰ κεφάλαια τοῦ τρίτου βιβλίου **M**: *om. cett.* 3 Prohemium ... ca. 1^{m}: α´ προοίμιον **V** (*Hüb.*)] *om. cett.* de ordine — precedentem] *om. codd.* 4 ca. 2^{m}: β´ **V**] α´ **D γ**: *om. cett.* 5 ca. 3^{m}: γ´ **V**] β´ **Υ D γ**: *om. cett.* 6 Quomodo oportet accipere horam ca. 4^{m}: *fort.* πῶς δεῖ λαμβάνειν τὴν ὥραν *legit* **G**: *cf.* ἐν ᾧ (*sc. cap.* 3) καὶ περὶ τοῦ «πῶς δεῖ λαβεῖν τὴν ὥραν» *post tit. cap.* 3 **γ**, *cf.* **M**; *cf. etiam infra* 109 *post* eleuationem «πῶς δεῖ λαμβάνειν ὥραν» *ut titulum inseruerunt* **D M**, *necnon Co* (90,30)] *om.* **V** *et cett.* 7 ca. 5^{m}: ε´ *legit* **G**] δ' **V** (*Hüb.*): γ´ **Υ D γ**: *om. cett.*; *et sic deinceps pro numeris capitulorum* 10 femininis (**Υ M** «*fort. recte*»)] + γενέσεων *cett.*

---

18 ca. 16^{m}] ca. 18. *Ox*

Prohemium
Capitulum primum

[1] Pertractata nobis in precedentibus theoria que est circa uniuersalia ˈ simptomata, tamquam principaliore et que in multis preualere potest super euentus qui contingunt circa unumquemque hominum secundum proprietatem nature, quorum pronosticam partem uocamus geneaticum sermonem, potentiam quidem unam et eandem ambarum specierum existimari oportet ˈ et circa theoricum. Quoniam quidem uniuersalium et eorum que penes unumquodque simptomatum causa quidem errantium stellarum et solis et lune motus, pronosticatiua autem scientionalis obseruatio uersionis suppositarum ipsis naturarum penes eos qui similis figure ˈ celestium transitus per continens; uerumtamen in quantum uniuersalis quidem effectus maior et perfectior, particularis autem non similiter.

[2] Principia non iam eadem ambarum esse putandum, a quibus celestium dispositionem supponentes attemptamus pronosticari ea que per ˈ figurationes tunc significantur. Sed uniuersalium quidem multa, eo quod non habemus unum totius, et hec non a suppositis semper accepta, sed etiam a continentibus et causas ferentibus; fere enim omnia a ˈ perfectioribus eclipsibus et annotabiliter concurrentibus stellis consideramus. Eorum que sunt circa unum quemlibet hominum et unum et multa; [et] unum quidem principium seminum commixtionis (et hoc enim habemus), multa autem, scilicet ˈ superuenientes ad primum principium significationes secundum consequentiam continentium; hic tamen merito precedit unum, quoniam hoc et alia efficit.

[3] Hiis autem sic habentibus, a primo quidem principio considerantur uniuersales proprietates concretionis, per alia autem ea que eueniunt per tempora ad magis et minus ˈ secundum dictas deinceps annorum diuisiones.

V (**α**=**YΣ** **β** =**DMS** **γ** =**ABC**) 19-20 Prohemium capitulum primum: α΄ προοίμιον **V** (*Hüb.*)] προοίμιον **D** **γ**: *om. cett.* 23 qui contingunt circa: τῶν περὶ **Σ M S γ** (*Hüb.*)] περὶ τῶν περὶ **V**: *aliter cett.* 24 geneaticum sermonem: γενεθλιαλογίαν **V** (*Hüb.*)] γενεθλιαλογικὸν **α β γ** («fort. recte») 25 et circa theoricum] καὶ περὶ τὸ ποιητικὸν *praem.* **V α β** (*Hüb.*), *add.* **γ** 28 ipsis naturarum: αὐτοῖς φύσεων **Ξ** (= Oxon. CCC gr. 100) (*Boer*)] αὐτῶν φύσεων **S γ**: αὐτῆς φύσεως **V**: αὐτῆς φύσεων **Y** (*Hüb.*): *diversimode cett.* 31 esse putandum: εἶναι νομιστέον **α**] *inv.* **V D A C** (*Hüb.*); *cf. cett.* (32 attemptamus: πειρώμεθα **α β B C**] πυρώμεθα **V**, *cf.* **A**) (37 seminum commixtionis] αὐτοῦ (αὐτὴν) τοῦ συγκρίματος *codd.*) 38 scilicet superuenientes ...: τὰς ... συμβαινούσας **Σ γ** (*Hüb.*)] τὰς ... συμβαίνειν **V Y β** 40 efficit: ἀποτελεῖ *cett.*] -εῖς **V**

---

21 Pertractata] *disiuncte Ox* uniuersalia] *ss. Ox, exp.* naturalia 22 principaliore] -ra *Ox* 23 qui] que (*abbr.*) *Ox* proprietatem] -te *Ox* 24 potentiam] -ia *Ox* 25 circa] contra *Ox, et sic deinceps, fere semper* 26 Quoniam] *ss. Ox, exp.* contra 30 quidem] quod (*abbr.*) *Ox* particularis] -iter *Ox* 33 significantur] -atur *Ox*

De seminatione et partu
Capitulum 2[m]

[1] Temporali autem principio generationum hominum existente per naturam quidem eo quod secundum seminationem, uirtute autem [et significationes] et | secundum accidens partu ex utero, in hiis quidem quorum scitur tempus seminationis uel casualiter uel per obseruationem, illi magis ad proprietates corporis et anime conuenit attendere, considerantes operationem | configurationis stellarum secundum tempus illud.

[2] Semel enim in principio sperma quale factum aliqualiter ex influentia continentis, quamuis differens fiat hoc secundum consequentia corporis formatiua tempora, ipsum conuenientem solam materiam naturaliter congregans ad se secundum incrementum, | adhuc magis assimilatur proprietati prime qualitatis.

[3] In non cognoscentibus autem, quod quidem quasi in omnibus accidit, principio quod est secundum partum attendere necessarium, tamquam maximo et ipso, solo hoc <a> primo | deficiente eo quod per illud et ea que ante partum possint precognosci. Et enim si hoc quidem dixerit utique aliquis principium, hoc autem uelut katarchen (id est optentionem principii), magnitudo ipsius tempore quidem fit posterior, equalis autem et perfectior magis uirtute; fere autem iuste illud quidem utique nominabitur | spermatis humani generatio, hoc autem hominis. Plurima enim tunc assumit infans, que non prius aderant ipsi quando in utero erat, et ipsa propria solius humane nature, corporalisque formatio.

[4] Quamuis nichil | ad hoc quod taliter sit uideatur ipsi conferre continens secundum partum, tamen ad ipsum prodeundum in lucem secundum conuenientem continentis figurationem confert, natura post perfectionem impetum exitus faciente ad consistentiam similis modi ei qui particulariter a principio formauit. Quare rationabile et | talium existimari significatiuam esse figurationem astrorum que est penes partum, non quasi factiuam tamen omnino, sed quasi ex necessitate et secundum naturam uirtute simillimam factiuo.

V (**α** =**YΣ** **β** =**DMS** **γ** =**ABC**) 45 Capitulum 2[m]: β´ **V** (*Hüb.*)] α´ **D M γ**: *om.* **α S** 46 generationum: γενέσεων **α M S** Procl. (151,6)] τέξεων **V D** (*Hüb.*), *cf.* **γ** (hominum] ἀνθρωπίνων *vel* ἀνθρωπείων *codd.*) 49 illi: ἐκείνῳ **β γ** Procl. (*Hüb.*)] ἐκεῖνο **V α** 51 secundum tempus illud: κατ' αὐτὸν **α D M γ** (*Hüb.*)] κατ' αὐτῶν **V S** (56 principio] + καὶ ταύτῃ **V D γ** (*Hüb.*), *diversimode cett.*; *fort. non redd.* **G**) 58 ante: πρὸ τῆς **Σ β γ** (*Hüb.*), *cf.* **Y**] πρὸς τῆς ἐκ τῆς **V** 59 si: εἰ *cett.*] ἡ **V**: ἠ **Y** (60 id est optentionem principii] *om. codd.*; *explicatio translatoris*) 63 aderant ipsi: προσῆν αὐτῷ *cett.* (*Hüb.*)] προς ἦν αὐτό **V** (**Y**) 64 humane (*cett.*)] *om.* **V** 65 uideatur: δοκῇ *cett.* (*Hüb.*)] δοκεῖ **V Y** (66 tamen] αὐτὸ *codd.*) 71 et (**Σ β γ**)] *om.* **V Y** (*Hüb.*) factiuo: τῷ ποιητικῷ **β γ** (*Hüb.*)] τὸ ποιητικόν **V α**

---

34 non a] *inv.* *Ox* 46 Temporali] Corporali *Ox* 56 non] + *spat. 5 litt.* *Ox* 57 attendere] ascendere *Ox* 58 illud] id (i[d] –> id) *Ox* 60 katarchen] et(?)katharchera *Ox* 67 figurationem] -ni *Ox* impetum] -us *Ox* 70 factiuam] -ua *Ox* omnino] -ia *Ox*

[5] Proposito autem secundum presens nobis existente et hanc | partem prosequi secundum protestatam in principio huius compilationis ratiocinationem de eo quod possibile est talis pronosticationis, antiquum quidem modum predicendi recusabimus propter usus difficultatem et inexplicabilitatem, illum uidelicet modum qui secundum speciem optentiuam astrorum omnium uel plurium, cum sit multiplex et fere | infinitus, si quis uelit omnimoda diligentia pertractare, et magis possit in particularibus adiectionibus eorum que naturaliter considerantur quam in traditionibus per singula speculari.

[6] Operationes autem ipsas, per quas conspiciatur quelibet | specierum secundum modum proficuum, et factiuas uirtutes secundum proprietatem et uniuersalem habitum astrorum ad singula, ut possibile est maxime, modo assecutibili simul et breui secundum naturalem coniectationem exponemus: loca ambientis, <ad> | que humanorum euentuum singula considerantur, presupponentes uelud signum ad quod <oportet> coniecturam fieri; eas etiam corporum que aptantur locis secundum dominium factiuas uirtutes, uelut emissiones sagittarum, in grosso adaptantes; illum <autem> qui ex commixtione | ex pluribus naturis circa suppositam speciem collectum euentum derelinquentes, quemadmodum sagittario bene coniecturatiuo, per ratiocinationem eius quod iam intuitus est.

[7] Primo autem de uniuersalibus que considerantur per principium secundum partum faciemus sermonem secundum conuenientem ordinis | consequentiam, omnibus quidem, ut diximus, naturam habentibus per hoc accipi possibilibus, cooperantibus <autem>, si quis amplius negotiari uoluerit, ad solas proprietates que sunt penes fetus concretionem, et earum que secundum tempus seminationis per eandem theoriam incidunt proprietatum.

V (**α** =**YΣ** **β** =**DMS** **γ** =**ABC**) 72 Proposito: προθέσεως *cett.*] -εων **V** autem secundum (*cett.*)] *spat. 12 litt.* **V** (75 recusabimus — inexplicabilitatem *hic* **G**] *in fine sententiae hab. codd.* [98°-99°]) 77 quis] + *αὐτὸν codd.* 78 possit: δυνάμενον **V β**] -*νων* **α γ** (*Hüb.*) 79 quam: ἢ *cett.*] *om.* **V** 81 proprietatem: ἰδιότροπον *cett.*] ἴδιόν ροπον **V** 81-2 uniuersalem habitum: ὁλοσχερὲς **V** (*Hüb.*)] ὁλοσχερέστερον **α β γ** 83 coniectationem exponemus (*cett.*)] *spat.* **V** 84 considerantur: θεωρεῖται *cett.*] θεωρεῖ **V** 85 ad quod: οὗ *cett.* (*Hüb.*)] οὐ **V Y** (94 fetus concretionem: αὐτὴν τὴν σύγκρισιν **V**, *Hüb.*: τὴν σύγκρισιν *plur.*: *ad sensum bene redd.* **G**)

---

73 ratiocinationem] -cinitantem *Ox* 76 qui] quid *Ox* 77 omnimoda diligentia] -modi -gentiā *Ox* 80 quelibet] quod- *Ox* 88 sagittario] -ari^{s} (-ius?) *Ox* 91 sermonem] sermocionem *Ox*

De gradu horoscopi
Capitulum 3^{m}

[1] Quoniam autem de primo et principalissimo, hoc est de parte horoscopi secundum partum, dubitatio euenit sepe, sola quidem ut ad omne dicatur consideratione horoscopiorum secundum partum ipsum per astrolabia | obseruantibus diligenter <- - -> percipere potente, aliis autem fere omnibus horoscopiis, quibus plurimi magis sollicitorum attendunt, multotiens mentiri a ueritate potentibus [2] (solaribus quidem propter mutationes <positionum> et propter incidentes gnomonum transmutationes, | aquaticis autem propter impedimenta et irregularitates fusionis aque a diuersis causis et casualiter), necessarium utique erit pretradi quo quis utique modo inueniat gradum zodiaci qui debet oriri secundum naturalem et consequentem rationem, | presupposito eo qui secundum datam propinquam horam inuento per opus eleuationum.

Quomodo oportet accipere horoscopum
Capitulum 4^{m} [3]

[3] Oportet itaque accipere coniugationem ante partum factam proxime, siue coniunctionalis sit siue pleniluniaris, et gradu diligenter considerato (coniunctione quidem existente | amborum luminarium, plenilunio autem gradum super terram existentis ipsorum) et secundum tempus partus uidere astra que habent ad ipsum dominatiuam rationem; modo uniuersali qui consideratur secundum do-

---

V (**α** =**YΣ** **β** =**DMS** **γ** =**ABC**) 97 3^{m}: γ′ **V** (*Hüb.*)] β′ **Y D γ**: *om.* **Σ M S** 98 Quoniam autem: ἐπεὶ δὲ **α M S**] ἐπειδὴ **V D γ** Heph. (*Hüb.*) 100 secundum partum ipsum: κατ' αὐτὴν τὴν ἔκτεξιν **α β γ** Heph. (*Hüb.*)] κατὰ τὴν τῆς ἔκτεξιν **V** 101 <...>*] τὸ λεπτὸν τῆς ὥρας *codd.* (*l.v.* **Y G Σ**) (103 mutationes] τὰς *codd.*) 106 modo: τρόπον **α M S** Heph. (*Hüb.*)] πον **V**; *cf.* **D γ** 109 Quomodo oportet accipere horoscopum: πῶς δεῖ λαμβάνειν ὥραν *ut titulum inseruerunt* **D M**, *cf.* Co le (90,30): πῶς δεῖ λαβεῖν τὴν ὡροσκοποῦσαν μοῖραν **γ**] *om.* **V α** (*Hüb.*) 110 Capitulum 4^{m} (*sic*), *cf. supra* 6] *om. codd.* 111 coniugationem: συζυγίαν **α β γ** Heph. (*Hüb.*)] -γίας **V** factam: προγενομένην **β γ** Heph. (*Hüb.*), *cf.* **α**] -μένης **V** (112 gradu ... considerato] τὴν μοῖραν ... διασκεψαμένους (-σκεψω-, -σκεψο-) *codd.*) (113 gradum*] τὴν *codd.*) 114 existentis: ὄντος **β γ** Heph. (*Hüb.*)] ὄντως **V**, *cf. cett.* et secundum: κατά τε **V Y** (*Robbins*), *cf.* Heph.] κατὰ **Σ M S γ** (*Hüb.*) que habent: ἔχοντας (*diversis locis*) *cett.*] *om.* **V**

---

97 Capitulum 3^{m}] cap^{o} tertio *Ox* 99 sola] *ss. Ox, exp.* idla quidem] pro *Ox* 100 horoscopiorum] -copiꞩ *Ox* astrolabia] -lobia *Ox* 101 <- - ->*] *spat. 3 litt. Ox, an spat. vel graece* **G**? 104 irregularitates] inecul- *Ox* 107 datam] dicam *Ox* 109 horoscopum] -um *dub. Ox* 113 plenilunio] plenissime *Ox* 113 gradum* *sic Ox*] *fort.* gradum (*pro* τὴν) *neglegenter loco* gradu (*vide supra* gradu considerato) *scr.* **G**

minium in hiis quinque: trigonalitate scilicet et domo et exaltatione | et terminis et apparentia aut configuratione; hoc est quando <unum - - - [4]- - - > unum ad hec omnia uel plurima conuenienter dispositum inuenimus, quem utique obtinuerit gradum iste certitudinaliter secundum quod peruagatur | signum in tempore partus, gradum equalis numeri cum ipso iudicamus oriri in signo proximo inuento per opus eleuationum. Si autem duos uel plures condominantes, cuiuscumque ipsorum qui secundum partum gradalis transitus | habuerit propinquiorem numerum gradui orienti secundum eleuationes, quantitate graduum huius utemur. Si autem duo uel etiam plures propinquum habeant numerum, ei qui magis habet rationem ad angulos et ad partitionem adherebimus.

[5] Si tamen amplior fuerit distantia graduum dominationis | ad horoscopum secundum totalitatem quam ad gradum secundum simile medium <celi>, eodem numero ad gradum mediantem celum utemur et per hunc reliquos angulos instituemus.

Diuisio geneatici sermonis
Capitulum 5m [4]

[1] Hiis itaque preexpositis, si quis ipsius ordinis gratia diuidat uniuersale geneatice theorie, inueniet utique eorum que secundum naturam et que possunt comprehendi hanc quidem que est solorum simptomatum ante natiuitatem, ut eam que est sermonis de parentibus; hanc autem et eorum que ante natiuitatem | et post natiuitatem, ut eam que est sermonis de fratribus; hanc autem [et] eorum que penes ipsam natiuitatem, que non iam ita una <est et> simplex; ultimam autem que eorum que post natiuitatem, multarum partium et hanc consideratam.

V (**α** =**YΣ** **β** =**DMS** **γ** =**ABC**) 116 terminis: ὁρίοις **M S**] ὁρίῳ *cett.* (*Hüb.*) 117 <unum - - - > unum] ἕν τι ἢ πλείονα — ἐὰν μὲν οὖν ἕνα *codd.*; *verba ante* unum² (ἕνα) *fort. in trad. latina exciderunt* 119 iste: οὗτος **Σ β γ** Heph. (*Hüb.*)] οὕτως **V Y** 120 cum ipso: αὐτῇ **α γ** (*Hüb.*)] αὐτοῦ **V**: αὐτὴν **β** signo: δωδεκατημορίῳ **β** (*Hüb.*)] -μόριον **V α γ** 121 condominantes: συνοικοδεσποτοῦντας **α** (*Hüb.*)] -δεσπότας **V**: -δεσποτῶσιν **β** 124 habeant numerum: ἔχοιεν τὸν ἀριθμὸν **Σ M S** Co (92,2)] εἶεν τῷ ἀριθμῷ **V Y D γ** (*Hüb.*) (128 et per hunc] διὰ ταύτης καὶ *codd.*) 128 instituemus: καταστησόμεθα **α β γ** (*Hüb.*)] διαστησ- **V** 131 Capitulum 5m [4]] δʹ **V D** (*Hüb.*): γʹ **Y γ**: *om. cett.*; *et sic deinceps* (134 solorum: μόνων **Σ β γ**] μόνον **V Y**, *Hüb.*) 134-5 ut — natiuitatem (*cett.*)] *om.* **V**

---

117 <unum - - ->] *cf. adn. gr.-lat.* 118 inuenimus] *an* -iamus *scrib.* ? 122 gradalis] *an* -dualis *scrib.* ? 125 adherebimus] -hēbimus *Ox* 126 dominationis] -tores *Ox* 127 medium] medii *Ox* 132 geneatice] -cis *Ox* 137 que non] non que *Ox* ita] ista *Ox* 138 multarum] -orum *Ox* consideratam] -ta *Ox*

[2] Est autem eorum que penes ipsam natiuitatem inquiruntur, sermo de masculinis et femininis et qui de gemellis uel etiam de pluribus nascentibus et qui de monstris et qui de non cibatis; eorum uero que post natiuitatem qui de temporibus uite, quoniam non copulatur cum sermone | de non cibatis. Deinde qui de forma corporis et de passionibus et de lesuris corporalibus. Consequenter autem qui de qualitate anime et de passionibus animalibus. Deinde qui de fortuna possessoria <et qui de fortuna> dignitatiua, post hec autem qui de qualitate operis.

[3] Deinde qui de conuictu nuptiali et de filiorum procreatione et qui de coniunctis et coaptatis et de amicis. Deinceps autem qui de peregrinatione, et ultimus qui de qualitate mortis, uirtute quidem coaptatus ei qui de temporibus | uite, ordine autem merito post hec omnia positus.

[4] De quorum unoquoque summatim faciemus introductionem, ipsas que speculationis operationes cum nudis effectiuis uirtutibus, ut diximus, exponentes ita, et ea que superflue a multis garriuntur et neque probabilem | habent rationem ad eas que a prima natura causas pretermittentes; ea <autem> que possunt comprehendi, non per sortes et numeros rationem cause non habentes, sed per ipsam theoriam configurationum ad conuenientia loca considerantes, uniuersaliter tamen et de omnibus | simpliciter, ut non secundum quamlibet speciem idem loquamur.

[5] Et primo quidem oportet considerare locum zodiaci conuenientem capitulo quod queritur circa natiuitatem secundum speciem, sicut, uerbi gratia, capitulo de operationibus locum medii celi, uel capitulo de patre locum solis. Deinde considerare | rationem habentes ad suppositum locum, astra scilicet dominantia, secundum <5> modos quos superius diximus, et si quidem unus sit qui secundum

---

V (**α** =**YΣ** **β** =**DMS** **γ** =**ABC**) 142 cum sermone: τῷ *cett.*] *om.* **V** non cibatis: ἀτρόφων *cett.*] ἀστρόφων **V** (143 de passionibus: περὶ παθῶν **S**: πάθῶν *cett.*] ὁ περὶ παθῶν **V Y**, *Hüb.*) 147 qui (*cett.*)] *om.* **V** de peregrinatione: περὶ ξενιτείας **Σ M S γ** Heph. (*Robb.*), *cf. tit.* IV.519] περὶ ξενιτειῶν **V D Y** (*Hüb.*) et ultimus: καὶ τελευταῖος **α β γ** Heph. (*Robb.*) «*fort. recte*»] τελευταῖος δὲ **V** (*Hüb.*) (152 ita] *om. codd.*) 153 pretermittentes: ἀποπεμπόμενοι **Σ β γ** (*Robb.*): παραπεμπόμενοι **Y** (*Hüb.*)] περιπεμπόμενοι **V** 156 idem loquamur: ταυτολογῶμεν **Σ β γ** (*Hüb.*)] ταῦτα λέγωμεν **V**, *cf.* **Y** (158 Et] *om. codd.*) (161 astra scilicet dominantia] τῶν ἀστέρων οἰκοδεσποτείας) 162 diximus: προείπομεν *cett.*] -πον **V**

---

139 sermo] + .1. *ss. Ox* 140 gemellis] + .2. *ss. Ox* nascentibus] -antibus *Ox* 141 monstris] + .3. *ss. Ox* cibatis] + .4. *ss. Ox* temporibus] + .5. *ss. Ox* 142 Deinde] + .2. (?) *ss. Ox* 143 Consequenter] + .4. *ss. Ox* qui] et *Ox* 144 Deinde] + .6. *ss. Ox* qui de] quidem *Ox* 145 dignitatiua (*cf.* IV.5 *et alibi*)] dignitatis (*ut uidetur*) *Ox* post] + .7. *ss. Ox* hec] hoc *Ox* 146 Deinde] + .8. *ss. Ox* et[1]] + .9. *ss. Ox* qui[2]] + .0. *ss. Ox* 147 Deinceps] + .11. *ss. Ox* 147-8 et ultimus — mortis] *it.* (*et exp.*) *Ox* 151 operationes] -nis *ante corr. Ox* exponentes] -nibamus *Ox* 160 locum[1]] loco (*dub.*) *Ox* 162 et si quidem unus] *mg. Ox*

omnes est dominus, ipsi dare illius gubernationem [dominium], si autem duo uel tres, hiis que plures habent dignitates.

[6] Post hec autem ad qualitatem quidem euentus considerare naturas ipsorum astrorum dominantium et naturas signorum, in quibus sunt ipsa et coaptata loca; ad magnitudinem autem et uirtutem considerare, utrum efficaciter sint | disposita et secundum id quod est mundi et secundum id quod ad generationem uel contrarie.

[7] Efficacissima quidem enim sunt, quando mundialiter quidem in propriis uel in conuenientibus fuerint locis, et iterum quando orientales extiterint <et> augmentatiui numeris; secundum natiuitatem autem, quando in angulis uel in | accedentibus sursum moram fecerint, et maxime in primis, dico autem in hiis que sunt secundum eleuationes super orizontem et secundum mediationes celi. Debilissima autem, mundialiter quidem, quando in alienis uel non conuenientibus locis fuerint, et occidentales uel retrogradi cursibus; secundum natiuitatem autem, quando declinauerint ab angulis.

[8] Ad uniuersale autem tempus euentus, utrum matutini sint uel uespertini ad solem et ad horoscopum, quoniam quidem principales utriusque ipsorum quarte et diametrales hiis matutine fiunt, relique autem et sequentes uespertine; et utrum in | angulis existunt uel in superaccedentibus: matutina quidem enim existentia uel angularia secundum principia fiunt efficacissima, uespertina autem uel in superaccedentibus tardiora.

[5.1] Principalis quidem igitur modus eius <que> secundum speciem speculationis, | cui semper oportet attendere, hunc habet modum. Incipimus autem iam secundum expositum ordinem a primo, eo uidelicet qui de parentibus sermone.

V (**α** =**YΣ** **β** =**DMS** **γ** =**ABC**) 163 ipsi: *αὐτῷ* **α M S** Heph.] *τούτῳ* **V D γ** (*Hüb.*) illius] + τὴς προτελέσεως (προτε *sequ. lac.* **V**) *codd.* 166 et²: τε καὶ *cett.*] περὶ καὶ **V** (168 id quod est mundi] *αὐτὸ τὸ κοσμικὸν* *codd.*) 175 mundialiter quidem, quando: κοσμικῶς μὲν, ὅταν **Σ** Procl., *cf.* Heph.] ὅταν κοσμικῶς μὲν *tr. cett.* (*Hüb.*) 176 cursibus] + ὦσιν *codd.* 185 cui: οὗ **Σ M S γ** Heph. (*Hüb.*)] οὐ **V D**: οὖν **Y**

---

163 illius] *an* effectus *post* illius *excidit* ? gubernationem dominium] *an duplex lectio* **G**? 165 hec] hoc *Ox* 172 augmentatiui] -tati *Ox* 173 moram] -a *Ox* 178 autem] *ss. Ox* 178 ad solem] *mg. Ox* 181 quidem enim] autem *praem. Ox* 185 habet] autem *Ox* Incipimus] *an* -piemus *scrib.* ?, *sed cf.* II.524

De parentibus
Capitulum 6m [5]

Sol quidem igitur et Saturnus persone patris coaptantur secundum naturam, luna autem et Venus matri, et | qualitercumque isti dispositi fuerint ad inuicem et ad alios, [et] talia oportet et que circa parentes existimare.

[2] Que quidem enim circa fortunam et possessionem ipsorum considerandum ex addextratione luminarium, quoniam circumdata quidem a beneficis et | ab hiis qui eiusdem partitionis, uel in eisdem signis uel etiam in consequentibus, eminentia et clara que circa parentes insinuant, et maxime quando solem quidem stelle matutine addextrauerint, lunam autem uespertine, bene et ipse diposite secundum modum quem diximus.

[3] Si autem | Saturnus et Venus fuerint et ipsi orientales <et> in propria dignitate uel etiam angulares, felicitatem preclaram existimandum secundum conueniens utriusque parentum. Econtrario autem, si luminaria uacue currentia fuerint et non addextrata extiterint, humiliationis et | ingloriationis parentum sunt significatiua, et maxime quando neque Venus uel Saturnus bene uideantur dispositi.

[4] Si autem addextrentur quidem, <non> tamen <ab> hiis qui sunt eiusdem partitionis (ut quando Mars adiungatur soli uel lune de die, Saturnus autem lune de nocte) uel non a bene dispositis beneficis | et secundum eandem partitionem, mediocritates et irregularitates circa uitam ipsorum existimandum. Et si quidem consona fuerit notata sors fortune in natiuitate addextrantibus solem uel lunam in bono, salua suscipientur que parentum, si autem inconsona | fuerit uel contraria, a benefico nullo uel a maleficis assumpta addextratione, parentum possessio erit ipsis inutilis et nociua.

V (**α** =**YΣ** **β** =**DMS** **γ** =**ABC**) 187 *titulum hic praebent* **V D γ** Co (95.28)] *ante* Principalis (184) **α M S** Heph. (*Hüb.*) 188 Capitulum 6m] ε′ **V D γ** 191 oportet: δεῖ **Σ M S γ** Heph. (*Hüb.*)] δὴ **V D**: γὰρ **Y** 192 Que: τὰ **Σ** Procl. (*Hüb.*), *cf.* Heph.] *om.* **V Y D γ** («*haud scio an recte*» *Hüb.*), *cf.* **M S** 193 circumdata: περιεχόμενα **Q** (*Marc. gr.* 594) Heph. (*Boer Hüb.*)] πῶς ἐχόμενα **V**: περιεχόμενοι **Σ β γ**, *cf.* **Y** 196 addextrauerint: δορυφορῶσιν **α β γ** Heph. (*Hüb.*)] διαφοροῦσιν **V** 202 neque ... bene: μηδ'... καλῶς **Y**] .... μὴ καλῶς **Σ M S γ** Heph. Procl. (*Hüb.*): ... κακῶς **V** (*Boer*); *cf.* **D** 204 (Mars] + μὲν *codd.*) adiungatur: ἐπαναφέρηται *cett.*] ἐπάνω φέρηται **V** uel lune de die, Saturnus autem lune de nocte (ἢ τῇ σελήνῃ ἡμέρας, Κρόνος δὲ τῇ σελήνῃ νυκτὸς)] ἢ τῇ σελήνῃ Κρόνος τῇ σελήνῃ **V**: ἢ τῇ σελήνῃ Κρόνος **D γ** (*Hüb.*): ὁ δὲ Κρόνος δὲ τῇ σ. **Y**: Κρόνος δὲ τῇ σ. **Σ M S** Heph. (*Boer*) (206 mediocritates et irregularitates] μετριότητα [-τατοι **V**] καὶ ἀνομαλίαν *codd.*) (207 notata] διασημανθησόμενος *codd.*) 209 a benefico nullo uel a maleficis (μηδενὸς τῶν ἀγαθοποιῶν ἢ τῶν κακοποιῶν)] μηδενὸς ἢ τῶν κακοποιῶν **V Σ β** (*Hüb.*): μηδενὸς τῶν ἀγαθοποιῶν **γ** Heph.: μηδενὸς τῶν κακοποιῶν **Y**

---

188 capitulum] *it.* *Ox* 193 ex addextratione] e9 ad dextrationem *Ox* beneficis] -ciis *Ox* 195 solem] -em *dub.* *Ox* 196 lunam] lumina *Ox* 200 uacue currentia] uacuo currentia *Ox* 201 extiterint] existerint *Ox*, *cf. supra* 171 203 addextrentur] adex- *Ox*, *et sic saepius* qui] que *Ox*, *vide supra* 194 204 adiungantur] adiug- *Ox* 206 irregularitates] -larites *Ox*

[5] De longeuitate autem ipsorum uel temporis paucitate considerandum ab aliis configurationibus. De ǀ p a t r e quidem, si Jupiter uel Venus configurentur qualitercumque soli uel Saturno (uel etiam si ipse Saturnus habeat consonam configurationem ad solem, uidelicet uel simul existens uel in sextili aspectu uel trino), in uirtute quidem existentibus ipsis, longeuitatem patris coniecturandum, ǀ debilibus autem non similiter, non tamen neque temporis paucitatem.

[6] Si autem hoc quidem non extiterit, Mars autem superauerit solem uel etiam Saturnum uel superinferatur ipsis, uel etiam si ipse Saturnus iterum non fuerit consonus ad solem, sed uel in quadrato uel diametrali aspectu, ǀ decidentes quidem ab angulis debiles solum patres efficiunt, angulares autem uel superaccedentes angulis, aut pauci temporis aut membrorum lesuram patientes: pauci quidem temporis quando fuerint in primis duobus angulis (oriente scilicet et eo qui medii celi) uel in horum accessibus, membrorum ǀ uero lesuram sustinentes uel languorosos, quando in reliquis duobus angulis fuerint (occidente scilicet et eo qui sub terra) uel in horum accessibus.

[7] Mars quidem enim solem superaspiciens secundum modum quem diximus, repentine perimit patrem uel lesuram circa oculos efficit, Saturnum autem aspiciens uel mortibus ǀ uel febribus cum rigoribus uel lesuris per incisiones uel adustiones inuoluit. Saturnus autem et ipse male configuratus ad solem mortes paternas languorosas preparat et passiones que eueniunt per molestationem ab humoribus.

[8] De m a t r e autem, si quidem Jupiter configuratus fuerit lune et Veneri qualitercumque uel etiam si ipsa Venus consone se habeat ad lunam, scilicet in sextili aspectu uel trino uel simul existens cum illa, in uirtute existentes longeuam insinuant matrem. Si autem Mars ǀ aspexerit lunam uel etiam Venerem, superallatus uel in quadrato uel in diametrali aspectu, uel Saturnus lunam solam eodem modo, ablatiui quidem existentes uel iterum decidentes inuoluunt aduersitatibus solum

---

V (**α** =**YΣ** **β** =**DMS** **γ** =**ABC**) 211 longeuitate ... ipsorum uel temporis paucitate: πολυχρονιότητος αὐτῶν ἢ ὀλιγοχρονιότητος **Y**] ... πολυχρ. ἢ ὀλιγοχρ. αὐτῶν **V β γ** (*Hüb.*): πολυχρ. αὐτῶν **G**° **Σ** Heph. 212 De: ἐπὶ *cett.* (*Hüb.*)] ἐπεὶ **V D** 213 uel[1]: ἢ **Σ β γ**] καὶ **V Y** Heph. (*Hüb.*) consonam: σύμφωνον *cett.*] ἄφωνον **V** 214 ad: πρὸς *cett.*] παρὰ **V** 215 debilibus: (*potius*) ἀδυναμούντων **α β γ** Heph. (*Hüb.*)] ἀδυνατούντων **V** (Boer) 220 angulares ... superaccedentes: ἐπίκεντροι ... ἐπαναφερόμενοι **α β γ** Heph. (*Hüb.*)] -κεντρος ... -φερόμενος **V** (223 uel] καὶ *codd.*) 226 Mars: Ἄρεως *cett.*] πυρόεντος **V D** superaspiciens (ἐπιβλέψας ?)] βλέψας **V Σ β γ** Heph. (*Hüb.*): ἐπιθεορήσας (*sic*) **Y** (228 uel[4]] καὶ *codd.*) 229 male (*cett.*, *Hüb.*)] *om.* **V** Heph. (*cod.* C) ad: πρὸς τὸν *cett.*] *om.* (*spat.*) **V** solem ἥλιον *cett.*] ἡλίου **V** 233-4 in sextili aspectu: ἑξάγωνος **Σ β γ** Heph.] + ὢν **V Y** (*Hüb.*) 235 uel etiam: ἢ καὶ **Y**] καὶ **V D γ** (*Hüb.*): ἢ **Σ M S** Heph.

---

212 patre] tempore *Ox* 218 etiam[1]] *an post* uel[1] *transponendum* ? non] *ss. Ox*, *exp.* ut 220 debiles] + non *ss. Ox* 224 languo-] lango- *Ox*, *etiam infra* 230 228 mortibus] moribus *Ox* 229 configuratus] -tur *Ox* ad] a 235 superallatus] super allegatus *Ox* 236 solam] et solem *Ox* 237 ablatiui] -tum *Ox* iterum] et (*abbr.*) *Ox*

uel debilitatibus, appositiui autem uel angulares | faciunt matres pauci temporis uel lesuram membrorum sustinentes. Pauci quidem temporis existentes in angulis orientalibus uel in accedentibus, lesuram autem patientes in occidentalibus.

[9] Mars quidem aspiciens lunam hoc modo orientalem quidem existentem mortes matrum | efficit repentine et lesuram circa oculos, deficientem autem mortes a partus doloribus uel talibus et lesuras per incisiones et per adustiones. Venerem autem aspiciens mortes febriles efficit et passiones per occultationes et obtenebrationes et concussiones repentinas. Saturnus autem lunam aspiciens | mortes et passiones facit, orientali quidem existente ipsa per febres cum rigore, depulsiua autem per deuastationes matricum et corrosiones.

[10] Coassumendum autem ad particulares species nocumentorum et passionum et mortium proprietates signorum in | quibus sunt causam efficientes, de quibus opportunius in hiis que de ipsa natiuitate elaborabimus. Et adhuc obseruandum de die quidem maxime solem et Venerem, de nocte autem Saturnum et lunam.

Reliquum autem de elaborationibus secundum speciem congruum et consequens utique erit locum paternalem uel maternalem qui fuerit secundum partitionem uelut horoscopum statuentes reliqua considerare ut in natiuitate ipsorum parentum secundum | regulas tradendas consequenter totaliorum specierum que pertinent ad actiones et corpus.

[11] Hic tamen et in omnibus reminisci oportet modi commiscitiui coniecturantem, si non uniformes sed diuersi uel contrariorum effectiui extiterint qui dominationes quesitorum locorum acceperunt planete, qui de habundantiis ad uirtutem circa unumquodque | accidentium plures dignitates habentes inueniuntur ad dominium euentuum, ut aut horum naturis consequentem faciamus considerationem uel, dignitatibus equalibus existentibus, quando quidem simul fuerint qui dominantur, quod colligitur ex commixtione | diuersarum naturarum bene coniecturabiliter ratiocinemur; quando autem distantes fuerint in parte altera unusquisque, per singula tempora conuenientia simptomata distribuamus, prioribus quidem matutinis magis, [et] posterioribus autem uespertinis.

V (**α** =**YΣ** **β** =**DMS** **γ** =**ABC**) 239 temporis] + ὁμοίως *codd.* 242 repentine: αἰφνιδίως **V**] -ίους *cett.* (*Hüb.*) 243 mortes: τε θανάτους *cett.*] τοῦ θανάτου **V** 245 et[1]: καὶ **Σ D M γ**] ἤ **V Y S** (*Hüb.*) et[2]: καὶ *cett.*] ἤ **V** 247 autem: δὲ *cett.*] μὲν **V** deuastationes: νομῶν *Par. suppl. gr.* 597 (*Robb., Hüb.*)] νόμων **V Y** Heph.: νόσων **Σ M S γ**; *cf.* **D** 257 corpus: *fort.* σωματικῶν *leg.* **G**] συμβατικῶν **V** (*Robb. Hüb.*): σημαντικῶν **Y**, *cf.* **G°**: συμβατικὰς **Σ β γ** (*Boer*) 262 ad (*cett.*)] *om.* **V** 266 distribuamus: ἀπομερίζωμεν **Σ β γ** (*Hüb.*)] -ζομεν **V** Procl., *cf.* **Y** 267 posterioribus: ὕστεροις *cett.*] -ον **V**

---

240 accedentibus] occidentalibus (*dub.*; oc- *ex* ac-) *Ox* 242 repentine] -tiue *Ox* 245 repentinas] -tiuas *Ox* 247 depulsiua] -pulsa *Ox* 255 natiuitate] -tatem *Ox* 256 tradendas] cadentes *Ox* 257 corpus (*sic*) *Ox* 258 coniecturantem] *an* -tes *scrib.*? 259 effectiui] defectiui *Ox*

[12] A principio quidem enim necesse coaptari | loco quesito planetam qui debeat circa ipsum operari; hoc autem si non factum fuerit, nichil magnum possibile est uniuersaliter disponere eum qui principio non communicauerit totaliter; temporis tamen euentus ipsius quod efficitur non iam erit causa que prime dominationis, sed | situs dominantis aliqualiter ad solem et ad mundi angulos.

De fratribus
Capitulum 7^m^ [6]

[1] Qui quidem igitur de parentibus locus fere et ab hiis utique nobis manifestus fiet. Qui autem de fratribus, si quis et hic | uniuersale solum exquirat et non ultra possibile numerum certitudinaliter et particularia inquirat, accipietur utique magis naturaliter qui de hiis qui eiusdem matris solum et a signo mediante celum et a materno loco (hoc est continente de die quidem | Venerem, de nocte autem lunam), quoniam hoc signum et superallatum ipsi fit matris qui de filiis locus, idemque debet loco geniti de fratribus.

[2] Si igitur benefici loco configurentur, bonam copiam fratrum dicemus, | coniecturam facientes ad multitudinem ipsorum astrorum, et prius si sint in signis unius forme uel bicorporibus. Si autem malefici superemineant ipsum uel etiam contrarientur secundum diametrum, raritatis fratrum sunt significatiui, maxime autem si solem coassument. Si | autem et in angulis contrarietas fiat, et maxime in ascendente, per Saturnum quidem et primogeniti et primoaliti fiunt, per Martem autem morte reliquorum rari fiunt fratres.

[3] Adhuc tamen dantibus astris, si quidem bene secundum mundum disposita extiterint, bene formatos | et gloriosos existimandum fratres qui dantur, si autem contrarie, deiectos et ingloriosos; si autem superauerint dantes uel superinferantur ipsis malefici, etiam pauci temporis. Dabunt autem masculos quidem qui mundialiter sunt masculinizati, feminas autem feminizati, | et iterum primos quidem orientaliores, posteriores autem occidentaliores.

V (**α** =**YΣ** **β** =**DMS** **γ** =**ABC**) 268 qui debeat] + τι *codd.* 269 (hoc autem] καὶ [ὡς **γ**] τούτου *codd.*) possibile est: οἷόν τε **Σ M S γ** (*Hüb.*)] οἴονται **V Y D** 270 temporis (*cett.*)] *om.* **V** 276 exquirat: ἐξετάζοι **α γ** (*Hüb.*)] -ζοιτο **V β** (*Boer*) 278 et a: καὶ ἀπὸ **α M S**] ἀπὸ **V D γ** (*Hüb.*) 279 et a: καὶ **c**, *suppl. Hüb., cf.* Heph.] *om. codd.* (282 igitur] μὲν οὖν *codd.*) 283-4 prius si sint ... uel[1] (*fort.* πρότερον *loco* πότερον *uel ambo verba legit* **G**)] πότερον ... τυγχάνουσι ἢ *codd.* 284 ipsum*: αὐτὸν **Σ M S** (*Hüb.*)] αὐτῶν **V Y D γ** (*Robb.*) 289 quidem: μὲν *cett.*] ὑπὸ μέντοι **V**

---

268 enim] *etiam ante* quidem *Ox* 275 nobis] uobis *Ox* 276 hic] hoc *Ox* 277 particularia] *an* -iter *scrib.* ? 284 ipsum] ipsi (*obsc.*) *Ox* 287 ascendente] -tem *Ox* primoaliti] primoalia *Ox* 288 reliquorum] -quarum *Ox* 290 extiterint] exist- *Ox* 293 masculinizati] -zati *ss.* *Ox*

[4] Ad hec autem si quidem qui dant fratres consone configurati existant dominanti super signum fratrum, datos fratres facient amicitia coniunctos, si | autem et sorti fortune, facient etiam communis uite; si autem extiterint in inconnexis uel in contrarietatis situ, discordes et inuidiosos et ut ad omne dicatur insidiosos. Reliquum autem, si quis amplius ea que pertinent ad unumquodque ipsorum uoluerit percunctari, coaptetur utique et hic iterum | supposito astro dante ad rationem ascendentis et reliquis ut in natiuitate consideratis.

De masculinis et femininis
Capitulum 8^m [7]

[1] Sub uisu iam nobis facto sermone de fratribus secundum congruentem et naturalem rationem, consequenter utique erit | incipiendum de hiis que penes ipsam natiuitatem et primo percurrendum rationem de masculinis et femininis. Consideratur autem hec non uno modo neque ab uno aliquo, sed a lunaribus ambobos et ab horoscopo et a stellis rationem habentibus ad ipsa, maxime quidem secundum | in seminatione dispositionem, grossius autem et secundum dispositionem in partu. Omnino autem obseruandum, utrum predicta tria loca et astra hiis dominantia si omnia uel plurima sunt masculina ad masculi procreationem uel feminina ad femine procreationem, et ita | pronuntiandum.

[2] Discernendum tamen masculina et feminina secundum modum que premisimus in tabularibus expositionibus in principio compilationis, a natura signorum in quibus sunt et a natura ipsorum astrorum et adhuc ab | habitudine ad mundum, quoniam orientales quidem existentes masculini fiunt, occidentales autem feminini; ad hec autem ab habitudine ad solem, matutini quidem enim existentes iterum masculini fiunt, feminini autem <uespertini>. Per que omnia predominationem generis secundum plurimum conuenit coniecturare.

V (**α** =**YΣ** **β** =**DMS** **γ** =**ABC**) 295 quidem: μὲν **Σ M S** Heph. (*Hüb.*)] *om.* **V Y D γ** (*Boer Robb.*) 297 communis uite: κοινοβίους **α β γ** Heph. (*Hüb.*)] κοινωνοὺς βίου **V** 299 amplius: ἔτι **Σ M S** Heph. (*Hüb.*)] ἐπι- **V Y D γ** (*Boer Robb.*) 302 De — femininis (**α C** Heph., *Hüb.*)] + γενέσεως **V D**, *cf.* **A B** (310 Omnino] ὅλον *codd.*) 311 si* omnia (εἰ πάντες ?)] ἢ πάντες **α β γ** (*Hüb.*): ἅπαντες **V** masculi procreationem: ἀρρενογονίαν *cett.*] ἀρρενολογίαν **V** 313 premisimus: ὑπεθέμεθα *cett.*] -τιθέμεθα **V** 317 ad[2] (*cett.*)] *om.* **V** 317-8 existentes iterum (**Σ M S γ**)] *inv.* **V Y D** Heph. (*Hüb.*)

---

296 amicitia] -cicā *Ox* 297 extiterint] exist- *Ox* 300 coaptetur] *an* coniectetur *scrib.* ? hic] hoc *Ox* 301 reliquis] -qu⁹ *Ox* 307 Consideratur] -antur *Ox* 311 si] *an abbr. pro* uel* («l'») *male legit Ox* ? 314 expositionibus] exposision- *Ox* 315 adhuc] ad hoc *Ox* 317 ad hec] ad hoc *Ox*

De gemellis
Capitulum 9m [8]

[1] De genitis autem similiter singulis duobus uel etiam pluribus eadem loca obseruare oportet, hoc est duo luminaria et horoscopum. Consequi enim consueuit tale simptoma circa commixtiones, quando duo uel | etiam tria loca contineant animalia bicorporea, et maxime quando planete dominantes ipsis idem patiantur, uel aliqui quidem in bicorporibus, aliqui autem fuerint ana duo positi uel etiam plures.

[2] Si autem in bicorporibus fuerint principalia loca et secundum idem plures planetarum configurati, | tunc accidit plures duobus procreari, multitudinis quidem numero a planeta euentum faciente coniectato, sexu autem a situ astrorum configuratorum soli et lune et horoscopo ad masculorum uel | femellarum procreationem secundum modos prius dictos.

[3] Quando autem talis dispositio non luminaribus coassumat angulum ascendentis sed angulum medii celi, tales matres duos genitos ut ad omne dicatur uel etiam plures pariunt. Singulariter autem tres quidem masculos proferunt | in Regum generatione, simul cum premissis locis in bicorporibus figurati Saturnus, Jupiter et Mars; tres autem femellas in generatione Kharitarum, Venus, luna cum Mercurio feminino; duos autem masculos et unam feminam sub ea que Iouis prolis, Saturnus, Iupiter, Venus; | duas autem feminas et masculum unum sub ea que Demetri et Puelle, Venus, luna et Mars.

[4] De quibus ut in pluribus accidere consueuit non ad perfectum perduci genita et cum defectibus quibusdam corporalibus procreari, adhuc in locis illis fieri aliqua mirabilia et inexspectata | per quasi insinuationem talium simptomatum.

V (**α** =**YΣ** **β** =**DMS** **γ** =**ABC**) 322 obseruare: παρατηρεῖν *cett.*] περιτηρεῖν **V** 323 enim: γὰρ **Σ β γ** Co[le], *sim.* Heph.] δὲ **V Y** (*Hüb.*) 324 tale simptoma: τὸ τοιοῦτον σύμπτωμα **α β γ**, *cf.* Heph.] τοῦτο τὸ σ. **V** (*Hüb.*) circa: περὶ **α β γ**] παρὰ **V** (*Hüb.*) commixtiones: συγκράσεις **α β γ** Procl. (*Hüb.*)] κράσεις **V** contineant: περιέχωσιν *cett.*] -χουσι **V** 330 euentum: *an* σύμπτωμα *legit* **G** ?] ἰδίωμα *codd.* 332 prius: *an* ἔμπροσθεν (*sine* ἐν τοῖς) *ut* **Y D A** (*cf.* **M**) *legit* **G** ?] ἐν τοῖς ἔμπροσθεν **V** *et cett.* (*Hüb.*) (333 coassumat: συμπεριλαμβάνῃ **α β A B**, *Hüb.*] συμπροσλαμβάνει **V**, *cf.* **C**) 337 in generatione: ὑπὸ τὴν *cett.*] τὴν *om.* **V** 339 unum: ἕνα **Σ β γ** (*Boer Robb.*)] μόνον **V Y** Heph. (*Hüb.*) 341 ut: ὡς *cett.*] ὅσα **V**

---

322 genitis] -tiuis *Ox* 334 sed] si *Ox* matres] matr[s] *Ox* genitos] gēios *Ox* 341 non] *ss. Ox, exp.* ut 342 procreari] -at *Ox*

De monstris Capitulum 10^m^ [9]

[1] Non alienus autem a premisssa speculatione neque de monstris sermo. Primo enim in talibus | luminaria quidem decidentia uel inconnexa horoscopo secundum plurimum inueniuntur, anguli autem occupati a maleficis. Quando igitur talis aliqua dispositio sub uisu ceciderit, quoniam fit sepe <circa> deiectas et infelices natiuitates, et si non monstruose sint, mox considerare | conuenit precedentem coniugationem coniunctionalem uel pleniluniarem et dominatorem ipsius et luminarium in partu.

[2] Si enim loca ipsorum in partu et locus lune uel etiam ascendentis omnia uel plura incolligata sint | loco coniugationis precedentis, genitum existimandum prodigiosum. Si quidem igitur hiis ita se habentibus luminaria in signis quadrupedibus uel bestialibus inuenta fuerint et maleficorum duorum angulares, neque ex hominibus erit quod procreatum est, sed nullo quidem benefico luminaribus testimonium reddente, | perfectum ferale et agrestem et malignam naturam habentium; testificantibus autem Ioue uel Venere, de hiis que ad solacium sunt, puta de canibus uel marturinis uel de simiis uel talibus; Mercurio [si] autem, in hiis que in humanum usum, puta de auibus uel porcis uel de bobus et capris | et talibus.

[3] Si autem in hiis que humane forme luminaria deprehendantur, aliis eodem modo se habentibus, ex hominibus quidem uel apud homines erunt genita, monstra autem et prodigiosa, proprietate secundum qualitatem et hic conspecta ex forma signorum, in quibus | malefici existunt qui luminaria et angulos optinent. Si quidem igitur et hic nullus beneficorum contestificetur alicui predictorum locorum, irrationalia et uere prodigiosa fiunt penitus.

V (**α** =**YΣ** **β** =**DMS** **γ** =**ABC**) 346 monstris: τεράτων *cett.* (*Hüb.*)] τερατωδῶν **V** 347 enim: γὰρ **Σ β γ**] μὲν γὰρ **V** (*Hüb.*): μὲν **Y** 350-1 mox — coniugationem (*cett.*)] *om.* **V** precedentem: προγενομένην **S** Heph. (*Hüb.*): -γινομένην **γ**] γενομένην **Σ D M**: [**V**] 353 locus lune: ὁ τῆς σελήνης **Σ β γ** (*Hüb.*)] οἱ τῆς σ. **V G**°; *cf.* **Y** (uel etiam] καὶ *codd.*) 357 sed (**α β γ**, *Robb.*)] *om.* **V** Heph. (*Hüb.*) 358 perfectum (**V γ**)] τῶν (δὲ) κακοποιῶν *praem.* **α M C** (*cf.* **G**°) (*Robb.*): ἀλλὰ τῶν κακοποιῶν *praem. Hüb. secundum Procl.* 359 uel: ἢ **V D γ**] μὲν ἢ **Σ M S** Heph. (*Hüb.*): μὲν **Y** de hiis que ad solacium sunt (τῶν ἐκπαιζομένων ?)] τῶν ἐκθειαζομένων **α β** Heph. (*Hüb.*): τῶν ἐκδιαζομένων **V**; *cf.* **γ** 360 uel de simiis: ἢ πιθήκων **Σ β γ**, *cf.* Heph. (*Hüb.*)] *om.* **V Y** (*Robb.*) (363 ex hominibus: ἐξ ἀνθρώπων **γ**] ὑπ' ἀν. **V α β** Heph. (*Hüb.*); *fort.* ὑπ' *per* ex *ad sensum redd.* **G**, *cf. supra* 356) 366 nullus: μηδεὶς **Σ β γ** Heph.] μηδὲ εἷς **V** (*Hüb.*), *cf.* **Y** beneficorum (**α M S** Heph.)] + ἀστέρων **V D γ** (*Hüb.*) contestificetur alicui: συμμαρτυρῇ μηδενὶ **Σ β γ**: προσμαρτυρῇ μηδενὶ **Y** Procl. (*Hüb.*)] πρὸς μαρτυρούμεναδενὶ (*sic*) **V**

---

346 alienus] -is *Ox* 351 coniunctionalem] coniuctio- *Ox* dominatorem] -res *Ox* 354 incolligata] in collegiata *Ox* 356 et maleficorum] in maleficio- *Ox* duorum] *an* duo *scrib.* ? 357 procreatum] *ex* procur- *corr. Ox* benefico] -cio *Ox* 360 in] *an* de *scrib.*? 361 humanum] -ne *Ox* 366 beneficorum] -ciorum *Ox*

[4] Si autem Jupiter aut Venus testificetur, pretiosum et | formosum erit monstri species, qualem consueuit accidere circa ermafroditas et circa uocatos arpocraticos et circa tales; si autem Mercurius testificetur, cum hiis quidem responsa dantes et per hec lucratiuos, solus autem edentulos | et surdos, aliter bene aptatos et astutos efficit.

De non cibatis
Capitulum 11^m [10]

[1] Reliquo autem existente de hiis que circa natiuitatem sermone de non cibatis, conuenit distinguere quod hac quidem | coheret modus ei qui de annis uite, quoniam species que inquiritur non aliena ab utroque, hac autem separatur, quia differt aliqualiter secundum considerationis potentiam. Qui quidem enim de annis uite in hiis qui totaliter habent annos sensibiles consideratur, hoc est non minores | reuolutione una solari; annus enim proprie talis regyratio accipitur, uirtute autem et qui minor hiis, menses et dies et hore.

[2] Qui autem de non cibatis in hiis qui neque omnino pertingunt ad premissum annum, sed in paucioribus numeris propter excessum | mali corrumpuntur. Vnde et ille quidem habet multipliciorem speculationem, iste autem magis grossam. Simpliciter enim, si in angulo fuerit alterum luminarium et alter maleficorum coextiterit uel etiam in diametrali situ, et hec gradualiter et secundum figuram equalium crurum, nullo quidem beneficorum | configurato, domino autem luminarium in maleficorum locis deprehenso: quod procreatum est non cibabitur sed continuo uite finem habebit.

[3] Si autem non secundum figuram equicruriam hoc acciderit, sed prope superferantur locis luminarium maleficorum | radii, duo autem sint malefici, <et> uel alterum luminarium uel etiam ambo impediant, uel secundum superallationem uel secundum diametrum, uel in parte [quam] alterum alter, uel alter

---

V (**α** =**YΣ** **β** =**DMS** **γ** =**ABC**) 368 formosum: εὔσχημον *cett.*] εὐσχήμων **V** 371 hec*: *an* τούτων *legit* **G**?] τοιούτων *codd.* autem (**α M S** Heph. «fort. recte»)] + ὁ (τοῦ) Ἑρμοῦ **V D γ** (*Hüb.*) 376 coheret: ἔχεται *cett.*] ἔρχεται **V** (modus] + οὗτος *codd.*) 377 uite (**Σ M S**, *cf.* Heph.)] + λόγου **V Y D γ** (*Hüb.*) 378 quia: παρὰ *cett.*] περὶ **V** 381 hiis] τούτου *codd.* menses] + εἰσι *codd.* 384 quidem: μὲν **α** Heph. (*Hüb.*), *cf.* **M S**] *om.* **V D γ** 386 uel etiam: ἢ καὶ **α β γ**] ἢ **V** Heph. (*Hüb.*) 390 acciderit: συμβαίνῃ **α β γ** (*Hüb.*)] συμβαίνει **V**

---

369 ermafroditas] eruntfronditas *Ox*: *an* ermafroditos *scrib.* ?, *cf. infra* 787 371 hec*] hoc *Ox* 376 hac] hanc *Ox* 377 hac] h^a^c *Ox* 380 regyratio] regryatio *ex* regriatio *Ox* 383 corrumpuntur] -piuntur *Ox* 385 maleficorum] -ciorum *Ox* 386 hec] hoc *Ox* 387 crurum] curium *Ox* beneficorum] -ciorum *Ox*

quidem diametralis sit, alter autem superallatus: etiam ita non annalia fiunt, [a] multitudine malorum exterminante | quod ex distantia superallationis ad permanentiam uite uidebatur amiciter.

[4] Impedit autem singulariter secundum superallationes quidem solem quidem Mars, lunam autem Saturnus, secundum diametrizationes autem et superationes econuerso solem quidem Saturnus, lunam autem | Mars, maxime autem si optineant localiter dominium luminarium et horoscopum. Si autem et due fuerint diametrises, luminaribus existentibus angularibus et maleficis secundum equicrurium, tunc et mortua uel semimortua pariuntur procreata.

[5] Hiis autem ita se habentibus, si quidem ab aliquo beneficorum extiterint defluxum habentia luminaria uel etiam si aliter ipsis sint configurata, partibus tamen se precedentibus radios superferant, uiuet natum pro numero graduum intermedio inter dimissionem et | propinquiores maleficorum radios, tot mensibus uel diebus uel etiam horis secundum magnitudinem mali et uirtutem causantium.

[6] Si autem maleficorum quidem radii ferantur in precedentia luminarium, beneficorum autem in | consequentia, quod procreatum est suscipietur et uiuet. Et iterum, si configurati quidem benefici obtineantur a maleficis, in malignitatem et subiectionem uiuet, si autem obtinuerint, in submissionem aliorum parentum. Si autem et beneficorum quidem aliquis uel ortum | uel contactum faciat ipsi lune, maleficorum autem sub occasu <sit aliquis - - [7] -> sit aliquid astrorum que secundum duo uel plura configurantur, semimortuum aliquid uel frustrum carnis et imperfectum | procreabitur quod nascitur; si autem a maleficis optinetur, sine cibo uel non annale erit quod procreatum est sub causa huius.

V (**α** =**YΣ** **β** =**DMS** **γ** =**ABC**) 394 diametralis sit: διάμετρος ᾖ **V**, *cf.* Heph.] διαμετρῶν **α β γ** (*Hüb.*) (398 et: καὶ **Y**] ἢ **V Σ** Heph., *Hüb.*: ἢ καὶ *cett.*) 399-400 maxime autem si optineant localiter dominium luminarium et horoscopum: καὶ μάλιστα ἐὰν κατάσχωσι τοπικῶς τῶν τὰ φῶτα καὶ τὸν ὡροσκόπον οἰκοδεσποτησάντων **Σ β γ**] κἂν κατάσχωσι τ. ἤτοι τὰ φῶτα ἢ τὸν ὡροσκόπον οἰκοδεσποτήσαντες **V Y**: *lect.* **Σ β γ** *ad sensum redd.* **G**: *cf. Hüb.* (*in initio* = **Σ β γ**; *in fine* = **V Y**) 400 et due: καὶ δύο **α β γ** Heph. (*Hüb.*)] δύο **V** (405 pro: ἀντὶ **V Y**] ἄχρι **Σ β γ**) 409 quod procreatum* est] *an* τὸ γεννώμενον ἐκτεχθὲν **Y** Heph. (2,28,4: *cod.* P) *legit et inde* ἐκτεχθὲν *non redd.* **G** ?: τὸ γεννώμενον ἐκτεθὲν *cett.* (*Hüb.*) 412 (quidem] *om. codd.*) uel[1] **Σ β** Procl.] *om.* **V Y γ** (*Hüb.*) 413 <sit aliquis ...> sit aliquid] τις εἴη ὑπ' αὐτῶν — ὑπὸ δύσιν τις ᾖ *codd.* (541°-543°), *fort. in trad. lat. exciderunt* 415 quod nascitur: τὸ γεννώμενον *cett.*] τὸν γ. **V**

---

394 annalia] almalia *Ox* fiunt] fuerit *Ox* 395 exterminante] -tia *Ox* 396 amiciter] amic[ꝭ] *Ox*: *an* amicabile *scrib.* ? 397 secundum] super *Ox* 398 lunam] lumina *Ox* 399 lunam] lumina *Ox* 401 diametrises] -trices *Ox* 402 pariuntur] -unt *Ox* 403 beneficorum] -iorum *Ox* 406 maleficorum] -iorum *Ox* 407 causantium] cauenter (*dub.*) *Ox* 409 beneficorum] maleficorum (!) *Ox* *an* expositum *post* procreatum* *excidit* ? 410 benefici] -cii *Ox* 410-1 malignitatem ... subiectionem] -te ... -ne *Ox*; *cf.* submissionem 412 et beneficorum] *inv.* *Ox* 413 <sit aliquis ... >] *verba fort. per hom. om. Ox* 414 frustrum *sic Ox*

De annis uite
Capitulum 12^m [11]

[1] Simptomatum autem que post natiuitatem precedit quidem qui de ˈ annis uite sermo, quoniam secundum antiquum ridiculum est per singulos euentus adaptare ei qui omnino non perueniet ex subsistentia uitalium annorum ad tempora ipsorum effectiua.

[2] Consideratur autem iste non simpliciter neque absolute, ˈ sed a predominio principalissimorum locorum multiformiter accipitur. Est autem qui maxime consonus nobis et alias coherens nature modus talis. Dependet quidem enim omne ex perceptione dimissoriorum locorum et ex perceptione ipsorum qui dominantur super dimissionem, ˈ et adhuc ex perceptione peremptiuorum locorum uel astrorum. Discernitur autem unumquodque sic.

[3] Loca primo quidem existimandum dimissoria, in quibus esse oportet modis omnibus accepturum dominium dimissionis: signum scilicet quod circa horoscopum, a quinque ˈ <gradibus> preeleuatis super orizontem usque ad reliquos uiginti quinque gradus qui eleuantur; et dextros sextiles hiis triginta gradibus qui dicuntur boni demonis; et quadrangulares medii celi super terram; et trigonales qui uocantur dei; et diametrales ˈ ipsius occidentis. Preiudicatis etiam in hiis secundum uirtutem dominationum primo gradibus acceptis secundum medium celi super terram; deinde hiis qui apud orientem; deinde illis qui sunt penes superallationem medii celi; deinde hiis qui apud occasum; deinde hiis <qui> precedunt medium celi.

[4] Omne enim quod sub terra merito repudiandum ad tantum dominium, exceptis solis prodeuntibus in lucem penes ipsam eleuationem. Eius autem quod super terram neque signum quod inconnexum est orienti congruit assumere, neque quod preortum est, quod uocatur mali ˈ demonis, quoniam quidem astrorum que in ipso defluxum ad terram, preter hoc quod deciderunt, obfuscat et quasi exterminat quod euaporat ex humiditatibus terre grossum et caliginosum,

---

V (**α** =**YΣ** **β** =**DMS** **γ** =**ABC**) 423 iste: οὗτος **Σ β γ** Procl. (*Hüb.*)] οὕτως **V Y** 424 Est (*cett.*)] ἔτι **V** 426 locorum **Y β γ** Heph. Procl. (*Hüb.*)] *om.* **V Σ** 434 ipsius occidentis: τοῦ δύνοντος *cett.*] οὐ δύνοντος **V** in: ἐν *cett.*] ἐκ **V** 435 dominationum: τῶν ἐπικρατήσεων **Σ**] (τῆς) ἐπικρατήσεως **V Y β γ** Heph. (*Hüb.*) (gradibus acceptis: τῶν **S γ**, *Hüb.*] *om. cett.*) 436 hiis: τῶν *cett.*] τὴν **V** 438 medium celi: τοῦ μεσουρανήματος *cett.*] μορίου **V** 440 eleuationem: ἀναφορὰν **α M S** (*Robb.*)] ἐπαναφορὰν **V D γ** (*Hüb.*) 442 quod² (**Y D γ**, *Hüb.*)] καὶ **V**: ὃ καὶ *cett.* quoniam quidem: ἐπειδήπερ **Σ β γ** (*Hüb.*), *cf.* ἐπείπερ **V**, ὅπερ **Y**] + κακοῖ **V Γ** (*Hüb.*): + κακῇ **Y** (443 preter hoc quod deciderunt] μετὰ τοῦ καὶ ἀποκεκλικέναι *codd.*)

---

422 effectiua] -tuū *Ox* 424 multiformiter] muttiferuntur *Ox* 433 demonis] hominis (hōīs) *Ox* 436 orientem] *ss. Ox*, *exp.* orizontem 437 precedunt] proc- *Ox* 442 preortum] prioratum *Ox* uocatur] -tus *Ox* demonis] hominis (hōīs) *Ox*

propter quod et coloribus et magnitudinibus non secundum naturam habentia uidentur.

[5] Post hec autem iterum dimissores accipiendum principalissimos quatuor (solem, lunam, horoscopum, sortem fortune) et dominatores horum. Sortem quidem fortune collectam semper nocte et die a numero qui est a sole ad lunam et | equalia ferente ab horoscopo secundum consequentia signorum, ut quam habet rationem et figurationem sol ad orientalem orizontem, hanc habeat et luna ad sortem fortune et sit quasi lunaris horoscopus.

[6] Forte autem idem hoc est et uult apud illum qui scripsit, | quod in natis de nocte numerandum est a luna ad solem et econuerso ab horoscopo, id est in precedentia, proicere. Et ita enim et illo modo idem locus sortis et eadem ratio figurationis eueniet.

[7] Preiudicandum autem ex hiis de die quidem solem, siquidem fuerit in locis dimissoriis, si autem non, lunam, si uero non, habentem plures rationes dominationis ad solem et ad precedentem coniunctionem et ad horoscopum; | pleniluniari autem, sortem fortune.

[8] Si autem luminaria uel etiam proprie partitionis dominus in dimissoriis locis extiterint, tunc luminarium quod in principaliori loco assumendum. Tunc autem | solum dominatorem amborum preiudicandum, quando principaliorem habuerit locum et ad ambas partitiones dominationis habuerit rationem.

[9] Dimissore autem diiudicato, adhuc dimissionum duos modos accipiendum, eum scilicet qui in consequentia | signorum solum sub uocata radiorum proiectione, quando dimissor fuerit in locis orientalibus (hoc est in hiis que a medio celi [ita se habentibus] ad horoscopum), et eum qui non solum in consequentia sed

---

V (**α** =**YΣ** **β** =**DMS** **γ** =**ABC**) 447 quatuor] + τόπους *codd.* 450 et equalia: καὶ ἴσα **α β** (*Hüb.*)] εἰς τὰ ἴσα **V**; *cf.* **γ** 451 orientalem orizontem: ἀνατολικὸν ὁρίζοντα **α M S** Heph.] ὡροσκόπον **V D** (*Hüb.*); *cf.* **γ** 453 est et uult (ἐστί τε καὶ θέλει ?)] θέλει τε (*om.* **Y**) καὶ δύναται **V Y** (*Hüb.*): ἐστί τε καὶ δύναται **Σ D M** 455 et illo modo: κἀκείνως **M**] κἀκεῖνος **V α D S γ** (*Hüb.*) 457 quidem] + πρῶτον *codd.* 459 horoscopum*] + τουτέστιν — τὸν ὡροσκόπον (*cum l. v.*) *codd.*, *vel in trad. gr. vel in trad. lat. per hom. exciderunt* 461 autem] + καὶ ἀμφότερα *codd.* 465 modos: τρόπους **α β γ** Heph. (*Hüb.*)] τόπους **V** (*cf. Boer*) 468 (horoscopum] + τόποις *codd.*, *fort. om.* **G**) 468-9 sed etiam (**α M S**, *Hüb.*)] + τὸν **V D γ**

---

445 quod et] *inv. Ox* 447 hec] hoc *Ox* iterum] inter *Ox* dimissores] + .i. yleg *ss. Ox*[2] (*e commentario arabo-latino in marg. inf., cf.* 458) 448 solem lunam] sole luna *Ox* 449 qui] quod *Ox* 450 ab horoscopo] ad horoscopum *Ox* 453 uult] uidelicet *Ox* 458 dimissoriis] -ris *Ox*, + .i. yleg *ss. Ox*[2] 459 coniunctionem] -iuct- *Ox* *an* post horoscopum* *verba fere haec per hom. exciderunt* <hoc est quando dominatiuorum modorum quinque existentium tres habet ad unum uel etiam plures dictorum; si autem non, ultimo horoscopum; de nocte autem primo lunam, deinde solem, deinde plures habentem rationes dominationis ad lunam et ad precedentem plenilunium et ad sortem fortune; si autem non, ultimo, coniunctionali quidem existente precedente coniugatione horoscopum> ? 463 solum] solem *Ox* 467 est] enim *Ox* 468 [ita se habentibus] *haec verba fort. ex alio loco perperam hic inserta sunt* (*cf. supra* 403, *infra* 471) ad] aliud *Ox*

etiam in precedentia, secundum eam que dicitur orimea, quando dimissor fuerit in locis decidentibus a ǀ medio celi.

[10] Hiis autem ita se habentibus, peremptorii fiunt gradus secundum dimissionem quidem que est in precedentia signorum: qui occidentalis orizontis solus, quia occultat dominum uite, gradus autem ǀ astrorum que sic occurrunt uel testificantur auferunt solum uel [occurrunt] apponunt annos collectis a dimissore usque ad occasum dimissoris, et non perimunt, quia ipsa non feruntur ad locum dimissorium, sed ille locis ipsorum. Et apponunt quidem benefici, auferunt enim malefici. Mercurio rursum ǀ apponente quibuscumque ipsorum configuretur.

[11] Numerus autem appositionis uel ablationis consideratur per graduum situm penes unumquodque. Quotcumque enim fuerint oraria [per] tempora gradus uniuscuiusque (existente quidem die que diei, nocte autem que noctis), tot annorum multitudo erit ǀ summa. Quod quidem ipsis in oriente existentibus supputandum, deinde secundum proportionem abscessionis auferendum, donec utique ad occasum in nichil decidat.

[12] Secundum dimissionem autem que est in consequentia signorum, perimunt maleficorum loca, Saturni scilicet et Martis, ǀ uel corporalibus occurrentibus uel radios inferentibus undecumque quadrangulares uel diametrales; aliquando autem perimunt et ab obedientibus uel aspicientibus secundum equipotentiam sextilibus. Et ipse etiam qui est dimissorio loco tetragonalis a consequentibus; aliquando autem et in signis multum tardantibus ǀ malignus factus qui sextilis, in signis autem parum tardantibus trigonalis. Luna autem cum dimiserit, solis locus perimit. Possunt enim secundum talem dimissionem occursus et perimere et saluare, quoniam hii loco dimissoris superferuntur.

[13] Non semper tamen hec ǀ loca perimere existimandum, sed solum quando fuerint malignata. Impediuntur enim, siue in terminos beneficorum inciderint, siue beneficorum aliquis radios intulerit tetragonales uel trigonales uel diametrales uel ad ipsum gradum peremptorium uel in consequentia ipsius ǀ (in Ioue quidem non super duodecim gradus, in Venere autem non super octo); siue autem corporum amborum, dimissoris scilicet et occurrentis, non eadem latitudo fuerit utrorumque.

V (**α** =**YΣ** **β** =**DMS** **γ** =**ABC**) 471 gradus: μοῖραι *cett.*] μόνον **V** 474 uel (**Σ β γ**)] καὶ **V Y** Heph. Procl. (*Hüb.*) a dimissore: ὑπὸ τοῦ ἀφέτου **Σ β γ**] *om.* **V Y** Heph. (*Hüb.*) (485 corporalibus*] σωματικῶς *codd.*) 487 ab: ἀπὸ **Σ** (ἐπὶ *ss.*) **β** Heph. (*cod.* P)] ἐπὶ **V Y γ** Procl. (*Hüb.*) 488 sextilibus: ἑξαγώνων **V Y** Procl. (*Boer Robb.*)] *-νον* **G**º **Σ β γ** Heph. (*Hüb.*) 489 in: ἐπὶ **S γ** Heph.] ἐπὶ μὲν **V Y** (*Hüb.*): ἀπὸ **Σ D M** 491 perimit: ἀναιρεῖ **Σ β γ** Heph. (*Hüb.*)] *om.* **V Y** Procl. 493 loca] + καὶ (*om.* **V**) πάντως *codd.* (494 terminos beneficorum] ἀγαθοποιοῦ ὅριον *codd.*)

---

469 que] qui *Ox* 472 que] qui *Ox* 473 sic] non *Ox* 474 dimissore] -orie *Ox* 476 enim] *an* autem *scrib.* ? 480 que^{1}] qui *Ox* 482 supputandum] suppit- *Ox* 485 corporalibus*] *an* -liter *scrib.* ? 489 qui] que *Ox* 490 solis] *ex* solus *corr.* *Ox* 492 hii] hoc *Ox* 497 non^{1}] ut *Ox*

[14] Cum igitur etiam duo uel plura fuerint utrimque, adiuuantia scilicet et per conrarietatem perimentia, considerandum | predominationem alterutrius specierum secundum multitudinem eorum que coassumuntur ipsis et secundum uirtutem; secundum multitudinem quidem, quando sensibiliter altera fuerint plura alteris, secundum uirtutem autem, quando adiuuantium uel perimentium astrorum hii quidem in propriis fuerint [in] locis, hii autem non; | maxime autem quando hii quidem fuerint orientales, hii autem occidentales.

[15] Vniuersaliter autem existentium sub radiis nullum assumendum neque ad peremptionem neque ad auxilium, excepto si, cum luna fuerit dimissor, ipse solis locus perimat, peruersus quidem a simul existente malefico et a nullo | benefico auxiliatus.

Numerus tamen annorum, quem faciunt gradus intermediarum distantiarum dimissorii loci et perimentis, neque simpliciter neque ut contingit debet accipi secundum multorum traditiones ex | temporibus eleuationum uniuscuiusque gradus, nisi solum quando uel ipse orientalis orizon dimissionem susceperit uel aliquis facientium ortum secundum ipsum.

[16] Proposito enim uno modo ex omnibus considerare oportet, post quot equinoctialia tempora | regulariter pertransire et orizontem et meridianum, ad quos ambos localium distantiarum similitudines accipiuntur; accipitur <autem> unumquodque temporum ualere | annum unum solarem merito [autem]. Quando quidem in orientali ipso orizonte fuerit dimissorius et principalis locus, eleuationum tempora graduum usque ad obuiationem conuenit accipere.

[17] Post tot enim equinoctialia tempora peremptor ad | locum dimissoris, hoc est ad orientalem orizontem, peruenit. Quando autem in ipso meridiano, eleua-

---

V (**α** =**YΣ** **β** =**DMS** **γ** =**ABC**) 500 etiam* ... uel: ἤ καὶ **α β γ**] καὶ **V** (*Hüb.*) 506 fuerint: ὦσιν *cett.*] εἰσὶν **V** 507 radiis: αὐγὰς *cett.*] αὐτὰς **V D** 509 et a nullo: ὑπὸ μηδενὸς δὲ **α**: καὶ ὑπὸ μηδενὸς Procl. (189.14)] διὰ μηδενὸς δὲ **V γ** (*Hüb.*), *cf.* **β** 510 auxiliatus: βοηθούμενος **D**] + ἤ ἀναλελυμένος **Σ M S γ**: ἀναλελυμένος **V Y** Heph. Procl. (*Hüb.*) 511 gradus: αἱ <μοῖραι> *suppl. Boer Hüb.*] αἱ *codd.* 512 loci (*cett.*)] *om.* **V** 513 eleuationum] + παντότε *codd.* 513-4 uniuscuiusque gradus: ἑκάστης μοίρας **α** Procl. (*Hüb.*)] ἑκάστη μοίρα **V**: -στας -ρας **β γ** 514 susceperit: ᾖ εἰληφὼς *cett.*] εἰληφὼς **V** 516 omnibus*] + τῷ φυσικῶς τοῦτο τὸ μέρος ἐπισκεπτομένῳ (*l. v.* **Y**) *codd.*; *an in trad. lat. exciderunt?* (516 oportet] *om. codd.*) 517 tempora* (**Y**)] + ὁ τοῦ — τοὺς ἰσημερινοὺς χρόνους **V Σ β γ** (684-7°); *per hom. vel in trad. gr.* (*cf.* **Y**) *vel in trad. lat. exciderunt* (518 accipitur] *om. codd.*) 521 graduum: μοιρῶν *cett.*] -ας **V**

---

500 etiam] *fort. transponend. post* uel 504 alteris] alterius *Ox* 505 hii[1,2]] hoc (h°) *Ox*; *an* hec *scrib.?*, *sed cf.* hii 506 507 autem] *an* enim *scrib.* ? 508 si] *ex* non *corr. Ox* 511 tamen] autem *ante corr. Ox* 513 uniuscuiusque] unuscuius- *Ox* 516 *an post* omnibus* *verba* <ei qui naturaliter hanc partem considerat> *exciderunt* ? 517 *an post* tempora* *verba translatoris per hom. exciderunt* ? 522 peremptor] perentior *Ox*

tiones in recta spera, in quot unaquaque decisio pertransit meridianum. Quando autem in occidentali ipso orizonte, in quot unusquisque | graduum distantie deorsum fertur, <hoc est> in quot diametrizantes hiis sursumferuntur.

[18] Principali autem loco non iam existente in tribus locis hiis, sed <in> intermediis distantibus, non iam premissarum eleuationum uel descensionum uel celi mediationum tempora | ferent loca sequentia super principalibus, sed diuersa. Similis quidem enim et idem locus est qui similem positionem et ad easdem partes habet simul ad orizontem et ad meridianum; hoc autem propinquissime accidit posito in uno semicirculo descriptorum | per decisiones meridiani et orizontis, quorum unumquodque secundum eandem positionem equalem propinquissime temporalem horam facit.

[19] Sicut autem, si circumducatur circa dictas incisiones, uenit quidem ad eandem positionem et orizonti et meridiano, | tempora autem pertransitionis zodiaci inequalia in utroque facit, eodem modo et secundum positiones aliarum distantiarum per inequalia illis tempora processus efficit.

[20] Vna autem quedam nobis ars erit talis, per quam — siue orientalem siue meridianam aut occidentalem siue aliam | aliquam habeat positionem principalis locus — accipietur proportio temporum ad ipsum ferentium sequentem locum. Preaccipientes enim gradum zodiaci mediantem celum et adhuc principalem et superuenientem, primo consideremus principalem | positionem, quot temporalibus horis distat a meridiano, numerantes eleuationes inter ipsum et gradum celum mediantem conuenienter siue <super terram siue> sub terra in recta spera, et partientes in multitudinem temporum horariorum ipsius gradus precedentis, si quidem | super terram fuerit, diurnalium, si autem sub terra, nocturnalium.

[21] Quoniam autem que secundum easdem temporales horas distant a meridiano decisiones zodiaci in uno et eodem fiunt predictorum semicirculorum, oportebit accipere post quot equinoctialia tempora et sequens | decisio per tot horas temporales distabit a meridiano eodem cum principali. Has autem cum ac-

---

V (**α** =**YΣ** **β** =**DMS** **γ** =**ABC**) 533 unumquodque (*sic*): ἕκαστον (*sc.* ἡμικύκλιον) **α M S** Heph.] ἕκαστος **V D γ** 537 utroque: ἑκατέρου **α β γ** Heph. (*Hüb.*)] -ον **V** (*Robb.*) 539 quedam nobis (**α β γ**)] *inv.* **V** erit: ἔσται **α β γ**] ἔστω **V** Procl. (*Hüb.*) 548 Quoniam autem: ἐπεὶ δὲ **Σ β γ**] ἐπὶ δὲ **Y**: ἐπειδὴ **V** Heph. (*Hüb.*) 550 oportebit (**Σ β γ** Heph., *Hüb.*)] καὶ *praem.* **V Y** tempora: χρόνους **Y** Heph. Procl. (*Hüb.*), *alio loco* **γ**] *om.* **V Σ β** et sequens decisio per tot (*cett.*)] *om.* **V** 551 Has autem cum acceperimus: ταύτας δὲ εἰληφότες (*coni. Hüb.*)] ταύτας δὲ ἃς εἰλήφαμεν **V Y** (*Boer*): ταύτας γὰρ διειληφότες **Σ M S**: ταύτας διειληφότες **D γ**, *cf. Robb.*

---

524 decisio] deficio *Ox* 525 occidentali] orientali *Ox* 527 non] ut *Ox* 528 descensionum] -tionum *Ox* 529 ferent] fierent *Ox* 530 Similis] simul *Ox* positionem] ratio- (*abbr.*) *Ox*[ac] 532 posito] *an* positis *scrib.* ? 533 positionem] ratio- (*abbr.*) *Ox* 536 pertransitionis] -transio- *Ox* 542 gradum] -uum *Ox* 543 principalem] *an* -palis *scrib.* ? 546 multitudinem] -e *Ox* 549 fiunt] sunt *Ox*, *cf.* I.938* 550 decisio] desisio *Ox*[ac]

ceperimus, considerabimus quot temporibus equinoctialibus secundum eam que a principio positionem distabat sequens gradus a gradu secundum idem celi medium, per eleuationes in recta iterum spera, | et quot quando faciebat temporales horas equales cum principali.

[22] Multiplicantes autem et ipsas in multitudinem horariorum temporum sequentis gradus (si quidem fuerit ad medium celi super terram rursum comparatio, temporalium horarum penes diurnalia, si autem ad id quod sub | terra, ad nocturnalia) et ea que fiunt ex excessu ambarum distantiarum sumentes, habebimus quesitorum annorum multitudinem.

[23] Vt autem manifestius fiat quod dicitur, supponatur quidem principalis locus principium Arietis exempli | gratia, sequens autem principium Geminorum, clima autem ubi maxima quidem dies est horarum 14, horaria autem magnitudo principii Geminorum 17 temporum equinoctialium proxime. Et oriatur primo principium Arietis, ut mediet celum principium Capricorni, et distet | a medio celi super terram principium Geminorum temporibus equinoctialibus 148.

[24] Quoniam igitur principium Arietis distat a medio celi per temporales horas 6, has multiplicantes in 17 tempora que quidem erant magnitudinis orarie principii Geminorum, | quoniam ad medium celi super terram est distantia 148 temporum, habebimus et huius distantie tempora 102; post excessus igitur tempora 46 sequens locus ad locum precedentis perueniet. Tot autem sunt proxime tempora et eleuationis et | Arietis et Tauri, quoniam dimissorius locus supponebatur ascendens.

[25] Sit autem in medio celi [sunt] similiter principium Arietis, ut distet secundum positionem <primam> principium Geminorum a medio celi super terram temporibus equinoctialibus 58. Quoniam igitur secundum | positionem secundam debet mediare celum principium Geminorum, habebimus distantiarum excessum ipsorum 58 temporum, in quot iterum, quia locus dimissorius mediat celum, pertransit meridianum Aries et Taurus.

V (**α** =**YΣ** **β** =**DMS** **γ** =**ABC**) 563 quidem (*cett.*)] *om.* **V** dies est horarum: ἡμέρα ὡρῶν ἐστὶν *cett.* (*Hüb.*)] ὥρα ἐστὶν ὡρῶν **V** 572 et[1]: καὶ **α M S** (*Hüb.*)] τοῦ **V D γ** 574 Sit ... in medio celi: μεσουρανείτω *cett.* (*Hüb.*)] si ... in medio celi sit*: μεσουρανήματος **V** 577 principium Geminorum: ἡ ἀρχὴ τῶν Διδύμων (Procl.) *coni. Hüb.*] ἡ ἀρχὴ τῶν **V**: ὁ ἀφετικὸς τόπος **Σ γ**: *om.* **Y β** habebimus: ἕξομεν *cett.*] ἕξω μὲν **V** 578 ipsorum: αὐτῶν τῶν **Y** (*Hüb.*)] τῶν τῶν **V**: αὐτὴν τὴν τῶν **Σ D M γ**, *cf.* **S**

---

556 in] et *Ox* 557 comparatio] operatio *Ox* 563 quidem] *ss. Ox* 564 principii] -pali *Ox* 568 quidem] -dam *Ox* 570 habebimus (ἕξομεν, *cf.* 559)] habemus *Ox* 574 Sit ... in medio celi] si ... in medio celi sunt *Ox*; *an* si ... in medio celi sit* *scrib.*? 575 principium] -iorum *Ox* 578 pertransit] per + *spat. 4 litt. Ox*

Occumbat autem eodem modo principium Arietis, ut celum quidem mediet principium Cancri, distet autem a medio celi super terram principium Geminorum in precedentia tempora equinoctialia 32. [26] Quoniam igitur rursum per 6 horas temporales distat a meridiano principium Arietis uersus | occasum, si has multiplicemus in 17, habebimus 102 tempora, quibus distabit a medio celi principium Geminorum, quando occubuerit. Distabat autem et secundum primam positionem ad eadem tempora 32; in septuaginta igitur temporibus excessus ad occasum peruenit, in quibus et deorsum feruntur quidem | Aries et Taurus et eleuantur diametralia signa, Libra scilicet et Scorpius.

[27] Supponatur igitur in nullo quidem angulorum esse principium Arietis, distans autem, exempli gratia, in precedentia meridiei per horas temporales tres, ut celum | quidem mediet Tauri gradus 18[us], distet autem secundum <primam> positionem Geminorum principium a medio celi super terram in consequentia temporibus equinoctialibus 13. Si igitur rursum <17 tempora per 3> horas multiplicemus, distat quidem et secundum <secundam> positionem | Geminorum principium a meridie <in> precedentia temporibus 51, omnia autem faciet tempora 64.

[28] Faciebat autem per eandem artem, quando quidem oriebatur dimissorius locus, tempora 46, quando autem mediabat celum, tempora 58, quando uero occumbebat, tempora 70. Differt quidem igitur qui secundum positionem mediam | inter medium celi et occasum numerus temporum ab unoquoque aliorum; est enim temporum 64; differt autem secundum proportionem excessus trium [annorum] horarum, quoniam hic in aliis quidem secundum angulos quartis 12 temporum erat, in distantia autem trium horarum | 6 temporum.

[29] Quoniam autem et in omnibus idem fere proportionaliter conseruatur, possibile erit et hoc modo simpliciori arte uti. Rursum enim in oriente quidem precedente gradu eleuationibus usque ad sequentem | coutemur, mediante autem celum eleuationibus in recta spera, occumbente autem hiis que secundum diametrum. Quando autem intermedio horum fuerit, ut uerbi gratia in exposita

---

V (**α** =**YΣ** **β** =**DMS** **γ** =**ABC**) 580-1 Arietis — principium (*cett.*)] *om.* **V** 583 6: ἓξ **Σ β γ** (*Hüb.*), *tr.* **Y**] καὶ **V** distat: ἀπέχει *cett.*] ἀπέχη (*sic*) **V** 589 quidem ... esse: μὲν οὖσα ... *cett.*] μένουσα ... **V** (594 distat*: ἀφίσταται **V**] ἀφέξει **α β γ**, *Hüb.*) 598 mediabat: ἐμεσουράνει *cett.* (*Hüb.*)] ἐμεσουράνῃ **V** 599 occumbebat: ἔδυνε **Σ β γ** (*Hüb.*)] ἔδεινε **V**: δύνῃ **Y** 600 medium celi: μεσουρανήσεως **α D M γ** (*Hüb.*)] -ούσης **V S** 605 et: καὶ **Y** (*Robb.*), *cf.* **V**] *om.* **Σ β γ** (*Hüb.*) 607 hiis que secundum diametrum] ταῖς καταφοραῖς *codd.*: *an* ταῖς καταφοραῖς *ad sensum redd. vel* ταῖς κατὰ διάμετρον *e glossa legit* **G**? 608 fuerit: ᾖ *cett.*] ἦν **V**

---

585 quando] *ss. Ox* Distabat] -bit *Ox* 591 mediet] -dūt *Ox* 592 positionem] pocionem (?, *e corr.*) *Ox* 593 13] 12 *Ox* 594 distat*] *fort.* distet *scrib.* 595 omnia] + sunt (*exp.*) *Ox* 598 mediabat] -iebat *Ox* 599 70] 10 *Ox* 602 annorum horarum] *an* **G** *primo perperam* annorum *loco* horarum *scr. et non delevit*? 605 arte] parte *Ox* 606 coutemur] conte- *Ox*

distantia Arietis, accipiemus primo tempora incidentia utrique continentium angulorum, inueniemus autem (quoniam | post medium celi super terram supponebatur principium Arietis, inter angulum medium celi et occumbentem) incidentia tempora usque ad principium Geminorum commediationum quidem celi 58, simul occasuum autem 70.

[30] Deinde addiscentes, ut premissum est, quot temporalibus | horis distat precedens decisio ab alterutro angulorum, quantacumque pars sint hee 6 temporalium quarte horarum, tantam partem excessus ambarum collectionum apponemus uel auferemus ab angulis comparatis; ut puta quoniam presuppositarum 70 et 58 partium | excessus est temporum 12 (supponebatur autem precedens locus distare ab utroque angulorum per equales temporales horas 3, que sunt 6 horarum media pars), accipientes igitur ipsorum 12 dimidium et aut 58 apponentes uel a 70 auferentes, inueniemus adiectionem 64 temporum.

[31] Si uero per | duas temporales horas distabat ab alterutro angulorum, que sunt 6 horarum tertia pars, tertiam partem rursum temporum 12 excessus, hoc est tempora 4, si quidem distantia duarum horarum a medio celi supponebatur, apponeremus utique 58 temporibus, si autem ab occasu, auferremus | utique a 70. Modus autem igitur quantitatis temporalium distantiarum ita nobis secundum consequentiam debet accipi.

[32] Diiudicabimus autem de cetero in unaquaque predictarum obuiationum uel etiam occasuum, secundum eum qui a parum tardantibus | ordinem, peremptorias et afflictiuas et aliter transitorias per malignizatam esse obuiationem uel per auxiliis subleuatam secundum predictum nobis modum et per temporales superingressus secundum unumquemque significatorum ex obuiatione.

[33] Malignizatis enim | simul locis et transitu astrorum ad ingressum annorum malefaciente principalissima loca, e regione mortes existimandum; altero autem horum amicabili, afflictiones magnas et <in>certas; ambobus autem, in-

---

V (**α** =**YΣ** **β** =**DMS** **γ** =**ABC**) 612 principium: τῆς ἀρχῆς **Y** Procl. (*Hüb.*)] *om.* **V Σ β γ** quidem: μὲν **Y S** (*Hüb.*)] μὲν μ° **V**: δὲ **Σ D M γ** 613 simul occasuum: συγκαταδύσεων **α β γ** (*Hüb.*)] καταδύσεων **V** 614 distat: ἀπέχει **Σ S γ** (*Hüb.*)] ἀπέχῃ **V Y D M** 615 angulorum (*cett.*)] *om.* **V** 617 presuppositarum: προυποκειμένων **V**] προκειμένων **α β γ** (*Hüb.*) 620 (igitur] καὶ **V Y D S γ**: δὲ καὶ **Σ M**) 12: ιβʹ **Y β** (*cf.* Heph. Procl.)] + δύο **V**: + χρόνων **Σ γ** (*Hüb.*) et aut: καὶ ἤτοι **Σ M S** (*Robb.*)] ἤτοι **V Y D γ** (*Hüb.*) 624 a medio celi: μεσουρανήματος **Σ β** Heph. (*Hüb.*), *cf.* **Y**] -οῦντος **V γ** 626 distantiarum: διαστάσεων **Y γ** (*Hüb.*)] -εως **V Σ β** 629 parum tardantibus: ὀλιγοχρονιωτέρων **Y S** Procl. (*Hüb.*), *cf.* **V**] -ωτάτων **Σ D M γ** 631 predictum: προειρημένον **α M S** Procl. (*Hüb.*)] εἰρημένον **V D γ**

---

612 commediationum] cum mediatione *Ox* 615 hee] *ss. Ox*, *exp.* hore 625 auferremus] -feremus *Ox* 626 autem] *an* quidem *scrib.* ? 627 consequentiam] continentiam *Ox* 628 obuiationum] obuicionum *Ox* 631 predictum] -tam *Ox* superingressus] *disiuncte Ox* 632 significatorum] sig[a]turum *Ox* 633 Malignizatis] -tatis *Ox* astrorum] temporum *Ox*

hertias solum uel nocumenta uel deiectiones | transitorias, assumpta et in hiis proprietate a coaptatione occursantium locorum ad res natiuitatis.

[34] Nichil autem aliquando prohibet, dubitantibus qui debeant accipere dominium peremptiuum, obuiationes per singulas ipsorum ratio<ci>nantes, uel | per consonantes maxime ad species simptomatum que euenerunt prosequi ad futuras, uel ad omnes secundum equalitatem uirtutis ualentes obseruatiue se habere, magis et minus ipsorum secundum eundem modum considerantes.

De forma et complexione corporali
Capitulum 13^m [12]

[1] Tradito autem negotio sermonis de annis uite, dicamus, principium particularium accipientes secundum conuenientem ordinem, de forma et corporali figuratione, quoniam et que corporis ante illa que | anime figurantur secundum naturam; corpore quidem propter materialius habente fere congenitas propriarum complexionum fantasias, anima autem post hec et paulatim manifestante ydoneitates a prima causa, exterioribus uero adhuc magis posterius secundum | tempus consequens superuenientibus.

[2] Obseruandum igitur uniuersaliter quidem orientalem orizontem et consequentes aut dominationem ipsius accipientes planetas, secundum modum quem diximus, particulariter autem et lunam eodem modo; per | naturam autem formatiuam horum amborum locorum et naturam dominantium et per commixtionem secundum utramque speciem, et per configurationem ipsis coorientium stellarum fixarum que circa configurationes corporum considerantur, antecedentibus quidem uirtute | astris habentibus dominationem, cooperante autem et proprietate ipsorum locorum. Id tamen quod secundum unumquodque et ut utique quis simpliciter ita assignabit, hunc habet modum.

V (**α** =**YΣ** **β** =**DMS** **γ** =**ABC**) 638 dubitantibus: δισταζομένων *cett.*] διαστ- **V** 640 ad[1] (*cett.*)] *om.* **V** species: εἴδη **γ**] ἤδη **V Σ β** (*Hüb.*): ἴδη **Y** 641 omnes: ἁπάσας *cett.*] ἃς πάσας **V** (ualentes] ὡς ... ἰσχυούσας *codd.*) 642 considerantes: ἐπισκεπτομένους *cett.*] ἐπισκεπτόμενα **V** (645 dicamus] λέγομεν *codd.*) 648 habente: ἔχοντος **Σ β A B** (*Hüb.*)] -τες **V D** 652 consequentes: ἕποντας **V Y**] ἑπόντας **Σ β γ** (*Hüb.*) (656 et] + ἔτι *codd.*) 660 habet: ἔχει *cett.*] ἔχῃ **V**

---

640 consonantes] -ans *Ox* 641 obseruatiue] -antie *Ox* 643 corporali] naturali *praem. et exp.* *Ox* 645 Tradito] *an* Tractato *scrib.* ? 648 corpore] corporee *Ox* 649 hec] hoc *Ox* ydoneitates] -ietates *Ox* 650 adhuc] adh° *Ox* 652 consequentes] -tia *Ox* 653 aut] autem *Ox* 654 autem[2]] *an* enim *scrib.* ? formatiuam] -atam *Ox*

[3] Primo enim in astris Saturnus quidem orientalis existens, forma quidem | mellei coloris efficit et mediocris pinguedinis et nigrorum capillorum et crispi capitis et pilosi pectoris et mediorum oculorum, et commensuratos magnitudinibus, complexione autem quod magis habentes in humido et frigido. Occidentalis autem existens, figura quidem nigros et subtiles et paruos et expansorum pilorum et | pauce barbe et paucorum pilorum et nigrorum oculorum, complexione autem quod magis habentes in sicco et frigido.

[4] Jupiter autem dominans in predictis locis orientalis, forma quidem facit albos ad bonam colorationem et medios in capillis et magnorum oculorum et bone magnitudinis | et honorabiles, complexione autem plus habentes in calido et humido. Occidentalis autem existens, colore quidem albos, non tamen ad <bonam> colorationem similiter, extense capillature autem et subcaluastros et mediocriter caluos et mediocres magnitudinibus, complexione autem plus habentes in humido.

[5] Mars autem similiter orientalis, forma quidem facit albedinem commixtam rubedini, bone magnitudinis et bone habitudinis et glaucorum oculorum et hispidos et medie capillature, complexione autem quod plus habentes in sicco et calido. Occidentalis autem existens, forma quidem rubeos simpliciter et mediocres magnitudinibus | et paruorum oculorum et pauce barbe et xantorum capillorum et tensorum, complexione autem quod plus habentes in sicco.

[6] Venus autem consimilia efficit Ioui, tamen ad formosius et gratiosius et ad | muliebrem decentiam magis et magis femininum et minus animosum et tenerosius; singulariter autem oculos facit cum bona decentia subgaudiosos.

[7] Mercurius autem orientalis, forma quidem facit mellei coloris et commensuratos magnitudinibus et bene proportionatos et | paruorum oculorum et medie capillature, complexione autem quod plus habentes in calido. Occidentalis autem existens, forma quidem albos, non tamen ad bonam colorationem, tense autem capillature, mixturam habentes ex nigro et uiridi et exilis uocis et graci-

---

V (**α** =**YΣ** **β** =**DMS** **γ** =**ABC**) 661 existens: ὢν *cett.*] τῶν **V** 662 mediocris pinguedinis: εὐεκτικοὺς **Σ β γ** (*Hüb.*)] εὐεκτίας **V Y** 665 subtiles: σπινώδεις **Σ β γ** (*Hüb.*)] -δης **V Y** 666 pauce barbe: ὑποψίλους **α β γ** (*Hüb.*)] ψίλους **V** paucorum pilorum (ὑπότριχας *redd.* **G**)] ὑπορίθμους **V Y**: ὑπορρύθμους **Σ β γ** (*Hüb.*) 669 medios in capillis: μεσότριχας *cett.*] μεσότειχον **V** 671-2 non tamen ... similiter, ... autem et: οὐκ ... δὲ ὁμοίως, ... τε καὶ *redd.* **G** (*Boer, Robb.*)] οὐκ ... δὲ; ὁμοίως δὲ ... τε καὶ **V Σ γ** (*Hüb.*), *cf.* **Y β** (677-8 sicco et calido] *inv. codd.*) 682 minus animosum (ἀθυμότερον *redd.* **G**)] εὐχυμότερον *codd.* tenerosius: τρυφερώτερον *cett.*] -φώτερον **V** 684 mellei coloris: μελλίχροας *cett.*] τελίχροας **V** 687 tamen*] + ὁμοίως **V Y** Procl. (*Hüb.*): [**Σ β γ**] 688 (autem, *cf.* 672] *om.* **V Y** Procl., *Hüb.*: [**Σ β γ**]) exilis uocis: σπινοὺς **Y S** Procl. (*Hüb.*), *cf.* 707] σπίρους **V**: σπανοὺς **Σ D M γ**

---

672 et[1]] etiam *Ox* subcaluastros] subcaluestres *Ox* 680 tensorum] tonsurerum *Ox*[ac] 682 femininum] fēim (*dub.*) *Ox* tenerosius] temerosus *vel* teneriosus (*abbr.*) *Ox* 683 autem] + angulos (*exp.*) *Ox* 687 *an post* tamen *excidit* similiter ? 688 ex] et *Ox* exilis] ex illis *Ox*

les et obliquorum oculorum et caprinos oculos et subrubeos, complexione autem quod plus | habentes in sicco.

[8] <Co>operantur autem unicuique horum configurati, sol quidem ad magnificentius et melioris habitudinis, luna autem, et maxime quando defluxum ipsius optineant, uniuersaliter quidem ad commensurationem et maiorem gracilitatem et ad | humidius complexione, particulariter autem proportionaliter proprietati illuminationum secundum complexionem in principio libri expositam.

[9] Rursum autem uniuersaliter matutini quidem existentes et apparitiones facientes magna faciunt corpora, stationarii autem primo | fortia et robusta, retrogradi autem incommensurata, secundo autem stationarii debiliora, occumbentes autem ingloria quidem penitus, toleratiua autem malorum et calamitatum.

[10] Et locis autem ipsis cooperantibus ad figurationes maxime plasmationum et ad complexiones, ut diximus. | Vniuersaliter rursum quarta que est a uernali equinoctio ad estiuam uersionem facit boni coloris, bone magnitudinis, bone habitudinis, bonorum oculorum, quod plurimum habentes in humido et calido. Quarta autem ab estiuo tropico usque ad autumpnale equinoctium medii coloris, commensuratos magnitudinibus, bone habitudinis, | magnorum oculorum, hispidos, crispos, quod plurimum habentes in calido et sicco. [11] Quarta autem ab autumpnali equinoctio usque ad yemalem tropicum mellei coloris, graciles, exilis uocis, amplarum spatularum, medie capillature, bonorum oculorum, quod plurimum habentes in sicco et frigido. Que autem ab hyemali tropico usque ad | uernale equinoctium nigri coloris, commensuratos magnitudinibus, tense capillature, pauce barbe, proportionatos, quod plurimum habentes in humido et frigido.

[12] Particulariter autem que quidem humane forme <signorum> que in zodiaco et eorum que extra, bene apta et commensurata | figuris corpora constituunt; que autem alterius forme transfigurant commensurationes corporis ad proprietatem sue forme et secundum quandam rationem assimilant conuenientes

---

V (**α** =**YΣ** **β** =**DMS** **γ** =**ABC**) 689 obliquorum oculorum: λοξοφθάλμους **Y** Procl. (*Hüb.*)] ληξοφ- **V**: ξηροφ- **Γ**: κοινοφ- **Σ**: κοιλοφ- **β γ** caprinos oculos: αἰγώπους **Y**: αἰγόπους Procl. (*cf. Robb.*)] αἰγόπλους **V**: αἰγόποδας **Σ β γ** 703 Quarta: τὸ (*sc.* τεταρτημόριον) *cett.*] τὰ **V** 708 amplarum spatularum: *an* σπαθηνούς *legit* **G** ?] παθηνούς **V** (*Robb., cf. Boer*): παθινούς **Y**: παθεινούς *Hüb.*: σπανούς **Σ γ**: σπανθινούς **M S**: *om.* **D** 715 forme: μορφώσεως *cett.*] φύσεως **V**

---

693 commensurationem] -suronem *Ox*; *an* maiorem commensurationem *scrib.* ? 694 humidius] habitudinis *Ox* complexione] -nem *Ox* 695 complexionem] -num *Ox* 698 occumbentes] Et cumbentes *Ox* 699 toleratiua] cole- *Ox* 701 complexiones] -nis *Ox* 702 estiuam] estiualem *ante corr. Ox* 706 crispos] cripos *Ox* 707 graciles] gracil' *Ox* uocis (*cf.* 688)] noctis *Ox* 712 forme] fortune *Ox* 713 commensurata] -anda *ante corr. Ox* constituunt] constitunt *Ox*

partes suis, uel ad maius uel ad minus uel ad robustius et debilius uel ad meliorem proportionem | et deteriorem.

[13] Ad maius quidem, uerbi gratia, Leo et Virgo et Sagittarius, ad minus autem ut Pisces uel Cancer et Capricornus; et iterum ut Arietis et Tauri et Leonis superiora quidem et anteriora ad meliorem habitudinem, inferiora autem et | posteriora ad debilius; contrarie autem ut Sagittarii et Scorpionis et Geminorum anteriora quidem ad gracilius, posteriora autem ad meliorem ualitudinem. [14] Similiter autem et Virgo quidem et Chele et Sagittarius ad commensurationem et bonam rithmizationem, Scorpio autem et Pisces | et Taurus ad malam rithmizationem et incommensurationem et in aliis similiter. Que quidem omnia conferentes et comparantes conuenit coniecturare eam que ex complexione colligitur proprietatem circa formationes et complexiones corporum.

De lesura et passionibus <corporalibus>
Capitulum 14^{m} [13]

[1] Consequente autem hiis sermone qui circa corporales lesuras et passiones, copulabimur ad ipsa secundum consequentiam speculationem penes hanc speciem existentem, ita se habentem. Et hic enim ad uniuersalem quidem acceptionem aspicere | oportet ad orizontis duos angulos, hoc est ad orientalem et occidentalem, maxime autem ad occidentalem ipsum et ad id quod preoccidit, quod est inconnexum orientali angulo; et obseruare maleficas stellas, quomodo figurate sunt ad ipsos.

[2] Si enim | ad eos qui eleuantur gradus dictorum locorum fuerint stantes uel corporaliter uel quadrangulariter uel diametraliter, aut alterutra ipsarum aut etiam ambe, lesuras et passiones corporales circa natos existimandum, maxime autem si et luminarium aut alterutrum aut ambo | angularia existant, secundum modum quem diximus, uel simul uel secundum diametrum.

V (**α** =**YΣ** **β** =**DMS** **γ** =**ABC**) 716-7 meliorem proportionem et deteriorem: εὐρυθμότερον καὶ ἀρρυθμότερον (*Boer Robb. Hüb.*)] *inv.* **Y G**^{o}: καὶ εὐρ- καὶ ἀρρ- **S:** εὐρ- **V Σ M γ**: ἀρρ- **D** 723 et[1]: καὶ **Σ M**] ὡς **V Y D S γ** (*Hüb.*) 726 comparantes: συνεπικρίνοντας **α β γ**: -ναντας Heph. (*cod.* P)] συνεπικίρναντας **V** (*Hüb.*), *cf.* Procl. (-κιρνᾶν) 734 id quod preoccidit: τὸ προδῦνον **Y** Co le (*Hüb.*), *cf.* Procl.] δῦνον **V M**: ἡγούμενον **Σ γ**, *cf.* **D G**^{o} **S**

---

720 autem] quidem *Ox* 724 rithmizationem[1]] rithimza- *Ox* 725 et incommensurationem] *om. in textu*, et commensurationem *in mg. Ox* 726 comparantes] -entes *Ox* 727 corporum *it. Ox* 728 lesura] *an* lesuris *scrib.* ? 740 existant] -ent *Ox*

[3] Tunc enim non solummodo si superferatur aliquis maleficorum, sed et si anteferatur luminaribus ipse angularis existens, sufficiens est disponere aliquid expositorum, qualecumque orizontis loca et animalium | insinuant, lesuram scilicet uel passionem, et qualecumque nature stellarum maleficarum insinuant et adhuc configuratorum ipsis.

[4] Partes enim uniuscuiusque signorum continentes partem orizontis infectam signabunt partem | corporis circa quam erit occasio, et utrum lesuram aut passionem aut ambo suscipere potest pars que insinuatur, natureque astrorum species et causas simptomatum efficiunt.

[5] Quoniam precipuarum hominis partium Saturnus quidem est dominus audituumque | dextrorum et splenis et uesice et flegmatis et ossium, Jupiter autem tactus et pulmonis et arteriarum et spermatis, Mars autem audituum sinistrorum et renunculorum et uenarum et partium, sol autem uisus et cerebri et cordis et neruorum et dextrorum omnium, | Venus autem <odoratus> et epatis et carnium, Mercurius autem sermonis et mentis et lingue et fellis et fundamenti, luna autem gustus et transglutionis et stomachi et uentris et matricis et sinistrorum omnium.

[6] Est autem uniuersalium lesuras quidem ut in pluribus euenire orientalibus existentibus maleficis qui occasionem efficiunt, passiones autem contrarie occidentalibus ipsis existentibus, quoniam et distinctum est horum utrumque eo quod lesura quidem semel efficiat et dolorem non habeat perseuerantem, | passio autem aut continue aut interpolate patientibus superueniat.

[7] Ad particularem enim adiectionem quales species future sunt, singularem sortiuntur obseruationem figurationes lesurarum illatiue et passionum, quia, ut in pluribus, secundum similis figure situs assecuntur | simptomata. Orbationes enim uisuum efficiuntur secundum alterum quidem oculorum, quando luna per se in predictis angulis existens et ad coniunctionem festinans uel ad plenilunium extiterit, [8] et quando in altera ratione habentium figurationem ad solem manens

---

V (**α** =**YΣ** **β** =**DMS** **γ** =**ABC**) 745 maleficarum: τῶν τε κακούντων] + καὶ τῶν κακουμένων **α β γ** (*Hüb.*): + καὶ τῶν κακούντων (*sic*) **V** 749 suscipere: ἐπιδέξασθαι **β γ**, *cf.* **Y**] ἐπιδείξασθαι **V** (*Hüb.*) 757 sinistrorum omnium: εὐωνύμων πάντων **Σ M S** Heph. Procl. (*Hüb.*)] εὐωνυμωτάτων **V Y D γ** 758 uniuersalium: τῶν καθόλου *cett.*] τὸ καθόλου **V** 762 superueniat (*cf. supra* efficiat ... habeat): ἐπισκήπτειν **Σ D M γ** (*Hüb.*), *cf.* **Y S**] ἐπισκεπτέον **V** 763 quales species future sunt: εἴδη τίνα] ἤδη τινὰ *codd.* 765 Orbationes: πηρώσεις *cett.*] πληρώσεις **V**

---

747 continentes] -tis *Ox* 748 signabunt] -bant *Ox* 749 natureque] nature que *Ox* 752 audituumque] auditum que *Ox* 753 audituum] auditurum *Ox* 754 renunculorum] recuniculorum (?) *Ox* 755 neruorum] niuernorum (?) *Ox* epatis] epatis *Ox* 757 gustus] guttus *Ox* transglutionis] tranglut- *ante corr.* *Ox* 760 distinctum] distict- *Ox* 763 enim] *an* autem *scrib.* ? 765 similis] simul (sil') *Ox* 767 coniunctionem] coniuct- *Ox*

extiterit | et coniungatur alicui nebulosarum in zodiaco conuolutionum (ut nubecule Cancri et Pliadi Tauri et cuspidi sagitte Sagittarii et aculeo Scorpionis et partibus penes nates Leonis aut ydrie Aquarie), et quando Mars aut etiam | Saturnus ipsi existenti angulari et occidentali — orientales ipsi existentes — superferantur, aut iterum sole existente angulari ipsi antesursumferantur.

[9] Si autem ambobus simul luminaribus uel penes idem signum uel secundum diametrum, ut diximus, configurentur, matutini quidem ad solem existentes, | ad lunam autem uespertini, circa ambos oculos occasionem efficient. Mars quidem enim a percussione uel pulsu aut ferro aut adustione facit orbationes, cum Mercurio autem configuratus in luctis aut exercitiis aut malignorum uiis, Saturnus autem per [suffocationes] | suffusiones aut frigidationes aut a glaucedinibus et talibus.

[10] Rursum si Venus fuerit in aliquo predictorum angulorum, maxime autem in occidentali, Saturno quidem coexistens aut configurata aut uicissitudinans loca, sub Marte autem superata aut diametraliter aspecta, | uiri quidem steriles efficiuntur, mulieres autem aborsibus et embrionum excisionibus inuoluuntur, maxime autem in Cancro et Virgine et Capricorno. Si luna ab oriente Marti copuletur, si autem et Mercurio secundum idem configuretur et cum | Saturno, Marte superante aut diametrizante: eunuchi aut ermafroditi aut sine folle aut sine foramine fiunt.

[11] Hiis autem <ita> se habentibus, si etiam sol configuretur et luminaribus quidem et Venere masculinizante, luna existente occidentali | et maleficis allatis superascendentes gradus: uiri quidem secti fiunt uel in genitalibus lesuras patientes et maxime in Ariete et Leone, Scorpione et Capricorno et Aquario, mulieres uero sine prole et steriles.

V (**α** =**YΣ** **β** =**DMS** **γ** =**ABC**) 768-9 in altera — extiterit *fort. ad sensum* **G**, *sed cf. adnotationes sequentes* in altera: ἐφ' ἑτέρου **V Y** (*Robb. Hüb.*)] ἐφ' ἑκατέρου **Σ β γ** (*Boer*) ratione (λόγου?)] λόγον **Σ γ**, *Boer Hüb.*: τῶν λόγων **V S**: τὸν λόγον **Y** Co le: τῶν λόγον **D M**, *Robb.* figurationem (σχῆμα ?)] σχήματος *codd.* manens extiterit (μένον ᾖ)] μὲν ᾖ **V** (*Hüb.*): μὲν ἢ **Γ**: μὲν **Y S**: *om.* **Σ D M γ**; *cf. adn. Hüb.* 769 et coniungatur: συνάπτῃ δὲ **Y** (*Hüb.*), *cf.* (-τῃ δὲ) **β** *et* Procl.] -πτει δὲ **V G**^{o}; *cf.* **Σ γ** (771 penes nates] περὶ τὸν πλόκαμον *codd.*) 775 existentes (**V Y** *Robb.*)] *om.* **Σ β γ** (*Hüb.*) 777 adustatione: κατακαύματος **Σ β γ** (*Hüb.*)] κατὰ καύματος **V**, *cf.* **Y** 779 a glaucedinibus: ἀπὸ γλαυκώσεων **M S**] ἀπογλαυκώσεων **V α D γ** (*Hüb.*) 784 aborsibus] + ἢ ὠμοτοκίαις ἢ *codd.* 785 Capricorno (**Y M S**, *Hüb.*)] + καὶ τοῖς νεφελοειδέσι συνάπτουσα ὀφθαλμὸν πηροῖ σελήνη **V D**, *simil.* **Σ γ** 786 copuletur: συνάπτῃ *cett.*] -πτει **V** (787 Marte] + πάλιν *codd.*) 790 masculinizante] ἠρρενομένων *codd.* allatis: ἐπιφερομένων **Σ M S**, *Hüb.*, *cf.* Heph.] ἐπανα- **V Y D γ**, *Boer*

---

769 in] ut *Ox* nubecule] nebecule *Ox* 770 cuspidi sagitte] cuspide sag^{t}tei (?)*Ox* 773 ante-] autem *Ox* 775 ut] uel *Ox* 782 uicissitudinans] uiscissi- *Ox* 783 autem] aut *Ox* superata] separata *Ox* aspecta] arespecta *Ox* 785 autem] aut *Ox* luna] lumina *Ox* 786 et^{2}] *an delend.* ? 787 ermafroditi] ermofodrin *in* ermofodriti *mut. Ox* 788 fiunt] fuerit *Ox* 789 configuretur] -gentur *Ox* 790 luna] lunam *Ox*

[12] Aliquando autem non sine lesura uisus tales perdurant, impediuntur autem secundum linguam et fiunt balbi aut uix loquentes, quicumque Saturnum et Mercurium simul existentes soli in predictis angulis habent, maxime autem si etiam occidentalis fuerit Mercurius et configurentur | ambo lune. Hiis autem Mars superueniens soluere consueuit <ut> in pluribus lingue impedimentum, a quocumque luna coobuiationem ad ipsum fecerit.

[13] Rursum <si> aut luminaria maleficis angularibus | afferantur simul uel secundum diametrum, aut si luminaribus malefici (et maxime luna existente aut in nodis aut in maximis latitudinibus aut in signis occasionantibus, puta Ariete, Tauro, Scorpione, Capricorno), fiunt impedimenta corporis a gibbositatibus uel incuruationibus uel claudicationibus | uel paralisibus, si quidem cum luminaribus fuerint malefici, a natiuitate ipsa, si uero in medio celi superantes luminaria aut prospicientes ex longinquo inuicem fuerint, a periculis magnis ut precipitationibus aut casibus aut a latronibus et quadrupedibus; Marte quidem enim | dominante ab hiis que per ignem aut uulnera aut per colericas egritudines aut per latrones, Saturno aut ab hiis que per casus aut nauigia aut spasmos.

[14] Vt in pluribus autem fiunt lesure luna existente circa tropica et equinoctialia signa, maxime autem circa equinoctiale quidem que per morpheas, circa estiuale autem | per impetigines, circa autumpnale autem que per lepras, circa hyemale uero que per lentigines et similia. Passiones autem accidere consueuerunt, quando in premissis sitibus malefici configurentur, secundum contrarium tamen, hoc est uespertini quidem ad solem, ad lunam autem matutini.

[15] Vniuersaliter | autem Saturnus quidem frigidos uentres facit et multi flegmatis et reumaticos demaceratos et debiles et hictericos et dissentericos et tussitiuos et spuitiuos et colicam patientes et elephanticos, mulieres autem adhuc histericas (id est matricis dolores patientes). Mars autem sanguinis emissiuos, | melancolicos, pneumonicos, scabiosos, adhuc etiam eos qui per incisiones

---

V (**α** =**YΣ** **β** =**DMS** **γ** =**ABC**) 802 in maximis latitudinibus: *fort.* ἐπικαμπίων **V D S** *legit et* in *sua sponte it.* **G**] ἐπὶ καμπίων **α M** Heph. (*Hüb.*): *cf.* κάμπια (= *flexiones*) 804 uel paralisibus (*cett.*)] *om.* **V** 805 ipsa (*cett.*)] *om.* **V** (811 equinoctiale] τὸ ἐαρινὸν *codd.*; *fort. perperam* **G**) 813 accidere (*cett.*)] συμβαίνοντα **V** (816-7 multi flegmatis: πολυφλεγμάτους **α β γ** (*Hüb.*)] πολυφλεγματώδεις **V**, *cf.* Heph.) 817 et reumaticos (**α M S**, *Hüb.*)] *om.* **V D γ** Heph. 818 elephanticos: ἐλεφαντιῶντας *cett.*] -τιῶν **V** 819 (id est matricis dolores patientes: *explicatio* **G**) sanguinis emissiuos: αἱμοπτυϊκοὺς **M S** Heph. (*Hüb.*), *fort. coni.* **G**] αἱμοπτοικοὺς **V α D γ** 820 pneumonicos: πνευμονικοὺς *cett.*] -ματικοὺς **V** etiam: τε **V Y D B C** (*Boer*)] δὲ **Σ M S A** Heph. (*Hüb.*)

---

794 autem] aut *Ox* uisus] usus *ante corr. Ox* 799 luna] lune *ante corr. Ox* coobuiationem] -ne *Ox* 800 aut] autem *Ox* afferantur] auf- *Ox* 806 ut] in *ante corr. Ox* 808 hiis] *it. Ox* colericas] coll- *Ox* 810 autem] aut *Ox* 812 autumpnale] autup- *Ox* 816 autem] *an* enim *scr.* ? quidem] quod *Ox* 817 demaceratos] dematratos *Ox* dissentericos] dist'ent'icos *Ox* 820 pneumonicos] puenmotucos *Ox*

aut adustiones occultorum locorum continue molestantur fistularum gratia aut etiam emoroidarum aut luporum aut cancrorum aut etiam igneorum ulcerum aut depascentium, mulieres etiam insuper et aborsibus aut embrionum excisionibus aut etiam ineptitudinibus ad escas consueuit inuoluere.

[16] Singulariter autem et secundum predictas configuratarum stellarum naturas ad partes corporis proprietates efficiunt passionum. Cooperatur autem ipsis maxime ad intensiones malorum Mercurius | Saturno quidem coaptatus ad frigidum et magis in motu continuo faciens reumatismos et humorum molestationes, precipue autem eorum que circa pectus et cannam et stomacum, Marti autem coaptatus ad maiorem exsiccationem ut in ulcerosis | crustulis et in apostematibus et erysipilis et impetiginibus siluestribus et in nigra colera aut mania aut morbo sacro et talibus.

[17] Et secundum signorum coaptationes continentium predictas in duobus angulis configurationes | fiunt quedam qualitates passionum. Proprie enim Cancer quidem et Capricornus et Pisces et uniuersaliter terrestria et piscium animalia faciunt passiones per depasturas uel corrosiones et per impetigines aut excoriantes aut scrophulas aut fistulas aut elefantias et talia, Sagittarius autem et Gemini eas que | per casus aut epilentias. Et in ultimis gradibus signorum existentes planete circa extremitates maxime lesuras et passiones faciunt per impedimenta aut reumaticos, a quibus et elefantie et ut ad omne dicatur ciragre et podagre eueniunt.

[18] Hiis autem ita se habentibus, si quidem nullus beneficorum configuretur maleficis occasionem facientibus aut luminaribus in angulis existentibus incurabiles et grauissime fient lesure et passiones, eodem autem modo, et si configurentur quidem, superentur autem a malignis in uirtute | existentibus; si autem et ipsi benefici in dominationis figuris existentes superauerint malignos occasionem

---

V (**α** =**YΣ** **β** =**DMS** **γ** =**ABC**) 822 (luporum aut cancrorum *duplex lect. Guill.*] κονδυλωμάτων [κονδυλο- **Y**] *codd.*) etiam[2]: καὶ **α M S** (*Hüb.*)] *om.* **V D γ** (824 aut etiam] ἢ *codd.*) 827 intensiones: ἐπιτάσεις **α M S** (*Hüb.*)] ἐπιστάσεις **V D γ** 828-9 molestationes ... eorum que: ὀχλήσεις ... τῶν **α β γ** (*Hüb.*)] ὀχλήσεως ... τὸν **V** 830 ulcerosis] + πιλλώσεων καὶ **V**: + πτηλώσεων καὶ **Y**: + ψιλώσεων καὶ **Σ β γ**: + πτιλώσεων καὶ *coni. Robb. et Hüb., cf.* Heph.: *an lect. corruptam non redd.* **G**? (833 coaptationes (ἐναρμογὰς ?)] ἐναλλαγὰς *codd.*) 839 circa: περὶ *cett.*] παρὰ **V** (lesuras et passiones] *inv. codd.*) 840 et[1] (**V Y** Heph.)] *om. cett.* (*Hüb.*) (844 malignis: κακούντων **V Y D γ**] κακοποιῶν **Σ M S**, *Hüb.*) (845 malignos (κακοῦντες ?)] κακοποιούς *cett.* (*Hüb.*): κακοποιοῦσι **V**)

---

824 escas] estas *Ox* 825 configuratarum] -orum *Ox* 827 intensiones] indiuisiones *Ox* 831 mania] in auia *Ox* 834 Cancer] cancri *Ox* 836 aut] ante *Ox* 838 que] -que *Ox* epilentias *sic Ox* 839 reumaticos] *an* -tismos *scr.* ? 840 ut ad] *inv. Ox* podagre] potagre *Ox* 841 beneficorum] -ciorum *Ox* 843 configurentur] -figentur *Ox* 845 dominationis] *fort.* -tiuis *scrib.*

inferentes, tunc lesure decentes et non probrose fiunt et passiones mediocres et facile consolabiles, aliquando autem facile permutabiles orientalibus existentibus beneficis.

[19] Jupiter quidem enim auxiliis humanis per diuitias aut dignitates lesuras occultare consueuit et passiones consolari, cum Mercurio autem et farmaciis aut bonorum medicorum sollicitudinibus. Venus autem per pronuntiationem deorum et responsionum lesuras quidem modo quodam decentes et gratiosas | instituit, passiones autem per ea que a diis medicamina facile consolabiles, Saturno tamen astante cum explicationibus et diuulgationibus et talibus, Mercurio autem cum negotiatione et acquisitione quadam prouenientibus per ipsas lesuras et passiones habentibus.

De qualitate anime
Capitulum 15m [14]

[1] De corporalibus quidem igitur simptomatibus modus speculationis talis quis utique erit. Qualitatum autem anime que quidem circa rationalem et intellectualem partem deprehenduntur per | condicionem consideratam circa Mercurium, que autem circa sensualem et irrationalem a materialiori luminarium, hoc est a luna, et configuratis astris quantum ad defluxus uel etiam coniunctiones ipsius.

[2] Multo autem formiori existente ea specie que secundum animales impetus, merito utique et | talem speculationem non simpliciter neque ut contingit faciemus, sed per plures et uarias [et] obseruationes. Et enim signorum continentium Mercurium et lunam aut eos qui habent dominium ipsorum planetas differentie multum possunt facere ad | animarum proprietates, et rationem habentium ad premissam speciem astrorum configurationes ad solem et ad angulos, et adhuc ea que secundum naturam ipsam uniuscuiusque astrorum proprietas ad animales motus.

V (**α** =**YΣ** **β** =**DMS** **γ** =**ABC**) (847 facile permutabiles: εὐαπάλλακτα **Σ β γ**, *cf.* Heph., *Hüb.*] εὐεπάλλακτα **V Y**) 855 prouenientibus: περιγιγνομένων (-γενο- **Y**) **V Y D γ**] περιγενομένου (-γινο- **Σ**) **Σ M S** (*Hüb.*) per ipsas: δι' αὐτῶν *Robb.* (*tacite*)] ἢ δι' αὐτῶν **Σ β A B** (*Hüb.*): διὰ τῶν **V Y**, *cf.* **C** 861 per condicionem: διὰ τῆς ... περιστάσεως **α γ** (*Hüb.*)] διὰ τὸν τῆς ... π. **V**, *cf.* **β** consideratam] + ἑκάστοτε *codd.* 864 ipsius: αὐτῆς **α γ** (*Hüb.*)] αὐτῶν **V D**; *cf.* **M S**; *cf.* Heph. 866 faciemus: ποιοίμεθα **Σ** (*Hüb., Robb.*)] -ούμεθα **V Y γ** (*Boer*), *cf.* **β** 867 sed: δὲ *cett.*] *om.* **V M** 868 differentie: διαφοραὶ *cett.*] -αῖς **V** 869 animarum: ψυχῶν **Σ D M γ**, *Hüb.*] ψυχικῶν **V G° S**, *cf.* **Y**

---

847 aliquando] aliū *Ox* 850 aut] autem *Ox* 852 modo (Bate, *Speculum Pars* XIX,19)] moto *Ox* 853 consolabiles (Bate, *loc. cit.*)] -lationes *Ox* 855 negotiatione] -em *Ox* per ipsas (δι' αὐτῶν)] pro ipsas (*dub.:* pro ipsis *e corr.* ?) *Ox* 859 quidem igitur] *inv.* *Ox* 863 configuratis] afiguratis *Ox* 863 quantum] quam *Ox* ad] *ss.* *Ox* 865 formiori] priori (p'ori *perperam pro* f'ori) *Ox* 868 aut] ad *Ox*

[3] Signorum quidem igitur uniuersaliter tropica quidem communiores | faciunt animas, popularium et ciuilium rerum concupiscitiuas, adhuc autem amatrices glorie et deo supplices, bene aptasque et boni motus, inquisitiuasque et inuentiuas, bene coniecturatiuas et astrologicas et diuinatiuas. Bicorporea autem uarias, facile permutabiles, male comprehensibiles, leues, | facile transponibiles, duplices, amatiuas, multorum modorum, amatrices musarum, uoluptate pigras, facile acquisitibiles, penitiuas. Fixa autem iustas, non adulatrices, immansiuas, firmas, intellectiuas, sufferentes, amatrices laborum, duras, continentes, memores malorum, persecutiuas, contentiosas, amatrices honoris, perstitiuas, | supergressiuas, duras, intransponibiles.

[4] Configurationum autem orientales quidem horoscopales, et maxime in propriis dignitatibus, liberales et expensas et liberas et fortes et bene aptas et acutas et non adopertas animas efficiunt. Matutine autem stationes | et celi mediationes superratiocinatiuas et immansiuas, memoratiuas, firmas, intellectiuas, magna sapientes, perfectiuas eorum que uolunt, non uertibiles, robustas, inseductibiles, iudicatiuas, operosas, castigatiuas, scientificas. Precessiones autem et occasus facile transponibiles et non firmas, debiles, | recusatrices laborum, passibiles, humiles, formidolosas, dubias, audentes magna mala cum formidine, hebetes, desides, difficile mobiles.

[5] Vespertine autem stationes et que sub terra celi mediationes, adhuc autem in Mercurio et Venere, de die quidem uespertini occasus, de nocte autem matutini: bene aptas quidem <et> sapientie intendentes, non | ualde autem memoratiuas neque superlaboriosas aut amatrices laborum, perscrutatiuas autem occultorum, inquisitiuas inuisibilium, puta magicas, misteriales, metheorologicas, organicasque et mechanicas, mirabilium factrices et astrologicas, philosophas, auguratiuas, sompniorum interpretatiuas et similes.

[6] Ad hec autem in propriis quidem aut etiam conuenientibus existentes locis et partitionibus habentes dominium anime operum, secundum modum quem in

---

V (**α** = **YΣ** **β** = **DMS** **γ** = **ABC**) 876 astrologicas: ἀστρολογικὰς **Y β γ** (*Hüb.*)] -λογίας **V**: *cf.* **Σ** 884 expensas (δαπάνας ?)] ἁπλᾶς *codd.* 886 (et] *om. codd.*) immansiuas — intellectiuas (*cett.*)] *om.* **V** 889 debiles: ἀσθενεῖς *cett.*] ἀθενεῖς **V** 892 autem[2] (*cett.*)] + καὶ **V** 896 inuisibilium (*cett.*)] *om.* **V** 897 -que et: τε καὶ **V**] *om. cett.* (*Hüb.*) philosophas auguratiuas (*cett.*)] *om.* **V** 899 Ad hec: πρὸς τούτοις **Y D γ** Heph. (*Hüb.*)] πρὸ τούτοις **V**; *cf. cett.*

---

874 concupiscitiuas (Bate, *Liber Servi Dei* 30, 26)] -scentias *Ox* 875 -que[2]] que *Ox* 876 diuinatiuas] dn̄atiuas *Ox*: dominatiuas Bate, *Liber Servi Dei* 30, 27 878 pigras] pigros *Ox* 880 firmas] ferimas *Ox* 881 contentiosas] -siosas *Ox* 882 in-] et *Ox* 885 adopertas] *disiuncte Ox* 887 que] qui *Ox* iudicatiuas] indi- *Ox* 890 cum] tamen *Ox* hebetes] habentes *Ox* desides] decides *Ox* 892 adhuc] Ad h' *Ox* 893 quidem] quem *Ox* 895 autem] aut *Ox* 898 sompniorum] suppremorum *Ox*

principio determinauimus, euidentes et inimpedibiles et expressas et efficaces faciunt proprietates et maxime quando iidem dominentur | simul duobus locis, hoc est quando fuerint qualitercumque Mercurio configurati et lune defluxum uel etiam coniunctionem optineant.

[7] Non sic autem dispositi sed <in> inconuenientibus existentes locis, que quidem ipsorum nature conueniunt ad animalem operationem ineuidentia et obscura et | imperfecta et sine profectu instituunt, que autem nature dominantium aut preualentium fortia et supernociua subiectorum. Vt quando qui propter maleficorum coaptationem iniusti et maligni, optinentibus quidem ipsis, bene cessibile et inimpedibile aut etiam sine periculo <et> super gloria | habent impetum ad male faciendum aliis, superatis autem a contraria partitione, decidentes et non perficientes et de facili punibiles fiunt.

[8] Qui autem rursum propter beneficorum coaptationem ad predicta loca boni et iusti, insuperatis quidem existentibus, | ipsi gaudent et bonam famam habent in aliorum beneficiis et a nullo iniustitiam patiuntur sed commodum reportantes iustitiam perficiunt; superatis autem a contrariis, econtrario propter hanc ipsam mansuetudinem et benignitatem et misericordiam facile decidentes et querelas tollerantes | aut etiam a pluribus iniuriam sustinentes existunt.

[9] Modus quidem igitur uniuersalis moralis speculationis talis quidam utique erit, particulares autem ab ipsa astrorum natura secundum dominium inditas proprietates deinceps summatim prosequemur usque ad | eam que secundum totalitatem consideratur commixtionem.

[10] S a t u r n u s quidem igitur solus dominationem anime accipiens et precedens Mercurio et lune, si quidem in gloria extiterit quantum ad mundum et angulos, facit corporis amatrices, rigidas in sententiis, profundi sensus, au-

V (**α** =**YΣ** **β** =**DMS** **γ** =**ABC**) 902 dominentur: ἐπικρατήσωσιν *cett.* (*Hüb.*)] -τήσουσιν **V D** 905 existentes locis (**α M S** Heph., *Hüb.*)] *inv.* **V D γ** 910 aut etiam: ἢ καὶ **V**] καὶ *cett.* (*Hüb.*) super gloria* (ἐπὶ δόξαν ?): ἐπίδοξον **Σ β γ** (*Hüb.*)] ἑτερόδοξον **V Y** habent: ἔχουσι **Σ β** Heph.] -ωσι **V Y γ** (*Hüb.*) 911 aliis: ἄλλους **Σ β γ** Heph. (*Hüb.*)] ἀλλήλους **V Y** superatis: κρατουμένων *cett.*] κρατοῦσι **V** 912 de facili punibiles: εὐτιμώρητοι **α β γ**] ἀτι- **V** (*Hüb.*) fiunt: γίνονται **α β** Heph.] γίνωνται **γ** (*Hüb.*): *om.* (*spat. rel.*) **V** 914 gaudent: χαίρουσι **V α β** Heph.] -ωσι **γ** (*Hüb.*) bonam famam habent: εὐφημοῦνται **V α S** Heph.] -ῶνται **D M γ** (*Hüb.*) 916 perficiunt: διατελοῦσι **α S** Heph.] -ῶσι **V D M γ** (*Hüb.*) hanc ipsam: αὐτὸ τοῦτο **β γ**, *cf.* **V**] αὐτό τε **Y** (*Hüb.*): τὸ **Σ** 917 facile decidentes (εὐκατάφοροι ?; *cf.* κατάφορος)] εὐκαταφρόνητοι *codd.* 918 existunt: τυγχάνουσιν **α β** Heph.] -ωσιν **V γ** (*Hüb.*) 919 quidem igitur (*cett.*)] *om.* **V**: ***abhinc usque ad* 984 *deficit cod.* Σ** 920 autem (*cett.*)] *om.* **V**: [**Σ**]

901 euidentes] emendantes *Ox* 902 iidem] idem *Ox* 905 sed] uel *Ox* 909 cessibile] sensibile *Ox* 910 super gloria] sic pro gloria (*vel* gloriam ?: gl̅'i̅a̅) *Ox* 913 propter] pro *Ox* beneficorum] -iorum *Ox* coaptationem] -e *Ox* 914 insuperatis] *disiuncte Ox* 919 quidam] quidem *Ox* 920 inditas] -dictas *Ox* 922 commixtionem] -i *Ox* 923 precedens] -cidens *Ox* 925 in] etiam *Ox*

steras, | singularis sententie, laboriosas, preceptiuas, castigatiuas, circumspectas, inuidiosas, pecuniarum amatrices, uiolentas, thesaurizatiuas.

[11] Contrarie autem et inglorie dispositus sordidas et paruorum sermonum et parui animi et indifferentes, male sententie et ingratos, formidolosos, sequestratos, maliloquos, | heremi amatores et planctus amatores, inuerecundos, demonibus obsequiosos, amatores malignitatis, sine affectione, insidiatiuos domesticorum, non letantes, corporis oditores.

[12] Ioui autem secundum expositum modum coaptatus in gloriosis quidem rursum dispositionibus facit bonos, honorabiles senioribus, astantes, | bone sententie, auxiliatiuos, iudicatiuos, possessionis amatores, magnanimos, redditiuos, bene electiuos, domesticorum <amatiuos>, mansuetos, intelligentes, toleratiuos, philosophie insistentes.

[13] In contrariis autem bonorum inexpertos, insaniosos, ad minima timidos, demonibus seruientes, sacra frequentatores, predicantes, suspiciosos, natorum oditores, | non amicos, occultatiuos sui, sine iudicio, infideles, in stultitia malos, ueneficos, ypocritas, immaturos, sine amore honoris, penitiuos, austeros, difficile precibus acquiescentes, difficile adibiles, reuerentiales, futuricios, stultosque similiter et non toleratiuos malorum.

[14] Marti autem coaptatus in gloriosis quidem dispositionibus facit | indifferentes, laboriosos, fidentes, molestatores, magna mala attemptantes cum formidine, austera operantes, immisericordes, contemptiuos, asperos, bellicos, proiectores periculi, tumultus amatores, dolosos, insidiatores, grauis ire, infrangibiles, populi turbatores, tirannicos, auaros, ciuium oditores, litigii amatores, mali memores, | profunde malignos, auctores, non sufferentes, superbos, graues, gloriatores, malefactores, iniustos, infrunitos, homines odientes, inconuertibiles, intransponibiles, multe actionis, bene conuersantes tamen et operosos et inconuincibiles et totaliter consummatiuos.

V (**Y** **β** =**DMS** **γ** =**ABC**) 926 circumspectas (περιθεατικούς ?)] περιδυσιαστικούς **V**: περιουσιαστικούς *cett.* (*Hüb.*) 927 inuidiosas *hic*] φθονερούς *post* θησαυριστικούς *codd.* 929 indifferentes: ἀδιαφόρους *cett.*] διαφόρους **V** 938 ad minima timidos: ψοφοδεεῖς *cett.*] ψοφοειδεῖς **V** 942 futuricios: μελ<λ>ητικούς **S**] *om.* **V D γ** (*Boer Robb. Hüb.*): [**YM**] 949 non sufferentes: ἀνυποίστους **β B C** (*Hüb.*)] ἀνυπίστους **V**, *cf. cett.* 950 infrunitos (καταφρονήτους ?)] ἀκαταφρονήτους *codd.*

---

926 sententie] lumine *Ox* laboriosas] -rosas *Ox* castigatiuas] castica- *Ox* 927 inuidiosas] timidiosas *Ox* thesaurizatiuas] -zantiuas 930 maliloquos] -logcos inuerecundos] -des *Ox* 933 in gloriosis] *coniuncte Ox, etiam* 944, 958 935 iudicatiuos] indi- *Ox* redditiuos] reddat- *Ox* 939 frequentatores] *an* -tantes *scrib.* ? 941 amore] amoris *Ox* 943 non] *an delend.* ?; *cf.* 936 toleratiuos 946 immisericordes] in miseri- *Ox* 947 tumultus] timul- *Ox* 949 mali memores] *coniuncte Ox* 951 operosos] -as *Ox* inconuincibiles] -conclusibiles *Ox*

[15] In contrariis autem ǀ raptores, latrones, adulteros, patientes malorum, turpiter lucrantes, sine deo, sine affectione, iniuriosos, insidiatores, fures, periuros, sanguinem facientes, illicite occisores, malignos, homicidas, uenenatores, sacrorum predatores, impios, sepulcrorum suffossores et uniuersaliter ualde malos.

[16] Veneri autem coaptatus ǀ in gloriosis quidem dispositionibus facit mulierum oditores, antiquorum amatores et heremi sectatores, indelectabiles ad intercessiones, non amatores honoris, oditores pulcrorum, inuidiosos, austeros ad conuentus, sine collatione cum aliis, singularis sententie, pro diuinatione purificantes se, superstitiosos, misteriorum et rituum concupiscitores, sacrorum conditores, ǀ indiuinatos, deo supplices, uenerabiles autem et facile uerecundantes, uerecundos, immiscentes se philosophie, fideles ad conuictus, continentes, superratio<ci>natiuos, reuerentes, indignatoresque <et> ad mulierum suspiciones zelotipos.

[17] In contrariis autem luxuriosos, impudicos, turpium operum, indifferentes ǀ et immundos ad coitus, incestuosos, insidiosos femininarum personarum et maxime propinquissimarum, confractos, ualde probrosos, precipites, oditores pulcrorum, exprobatores, maliloquos, ebriosos, latreuticos, subcorruptores, nefarios ad conuentus, supponentes et suppositos, ǀ non solum ad ea que secundum naturam, sed etiam ad ea que preter naturam, seniorum et uilium et contra legem et bestialium commixtionum appetitiuos, impios, deorum contemptores, misteriorum et sacrorum dissipatores, turbatores, criminatores, ueneficos omnimodo.

[18] Mercurio autem ǀ coaptatus in gloriosis quidem dispositionibus facit operosos, delectantes in addiscere, legitimorum inquisitores, amatores medicinalium, misticos, participes occultorum <et> inenarrabilium, prodigiorum conditores, paralogizatores, paucis pro uita contentos, adaptatores, discussores rerum et sollertes, amaros et diligentes, uigiles, ǀ sapientiam sectantes, amatores operationis, consummatiuos.

[19] In contrariis autem tediosos, memores malorum, duros animabus, superlaboriosos, suos odientes, tormentorum amatores, non letantes, nocte errabundos, insidiatores, proditores, incompatientes, fures, magicos, uenenatores, litterarum falsatores, traiectatores, non ad consummationem pertingentes ǀ et facile incidentes.

V (**Y** **β** =**DMS** **γ** =**ABC**) 964 ad conuictus: πρὸς συμβιώσεις **Y M S** (*Hüb.*)] ἐνσυμβιώσεις **V D**, *cf.* **γ** 968 incestuosus: ἀνάγνους *cett.*] ἀναγώους **V** 974 turbatores: *an* καναπιστάς **V Y** *sic redd. vel* κυκητάς (*e glossa*) *legit* **G** ? (*cf. infra* 1081)] πάμπαν ἀπίστους **β γ** (*Hüb.*) omnimodo: παντοίους **V**] παντοποιούς **β γ** (*Hüb.*), *cf.* **Y** 976 delectantes in addiscere (*cett.*)] *om.* **V** 981 duros: νηλεεῖς **β γ** (*Hüb.*)] μηλίους **V Y** 984 pertingentes: ἀποτευκτικούς *cett.*] ἀποικτικούς **V**: ***abhinc Σ denuo adest***

---

953 raptores] -oris *Ox* 962 superstitiosos] superfaciosos *Ox* 964 philosophie] prophetie *Ox* 970 latreuticos] latrenti- *Ox* 974 ueneficos] uenif- *Ox* 980 consummatiuos] contuma- *Ox* 982 non] si *Ox* 984 pertingentes] -gintes *Ox* incidentes] insidantes *Ox*

[20] J u p i t e r autem solus dominationem anime accipiens in gloriosis quidem dispositionibus, facit magnanimos, largitiuos, dei cultores, honorabiles, fruitores, benignos, magnificos, magna sapientes, | uenerandos, proprio operi insistentes, misericordes, rationis amatores, affectuosos, presides.

[21] In contraria autem dispositione existens similes quidem fantasias imprimit animabus, ad humilius tamen et ineuidentius et indiscretius, puta pro magnanimitate quidem | prodigalitatem, pro dei cultu demonum obsequium, pro uerecundia formidinem, pro reuerentia gloriationem, pro benignitate indifferentem beneuolentiam, pro amore pulcritudinis amorem delectationis, pro magna sapere desidiam, pro liberalitate autem indifferentiam et quecumque hiis similia.

[22] Marti autem coaptatus in gloriosis quidem dispositionibus facit asperos, pugnaces, militares, rerum consumptores, motos, insubditos, calidos, parabolizantes, fidentes, operatores, arguitores, augmentatores operis, uictorie amatores, principatores, magnanimos | autem et amatores honoris et animosos et iudicatiuos et consummatiuos.

[23] In contrariis autem iniuriatiuos, indifferentes, insociales, inexorabiles, seditiosos, contentiosos, singulariter laborantes, criminatores, superbos, supergressores, rapaces, cito transmutabiles, leues, penitiuos, instabiles, | precipites, infideles, sine iudicio, sine sententia, excessiuos, operosos, querulosos, prodigos, tediosos et uniuersaliter irregulares et ultramotos.

[24] Veneri autem coaptatus in gloriosis quidem dispositionibus facit mundos, fruitores, amatores speculationis, | fistularum amatores, cantatiuos, amatores gesticulationis, in excessu beneuolos, beneficos, miseratiuos, sine malitia, dei amatores, uirtuti studentes, agonum amatores, prudentes, osculatiuos, uenereos in reuerentia, fulgentis anime, bone sententie, amatores litterarum, redditiuos, iudicatiuos, moderatiuos et bene institutos ad uenerea, | domesticorum amatores, pios, amatores iustitie, amatores communitatis, glorie amatores et uniuersaliter bonos et optimos.

V (**α** =**YΣ** **β** =**DMS** **γ** =**ABC**)   988 magnificos (**V**, *cf.* **γ**)] + ἐλευθερίους δικαίους **Y** Procl. (*Hüb.*), *cf.* **Σ β**   989 rationis amatores: φιλολόγους **Σ M S** Procl. (*Hüb.*)] ἐμφιλολόγους **V D γ**   affectuosus] εὐεργετικούς *praem.* **V Y** *Procl.* (*Hüb.*), *cf.* **Σ β γ**   994 magna sapere: μεγαλοφροσύνης **α M S** Heph. Procl. (*Hüb.*)] σοφροσύνης **V**, *cf.* **D γ**   997 motos: κεκινημένους *cett.*] κε + *lac.* **V**   999 principatores (**V**)] + εὐεπιβούλους, ἐπιεικεῖς, ἐπάνδρους, νικητικούς *cett.* (*Hüb.*)   1002 singulariter laborantes: μονοπόνους **V Y D γ** (*Boer*)] μονοτόνους **M S** Procl. (*Hüb.*), *cf.* **Σ**   1004 operosos: ἐμπράκτους Procl. (*Robb.*)] ἐμπατάκτους **V Y** (*Hüb.*): ἐπιτακτικοὺς εὐπατάκτους **β γ**, *cf.* **Σ**   1007 fruitores] + φιλοκάλους, φιλοτέκνους (*cum l.v.*) *codd.*   amatores gesticulationis: *an* φιλοστρόφους *vel* φιλοκώμους (*e glossa, cf.* **V** *infra ad* 1012) *legit* **G**?] φιλοτρόφους *codd.*   1010 amatores litterarum, redditiuos] *inv. codd.*   1011 domesticorum amatores: φιλοικείους **S** (*Hüb.*), *cf.* φιλικείους **V**, φιλοικίους **Y**] φιλονείκους **Σ M γ**; *cf. cett.*   1012 amatores communitatis (φιλοκοίνους ?)] φιλοκώμους **V** (*Boer*): φιλοτίμους *cett.* (*Hüb.*)

---

987 largitiuos] largitios *Ox*   997 parabolizantes] per aboliz- *Ox*   1002 criminatores] criminaciones (*abbr.*) *Ox*   1007 amatores[1]] amores *Ox*   cantatiuos] canitiuos *Ox*   1008 beneuolos] vn̄ (*del.*) beni- *Ox*   1010 iudicatiuos] indi- *Ox*   1012 communitatis] 9m̄tatis *vel* 9n̄itatis *abbr.* *Ox*

[25] In contrariis autem teneros, deliciose uite, femininos animo, saltatrices, muliebres animo, expensores, in malum muliebres, amorosos, luxuriosos, precipites, maledicos, ornatus amatores, submolles, | desides, prodigos, probrosos, compassiuos, ornatiuos, muliebris intellectus, sacrorum fanaticos, productiuos, misteriales, fideles tamen et sine nequitia et superlargientes et facile adibiles, et bene deductibiles et ad calamitates liberaliores.

[26] Mercurio autem coaptatus | in gloriosis quidem dispositionibus facit multum negotiosos, amatores sermonis, geometras, mathematicos, poeticos, contionatores, bene aptos, temperatos, boni sensus, bene consiliatiuos, politicos, beneficiatiuos, procuratiuos, dulcis moris, amatores dationis, amatores populi, facile se ingerentes, bene consummatiuos, | principatiuos, pios, dei amatores, bene pecuniosos, affectuosos, domesticorum amatores, facile erudibiles, philosophicos, dignatiuos.

[27] In contrariis autem excessiuos in beneuolentia, tediosos, deceptibiles, facile despectibiles, reuelationibus attendentes, deo supplices, stultiloquos, subamaros, fictione sapientes, | insensatos, iactatores, magicos, subtus motos, multarum autem historiarum et memoratiuos et magistrales et mundos concupiscentiis.

[28] M a r s autem solus dominationem anime accipiens in gloriosa quidem dispositione facit generosos, | principatiuos, animosos, armorum amatores, multum agiles, fortes, parabolizantes, proiectores periculi, insubicibiles, indifferentes, robustos, acutos, liberos, contemptiuos, tirannicos, operatiuos, iracundos, precessores. In contraria autem insociales, iniuriosos, sanguinis amatores, tumultus amatores, sumptuosos, | clamosos, percussores, precipites, ebriosos, raptores, immisericordes, astutos, perturbabiles, insanos, domesticorum oditores, sine deo.

V (**α** =**YΣ** **β** =**DMS** **γ** =**ABC**) 1014 deliciose uite: ἡδυβίους *cett.*] *om.* **V** 1015 muliebres animo: θηλυψύχους **α M S** Heph. (*Hüb.*)] μεγαλοψύχους **V D γ** in malum muliebres: κακογυναίους Procl. (*Robb.*)] καταγυναίους **V** (*Hüb.*): κατὰ γυναίων **Y**: καὶ γυναίους **Σ β γ** 1016 maledicos: λοιδόρους **V D B C**, *cf.* **A**] μοιχούς **α M S** (*Boer*): λοιδόρους, μοιχούς Procl. (*Robb. Hüb.*) 1016 probrosos: ἐπιμώμους **Y** Procl. (*Hüb.*)] ἐπιβώμους **V**: φιλομώμους ὑπομώρους **Σ M S**, *cf.* **D γ** 1021 multum negotiosos: πολυπραγμάτους (*cf.* **Y**)] φιλογραμμάτους πολυπραγμάτους φιλοπράκτους **Y**: πολυγραμμάτους **V D γ** (*Hüb.*): φιλογραμμάτους **α M S** 1024 facile se ingerentes: εὐεπιβόλους **γ** (« fort. recte»), *cf.* **M S**] ἐπιβόλους **V**: εὐεπιβούλους **α D** (*Hüb.*) bene consummatiuos (εὐεπιτεύκτους *vel* εὐεπιτευκτικούς ?)] εὐεπιτέκτους **V**: ἐπιτευκτικούς *cett.* (*Hüb.*) 1025 facile erudibiles: εὐπαιδεύτους *cett.*] εὐπεδεύτας **V** 1027 excessiuos in* beneuolentia: εὐήθεις *cett.*] συνήθεις **V** 1028 deo supplices: θεοπροσπλόκους **β γ** (*Hüb.*)] θεοπλόκους **V**; *cf.* **Y** *et* **Σ** 1029 fictione sapientes: προσποιησισόφους **M S** (*Hüb.*)] προσποιήσεις σοφούς **V**; *cf. cett.* iactatores] + ἐπιτηδευτάς *codd.* 1030 et magistrales (*cett.*)] *om.* **V**

1014 animo] -os *Ox* 1023 dationis] dominationis *Ox* 1026 dignatiuos *sic Ox*] *an* dignitatiuos *scrib.* ? 1027 in[2]] ut *Ox* 1028 despectibiles] deceptibiles *Ox* 1029 fictione] siccione *Ox* 1031 dispositione] -ponere *Ox* 1032 generosos] gloriosos *Ox* 1034 acutos] actutos *Ox* 1035-6 tumultus amatores] et amultus amatores (*in mg. inf.*) *Ox*

[29] Veneri autem coaptatus in gloriosis quidem dispositionibus facit superlargitores et bene degentes et sociorum amatores, delectabilis uite, bene sapientes, lusiuos, integros uel illesos, | bene rithmizatos, amatores saltationis, amorosos, artis amatores, imitatores, fruitiuos, preparatores, uiriles, bene procliues quidem ad uenerea peccata, [et] consummatores autem et facile uestitu ornabiles et intellectu acutos et difficile redarguibiles et diiudicatiuos, adhuc autem iuuenum concupiscitiuos et masculorum | et feminarum, sumptuososque et acute ire.

[30] In contrariis autem iniectores oculorum, luxuriosos, precipites, dissipatores, adulteratiuos, iniuriatores, mendaces, dolorum implicatores, subcorruptores domesticorum et alienorum, acutos simul et non gaudentes ad concupiscentias, | stupratores mulierum et uirginum, parabolizatores, calidos, inordinatos, insidiatores, periuros, facile incidentes, sensum ledentes, aliquando autem et prodigos, amatores ornatus et audaces et subactos et impudica agentes.

[31] Mercurio autem coaptatus in gloriosis quidem dispositionibus facit militares, | industrios, operosos, bene mobiles, non despicabiles, multorum modorum, inuentiuos, sophistas, astutos, prelinguosos, impositorios, dolosos, instabiles, methodicos, in malum artificiosos, acute sentientes, seductores, simulatores, insidiatores, mali modi, multorum negotiorum, amatores malignorum, | consummatoresque aliter et ad similes bene conuenientes et bene condextrabiles et uniuersaliter inimicorum lesiuos, amicorum autem benefactiuos.

[32] In contrariis autem sumptuosos, supergressores, insociales, parabolicos, audaces, penitiuos, promptos percutere, ultramotos, mendaces, fures, | sine deo, periuros, impostores, seditiosos, incensores, theatrorum inuasores, superiniuriosos, latrocinantes, murorum suffossores, turpes occisores, falsarios, traiectatores, demonum inuocatores, magos, uenenosos, uirorum interfectores.

V (**α** =**ΥΣ** **β** =**DMS** **γ** =**ABC**) (1040 integros uel illesos] ἀφελεῖς *codd.*: *fort.* ἀφελεῖς *sic redd.* **G**) 1041 artis amatores: φιλοτέχνους **Y** Procl.] φιλοτέκνους *cett.* (*Hüb.*) uiriles: ἐπάνδρους *cett.*] ἀνάνδρους **V** 1042 consummatores: ἐπιτευκτικούς *cett.*] εὐεπιτευκτικούς **V** 1044 acute ire] + καὶ ζηλοτύπους *codd.* 1045 dissipatores: διασύρτας **V D γ** (*Boer*)] ἀδιαφόρους διασύρτας **α M S** (*Hüb.*) 1047 (non gaudentes] προσκορεῖς *codd.*; *fort. male intell.* **G**) concupiscentias: ἐπιθυμίας **α β** (*Hüb.*)] ἐπιθυμιτάς **V**; *cf.* **γ** 1048 stupratores: διαφθορεῖς *cett.*] διαφθορὲ **V** 1049 facile incidentes: εὐεμπτώτους (+ τε καὶ) *cett.*] εὐεπτώτους (+ τε καὶ) **V**; *cf.* **S** 1053 inuentiuos: εὑρετικούς **Y M S γ** (*Hüb.*)] ἐρετικούς **V**, *cf.* **Σ D** sophistas] + ἐπιπόνους (-μόνους **Y**) *codd.* 1059 promptos percutere (εὐπατάκτους ? *cf. adn. Hüb. ad* 1321): ἐμπατάκτους **Σ M S** (*Hüb.*)] ἐμπαράκτους **V**: ἐμπράκτους **Y D γ**: εὐπαράκτους *temptavit Robb.*

1039 bene degentes] *coniuncte Ox* 1040 lusiuos] luciuos *Ox* 1041 fruitiuos] funiti- *Ox* 1042 autem] aut *Ox* 1043 iuuenum] inuenum *Ox* 1048 in-] et *Ox* 1049 ledentes] lendentes *Ox* 1051 quidem] quem *Ox* 1053 prelinguosos] -gosos *Ox* impositorios] -terios *Ox* 1055 multorum] malum *Ox* 1056 condextrabiles] -trales (?)*Ox* 1057 lesiuos] lestiuus *Ox* benefactiuos] -uus *Ox* 1060 impostores] impest- *Ox*

[33] Venus autem sola dominationem | suscipiens in gloriosa <quidem> dispositione facit mansuetos, bonos, teneros, rationabiles, mundos, prudentes, bonorum amatores, deo carissimos, mali zelatores, malignorum oditores, et artium amatores, bene compositos, bene se habentes, facile sompniantes, affectuosos, beneficiatiuos, misericordes, indelectabiles, bene communicatiuos, | consummatiuos et uniuersaliter uenereos. In contraria autem desides, amorosos, femininos, muliebres, non audaces, indifferentes, precipites, probrosos, ineuidentes, uituperosos.

[34] Mercurio autem coaptata in gloriosis quidem dispositionibus facit artium amatores, philosophantes, scientionales, | bene aptos, poeticos, fistularum amatores, dulcis moris, fruitiuos, delicate diete, bone moderationis, amicorum amatores, pios, intellectiuos, multum ingeniosos, meditatiuos, directiuos, bene ingressiuos, cito discentes, per se dociles, zelatores optimorum, imitatores bonorum, | boni oris et gratiosos, bene compositos moribus, studiosos, premii amatores, rectos, iudicatores, magna sapientes, uenereorum autem ad mulierculas quidem euitatiuos, ad pueros autem magis motos, et zelotipos.

[35] In contrariis autem impostores, multum ingeniosos, | mali oris, aliud simulantes, male sententie, seductores, mendaces, criminatores, periuros, profunde malignos, turbatores, obseruatores, incompositos, sine federe, peruersores, mulierum corruptores et puerorum, adhuc autem comptos, submolles, probrosos, male fame, multum notatos, | omnia aggredientes et aliquando quidem in corruptione talia simulantes, aliquando autem et in ueritate dispositos et turpia agentes et uariis passionibus inquinatos.

V (**α** =**YΣ** **β** =**DMS** **γ** =**ABC**) 1063 dominationem **V** Procl.] + τῆς ψυχῆς *cett.* (*Hüb.*) 1065-6 bonorum amatores (φιλοχρήστους)] φιλορχηστάς *codd.* deo carissimos *hic*] φιλοθεωτάτους *ante* εὐσχήμονας (bene se habentes, 1066) **V D γ** (*Hüb.*): *post* εὐσχήμονας **α M S** mali zelatores: κακοζήλους **Σ M S** Procl.] καλοζήλους **V Y D γ** (*Hüb.*) et artium amatores: καὶ φιλοτέχνους **V D**] φιλοτέχνους *cett.*: φιλοτέκνους *coni. Hüb.* 1066 facile sompniantes: εὐονείρους *cett.*] ἐνονείρους **V** 1072 fistularum amatores] + φιλοκάλους *codd.* 1074 directiuos, bene ingressiuos: κατορθωτικούς εὐεπιβούλους] *inv. codd.* 1075 cito discentes: ταχυμαθεῖς **Y D γ** (*Hüb.*)] ταχ᾽ ἡμαθεῖς **V**; *cf. cett.* 1076 gratiosos] + τῷ λόγῳ ἐρασμίους *codd.* 1079 contrariis: τῶν ἐναντίων **γ**] τῆς ἐναντίας **V α β** (*Hüb.*) 1081 turbatores *hic*] κυκητάς *ante* ψεύστας (mendaces, 1080) *codd.* obseruatores: *an* ἐπιβόλους *legit* **G** ? (*cf. supra* II.83)] ἐπιβουλευτικούς (-τικτούς **V**) *codd.* 1083 in: ἐπὶ *cett.*] ὑπὸ **V** 1085 inquinatos: ὑβριζομένους **α D M C** (*Hüb.*)] μεριζομένους **V**: *cf. cett.*

1067 beneficiatiuos] -catiuos *Ox* 1068 desides] decides *Ox* amorosos] -as *Ox* 1074 meditatiuos] medicat- *Ox* 1076 premii] primi *Ox* 1078 motos] motiuos *Ox* 1079 impostores] -posteros *Ox* 1079 aliud] ad *Ox* 1080 mendaces] ui^{s}daces *Ox* periuros] *disiuncte* *Ox* 1081 malignos] + compos *Ox* 1084 simulantes] simil- *Ox*

[36] Mercurius autem solus dominationem | anime suscipiens in gloriosa quidem dispositione facit natos perspicaces, sollertes, intelligentes, multum historiosos, inuentiuos, expertos, logicos, fisiologos, speculatiuos, bene aptos, zelatiuos, beneficiatiuos, ratiocinatiuos, mathematicos, misteriales, bene coniecturatiuos, disciplinales, adeptiuos. In | contraria autem astutos, preuolantes, obliuiosos, impetuosos, leues, facile permutabiles, penitiuos, cum malignitate stultos, imprudentes, peccatores, mendaces, indifferentes, instabiles, infideles, auaros, iniustos et uniuersaliter falsos mente et procliues ad peccata.

[37] Hiis autem ita se habentibus cooperatur quidem et ipsa lune consistentia, quoniam in flexionibus existens australis et borealis termini cooperatur anime proprietatibus ad multiformius, in | nodis autem ad acutius et operosius et facilius mobile. Adhuc autem in orientalibus et luminarium augmentationibus ad maiorem habilitatem et maiorem euidentiam <et> maiorem firmitatem et maiorem fiducialitatem, in diminutionibus autem luminarium aut occultationibus ad maiorem tarditatem et maiorem <h>ebetudinem | et penitudinem ampliorem et maiorem formidinem et ineuidentiam ampliorem.

[38] Cooperatur autem aliqualiter sol coaptatus aliquo modo dominanti complexioni anime, secundum gloriosam quidem dispositionem rursum ad iustius et ad eminentius operationis et honorabilius et reuerentius et | ad maiorem dei cultum. Secundum contrarium autem et inconueniens ad humilius et laboriosius et minus insigne et insocialius et singularioris sententie et austerius et ad id quod peioris regiminis et uniuersaliter ad id quod peius dirigibile.

V (**α** =**YΣ** **β** =**DMS** **γ** =**ABC**) 1089 bene coniecturatiuos *hic*] εὐστόχους *ante* μαθηματικούς *codd.* 1090 disciplinales (μαθητικούς ? *Vel alia translatio pro* μαθηματικούς, 1089 ?)] *om. codd.* 1095 flexionibus: τοῖς καμπίοις **γ** Heph. (*Robb.*: «fort. recte»): ταῖς καμπίαις **V** Procl. (*Boer*)] τοῖς ἐπικαμπίοις **Σ β** (*Hüb.*) 1096 multiformius] + καὶ τὸ πολυμεχανώτερον καὶ εὐμεταβολώτερον *codd.*; *in trad. gr. vel in trad. lat. excid.* 1099 diminutionibus: ταῖς μειώσεσιν **G**° **β γ** Heph. Procl. (*Hüb.*)] ταῖς βιώσεσιν **V**: *cf. cett.* 1099-100 maiorem tarditatem: τὸ νωχελέστερον *cett.* (*Hüb.*)] τὸν ὠλοσχερέστερον **V** 1102 autem aliqualiter: δέ πως **V Y D γ**] δὲ **Σ M S** Heph. (*Hüb.*) sol: ὁ ἥλιος *cett.*] ὅλως **V** 1106 minus insigne: ἀσημότερον *redd.* **G** (*coni. Robb. Hüb.*), *cf.* ἀσημώτερον **V**: ἀσημιότερον **Y**] ἀσεμνότερον **Σ β γ** (*Boer*)

1088 fisiologos] silogos *Ox* 1089 ratiocinatiuos] rem conatiuos *Ox* 1090 adeptiuos] adempt- *Ox* 1091 stultos] stratos *Ox* 1094 lune] lumine *Ox* 1095 in flexionibus] influxionibus *Ox* 1096 *an post* multiformius* *quaedam verba excid.?* 1101 formidinem] formed- *Ox* 1105 laboriosius] -rosius *Ox* 1106 et[3]] in- *Ox*

De passionibus animalibus
Capitulum 16^{m} [15]

[1] Quoniam autem anime proprietatibus consequitur aliqualiter qui de precipuis ipsius passionibus sermo, uniuersaliter quidem rursum notari et obseruari oportet Mercurium et lunam, quomodo se habent ad inuicem et ad angulos et ad alias | stellarum que sunt apte ad malignationem; quod siue ipsi inconnexi fuerint ad inuicem siue ad orientalem orizontem superentur aut habitudine inuoluantur aut longitudinis aspectu impediantur a figuratis inconuenienter et nociue, circa propietates animales incidentium uariarum passionum <sunt> effectiui, distinctione | ipsarum rursum considerata a proprietate preaccepta stellarum locis coaptatarum.

[2] Plurima quidem igitur magis mediocrium passionum fere prius in dictis de anime proprietatibus distincta sunt, intentione ipsarum possibili considerari ex | excessu malignantium, quoniam aliquis merito extrema morum et aut deficientia a medietate aut excedentia dicet utique passiones. Que autem precipuum habent excessum et sunt quasi egritudinales et preter totam naturam, circa ipsam ratiocinatiuam anime | partem et circa passiuam, ut in tipo talem habent obseruationem.

[3] Epilentici quidem enim ut in pluribus fiunt quicumque, luna et Mercurio, ut diximus, inconnexis existentibus aut ad inuicem aut orientali orizonti, | Saturnum quidem de die, Martem autem de nocte habent in angulo et aspectu impedientem premissam figuram. Insani autem, quando in eisdem econtrario Saturnus quidem de nocte, Mars autem de die dominatus fuerit figure, et maxime in Cancro aut Virgine aut Piscibus. Demoniaci | autem et humidi capitis, quando malignantes sic se habentes optineant lunam existentem in apparentia, Saturnus quidem festinantem ad coniunctionem, Mars autem festinantem ad plenilunium, maxime quidem in Sagittario et Piscibus.

[4] Soli quidem igitur malefici secundum predictum modum dominium | figure suscipientes, incurabiles quidem, secretas <tamen> et non productas in exemplum faciunt premissas ratiocinatiue anime egritudines. Coaptatis autem beneficis, Ioue

---

V (**α** =**YΣ** **β** =**DMS** **γ** =**ABC**) 1117 considerata: θεωρουμένης *cett.*] δεω- **V** (1121 quoniam: ἐπειδήπερ **V**] + ἤδη *cett., Hüb.*) (1122 aut: ἢ **S**] ἢ καὶ *cett.*) 1127 inconnexis existentibus *hic*] *post* ὁρίζοντι **V Y** (*Hüb.*): *diversis locis* **Σ β γ** 1132 existentem in apparentia: ἐπὶ φάσεως οὖσαν (*Hüb.*), *cf.* ἐπιφάσεως οὖσαν **V** (*cf.* Procl.)] ἐπιφάσεως ἰοὺς ἂν **Y**: *diversimodo cett.*

1113 inconnexi] connexi *Ox*; *fort.* in *post* ipsi ($\overline{\text{ipi}}$) *excidit, cf. infra* 1127 1114 siue] sine $\overline{\text{coe}}$ (?) *Ox* 1122 precipuum] principium *Ox* 1129 econtrario] contrario *Ox* 1132-3 festinantem ad coniunctionem, Mars autem (Bate, *Speculum* XIX,19)] *om. Ox* 1133 plenilunium (Bate, *loc. cit.*)] plenissimum *Ox* 1137 ratiocinatiue] rationatiue *Ox*

et Venere, in occidentalibus partibus existentes ipsi, beneficis autem in orientalibus | angulariter, curabiles quidem, facile tamen productibiles in exemplum faciunt passiones; sub Ioue quidem per curas medicinales et aut per regimen diete aut farmacie, sub Venere autem per responsa et auxilia a diis.

[5] In orientalibus autem ipsi in angulo existentes, beneficis | occumbentibus, incurabiles simul et multum diuulgatas et euidentes faciunt egritudines; secundum epilentias quidem eos qui patiuntur inuoluentes constrictionibus et clamoribus et periculis mortalibus, secundum insanias autem et extases inquietudinibus et abalienationibus a domesticis et | denudationibus et blasfemiis et talibus, secundum demonizationes autem aut humorum molestationes reuelationibus et prophetiis et plagis et similibus exemplarium.

[6] Singulariter autem et locorum figuram continentium que quidem solis et que Martis ad manias maxime | cooperantur, que autem Iouis et Mercurii ad epilempsias, que autem Veneris ad dei illustrationes et elocutiones, que autem Saturni et lune ad humorum molestationes et in demonizationes.

[7] Egritudinalis quidem igitur uicissitudo circa operatiuum anime secundum totales | naturas fere in talibus speciebus et per tales figurationes efficitur; que autem circa passiuum rursum precipue consideratur, manifestatur maxime circa eos qui secundum genus masculi et feminine excessus | et defectus eius quod secundum naturam, deprehenditur autem speculatiue secundum modum premisso similem, sole quidem cum luna pro Mercurio assumpto et coaptatione Martis cum Venere ad ipsos.

[8] Hiis enim sub uisu cadentibus, si | sola quidem luminaria in masculinis sint signis, uiri quidem excedunt id quod secundum naturam, mulieres autem id quod preter naturam, ad uirile simpliciter anime et operosius; si autem et Mars et Venus et alteruter uel etiam ambo sint masculinizati, uiri quidem | ad coitus secundum naturam fiunt procliues et adulteratiui et sine saturitate et in omni tempore prompti ad turpia et ad illegitima uenereorum, mulieres autem ad eos

---

V (**α** =**ΥΣ** **β** =**DMS** **γ** =**ABC**) (1138 autem] *om. codd.*) 1140 diete: διαιτητικῆς *legit vel coni.* **G**, *cf. Boer Boll Hüb.*, *cf.* διαιτικῆς **V**] διαγητικῆς **Υ**: ἰατρικῆς *cett.* 1146 denudationibus: γυμνητείας *cett.*] γυμνασίας **V** 1154 talibus (**V γ** Procl., *Robb.* «fort. recte»)] + ἐστὶν **α β** (*Hüb.*) 1155 rursum precipue: πάλιν ἐξαιρέτως **V**] κατὰ τοῦτο (**Υ**: κατ' αὐτὸ **Σ β γ**) πάλιν τὸ ἐξαίρετον **α β γ** (*Hüb.*) 1157 modum: τρόπον *cett.*] τρόπῳ **V** 1161 sint: ἦ̣ **Σ β** (*Hüb.*)] ἦ **V**: *om.* **Υ γ** 1165-6 eos* qui (*sc.* coitus) ... similiter: τὰς (*sc.* συνουσίας) ... ὁμοίως **V Υ** Procl. (240.5)] τὰς ... ὁμιλίας **Σ β γ** (*Hüb.*)

1139 angulariter] *ss. Ox* 1140 faciunt (Bate, *loc. cit.*)] *om. Ox* et aut (Bate, *loc. cit.*)] *inv. Ox* 1145 insanias] insanies *ante corr. Ox* 1146 abalienationibus] *disiuncte Ox* 1147 demonizationes (Bate, *loc. cit.*)] -ionibus *Ox* 1152 humorum] humanorum *Ox* demonizationes] -is *Ox* 1155 precipue] passiue *praem. et exp. Ox* 1157 modum] modo[s] *Ox* 1158 sole] -em *Ox* 1162 operosius] *in* -roisius *mut. Ox* 1163 et[2]] *an* uel *scrib.* ? masculinizati] -liniti *ante corr. Ox* 1164 sine saturitate (ἀκόρεστοι)] sine firmitate (ſa –> fi ; turi –> rmi ?) *Ox* 1165 tempore] -poree *Ox* 1165-6 eos qui] eas qui (*sic*) *Ox*

qui preter naturam similiter luxuriose et iniectrices oculorum et uocate confricatrices. [9] Subiciunt autem | femellas uirorum opera facientes, et si quidem sola Venus masculinizata sit, latenter et non in palam, si autem et Mars, econtrario, ita ut aliquando tamquam legitimas mulieres subactas ostendant.

[10] Econtrario autem luminaribus secundum expositam configurationem | in femininis signis existentibus solis, mulieres quidem excedunt id quod secundum naturam, uiri autem id quod preter naturam ad id quod facile fragile et feminizatum anime. Si autem et Venus fuerit masculinizata, mulieres procliues et adulteratrices et luxuriose fiunt | ad hoc quod subiciantur secundum naturam in omni tempore et ab omni quocumque, ita ut a nullo simpliciter uenereorum abstineant, siue turpe sit siue contra legem, uiri autem molles et proni ad coitus preter naturam et subacti in modum mulierum passiue, | secrete tamen et latenter. [11] Si autem et Mars fuerit feminizatus, econtrario et cum fiducia sine uerecundia sunt, premissa secundum utramque speciem efficientes, quod fornicarium et multum commune et quod multum probrosum et omnimode turpe scema defundentes, usque ad diuulgationem secundum maledictionem et | usus ignominiam.

Comparantur autem et orientales et matutine figurationes Martis et Veneris et ad masculius et facile diuulgabilius, occidentales autem et uespertine ad feminius et secretius. [12] | Similiter et Saturnus quidem configuratus ad magis impudicum aut immundius aut etiam probrosius unicuique expositorum natus est cooperari, Jupiter autem ad decentius, Mercurius autem ad uulgabilius | et ad id quod passionum facile mobilius et multiformius et quod facile profectibilius.

Explicit liber tertius, incipit quartus.

V (**α** =**YΣ** **β** =**DMS** **γ** =**ABC**) 1166 preter: παρὰ *cett.*] περα (*sic*) **V** 1167 autem: δὲ **V Y** (*Robb.*, «*forte recte*»)] γὰρ **Σ β γ** (*Hüb.*) 1168 sit: ᾖ **Σ** (*Hüb.*)] ἦν **V**: καὶ **Y**: *om.* **β γ** Heph. et² (*cett.*)] *om.* **V** 1170 Econtrario autem: τὸ δὲ ἐναντίον *cett.* (*cf.* **γ**)] τῶν δὲ ἐναντίων **V** 1174 hoc quod subiciantur: τὸ διατίθεσθαι *cett.*] τὸ διαθεῖναι **V** 1176 proni: θρασεῖς **β γ** Heph., *cf.* **Σ c**] σαθροὶ **V Y** Procl. (*Hüb.*) 1180 quod multum probrosum (τὸ πολύψογον ?): πολύψογον *cett.*] ἀπολύψογον **V** (defundentes] περιβαλλόμενοι *codd.*) diuulgationem: δημοσιώσεως **S**: δημοσίως ἕως **V**, *cf.* **Y Γ**] σημειώσεως **Σ D M γ** (*Hüb.*) 1182 Martis (*cett.*)] *om.* **V** 1186 decentius] + καὶ φυλακτικώτερον καὶ αἰδημονέστερον **V Y D γ** (*Hüb.*), *inv. cett.* 1187 facile profectibilius: εὐπροκοπώτερον **V** (*Boer*)] εὐπροσκοπώτερον **α β A C** (*Robb. Hüb.*), *cf.* **B** 1188 Explicit liber tertius, incipit quartus: τέλος τοῦ γʹ βιβλίου κλαυδίου πτολεμαίου **V**: τέλος τοῦ τρίτου **Y**: τέλος τοῦ γʹ βιβλίου **D A**: *om. cett.*

1169 aliquando] aliū *Ox* 1172 feminizatum] feminitatem (*sic*) *Ox* 1177 subacti]–tū *Ox* 1178 sine] siue *Ox* 1180 scema] stema *Ox* 1181 maledictionem] -dicio- *Ox* 1183 masculius] m[a]lius *Ox* 1184 configuratus] -atur (? *abbr.*) *Ox*

<LIBER IV>

<Hec sunt capitula libri quarti>

Prohemium — ca. p^{m}
De fortuna possessoria — ca. 2^{m}
De fortuna dignitatiua — ca. 3^{m}
De qualitate operationis — ca. 4^{m}
De coniugiis — ca. 5^{m}
De natis — ca. 6^{m}
De amicis et inimicis — ca. 7^{m}
De peregrinatione — ca. 8^{m}
De qualitate mortis — ca. 9^{m}
De temporum diuisione — ca. 10^{m}

Prohemium
Capitulum primum

[1] Que quidem igitur ante natiuitatem <et que penes ipsam natiuitatem> | possunt considerari et adhuc que post natiuitatem quecumque sunt propria consistentie, insinuantia uniuersaliter qualitatem commixtionum, fere hec erunt utique. Eorum autem que penes id quod extra accidunt et que consequenter debent pertractari precedit quidem qui de fortuna possessoria et dignitatiua | sermo, connectitur autem sicut possessoria quidem corporis conuenientiis, ita dignitatiua hiis que anime.

V (**α** =**YΣ** **β** =**DMS** **γ** =**ABC**) 1 *tit. om.* (**V S B C**)] ΒΙΒΛΙΟΝ Δ′ *Hüb., cf. cett.* 2 <Hec sunt capitula libri quarti>, *cf. lib.* II *et* III] Τάδε ἔνεστιν ἐν τῷ δ′ βιβλίῳ **V** (*Hüb.*), *cf.* **D B C**: τὰ κεφάλαια τοῦ δ′ βιβλίου **M**, *totum indicem om.* **S** (**Y Σ A** non citantur) 3 Prohemium ca. p^{m}: α′ προοίμιον **V** (*Hüb.*)] *om.* **α β γ** 4 ca. 2^{m}: β′ **V** (*Hüb.*)] α′ **Y D γ**: *om.* **Σ M S** *et sic deinceps* 13-14 Prohemium capitulum primum: α′ προοίμιον **V D** (*Hüb.*)] *cf. cett.* (**20** ita ...] ... δὲ *codd.*)

---

6 qualitate] fortuna *Ox* 16 quecumque] qui- *ante corr. Ox* 17 qualitatem] -te *Ox* 18 accidunt] -it *Ox* consequenter] conuenienter (9^{i}enter) *Ox* 20 connectitur] coniectutur *Ox*

De fortuna possessoria
Capitulum 2m

[1] Que quidem igitur possessionis qualia quedam erunt, accipiendum a uocata sorte fortune, solo tamen modo secundum quem | semper distantiam que est a sole ad lunam proicimus ab horoscopo et in natis de die et in natis de nocte, propter causas quas diximus in hiis que de temporibus uite. Considerare igitur oportebit, hac instituta secundum hunc modum, signi | dominationem habentes, et qualiter se habent hii quantum ad uirtutem et quantum ad conuenientiam, secundum modum quem in principio determinauimus, adhuc autem configuratos ipsis uel superantes | hos qui eiusdem uel contrarie partitionis.

[2] In uirtute enim existentes qui sortis dominium susceperunt | faciunt multarum possessionum, et maxime quando a luminaribus extiterint conuenienter testificati; sed Saturnus quidem per edificia et agriculturas et nauigia, Jupiter autem per fidem et per prefecturas et per sacerdotalia, Mars autem per militaria et per ducatus, Venus uero per amicitias | et mulierum dona, Mercurius autem per sermones et negotiationes.

[3] Proprie autem Saturnus fortune possessorie coaptatus, si Ioui configuretur, hereditates acquirit et maxime quando in superioribus angulis hec accident, Ioue in bicorpore signo existente aut etiam | copulationem <lune> optinente; tunc enim in filiorum procreationem reducti aliena hereditabunt.

[4] Et si quidem <qui eiusdem> partitionis cum ipsis dominatoribus testimonia dominationis fuerint facientes ipsi, possessiones inuiolatas conseruabunt; si autem qui contrarie partitionis | superexaltati fuerint principalibus locis aut superillati ipsis, dissipationes faciunt existentium, uniuersali tempore accepto per prosneusim (id est adnuitionem) occasionem inferentium ad angulos et ad superalleuationes.

V (**α** =**YΣ** **β** =**DMS** **γ** =**ABC**) (25 modo] *om. codd.*) 28 hac: τούτου **V Y** Procl. (*Robb.*)] τοὺς τοῦ **Σ β γ** (*Boer Hüb.*) 30 autem (*cett.*)] *om.* **V** 32 enim: γὰρ **Y**] μὲν γὰρ **V D γ** (*Hüb.*): *om.* **Σ M S** (34 et[1,2]] ἢ *codd.*) 35 fidem: πίστεως **V** (*Robb.*)] πίστεων *cett.* (*Hüb.*) (et[1,2]] ἢ *codd.*) Mars: Ἄρεως (*cett.*)] τοξότου **V** 36 amicitias: φιλίων **G**° Heph.] φιλικῶν **V** (*Hüb.*), *cf.* **Y**: φιλῶν *cett.* 39 superioribus: ἄνω **α M S γ** Heph. (*Hüb.*)] ἄλλων **V**, *cf.* **D** hec (accident): ταῦτα] τοῦτο *codd.* 41 reducti: ἀναχθέντες **Σ β γ** (*Hüb.*)] ἀνεχθέντες **V Y** (hereditabunt] κληρονομοῦσιν) (42 cum ipsis] *om. codd.*) 43 ipsi: αὐτοὶ **Σ β γ** (*Hüb.*)] *αὐτοῖς **V**: *om.* **Y** Heph. possessiones: κτήσεις **Σ β γ** (*Hüb.*)] κτίσεις **V Y** (conseruabunt] διαφυλάττουσιν *codd.*) 45 accepto: λαμβανομένου **α β γ** Heph. (*Hüb.*)] -μένων **V** (*Boer*)

---

24 a] et *Ox* 26 proicimus] proximus *Ox* 31 partitionis] petitionis *Ox* 32 qui] quod *Ox* 33 possessionum] *ss. Ox* extiterint] exsti- *Ox* 38 possessorie] -re *Ox* 39-40 Ioue in] Iouem *Ox* 41 reducti] -duci *Ox* 42 partitionis] -cipationis *Ox* 43 ipsi] *ipsis *Ox* (*cf.* 42) 44 superillati] -tis *Ox* 46 prosneusim] prouensim *Ox* adnuitionem] adnut- *Ox* superalleuationes] *an* -allationes *scrib.* ?

De fortuna dignitatiua
Capitulum 3[m]

[1] Que autem dignitatis et talis felicitatis oportebit speculari a dispositione luminarium et ab addextrantium stellarum conuenientia cum ipsis. In masculinis enim signis existentibus ambobus luminaribus et in angulis aut ambobus rursum aut etiam altero, maxime autem eo quod | partitionis, et addextratis <a> 5 planetis, sole quidem a matutinis, luna autem a uespertinis: qui nati reges erunt.

[2] Et si quidem addextrantes planete uel in angulis ipsi fuerint uel ad angulum qui super terram configurentur, magni et potentes | et rectores mundi perseuerabunt, et adhuc magis felices, si addextrantes dextri hiis qui super terram angulis configurentur. Si autem aliis sic habentibus, solus sol fuerit in masculino, luna autem in feminino, angulare autem alterum luminarium: presides solum erunt uite et | mortis domini.

[3] Si autem <cum> hiis neque addextrantes planete angulares fuerint uel testificentur angulis, magni solum erunt et in dignitatibus procuratoriis uel exercitus ductiuis uel sacerdotalibus et non principalibus.

[4] Si autem luminaria | non fuerint in angulis, addextrantium autem planetarum plurimi aut in angulis sint aut configurentur angulis, in dignitatibus quidem insignioribus non erunt, in preminentiis tamen ciuilibus et mediocritate circa preacceptationes secundum uitam; neque addextrantibus tamen | coaptatis ad angulos, non insignes operibus et sine profectu constituuntur; tandem autem humiles et infelices fiunt fortunis, quando neutrum luminarium neque in angulo neque in masculino signo extiterit neque addextretur a beneficis.

[5] Vniuersalis quidem igitur modus proposite speculationis dignitatum talem quandam [paulationem] paulatiuam augmentationem habet, intermedias autem horum consistentias ualde multas existentes coniecturandum a particulari immutatione luminarium et ipsorum addextratione circa ipsam speciem et a | dominio addextrationum.

V (**α** =**YΣ** **β** =**DMS** **γ** =**ABC**) 52 enim: γὰρ **V D γ**] μὲν γὰρ *cett.* (*Hüb.*) 56 configurentur*: συσχηματίζωνται **Σ β B C** (*Hüb.*)] -ζονται **V Y A** 58 configurentur: συσχηματίζωνται **Σ β B C** (*Hüb.*)] -ζονται **V Y A** 62 dignitatibus] + τοῖς ἀπὸ μέρους στεμματηφορικοῖς (*l.v.* **Σ β γ**; *cf.* **Y**) ἢ *codd.* (*Hüb.*) 63 uel sacerdotalibus: ἢ ἱερατικοῖς **Y D γ** (*Hüb.*)] ἱερατικοῖς **V**: *om. cett.* 64 luminaria (**V Y** Heph. Co le Procl.)] + μὲν **Σ β γ** (*Hüb.*) 66 erunt: γενήσονται *cett.*] γεννήσονται **V** 69 in angulo (**V β γ** Heph., *Boer*)] + ἢ **α** (*Hüb.*) 70 extiterit: τυγχάνῃ **α β γ** (*Hüb.*)] τυγχάνει **V** Heph.

---

50 speculari] speculatori speculari *Ox* 51 addextrantium] -tutium *Ox* 52 enim] autem *Ox* 53 addextratis] adextrans *Ox* 54 qui nati] a nati (*sic*) *Ox* 55 fuerint] -erunt *Ox* 56 configurentur*] -retur *Ox* 58 sic] se *Ox*: *an* sic se *scrib.*? masculino] + et *Ox* 62 magni] magis *Ox* exercitus] excercus (*sic*) *Ox*, *sed* -us *cum obscuris correctionibus* (-itis, -itus ?) 64 autem[2]] aut *Ox* 68 constituuntur] -tuntur *Ox* 70 beneficis] uenef- *Ox* 72 quandam] quadam *Ox* 73 immutatione] in mut- *Ox*

[6] Hac enim circa habentes quidem partitionem et circa beneficos consistente, maior dominabilitas et minor decidentia dignitates assequitur, circa contrarios autem aut maleficos, submissio et instabilitas. | Species autem dignitatis future consideranda a proprietate addextrantium planetarum, quoniam quidem Saturnus dominium addextrationis habens potentatus facit super possessionem et pecuniarum collectionem, Jupiter autem aut Venus super gratias et dona et honores aut magnanimitates, | Mars autem super militias et uictorias et timores subiectorum, Mercurius autem propter intellectum aut eruditionem aut curam et rerum dispensationem.

De qualitate operationis
Capitulum 4m

[1] Habens autem dominium operationis accipitur secundum duos modos, a sole et a signo medii celi. Considerare enim oportebit facientem apparitionem matutinam propinquissime ad solem et eam que in medio celi, quando maxime coniunctionem lune habeat. Et si quidem eadem | fuerit stella habens ambo predicta, hac sola utendum; similiter autem et si alterum nullum habeat, ea sola que alterum accepit. Si autem altera fuerit que propinquissimam apparitionem fecit et altera que medio celi et lune coaptatur, ambabus utendum, primoritatem | dantes habenti secundum dominium plures rationes dignitatis, secundum modum quem preexposuimus. [2] Si autem nulla inueniatur neque apparitionem faciens neque in medio celi, dominum ipsius assumendum, ad ydoneitates tamen temporales; nam sine opere ut | in toto tales fiunt.

Planeta quidem igitur operationis dominium suscipiens ita nobis diiudicabitur, qualitas autem operationum ex proprietate trium planetarum, scilicet Martis et Veneris et Mercurii, et a proprietate signorum et in quibus utique extiterint | gradientes.

V (**α** =**YΣ** **β** =**DMS** **γ** =**ABC**) 82 aut: ἢ **V**] καὶ *cett.* (*Hüb.*) 88 apparitionem (*cett.*)] *om.* **V** 89 et (*cett.*)] *om.* **V** 91 nullum (*an* μηδὲν *legit vel coni.* **G** ?)] μηδεὶς **V Y D γ**: μηδὲ εἷς **Σ M S**: εἷς *Hüb. scr.* (*cf.* Heph.) 93 et: καὶ **V A B** (*Boer, Robb.*)] ἢ καὶ **α β C** (*Hüb.*) (94 rationes dignitatis] ψήφους οἰκοδεσποτείας *codd.*) 95 nulla: μηδεὶς **V D γ**] μηδὲ εἷς **α M S** (*Hüb.*) 98 ita: οὕτως *cett.*] οὗτος **V** 100 et in: καὶ ἐν **V**] ἐν *cett.* (*Hüb.*) utique: ἂν *cett.*] ἐὰν **V**

---

76 circa habentes] contrahentes *Ox* et] *an* aut *scrib.* ? consistente] -tem *Ox* 81 pecuniarum] -cuniam *Ox* 82 militias] iustitias *Ox* 89 matutinam] -ne *Ox* 93 primoritatem] -te *Ox* 95 preexposuimus] -posimus *Ox* 96 dominum] -nium *Ox* ydoneitates] ydometates *Ox*

[3] Mercurius quidem enim agere exhibens, ut figuraliter utique dicat quis, facit scriptores, negotiatores, logistas, doctores, mercatores, trapezitas, diuinatores, astrologos, immolatores, et uniuersaliter operantes in litteris et interpretatione et datione et acceptione; | et si quidem Saturnus ipsi testimonium ferat, alienorum dispensatores aut sompniorum iudicatores, aut in sacris conuersationes facientes occasione diuinationum et responsionum; si autem Jupiter, legis scribas, rethores, sophistas, cum personis maioribus habentes | conuersationes.

[4] Venus uero agere exhibens facit habentes operationes circa odores florum aut unguentorum aut circa uina aut colores aut tincturas aut aromata aut ornamenta, uelud unguenti uenditores, sertorum complicatores, aromatum uenditores, balneatores, tabernarios, ueneni uenditores, | textores, pictores, tinctores, uestimentorum uenditores; et si quidem Saturnus ipsi testificetur, negotiatores eorum que ad fruitionem et ornatum, magos autem et ueneficos et precessores et acquirentes ex similibus hiis; si autem Iupiter, athletas, | coronarum gestatores, honoribus dignificatos, a muliebribus personis promotos.

[5] Mars autem cum sole quidem figuratus facit per ignem operantes, uelud cocos, calcem coquentes, carbonarios, erarios, metallorum fusores; existens autem sine sole | eos qui per ferrum, uelud naues compingentes, carpentarios, agricultores, latomos, lapidum sculptores, historiarum exaratores, lapidarios, lignorum fissores, politores; et si quidem Saturnus ipsi testificetur, nauticos, munitores, irrigatores, pictores, bestiarum alitores, cocos, portionatores; si autem Jupiter, milites, | ualetos, telonearios, receptores hospitum, remigatores, sacrificatores.

[6] Rursum autem inuentis duobus qui operationes exhibent, si quidem Mercurius et Venus susceperint dominium, a musa et organis et melodiis et poematibus et rithmis faciunt operationes, et maxime | quando loca sibi commutauerint; efficiunt enim ioculatores, simulatores, corporum locatores, organorum factores, cordarum tortores, pictores, saltatores, textores, cere plasmatores. Et si

---

V (**α** =**YΣ** **β** =**DMS** **γ** =**ABC**) 111 aromatum uenditores *hic*] ἀρωματοπώλας *inter* ὑφάντας (textores) *et* ζωγράφους (pictores) **V α M S** (*Hüb.*), *cf.* **D γ** (112 balneatores] ἐκδοχέας *codd.*) 115 acquirentes: πορίζοντας *cett.*] προρίζ- **V** (120 historiarum exaratores] λαοξόους **V β A C** (*Hüb.*): *om.* **Y G°** Procl., *Robb.*) 121 munitores, irrigatores] *tr.* (ἀντλητάς, ὑπονομευτάς) *codd.* 122 pictores: ζωγράφους **V** Procl. (*Hüb.*)] στρατιώτας **Σ D M γ**: *om.* **Y S** Heph. portionatores: παρασχίστας (-σχηστάς **Y**) **V Y** Procl. (*Hüb.*)] περιχύτας **Σ β γ** Heph., Co le 128 pictores *hic* (**V D γ**)] *post* cere plasmatores **α M S** Heph. Procl. (*Hüb.*)

---

102 exhibens] exibens *Ox* 103 trapezitas] traperitas *Ox* 106 sompniorum] suppremorum *Ox* 109 odores] ordines *Ox* 110 aut ornamenta] *ss.* *Ox* 114 fruitionem] finit- *Ox* 114 ueneficos] uenif- *Ox* 116 muliebribus] mulieribus *Ox* 118 cocos] cecos *Ox* 120 latomos] latrones (lat°nes) *Ox* 121 politores] polli- *Ox* munitores] Inturitores *Ox* 123 telonearios] thelenearius *Ox* 124 autem] aut *Ox* 126 faciunt] facit *Ox* 127 simulatores] simil- *Ox* 128 pictores] pactores *Ox*

quidem Saturnus iterum ipsis testificetur, facit eos qui circa genera predicta et muliebres | ornatus negotiantes; si autem Jupiter, causidicos, ratiocinationibus presidentes, in publicis occupatos, puerorum doctores, turbarum capitaneos.

[7] Si autem Mercurius et Mars simul susceperint dominium operationis, statuificos, | armorum factores, sacrorum sculptores, animalium fictores, ad palmam mensuratores, medicos, chirurgicos, accusatores, adulteratiuos, malorum actores, falsarios. Et si quidem Saturnus ipsis testificetur, occisores, raptores, latrones, iniuriosos, traiectatores; si autem Jupiter, amatores armorum, amatores monachorum, | actitores, industrios, amatores operationis, alienorum deportatores, et per talia acquirentes.

[8] Si autem Venus et Mars simul susceperint dominium operationis, faciunt tinctores, unguentarios, stangnifices, plumbi fusores, aurifices, argenti laboratores, | agricolas et armis gesticulatores, farmacorum compositores, medicos per farmacias curis utentes. Et si quidem Saturnus ipsis testificetur, sacrorum animalium curatores, hominum tumulatores, lamentatores, in sepulcris carmina decantantes, deis deditos, ubi misteria et lamentationes et purificationes | conuersationes facientes; si autem Iupiter, sacris supplices, augures, sacrorum gestatores, mulierum prefectos, nuptiarum et colligationum interpretes et per talia uiuentes uoluptuose simul et pericula proiciendo.

[9] Et signorum autem, in quibuscumque fuerint qui operari | exhibent, proprietates secundum speciem coaccipiuntur et faciunt aliquid ad uarietatem operationis. Que quidem enim humane forme cooperantur aliqualiter ad omnes scientificas [et circa humanam oportunitatem necessarias] et circa humanam indigentiam oportunas, quadrupedia autem ad metallicas et adquisitiuas et edificatiuas | et fabricatiuas, tropica autem et equinoctialia ad interpretatiuas et permutatiuas et mensuratiuas et agricultiuas et sacerdotales, terrestria autem et que in humido ad humidas et eas que per humida et herbarias et nauales, adhuc autem circa sepulturas aut fixuras aut sallituras aut commassationes.

V (**α** =**YΣ** **β** =**DMS** **γ** =**ABC**) 133 statuificos] ποιοῦσιν *praem. codd.* (136 raptores] λωποδύτας ἅρπαγας *codd.*: λωποδύτας *fort. non reddit* **G**) 137 amatores monachorum: ἢ φιλομονάχους **D γ**] ἢ φιλομονομάχους **V M S** (*Hüb.*); *cf.* **Y** Heph.: *om.* **Σ** (141 et] *om. codd.*) 142 ipsis: αὐτοῖς **Σ M S γ** Procl. (*Hüb.*)] αὐτὸς **V D**; *cf.* **Y** 148 quibuscumque: οἷς ἂν **Σ β γ** (*Hüb.*)] οἷς ἐὰν **V Y** Heph. (149 et faciunt] *om. codd.*) 150 operationis: τῆς πράξεως **V**] τῶν πράξεων *cett.* (*Hüb.*) 155 et[1]: καὶ **Y** (*Boer*)] ἢ **Σ M S γ** Heph. (*Hüb.*): *om.* **V D** autem: δὲ **Σ**] τε *cett.* 156 aut commassationes] ἢ ψυλλίας («vox inaudita» *Hüb.*) *codd.* (*exc. om.* **Y**); *fort. sua sponte* **G**

---

129 qui] quod *Ox* 131 in] et *Ox* publicis] puplicis *Ox* 132 susceperint] -erunt *Ox* 134 chirurgicos] -rur- *ss. Ox* 137 industrios] -tros *ante corr. Ox* 138 deportatores] -titores *Ox* 143 carmina] caruma *Ox* 144 misteria] missia *Ox* 151-2 et circa (contra *Ox ut fere semper*) humanam indigentiam oportunas] *non hic sed post* fabricatiuas (153) *Ox*: *haec versio fort. correctio pro* et circa humanam oportunitatem necessarias (151) *a Guillelmo in margine adposita, deinde a quodam librario false inserta et sic a librario codicis Ox descripta erat*

[10] Proprie autem rursum luna si operatiuum locum optineat a coniunctione cursum faciens cum Mercurio, in Tauro quidem et Capricorno et Cancro facit diuinatores, immolatores, in ollis diuinantes, in Sagittario et Piscibus in mortuis diuinatores et demonum commotores, in Virgine autem | et Scorpione magos, astrologos, pronuntiatores, prescientias habentes, in Libra autem et Ariete et Leone raptos, sompniorum iudices, coniuratores.

[11] Species quidem igitur ipsarum operationum per talia secundum commixtiuum modum oportebit coniecturare, | magnitudinem autem ipsarum ex uirtute astrorum dominantium. Orientales quidem enim existentes aut in angulis faciunt operationes [et] principales, occidentales autem aut decidentes ab angulis secundarias operationes, et a beneficis quidem magistrati magnas et gloriosas et lucrosas et | sine offensione et complacentes, [et] a maleficis autem humiles, inglorias, sine lucro et incertas.

[12] Saturno quidem contrariante secundum infrigidationes et discolorationes, Marte autem secundum proiectiones periculorum et uociferationes, ambobus autem secundum perfectas exsurrectiones; uniuersali tempore augmentationis | aut humiliationis iterum considerato per dispositionem astrorum effectum causantium ad matutinos et uespertinos angulos semper.

De coniugiis
Capitulum 5m

[1] Existente autem hiis consequenter eo qui de coniugiis sermone, de | conuictu quidem uiri et mulieris secundum leges ita considerandum. In uiris quidem aspicere oportet lunam ipsorum, quomodo disponitur. Primo quidem in orientalibus existens quartis facit uiros nubere iuuenes uel iunioribus secundum etatem connubentes, in | occidentalibus autem tarde nubere aut uetulis coniungi, si autem sub radiis fuerit et Saturno configurata, penitus innuptos.

V (**α** =**YΣ** **β** =**DMS** **γ** =**ABC**) 159 immolatores (**α M S** Heph., *Hüb.*)] *om.* **V D γ** 163 Species ... ipsarum: τὸ ... αὐτῶν ... εἶδος **Σ M S γ** Heph. (*Hüb.*)] τῶν ... αὐτῶν ... εἶδος **V Y** 167 secundarias operationes: *an* ὑποπρακτικάς **V Y D γ** *sic redd. vel* ὑποτακτικάς *ut* Heph., Procl. (*Hüb.*) *legit et* operationes *add.* (*cf.* 166) **G** ?] ὑπὸ τὰς πρακτικάς **Σ M S** (magistrati] καθυπερτερούμενοι *codd.*) 170 secundum infrigidationes (κατὰ ψύξεις)] καταψύξεις *codd.* 171 secundum proiectiones periculorum (κατὰ ῥιψοκινδυνίας)] καταρριψοκινδυνίας **V** (*Hüb.*); *cf.* **Y** *et cett.* 177 conuictu: τῆς ... συμβιώσεως **Y** Procl.] τῶν ... συμβιώσεων **V D γ** Heph. (*Hüb.*); *cf.* **Σ S M**

---

157 luna (Bate, *Speculum* XIX,19)] lumina *Ox* 162 raptos *sic Ox, etiam* Bate, *loc. cit.*] *an* deo raptos *scrib.* ? 163 talia] taliū *Ox* 167 beneficis] -ciis *Ox, etiam* Bate, *loc. cit.* magistrati *sic Ox, etiam* Bate, *loc. cit.*] *an* superati *scrib.* ? 168 a] al *praem. et exp. Ox* 172 augmentationis] -nes *Ox* 174 causantium] cannentium *Ox* 177 conuictu] coniunctu *Ox*

[2] Deinde si in signo quidem unius speciei aut etiam uni planetarum coniuncta extiterit, semel nubentes efficit, si autem in bicorpore aut etiam multiformi uel etiam pluribus in | eodem signo coniunctionem habens, pluries nubentes. Et si quidem astrorum que coniunctiones suscipiunt uel secundum adherentias uel secundum testimonia benefica extiterint, accipiunt uxores bonas, si autem malefica, contrarias.

[3] Saturnus quidem enim suscipiens coniunctionem procurat mulieres laboriosas | et austeras, Jupiter autem reuerendas et yconomicas, Mars autem audaces et insubiectas, Venus autem hilares et formosas et gratiosas, Mercurius autem perspicaces et acutas. Adhuc autem Venus cum Ioue quidem aut Saturno aut cum Mercurio uite proficuas et uirorum amatrices et natorum, cum | Marte autem iracundas et dissidentes et sine sententia.

[4] In mulieribus autem aspicere oportet solem ipsarum, quoniam quidem et ipse in orientalibus rursum existens quartis facit habentes ipsum ita dispositum aut iuuenculas nubere aut cum iunioribus contrahere, in occidentalibus autem | tarde nubere aut cum senioribus secundum etatem contrahere; et in signo quidem unius speciei existens aut in uno matutino planetarum, semel nubere; in bicorpore autem aut multiformi iterum aut etiam pluribus matutinis configuratus, pluries nubere.

[5] Saturno quidem igitur eodem modo soli quidem configurato | accipiunt uiros constantes et utiles et laboriosos, Ioue autem reuerendos et magnanimos, Marte autem actitores et sine affectione et insubicibiles, Venere autem mundos et formosos, Mercurie autem uite proficuos et negotiosos, Venere autem cum Saturno quidem parum coeuntes | et reuerentiores in uenereis, cum Marte autem calidos et procliues <et> adulterinos, cum Mercurio autem circa pueros pauidos pro affectu.

[6] Dicamus autem nunc orientales quartas in sole quidem que precedunt oriens signum zodiaci et occidens, | in luna autem que a coniunctione et plenilunio usque ad dichotomias, occidentales autem que dictis contrarie.

V (**α** =**YΣ** **β** =**DMS** **γ** =**ABC**) 184 in (*cett.*)] *om.* **V** 185 habens: ἐπέχουσα **Y D** Heph. (*Hüb.*), *cf. cett.*] ἐπέχοντα **V** 186-7 secundum testimonia: κατὰ μαρτυρίας **Y D γ** (*Hüb.*)] καταμαρτυρίας **V**: *cf. cett.* 199 matutino: ἑῴου **Σ M S c**] ἑῴων **V D γ** (*Hüb.*), *cf.* **Y** planetarum (**α M S**)] + συνάπτων **V D γ** (*Hüb.*) 200 configuratus: *fort.* συσχηματισθεὶς **α M S** Procl. (*Hüb.*) *legit* **G**] σχηματισθεὶς **V D M γ** 209 Dicamus: λέγωμεν] λέγω μὲν **V**: λέγομεν *cett.* (*Hüb.*) 211 dictis: τοῖς εἰρημένοις **Σ β γ** Procl. (*Hüb.*)] τῆς εἰρημένης **V Y**

---

184 semel] sol' *Ox* 185 etiam] + ex *Ox* 190 austeras] asteras *ante corr. Ox* reuerendas] uenerem *praem. et exp. Ox* 191 hilares] hill- *Ox* 192 aut[1]] autem *Ox* 195 quidem] quoddam *Ox* 202 quidem igitur — quidem] *mg. Ox*[2]: equidem *Ox*[1] 209 que] qui *Ox* 211 occidentales] -lis *Ox*

Perseuerant quidem igitur ut ad omne dicatur conuictus, quando utrarumque natiuitatum luminaria configurata extiterint | concorditer, hoc est trigona inuicem uel sextilia, et maxime quando uicissim hoc accidit, multo autem plus quando luna uiri soli mulieris.

[7] Dissoluuntur autem ex contingentibus et abalienantur penitus, quando predicti luminarium situs extiterint in | inconnexis signis aut in diametralibus aut in tetragonalibus. Et si quidem concordes luminarium configurationes benefici planetarum aspiciant, delectabiles <- - - et indelectabiles> et dampnosas.

[8] Similiter autem et in sitibus non concordantibus benefici quidem luminaribus testificantes non penitus abrumpunt conuictus, sed faciunt regressus et rememorationes conseruantes placabilitatem et affectionem, maligni autem cum quadam inuectione et iniuria faciunt | dissolutiones; Mercurio quidem igitur solo cum ipsis facto litigiis et inclamationibus inuoluuntur, cum Venere autem adulteriis aut farmaciis aut talibus.

[9] Conuentiones autem que fiunt secundum alium quemque modum | discernendum aspicientes ad Venerem et ad Martem et ad Saturnum. Coexistentibus enim ipsis cum luminaribus conuenienter, et conuictus conuenientes et legitimos; habet enim Venus quasi cognationem ad utrumque dictorum planetarum, et | ad Martem quidem secundum faciem perfecti status, quoniam in trigonalibus < - - - in trigonalibus> inuicem habent domos.

[10] Vnde Venus cum | Marte quidem simpliciter amorosas dispositiones facit, accedente autem Mercurio et acclamationes; in communibus autem et coaptatis signis, Capricorno et Piscibus, fratrum aut cognatorum implicationes. Et si quidem in uiris lune coextiterit, facit duabus sororibus aut cognatis conuenire, | si autem in mulieribus Ioui, duobus fratribus aut cognatis,

[11] Cum Saturno autem rursum Venus existens simpliciter quidem delectabiles et bene stabiles < - - | - instabiles> et nociuos et zelotipos. Et si quidem similis

---

V (**α** =**YΣ** **β** =**DMS** **γ** =**ABC**) (213 trigona] ὅταν ἤ τρίγονα ᾖ *codd.*) 214 accidit: συμβαίνει **V**] συμβαίνῃ **α β γ** (*Hüb.*) 217 extiterint (**V D γ**)] + ἤ **α M S** (*Hüb.*) (218 in: ἐν **Σ M S**] *om.* **V Y D γ**, *Hüb.*) 219 delectabiles: ἡδείας **Σ β B C** Heph. (*Hüb.*)] ἰδίας **V Y**: *def.* **A** *<- - - et indelectabiles>] καὶ προσηνεῖς — καὶ ἀηδεῖς *codd.* (*cf.* 277°-279°); *fort. in trad. lat. per hom. exciderunt* 224 facto: γενομένου (γίνο-) *cett.*] γεννωμένου **V** inuoluuntur: περικυλίονται *cett.*] -ίοντες **V** 228 et[1] (*cett.*)] *om.* **V**: *def.* **Σ** legitimos (**V α**)] + ποιοῦσι *cett.* (*Hüb.*) 230 *< - - - in trigonalibus >] ἀλλήλων ζῳδίοις ἔχουσι — ἐν τριγωνικοῖς *codd.* (*cf.* 296°-298°); *an in trad. lat. per hom. exciderunt?* 233 accedente (προσιόντος ?)] προσόντος *codd.* 237 delectabiles: ἡδείας **Σ D M γ** Heph. (*Hüb.*)] ἰδίας **V Y S** 238 bene stabiles *<- - - instabiles>] εὐσταθεῖς ποιεῖ τὰς συμβιώσεις — τοῦ Ἄρεως ἀστάτους *codd.* (*cf.* 308°-310°); *fort. in trad. lat. per hom. exciderunt*

---

218 benefici] uene- *Ox* 219 <- - - et indelectabiles>] *cf. adn. gr.-lat.* 225 farmaciis] forma- *Ox* 230 *an post* trigonalibus* *verba exciderunt fere haec*: <signis inuicem habent exaltationes, ad Saturnum autem secundum faciem senioris status, quoniam iterum in trigonalibus>? 238 *an post* stabiles* *verba exciderunt fere haec*: <facit conuictus, accedente autem Mercurio et proficuos, accedente autem et Marte instabiles> ? zelotipos] -tipoos *ante corr.* *Ox*

figurationis fuerit cum ipsis, ad coetaneas facit implicationes; si autem orientalior ipsis, ad iuniores masculos aut iuniores feminas; si autem occidentalior ad seniores feminas uel seniores masculos.

[12] Si autem in communibus signis | fuerint Venus et Saturnus, hoc est in Capricorno et Libra, cognatorum faciunt conuentus. Ascendente autem <aut> ad celum mediante predicta figura, luna quidem accedens facit masculos quidem matribus aut matrum sororibus aut consanguineis ex parte matris conuenire, | feminas autem filiis aut filiis fratrum aut filiarum uiris. Sol autem, occidentalibus maxime existentibus planetis, masculos quidem filiabus aut filiarum sororibus aut uxoribus filiorum, feminas autem patribus aut patrum fratribus aut cognatis ex parte patris.

[13] Si autem premisse figurationes cognatorum quidem signorum non extiterunt, in femininis autem fuerint locis, faciunt et ita procliues et ad subigi et subigere per omnem modum promptos, in quibusdam autem formationibus et impudicos, ut in anterioribus et posterioribus | Arietis et in Yade et in natibus et in posterioribus Leonis et in facie Capricorni.

[14] In angulis autem existentes: in primis quidem duobus angulis, scilicet orientali et meridionali, omnino patefaciunt passiones <et> in publicis locis producunt, in | extremis autem duobus, scilicet occidentali et boreali, faciunt spadones et canales habentes aut steriles aut imperforatas; Marte autem astante, abscisos aut fricatores.

[15] Et uniuersaliter autem omnimodam quandam dispositionem ad uenerea habebunt, in uiris quidem a Marte | consideramus; a Venere quidem enim et Saturno separatus, testificatus autem a Ioue, mundos et reuerentes circa uenerea facit et coniectantes solam naturalem oportunitatem; cum Saturno autem solo quidem existens, reuerentes et pigros et infrigidatos efficit.

V (**α** =**YΣ** **β** =**DMS** **γ** =**ABC**) 240 iuniores masculos aut iuniores feminas: νεωτέρους ἢ νεωτέρας **Σ M** (*Hüb.*)] *inv.* **Y S γ**: νεωτέρας **V** 242 communibus: ἐπικοίνοις **α β γ** Heph. (*Hüb.*)] ἐπικινδύνοις **V** 245 aut matrum: ἢ μητέρων *cett.*] ἡμετέρων **V** 251 in femininis: ἐν θηλυκοῖς **Σ β γ** Heph. Procl. (*Hüb.*)] ἐν θηλυκῶν **V Y** 251-2 subigi ... subigere (διατεθῆναι ... διαθεῖναι ?)] *inv.* **α S** (*Hüb.*), *cf.* **D M γ**: διαθεῖναι ... διαθεῖναι **V**; ***hic desinit* Σ**

V (**Y** **β** =**DMS** **γ** =**ABC**) 253 in[1]: ἐπί τε *cett.*] ἐπίπαν **V** 253-4 et in natibus et in posterioribus (Leonis): *fort. dupl. lect.* **G** *pro* καὶ τῶν ὀπισθίων] καὶ τῶν ὀπισθίων **M S** Heph. (*Robb.*: «*fort. recte*»): καὶ τῆς (*om.* **D**) κάλπιδος καὶ τῶν ὀπισθίων (-θέων **V**) **V Y D γ** (*Hüb.*) 255 primis (*cett.*)] τὸν πρῶτον **V** 258 canales habentes: αὐλίσκους **Y**, Co (*sic*, p. 157)] αὐλικοὺς **V** *et cett.* Co le (p. 157) (*Hüb.*) 260 omnimodam: παντοδαπήν **V Y D γ**] ποδαπήν (*Hüb.*), *cf. cett.*

---

239 implicationes] ampli- *Ox* 244 matribus] *ex* mulieribus *corr.* *Ox* 250 non] in *ante corr.* *Ox* 251 autem] aut *Ox* 254 Capricorni] -nii *Ox* 256 publicis] puplicis *Ox* 260 quandam dispositionem] quadam -tioni *Ox* 264 reuerentes] reuentes *Ox*

[16] Configuratis autem Venere et Ioue facile mobiles quidem et concupiscitiuos, continentes autem et supersedentes et turpitudinem cauentes, [autem] cum sola autem Venere aut etiam cum Ioue cum ipsa existente, Saturno <absente>, luxuriosos et uoluptati inherentes et undecumque sibi ipsis delectationes | procurantes; et si hic quidem planetarum uespertinus fuerit, hic autem matutinus, et ad masculos et femellas conuenienter se habentes, non ualde passionabiliter tamen ad neutras personas.

[17] Si autem ambo uespertini, ad femininas solas precipites, femininis autem existentibus signis et ipsos subactos, | si autem ambo matutini, ad pueros solos egritudinales, masculinisque existentibus signis ad omnem masculorum etatem. Et si quidem Venus occidentalior fuerit, humilibus femininis aut seruilibus aut alienigenis conuenientes, si autem Mars, excellentibus aut uirilibus aut dominabus.

[18] In mulieribus autem Venerem considerandum; configurata autem Ioui aut etiam Mercurio temperatas et mundas facit circa uenerea, et Saturno autem absente Mercurio configurata motas quidem et appetitiuas, <re>uerentes tamen et pigritantes | secundum plurima et turpitudinem cauentes.

[19] Marti autem solo quidem coexistens aut etiam configurata Venus facit luxuriosas et pronas et magis quiete letatrices; si autem et Jupiter ipsis assistat, et si quidem Mars sub radiis fuerit, coeunt cum seruis humilioribus aut alienigenis, | si autem Venus, cum excellentibus aut dominis, meretricum aut adulterarum more se habebunt; et si quidem feminizati fuerunt locis aut figuris planete, ad subigi solum pronas, si autem masculinizati, et ad subigere mulieres.

[20] Saturnus tamen | premissis figuris coaptatus, si quidem et ipse fuerit feminizatus, solarum impudiciarum causa fit, si autem orientalis et masculinizatus, scortatrices instituit aut efficit amatrices scortatorum. Ioue quidem rursum semper coassumpto ad decentius passionum, | Mercurio autem ad diuulgabilius et facile peccabilius.

V (**Y** **β** =**DMS** **γ** =**ABC**) 272 ad femininas: πρὸς θηλυκὰ *cett.*] πρὸς καθολικὰ **V** (277 autem^{2}] γὰρ *codd.*) 283 alienigenis: ἀλλοφύλοις *cett.*] -φίλοις **V** 285 se habebunt: ἐπέχουσι **V** **D** **γ**, *cf.* **Y**] -ουσαι **M** **S** (*Hüb.*) 288 solarum (μόνων)] μόνος (*diversis locis*: -ον **Γ**) *codd.* (*Hüb.*): *om.* *Heph.* (289 scortatrices instituit: ἐπὶ τέγους ἵστησιν Heph. (*Hüb.*)] ἐπιτέγους ἵστησιν **V** **D**, *cf.* *cett.*) 290 coassumpto: συλλαμβανομένου *cett.*] -μένων **V**

---

267 autem] + cum (*exp.*) *Ox* 268 sibi] sub *Ox* 269 hic^{2}] hoc *Ox* matutinus] -is *Ox* 275 seruilibus] secundum libet *Ox* 278 facit] -iunt *Ox* 281 aut] autem *Ox* 282 quiete] quidem *Ox* 283 Mars] + exultat (*exp.*) *Ox* 284 Venus] Veus *Ox* 287 tamen] cum *Ox* 288 impudiciarum] *an* impudicitiarum *scrib.*?

De natis
Capitulum 6^m^

[1] Quoniam autem ad locum de nuptiis qui de natis <consequitur>, considerare oportebit eos qui apud locum qui secundum uerticem uel apud | allatum locum, hoc est locum boni demonis uel fortune, coexistunt uel configurantur, si autem non, eos qui apud diametros ipsorum; et lunam quidem et Iouem et Venerem ad dationem natorum accipere, solem autem et Martem et Saturnum ad defectum natorum aut paucitatem natorum; Mercurium autem, | ad quodcumque ipsorum extiterit configuratus, communem, et datorem quidem quando orientalis fuerit, ablatorem autem quando occidentalis.

[2] Datores quidem igitur simpliciter ita positi et soli existentes unicos dant natos, in bicorporibus duplices, et in | femininis signis, similiter autem et in animalibus multi spermatis, uelud in Piscibus et Cancro et Scorpione, duplices aut etiam plures. Et masculinizati quidem signis et ad solem configurationibus masculinos, feminizati autem femininos; superati autem a maleficis aut etiam in sterilibus | locis existentes, uelut Leone aut Virgine, dant quidem, non autem in bonum neque ad permanendum.

[3] Sol autem et malefici optinentes dicta loca, si quidem in masculinis signis fuerint aut sterilibus et a beneficis non superati, sunt significatiui perfecte natorum priuationis, | in femininis autem aut in signis multi spermatis existentes aut a beneficis testificati, dant quidem, sed lesuris infectos aut pauci temporis.

[4] Partitionibus autem ambabus rationem habentibus ad signa natorum factiua, datorum natorum abiectiones fiunt aut omnium aut paucorum existentium, secundum excessus | testificantium penes utramque partitionem, quoscumque utique inuenerimus aut plures aut potentiores in existendo orientaliores aut magis angulares aut in superando aut in superferri.

V (**Y** **β** =**DMS** **γ** =**ABC**) 298 Venerem (*cett.*)] Ἀφροδίτι **V** 298-9 solem ... Martem ... Saturnum (*cett.*)] ἡλίου ... Ἄρεως ... Κρόνου **V** 304 duplices: δισσὰ **V D**, *cf.* διδυμογωνίαν δωροῦνται Procl. (264, -2)] δὲ **M S γ** Heph. (*Robb. Hüb.*): δὲ δισσὰ **Y** (*Boer*) 306 masculinizati: ἠρρενωμένοι **Y β B C** (*Hüb.*)] -μένα **V**: -μένοις **A** 307 feminizati: τεθηλυσμένοι **Y D C** (*Hüb.*)] -μένα **V**: -μένοις *cett.* 308 locis: τόποις **V** (*Robb.*)] ζῳδίοις **β γ** Heph. (*Hüb.*): τόποις ἢ ζῳδίοις **Y** («fort. recte»): ἢ ἐν τόποις ... ἢ ἐν ζῳδίοις Procl. (265,9-10) uelut: οἷον **Y M S γ** (*Hüb.*)] οἷόν ἐστι **V D** Procl. (*Robb.*) 315 existentium: ὄντων **V**] *om. cett.* (*Hüb.*)

---

300 configuratus] -tur *Ox* 304 duplices] -plic' (dupliciter ?) *Ox* 306 aut] autem *Ox* 307 autem[1]] + et *Ox* 314 autem] aut *Ox* rationem] *ex* ratione *corr. Ox* 315 fiunt] *an* fient *scrib.* ? 318 in superando ... in superferri] *coniuncte Ox*

[5] Si quidem igitur dominatores dictorum signorum | orientales extiterint, existentes datores natorum, aut in propriis fuerint locis, gloriosos et insignes faciunt datos natos; si autem occidentales et in locis aliene partitionis, deiectos et uiles.

[6] Et si quidem concordes hora et sorte fortune deprehendantur, | amicabiles parentibus et acceptos hereditantes ipsorum substantias, si autem inconnexi aut oppositi, rebelles et inimicitias facientes et nociuos et non recipientes parentum substantias. Similiter autem et si quidem inuicem fuerint configurati concorditer qui natos <dant>, | permanent dati in amicitia fraterna et honoratiui inuicem, si autem inconnexi aut in dyametrali situ inimicabiliter et insidiose dispositi.

[7] Particularia autem rursum utique aliquis coniecturabit utens <in> unoquoque planetarum dationem faciente pro | ascendente, et <a> reliqua dispositione sicut in natiuitate faciens considerationem de totalioribus.

De amicis et inimicis
Capitulum 7m

[1] Amicabilium autem dispositionum et contrariarum, quarum maiores quidem et durabiliores uocamus compassiones et | inimicitias, minores autem et minoris durabilitatis sinastrias (uel beneuolentias) et obsistentias, consideratio nobis erit secundum modum hunc. In hiis quidem que considerantur penes magna simptomata obseruare oportet loca principalissima ambarum natiuitatum, hoc est solarem et lunarem et | horoscopum et locum sortis fortune.

[2] Quoniam quidem in <eisdem> signis existentia aut uicissim permutantia loca aut omnia aut plura, et maxime quando horoscopi circa 17 gradus ab inuicem destiterint, faciunt <compassiones> indeci<di>biles et indissolubiles | et inculpabiles. In <in>connexis aut in diametrizantibus constituta, inimicitias maximas et contrariationes multi temporis.

V (**Y** **β** =**DMS** **γ** =**ABC**) 319 existentes: ὄντες **V Y** Procl. (*Robb.*)] ἔσονται **β γ** (*Hüb.*) 320 aut: ἤ **V Y**] εἰ δὲ **D M γ** (*Boer*): εἰ **S** Heph. (*Robb.*): ἐὰν δὲ *coni. Hüb.* 322 deprehendantur: καταλαμβάνωνται **M S B**pc **C** (*Hüb.*)] -βάνονται **V Y D A B**ac 323 acceptos (**V**)] + καὶ *cett.* (*Hüb.*) 325-6 concorditer qui: συμφώνως οἱ **G**o (*Hüb.*)] ἢ συμφώνως **V**: συμφῶνος (*sic*) οἱ **Y**: ἢ σύμφωνα **γ**; *cf. cett.* 340 17: ιζ′ **Y β**] ζι′ **V**: ζ′ **γ**

---

320 datores] -rum *Ox* 322 hora et] horret *Ox* 323 inconnexi] inde connexi *Ox* 326 fraterna] superna (frat- → ſup-) *Ox* 327 inconnexi] *disiuncte Ox* inimicabiliter] inmutabiliter *Ox* 336 erit] *ex* erat *corr.* (?) *Ox* hunc] habens *Ox* considerantur] -atur *Ox* 341 destiterint] *fort.* dist- *scrib.* indecidibiles, *cf. supra* 77] indecibiles *Ox* 342 aut] autem *Ox* inimicitias] in inimicitias *Ox*

[3] Neutro autem modo existentia, sed solum in signis configuratis, si quidem in trigonalibus aut sextilibus, minores faciunt compassiones, | si autem in quadrangularibus, minores contrariationes, ita ut fiant per tempora in amicitiis quidem collocutionis euitationes et parui affectus uerba, quando malefici configurationem obsedeant, in inimicitiis autem compactiones et inconstantie secundum | accessus beneficorum in configurationibus.

[4] Sunt autem amicitie et inimicitie species tres: aut enim propter electionem ita se habent ad inuicem aut propter utilitatem aut propter delectationem et tristitiam; quando quidem omnia aut plura predictorum locorum coaptantur ad inuicem, | amicitia congregatur ex omnibus speciebus, sicut, quando inconuenienter, inimicitia. Quando autem loca luminarium solum, propter electionem que est optima amicitia et infallibilissima et inimicitia pessima et infidelis; similiter autem, quando loca sortis fortune, propter utilitates, quando autem horoscopi, propter delectationes et | tristitias.

[5] Obseruandum autem configuratorum locorum superationes et planetarum aspectus; in quibuscumque quidem enim natiuitatibus fuerit figurationis superioritas, si idem uel propinquissime sit signum aut superillatio, | illi principalius amicitie aut inimicitie attribuendum; in quibuscumque autem aspectus planetarum melior ad beneficentiam et uirtutem, illis attribuendum ex amicitia magis commodum et inimicitie directibilius.

[6] In hiis autem que per tempora eueniunt aliquibus, temporales sinastrie (uel beneuolentie) et obsistentie, attendendum ad motus planetarum secundum utramque natiuitatem, hoc est secundum que tempora dimissiones planetarum alterius natiuitatis superueniunt locis planetarum alterius natiuitatis; | fiunt enim secundum hec amicitie particulares ad tempus durantes, minimum quidem quod

---

V (**Y** **β** =**DMS** **γ** =**ABC**) 345 trigonalibus (**V** Procl.)] + εἰσὶ **Y**: + εἶεν **β** **γ** Heph. (*Hüb.*) (346 fiant] + τινας *codd.*) 348 compactiones: σπονδὰς **β** **γ** Heph. (*Hüb.*)] σπουδὰς **V** **Y** inconstantie: ἀκαταστασίας **V** **D** **γ**] ἀποκαταστάσεις **Y** **M** **S** (*Hüb.*) 350 (Sunt* autem] ἐπεὶ δὲ **V** **Y** **D** **B** **C**: ἐπειδὴ **M** **S** **A**) tres aut: τρία ἢ *cett.*] τὴν εἰ **V** 351 habent (*cett.*)] ἔχωσιν **V** 354 loca: οἱ **M** **S** (*Hüb.*), *cf.* οἱ ... τόποι Heph.] ἡ **V** **Y** **D** **γ** 356 (sortis: τοῦ κλήρου] τὸν κλῆρον **V** **Y**: κλῆροι **β** **γ** Heph.: τῶν κλήρων **G**° Procl., *Boer Robb. Hüb.*) utilitates: χρείας **V** **Y** **D** **γ**] χρείαν **M** **S** Heph. Procl. (*Hüb.*) (horoscopi] τῶν ὡροσκόπων *codd.*) 359 fuerit: ᾖ *cett.*] ἤτοι **V** 360 uel propinquissime (ἢ ἔγγιστα ?): ἢ τὸ ἔγγιστα **V** (*Hüb.*)] ἔγγιστα *cett.* sit: ᾖ **γ** (*Hüb.*), *cf.* Heph.] ἦν **V** **β** (*Boer*): ἢ **Y** aut superillatio: ἢ ἐπαναφορά **V** **γ** (*Boer*)] τῇ ἐπαναφορᾷ **M** **S**, *cf.* Heph. (*Hüb.*); *cf.* **Y** **D** principalius (**M** **S**)] + καὶ ἐπιστατικώτερον *cett.* (*Hüb.*) (361 inimicitie] ἐκ τῆς ἔχθρας) 365 attendendum: προσεκτέον *cett.*] *om.* **V** 368 amicitie] + καὶ ἔχθραι *codd.*

---

344 existentia] -ias *Ox* 347 euitationes] -dationes *Ox* et] *it.* *Ox* parui] parus *Ox* 348 autem] aut *Ox* 349 beneficorum] -ciorum *Ox* 350 Sunt] *an* *quoniam (qm̄) *falso ut* sunt (sūt) *legit Ox* ? 352 quando] qui *Ox* 357 tristitias] -ia *Ox* 368 amicitie] *an* et inimicitie *addendum* ? minimum] nimium *Ox*

usque ad dissolutionem ipsam, plurimum autem quod usque ad apprehensionem alterius alicuius superuenientium planetarum.

[7] Saturnus quidem igitur et Jupiter aduenientes super loca inuicem faciunt amicitias per confederationes | aut agriculturas aut hereditates, Saturnus autem et Mars pugnas et insidias que secundum electionem, Saturnus autem et Venus colligationes per personas connatas, cito tamen remissas, Saturnus autem et Mercurius conuictus et communiones dant, et acceptationem et negotiationem aut secreta.

[8] Jupiter autem et Mars sodalitates per presulatus et dispensationes, Jupiter autem et Venus amicitias per personas femineas aut per religiones in sacris aut per reuelationes aut talia, Jupiter autem et Mercurius simul conuersationes propter sermones et scientias et electionem philosophicam. Mars autem et Venus | colligationes propter amores et adulteria aut quietes fallibiles et ad multum deficientes, Mars autem et Mercurius inimicitias et conclamationes et lites propter negotia aut farmacorum occasiones. Venus autem et Mercurius conuictus qui propter artem aliquam aut musam aut compactionem a litteris | aut a personis femininis.

[9] Eam quidem igitur que ad magis et minus intentionem et remissionem sinastriarum aut contrariationum discernendum ex dispositione acceptorum locorum ad prima et principalissima quatuor loca (quoniam in | angulis quidem uel sortium uel luminarium existentia euidentiores faciunt significationes, alienatis autem ipsis, ineuidentes), eam autem intentionem que ad maius nocumentum uel maiorem profectum aliis ex proprietate planetarum aspicientium dicta loca in bonum uel malum.

[10] Singulariter autem de seruis locus aut sermo et de compassione dominorum ad ipsos aut duritia ex mali demonis (uel infortunii) signo accipitur et ex naturali ydoneitate planetarum aspicientium locum secundum ipsam natiuitatem | et secundum accessus aut diametrizationes, et maxime, quando dominatores signi uel concordes fuerint locis natiuitatis dominatoriis uel contrarias fecerint configurationes.

V (**Y** **β** =**DMS** **γ** =**ABC**) 372 confederationes: συστάσεις **V** Procl. (*Robb.*)] συστάσεως **β** **γ** (*Hüb.*), *cf.* **Y** 375 dant: διδόασιν **B**: -δώασιν **Y**] -δόασιν καὶ διὰ δόσιν **M S**: διὰ δόσιν **V D A C** (*Hüb.*) 381 aut quietes fallibiles: ἢ νωθείας ἐπισφαλεῖς δὲ **S γ** (**C** *ante corr. ut vid.*) **c**] ἢ νοθείας ἐπισφαλεῖς δὲ **V Y D M** (*Hüb.*) et ad multum deficientes (*sic*): καὶ οὐκ ἐπὶ πολὺ διευθυνούσας **Y M** (*Hüb.*)] καὶ ἐπὶ πολὺ διευθυνούσας **V D S γ** 383 conuictus (*cett.*)] + ποιεῖ **V** 387-8 angulis ... sortium: κέντρων... κλήρων *cett.*] κέντρον ... κλῆρον **V** 388 alienatis: ἀλλοτριωθέντων **V**] -θέντες *cett.* (*Hüb.*) 396 concordes fuerint: συμφωνῶσι **V** (*Robb.*)] -νως **Y β** Heph. (*Hüb.*), *cf. cett.* locis natiuitatis (**V D γ**, *Boer Robb.*)] *tr.* **Y M S** (*Hüb.*) contrarias: ἐναντίους **Y β γ** (*Robb.*)] -ως **V** Heph. (*Hüb.*)

---

377 sodalitates] solidationes sodalitates (sodali- *ex* solida-) *Ox* 383 occasiones] -ne *Ox* 385 sinastriarum] sinistri- (*ex* sinisteri-) *Ox* 390 aspicientium] aspientium *Ox* 394 ydoneitate] -nietate *Ox*

De peregrinatione
Capitulum 8^m^

[1] Qui uero de peregrinatione locus deprehenditur per situm luminarium ad angulos, amborum quidem, maxime autem lune. Occumbens enim aut cadens ab angulis peregrinationes et locorum permutationes efficit; ualet autem consimile aliquando et Mars, aut occumbens aut etiam | cadens ipse ab eo qui secundum uerticem, quando diametralem aut tetragonalem habuerit situm ad luminaria.

[2] Si autem et sors fortune in signis facientibus peregrinationem ceciderit, habentes totam uitam et conuersationes et operationes in loco extraneo perseuerant. Beneficis quidem igitur aspicientibus | dicta loca aut illatis super ipsa, gloriosas habebunt et lucrosas in peregrinatione operationes et reditus celeres et sine impedimento; maleficis autem, laboriosas et nociuas et periculosas et uix regressiuas, consideratione commixtiua ubique coassumpta | secundum dominium configuratorum ad eadem loca, sicut in primis determinauimus.

[3] Vt in totum autem dicatur, in regionibus quidem matutinalium quartarum incidentibus luminaribus, ad partes habitationum ad orientem | et meridiem peregrinationes fieri accidit, in hiis autem que ad occasus uel etiam in ipsa occumbente, ad boreales et occidentales. Et si quidem unius forme extiterint signa facientia peregrinationem aut ipsa aut planete dominatores ipsorum, diutinas et | secundum tempora faciunt peregrinationes; si autem bicorpora aut duplicis forme, continue et ad plurimum tempus.

[4] Iupiter quidem igitur et Venus domini existentes locorum et luminarium facientium peregrinationem, non solum sine periculo, sed etiam complacentiores faciunt peregrinationes: aut enim | a presidentibus in regionibus aut per occasionem amicorum proficient, cooperante ipsis bona consistentia aeris et oportunorum habundantia. Accedente autem ipsis et Mercurio, sepe et per ipsum dictum occursum profectus | et honores eueniunt.

V (**Y** **β** =**DMS** **γ** =**ABC**) 403 aut etiam ... ipse: ἢ καὶ αὐτὸς ... **V** (*Robb.*)] καὶ αὐτὸς ἢ ... **M S γ** (*Hüb.*), *cf. cett.* 406 perseuerant: διατελοῦσιν *cett.* (*Hüb.*)] -ῶσιν **V** (408 in: ἐπὶ **Y M S** Heph. (*Hüb.*)] ἀπὸ **V D γ**) 416 aut ipsa: ἤτοι αὐτὰ **Y β γ** Heph., *cod.* P (*Hüb.*)] ἢ τοιαῦτα **V** Heph., *cod.* C 417-8 duplicis forme: δίμορφα *cett.*] διάμορφα **V** 422 proficient*: παραπέμπονται *cett.*] περι- **V** 424 profectus] + καὶ προκοπαὶ καὶ δωρεαὶ *codd.*

---

404 diametralem] -e *Ox* 407 Beneficis] -ciis *Ox* 408 peregrinatione] -es *ante corr. Ox* celeres] -lebres *Ox* 409 laboriosas] -rosas *Ox* 413 ad[2]] et *Ox* 414 occumbente] -es *Ox* 416 facientia] -ienda *Ox* planete] -ta *Ox* 419 domini] -nis *Ox* 421 occasionem] *an* -nes *scrib.* ? 422 proficient] *an* proficiscuntur *scrib.* ?

[5] Saturnus autem et Mars assumentes luminaria, et si maxime per longitudinem distant inuicem, faciunt ea que eueniunt inutilia et magnis periculis inuoluunt: in aquosis quidem existentes signis per difficiles nauigationes et nauigia aut iterum per immeabilia | loca et deserta, in fixis autem per precipitia et immissiones spirituum, in tropicis autem et equinoctialibus per indigentiam oportunorum et egritudinales consistentias, in hiis autem que humane forme per latrocinia et insidias et spoliationes, in terrestribus autem propter bestiarum incursus aut terremotus. [6] | Mercurio autem accedente, propter metheora et accusationes incertas, adhuc autem et propter reptilium et aliorum uenenosorum percussuras.

Obseruata principalitate in proprietate quidem symptomatum, siue proficua siue nociua sint (et hoc est secundum differentiam que attenditur penes causam) et ex dominio | locorum causatiuorum operationis aut possessionis aut corporis aut dignitatis secundum eam que e principio dispositionem; temporum autem facientium significationes ex qualitate accessuum quinque planetarum secundum tempora. Et hec quidem nobis usque tantum figuraliter dicta sint.

De qualitate mortis
Capitulum 9[m]

[1] Residua autem post omnia ea que circa qualitatem mortis speculatione, quam predistinximus per tractata in hiis que de temporibus uite, utrum per dimissionem radii peremptio efficietur uel per | descensum dominatoris ad gradum occumbentem. Si quidem enim secundum dimissionem et obuiationem peremptio fiat, locum obuiationis ad qualitatem mortis conuenit obseruare, si autem secundum descensum ad gradum occumbentem, ipsum occumbentem locum.

V (**Y β** =**DMS γ** =**ABC**) 425-6 et si maxime: κἂν μάλιστα **V Y**] καὶ μάλιστα ἂν **β γ** (*Hüb.*) per longitudinem distant: διαμηκίζουσιν **V**] -κίσωσιν **D M γ** Heph. (*Hüb.*), *cf. cett.* 428 immeabilia (*sc.* loca): δυσοδίων *legit vel intell.* **G**: δυσοδιῶν **D γ** (*Hüb.*)] δυσωδιῶν **V Y M S** immissiones: ἐμβολῶν **Y M S** (*Hüb.*)] + καὶ **V D γ** 435 (principalitate] *om. codd.*) in: ἐπὶ **V Y**] ἔτι **β γ** (*Hüb.*) 436 et hoc est: τοὐτέστι(ν) **V Y** (*Robb.*)] ἐκ **β γ** (*Hüb.*) (secundum differentiam que attenditur* penes causam] τῆς περὶ τὸ αἴτιον ἔσται διάφορᾶς **V Y D**: τῆς περὶ τὸ αἴτιον διαφορᾶς **M S γ**, *Hüb.*) 437 possessionis: κτήσεως *cett.*] κτίσεως **V** 438 e principio: ἐξ ἀρχῆς *cett.*] ἐξῆς **V** 439 facientium] μάλιστα *praem. codd.* quinque planetarum: ε' πλανωμένων **V** Procl. (*Robb.*)] *om. cett.* Heph. (*Hüb.*) (444 quam* predistinximus] προδιαληψόμεθα *codd.* [**Y**]) 445 dominatoris: ἐπικρατήτορος **Y A** (**C**) (*Hüb.*), *cf.* **M S**] ἐπικρατῆρος **V D B** 447 fiat: γένοιτο **Y β γ** (*Hüb.*)] γίνεται **V** Heph.

---

428 immeabilia] in mea- *Ox* precipitia] -pua *Ox* 429 per] *an* propter *scrib.* ?, *etiam* 430 430 egritudinales] -nes *Ox* 436 attenditur] attentior *Ox* 438 autem] aut *Ox* 439 significationes] -tionum *Ox* accessuum] et cessium *Ox* 440 hec] hoc *Ox* 443-4 speculatione quam] -tionem quod *Ox*

[2] Qualescumque enim utique fuerint, uel sequentes ad | loca predicta uel, si non sequantur, primi aliorum ipsis illati, tales et mortes fore distinguendum, coacceptis cum naturis ipsorum ad uarietatem simptomatum et planetis configuratis et proprietate ipsorum peremptiuorum dictorum | locorum, cum per proprietatem animalis tum per naturam terminorum.

[3] S a t u r n u s quidem dominium mortis suscipiens facit fines per egritudines diuturnas et pthises et per reumatismos et contabescentias et rigores febriles et per egritudines splenis | et uentris et per ydropises et per matricales dispositiones, et quecumque secundum habundantiam frigidi existant.

[4] I u p i t e r autem facit mortes a squinanchia et peripneumonia et apoplexia et a spasmis et cardiacis dispositionibus, et quecumque <secundum> spiritus | immoderationem aut prauitatem odoris copulantur.

[5] M a r s autem a febribus continuis et emitriticis et repentinis percussionibus nefreticis et a dispositionibus sanguinem spuere facientibus et a sanguinis fluxibus et aborsibus et partubus et ab erisipelis et aliis molestiis, et quecumque egritudinum secundum incensionem | et immoderantiam calidi mortes inferunt.

[6] V e n u s autem per stomaticas et epaticas et per impetigines et dissentericas dispositiones facit mortes, adhuc autem per deuorationes et fistulas et farmacorum dationes, et quecumque humore habundante aut corrupto | simptoma efficiunt.

[7] M e r c u r i u s autem per manias et extases et per melancolias et per casus et per epilempses et per tussiuas et euaporatiuas egritudines, et quecumque sicco habundante aut corrupto [quod] consistunt.

[8] Propriis quidem igitur mortibus moriuntur secundum dictum | modum transeuntes a uita, quando qui dominium mortis susceperunt in propria uel conuenienti naturali proprietate fuerint existentes, superati a nullo potentium malignari et fallaciorem facere finem.

V (**Y** **β** =**DMS** **γ** =**ABC**) 449 enim (*cett.*)] *om.* **V** sequentes (ἕποντες ?)] ἐπόντες *codd.* 450 sequantur: ἕποσιν **V Y**] ἐπῶσιν **β γ** (*Hüb.*) (456 et uentris et per ydropises] *inv. codd.*) 459 spasmis] + καὶ κεφαλαλγίας *codd.* 460 copulantur: ἐπισυνάπτουσιν **V Y** (*Robb.*)] ἐπισυμπίπτουσιν **β γ** (*Hüb.*) (461 percussionibus] + καὶ *codd.*) (463 aliis molestiis] ὀλέθρων *codd.*)

---

450 sequantur] -atur *Ox* illati] allati *ante corr. Ox* fore] fere *Ox* 451 cum] $\overline{\text{tn}}$ *Ox* 452 proprietate] -tem *Ox* cum] $\overline{\text{tn}}$ $\overline{\text{cu}}$ *Ox* 455 pthises *sic Ox* 457 existant] *an* -unt *scrib.* ? 458 peripneumonia] peripleum- *Ox* 462 sanguinis] -ne *Ox* 464 incensionem] intensionem *Ox* 465 stomaticas] *an* stomachicas *scrib.* ? 466 adhuc] ad hoc *Ox* deuorationes] deuort- *Ox* 467 humore habundante] humano rehabundate *Ox* 468 manias] manies *Ox* melancolicas] malen- *Ox* 469 quecumque] qui- *Ox* 471 quidem igitur] igitur quoddam *Ox* 474 facere] faciunt *Ox*

[9] Violentis autem et notabilibus, quando ambo | malefici dominantur peremptiuis locis, uel simul existentes uel in quadrato aspectu uel in maxime remoto, uel alteruter ipsorum uel etiam ambo solem aut lunam aut etiam ambo luminaria comprehendant. Malitia quidem igitur mortis constituta [ab ipsa] ab ipsorum conuentu, | magnitudine autem ab attestatione luminarium, qualitate autem rursum ab aspectu reliquorum planetarum et a signis maleficos continentibus.

[10] S a t u r n u s quidem enim solem preter partitionem aspiciens quadrangulariter aut diametraliter, in fixis quidem facit | per oppressionem turbarum aut constrictionibus aut strangulationibus pereuntes (similiter autem et si occumbat luna superillata); in bestialibus autem locis aut signis, a bestiis corruptos; et si Iupiter attestetur impeditus et ipse, in publicis locis et publicis diebus cum bestiis pugnantes; | in <opposito autem> oroscopi alterutro luminarium, in carceribus pereuntes. Mercurio autem configuratus, et maxime circa eos qui in spera serpentes aut circa terrestria animalium, a morsibus uenenosis morientes.

[11] Venere autem ipsis superueniente, a farmaciis et | mulierum insidiis; in Virgine autem et Piscibus aut in humidis signis, luna configurata, inundationibus et in aquis suffocatos; circa Nauem autem, et nauigiis incidentes; in tropicis autem aut quadrupedibus, soli existens aut diametrizans aut <pro> sole | Marti, a simptomatibus comprehensos; si autem in medio celi fuerint uel in opposito medii celi, ab alto precipitatos.

[12] M a r s autem soli preter partitionem aut lune quadrangularis aut diametralis existens, in humanis quidem | signis, in seditionibus sue gentis aut ab aduersariis facit ingulatos aut sibi ipsis manus inicientes; propter mulieres autem et mulierum inuidias, quando Venus ipsis testificetur; et si Mercurius autem ipsis testificetur, a piratis, latronibus aut a malignis pereuntes; | in animalibus autem imperfectis et truncatarum partium aut secundum Gorgonum Persei, decapitatos aut membris detruncatos; in Scorpione autem et Tauro, adustionibus uel incisionibus aut decisionibus medicorum et spasmis morientes.

V (**Y** **β** =**DMS** **γ** =**ABC**) (485 impeditus] κεκακωμένος *codd.*; *cf. infra* 506) 487 circa (**β** **γ**, *Hüb.*)] *om.* **V Y** 489 Venere ... superueniente (**Y** **β** Heph., *Hüb.*)] Ἀφροδίτη ... προσγενομένη **V**; *cf.* **γ** 490 luna configurata: τῆς σελήνης συσχηματισθείσης *cett.*] τῇ -νῃ -θείσῃ **V** 491 tropicis: τροπικοῖς (*cett.*)] τρεπτικοῖς **V** 493 comprehensos: καταλαμβανομένους *cett.*] -ας **V** 496 aut (*cett.*)] *om.* **V** 497 sibi ipsis: ἑαυτῶν *cett.*] αὐτῶν **V** 498 inuidias: φθόνους (?)] φονέας *codd.* 499 (testificetur*] συσχηματισθῇ *codd.*) 500 imperfectis et truncatarum partium] *inv. codd.* (truncatarum partium: μελοκοπουμένων *Robb. Hüb., fort. coni.* **G**)] μελεοκοπ- *codd.*) (501 membris detruncatos: μελοκοπουμένους *Robb. Hüb., fort. coni.* **G**] μελεοκοπ- *codd.*)

---

484 autem] aut *Ox* corruptos] -tus *Ox* 486 pugnantes] pung- *Ox* 487 configuratus] -tur *Ox* 490 inundationibus] in nudationibus *Ox* 492 quadrupedibus] quadri- *Ox* 496 humanis] humidis *Ox* 497 ingulatos] *ut* inguliatos *vel* ingloriatos *abbr. Ox* autem] aut *Ox* 499 testificetur*] *an perperam ex linea praeced. repetitum est* ? 500 imperfectis] -tas *Ox* 501 in] que (q̄) *Ox*

[13] In medio autem celi uel in opposito medii celi, crucibus | erectos, et maxime circa Cepheum et Andromedam; in occumbente uel contraascendente, uiuos aduri; in quadrupedis autem, a casibus et confractionibus et simptomatibus morientes. [14] Ioue autem huic testificante et coimpedito, | notabilibus iterum pereunt condempnationibus et iris presidum aut regum.

Si simul autem sint inuicem malefici et ita diametrizantes in aliqua dictarum dispositionum causatiuarum, cooperantur adhuc magis ad malitiam mortis, | eo quod penes qualitatem dominio facto circa ipsius peremptiui locum existentem; et hoc aut per simptomata mortis duplicia aut secundum qualitatem effecta, quando ambo rationem habeant ad peremptoria loca.

[15] Tales autem et sepultura expertes | deprehenduntur, deuorantur autem a bestiis uel auibus, quando circa conformia signorum malefici extiterint, nullo beneficorum testimonium reddente loco sub terra aut locis peremptiuis. Peregre autem mortes fiunt planetis optinentibus peremptoria loca in | peregrinis locis incidentibus, et maxime quando luna assistens aut quadrangulariter aut diametraliter aspexerit dicta loca.

De temporum diuisione
Capitulum 10m

[1] Pertractato autem a nobis summatim modo | speculationis secundum unamquamque speciem usque ad sola ipsa, sicut in principio premisimus, negotia que accipiuntur in grosso, reliquum utique erit adicere secundum eundem modum quecumque et circa diuisiones temporum debent considerari naturaliter et consequenter hiis que in parte expositis | negotiis.

V (**Y** **β** =**DMS** **γ** =**ABC**) 505 et simptomatibus: καὶ συμπτωμάτων **V Y** (*Robb.*), *cf.* Procl. (ἀπὸ συμπτωμάτων καὶ κλασμάτων 281.3)] ἢ συρμάτων **β γ** (*Boer*): καὶ συρμάτων Heph. (*Hüb.*) (506 coimpedito] συγκακωθέντος *codd.*) 507 et ... aut (*cett.*)] *tr.* **V** 510 locum: τόπον **V Y D**] τόπου **M S γ** Heph. (*Hüb.*) (510-1 et hoc aut] ἢ καὶ *codd.*) 511 per: διὰ τῶν **V Y** (*vel* διὰ τῶν τῶν ?)] πόλλων τῶν **β γ** (*Boer Robb. Hüb.*) duplicia (**V Y**)] ἢ *praem.* **β γ** (*Hüb.*) qualitatem] + ἢ κατὰ τὸ πόσον *codd.* 522 sola ipsa: μόνων αὐτῶν **β B C** («fort. recte»)] μόνον αὐτῶν **V Y A** (*Hüb.*)

---

503 crucibus] cruribus *ante corr.* *Ox* 504 Andromedam] -dem *Ox* 505 autem] aut *Ox* 507 iris] ius *Ox* 508 Si simul] similiter *Ox* diametrizantes] -zates *Ox* 514 uel] ut *Ox* conformia] -for[a] *Ox* 524 quecumque] quem- *Ox* debent] dn̄t *Ox* 525 expositis] exponetis *Ox*

[2] Sicut igitur in omnibus simpliciter geneaticis sermonibus seu speciebus ante particularia preexistit quoddam fatum maius, ipsum uidelicet quod r e g i o n u m ipsarum, cui que secundum unumquemque in grosso considerantur circa natiuitates subesse nata sunt, ut corporum | forme et animarum proprietates et morum <et> legitimorum uicissitudines.

[3] Et oportet naturaliter speculantem semper primam et principaliorem <causam> scire, ut non lateat aliquando secundum consimilitudinem natiuitatum eum quidem qui in Ethiopia nascitur, uerbi gratia dicere albi coloris aut dependentes | crines habentem, Germanum aut Galatam nigri coloris aut crispi capitis, aut hos quidem [in] domesticos moribus et rationis amatores aut amantes speculari, eos autem qui in Grecia animabus siluestres et ratione ineruditos; aut rursum secundum consuetudinum et legitimorum proprietatem in conuictibus, | uerbi gratia Ytalico secundum genus sororis connubium proponens, quod oportunum erit Egiptio, huic autem matris, oportunum <ei> qui de Perside; et uniuersaliter preaccipere circumstantias uniuersales fati, deinde particulares ad maius uel minus adaptare.

[4] Eodem modo et in | t e m p o r u m diuisionibus etatum differentias et ydoneitates ad singulos effectuum necessarium presupponi et considerare, quatinus non secundum communem et simplicem percuntatorem eorum que ad speculationem considerantur lateat nos aliquando infanti quidem predicentes agere et | sponsas ducere, aut aliquid magis perfectorum, decrepito autem filiorum procreationem aut aliquid minorum, sed semel que considerantur per successus temporales secundum consimilitudinem et contingentiam conformium etatibus adaptare.

V (**Y β** =**DMS γ** =**ABC**) (526 geneaticis sermonibus seu speciebus] τῶν γενεθλιαλογικῶν τόπων *codd.*) (529 ut … forme] ὡς τά τε … μορφὰς **V**: ὥσπερ τὰς … μορφὰς **Y**: ὡς τά τε περὶ τὰς … μορφὰς **β γ**, *Hüb.*) (534 habentem] εἰπῶν *codd.*; *an sua sponte* **G** ?) 535 domesticos: ἡμέρους **β γ** (*Hüb.*), *cf.* **Y**] ἡμέρως **V** 538-9 proponens: προθέμενος **β γ** (*Hüb.*)] μὴ *praem.* **V Y** (quod oportunum erit] δέον **β γ** (*Hüb.*): οὓς δέον **V Y**) 542 temporum (*cett.*)] *om.* **V** 544 secundum communem et simplicem percuntatorem …: κατὰ τὸν (*sic*) κοινὸν καὶ ἁπλοῦν τῶν … συμβατικῶν **V**] κατὰ τὸ κοινὸν καὶ ἁπλοῦν τῶν … συμβατικῶν **D γ** (*Hüb.*), *cf.* **Y M S**: *fort. lectionem codicis* **V** (τὸν) *legit et* τῶν … συμβατικῶν *sua sponte redd.* **G** (545 lateat nos: λάθωμεν αὐτούς **Y B C**, *Robb. Hüb.*] λάθωμεν αὐτούς **V β A**, *Boer*) 546 aliquid: τι *cett.*] τοι **V** 548 adaptare: ἐφαρμόζειν **V**] -ωμεν **γ** (*Hüb.*), *cf.* **Y β**

---

528 cui que] cuique *Ox* 531 primam] primum (p^i m̄) *Ox* 534 Galatam] Gall- *Ox* 535 amatores aut amantes] amatores *in* amantres (*sic*) *mut. et* aut amatores aut *mg. add. Ox* 537 conuictibus] coniunctibus *Ox* 540 uniuersales] naturalis *Ox* 542 Eodem] eorum (*exp.*) eodem *Ox* ydoneitates] ydonyetates (?) *Ox* 543 ad singulos effectuum] adsi^s effecti^m *Ox* 544 percuntatorem] -ciintatorem *Ox* eorum que] eorumque *Ox* 546 autem] aut *Ox* 548 contingentiam] -tium (*sed* u *e corr.*) *Ox*

[5] Est enim quedam adiectio una et eadem omnium de | temporis differentiis, communi nature hominum assequens secundum similitudinem et consequentiam ordinis VII erraticorum, incipiens quidem a prima etate et a spera prima a nobis, hoc est a lunari, terminans autem ad nouissimam etatem et ad | ultimam sperarum planetarum, uocatam Saturni; et accidunt ut uere unicuique etatum conformia nature adiecti planetarum, que oportebit obseruare, quatinus uniuersalia quidem temporalium hinc consideremus, particulares autem | <differentias> a proprietatibus inuentis in natiuitatibus.

[6] Vsque quidem enim ad primos fere 4 annos, secundum conuenientem numerum quadriennii, infantis etatem l u n a sortiens humiditatem et | incompactionem corporis et acumen augmentationis et nutriens quasi in totum aquosum et facile permutabilitatem habitus et anime imperfectionem et inarticulationem operata est, hiis que circa efficientiam ipsius accidentibus conformiter.

[7] Ad consequens autem decennium, puerilem uidelicet etatem, secundus | et secundam sortitus speram M e r c u r i u s secundum medietatem numeri XXti annorum, intellectiuum et logicum anime incipit dearticulare et quasi plasmare et theorematum imponere semina quedam et elementa, morumque et ydoneitatum insinuare | proprietates, doctrinis iam et instructionibus et primis exercitiis excitans animas.

[8] V e n u s <autem> adolescentialem et tertiam etatem assumens ad VIII annos consequentes secundum numerum equalem proprie peryodi, inchoat merito efficere motum spermaticorum | meatuum et secundum repletionem ipsorum impetum ad uenereorum coitum, quando maxime nociuus appetitus quidam fit in animabus et intemperantia et ad contingentia uenereorum amor et profluentia et deceptio et precipitis inconsideratio.

V (**Y β** =**DMS γ** =**ABC**) 549 quedam: τις **Y**] *om. cett.* (*Hüb.*) eadem: ἡ αὐτὴ **M S γ** (*Hüb.*)] αὐτῶν **V D**: αὐτὴ **Y** 552 etatem: ἡλικιῶν **Y M S** Heph. (*Hüb.*)] ἡλιακῶν **V D γ** 555 autem (*cett.*)] *om.* **V** 559 (nutriens] τὸ τῶν τροφῶν *codd.*) aquosum (*cett.*)] ὑδατώδεσει **V** 560 inarticulationem: ἀδιάρθρωτον **Y S γ** (*Hüb.*), *cf.* **D M**] ἀρθρωτον (*sic*) **V** 561 hiis: τοῖς **Y S B C** Heph. (*Hüb.*)] τῆς **V D M A** (563 speram] *om. codd.*) 566 insinuare: ἐμφαίνειν **Y β γ**, *cf.* Heph. (*Robb. Hüb.*)] ἐμφαίνει **V** (*Boer*) 568 Venus (*cett.*)] *om.* **V** 570 meatuum: πόρων **Y D γ** Heph. (*Hüb.*)] σπόρων **V M S** et secundum] κατὰ *cett.* (*Hüb.*): καὶ **MS** impetum (**V Y**)] καὶ *praem.* **β γ** (*Hüb.*)

---

550 hominum] omnium *Ox* consequentiam] *fort.* comparationem (παραβολὴν) *scrib.* (*cf. supra* I.40, *Rhet.*, Simpl. *In Cat.*, *e.a.*) 554 conformia] cumfor[a] *Ox* 555 autem] aut *Ox* 557 ad primos] a[i]d plurimos *Ox* conuenientem] 9[i]tatem (communitatem ?) *Ox* 561 efficientiam] -cent- *Ox* 562 decennium] decentium *Ox* 565 semina quedam] scientia ($\overline{\text{scia}}$) quadam *Ox* ydoneitatum] ydoniet- *Ox* 566 insinuare] sinuare *post corr.* (in- *exp.*) *Ox* 569 proprie peryodi] proprio peryoda *Ox* efficere] afficere *Ox* 571 fit] sit *Ox* 572 uenereorum] uenerorum *Ox* 573 precipitis] -tii *in* -tis *corr.* (?) *Ox*

[9] Quartam autem et ordine mediam etatem iuuenilem accipiens medie spere dominus s o l ad annos x et ix, dominatiuum iam et liberum operationum facit in anima, uite et glorie et status appetitum et transitum a puerilibus | et fictis peccatis ad attentionem et uerecundiam et amorem honoris.

[10] Post solem autem M a r s quintus suscipiens quod uirile etatis ad equalia sue periodi quindecim annos austeritatem et sufferentiam mali in | uita inducit, sollicitudinesque et anxietates efficit in anima, quasi †ad hec† quendam iam et intellectum parakmes (id est transacti status), imprimens et conuertens ad hoc quod, antequam ueniat prope finem, excellere agendo aliquid sermone dignum cum labore conantium.

[11] Sextus autem I u p i t e r seniorem etatem sortiens, ad duodennium iterum secundum propriam peryodum, corporaliter agere et laborare et turbatione repleri et periclitari in operationibus abicere facit, decentiam autem et prouisionem et a tumultibus eleuationem, adhuc | de omnibus ratio<ci>nationem et consolationem inducit econtra, honoris tunc maxime et laudis et liberalitatis gratia agere instituens cum uerecundia et reuerenda decentia.

[12] Vltimus autem S a t u r n u s extremam et senilem etatem sortitus est usque ad residua uite tempora, | infrigidatis et impeditis corporalibus et anime motibus in impetu et fruitionibus et concupiscentiis et celeritatibus, facta in naturam declinatione in uita indurata et inanimosa et debili et facile fatigabili et ad omnia displicentiam habenti | secundum conformitatem tarditatis motuum.

[13] Proprietates quidem igitur temporum secundum commune et uniuersale nature <considerate> hoc modo figuraliter premisse sint; earum autem que in particulari secundum proprietatem natiuitatum debent accipi, has quidem se-

---

V (**Y β** =**DMS γ** =**ABC**) (575 annos x et ix: τὰ τῆς ἐννεακαιδεκαετηρίδος ἔτη **Y**, *cf.* **β** *et* **γ** (*Hüb.*)] τὰ τῆς ἐνάτης καὶ δεκάτης ἐτηρίδος ἔτη **V**) 576 puerilibus: παιγνιωδῶν **Y M S**] παιγωωδῶν **V**: παιδικῶν ἀταξιῶν **D γ** fictis: ἀνεπιπλάστων **V Y γ** (*Robb.*)] ἀνεπιστάτων **D** Heph. (*Hüb.*): ἀνεπιστάτων πλάστων **M S** 578 quintus: πέμπτος **β γ** Heph. (*Hüb.*)] πέμπτην **Y**: *om.* **V** 580 anxietates: σκυλμοὺς **β B C** (*Hüb.*), *cf.* **A** *et* **Y**] κυλμοῦς **V**: κλυσμοὺς Heph. anima] + καὶ τῷ σώματι *codd.* †ad hec†] αἴσθησιν *codd.* 583 conantium: τῶν μεταχειριζομένων **Y β** (*Hüb.*)] τῷ -μένῳ **V γ** 584 Sextus (**β γ** Heph.)] ἐκτὸς **V**: ἕκτην **Y** 587 de omnibus ratio<ci>nationem et consolationem: ἐπιλογιστικὸν πάντων καὶ παραμυθητικὸν] ἐπιλογιστικὸν (-κῶν **V**) καὶ παραμυθητικὸν **V D γ**: ἐπιλογιστικὸν πάντων καὶ νουθητικὸν καὶ παραμυθητικὸν **Y M S** Heph. Procl. (*Robb. Hüb.*) 589 cum uerecundia (*cett.*)] μετὰ ἰδοὺς **V** 592 et celeritatibus: καὶ ταχείαις **V Y**] καὶ -ας **γ** (*Boer Pingree*): -ας **β**: -αις *coni. Hüb.* 593 in uita: ἐν βίῳ **V** Heph. («haud scio an recte»)] τῷ βίῳ *cett.* (*Hüb.*)

---

574 ordine] -em *Ox* 576 fictis] ftis (*pro* fictis) *Ox* 577 attentionem] -sionem *Ox* 578 equalia] qualia *Ox* 580 †ad hec† (*vel* adhuc): *locus corruptus* (*pro* sensum ?) *Ox* 582 ueniat] + a *Ox* 584 duodennium] -uum *Ox* 585 peryodum] -dem *Ox* 590 sortitus] fortius *Ox* 591 infrigidatis] -ditatis *Ox* 592 celeritatibus] sceler- *Ox* 594 displicentiam] dispic- *Ox* 596 earum] eorum (*dub.*) *Ox*

cundum | presuppositum rursum et secundum totalius faciemus a principalissimis dimissionibus, omnibus quidem et non ab una, sicut in hiis que de temporibus uite. [14] Sed dimissionem quidem que ab ascendente ad corporalia simptomata et peregrinationes, illam autem que a sorte fortune ad | ea que possessionis, eam autem que a luna ad anime passiones et conuictus, illam uero que a sole ad ea que penes dignitatem et gloriam, eam autem que a medio celi ad reliquas et secundum partem uite deductiones, uelut sunt actiones, ad amicitiasque et filiorum procreationes.

[15] Ita enim utique in hiisdem temporibus | non unus erit siue beneficus siue maleficus dominus ipsorum, multis, ut in plurimis, accidentibus sub eisdem temporibus contrariis simptomatibus, quando aliquis amittens personam oportunam acceperit hereditatem, aut etiam egritudine affectus fuerit secundum idem tempus et sortitus sit dignitatem aliquam et profectum, | et sine opere existens natorum pater fiat, et quecumque talia accidere consueuerint.

[16] Non enim secundum idem corporis et anime et possessionis et dignitatis et euenientium bonorum aut malorum de neccessitate in omnibus hiis fortunari quosdam aut iterum infortunari, sed accidit utique forte | et tale in temporibus perfecte felicibus aut lamentatis, quando in omnibus aut plurimis dimissionibus concurrant obuiationes beneficorum omnium aut iterum maleficorum. Raro autem propter imperfectionem nature humane ad utramque extremitatum, facile | autem procliue ad mediocritatem ex uicissitudine bonorum et malorum.

[17] Et dimisssoria quidem loca secundum dictum modum discernemus, obuiantes autem dimissionibus non solum rursum peremptores sicut in | temporibus uite, sed omnes simpliciter assumendum, et similiter non eos qui corporaliter solum aut secundum diametrum aut quadrangulariter situm occursantes, sed etiam eos qui secundum trigonalem et sextilem configurationem.

V (Y β (= DMS) γ (= ABC) 599 temporibus ...: τῶν ... χρόνων *cett.* (*Hüb.*)] τὸν ... χρόνον **V D** 600 dimissionem quidem: τὴν (*sc.* ἄφεσιν) μὲν *cett.*] τῆς μὲν **V** 601 possessionis: κτήσεως *cett.*] κτίσεως **V** 602-3 ***post* ad ea que (πρὸς τὰ 797°) *deficit* V *media in pagina***

Y β (= DMS) γ (= ABC) (Γ) 606 enim utique: γὰρ ἂν **β γ** (*Boer*)] γὰρ **Y** (*Robb. Hüb.*) unus (**β γ**)] *om.* **Y** siue[1] (**Y β**)] *om.* **γ Γ** 608 quando (**Y**)] ὡς *praem.* **β γ** (*Hüb.*) 609 affectus fuerit: καταληφθεὶς **Y** (*Hüb.*)] κατακλιθῇ **Γ β γ** (*Robb.*) 610 (et[3]] ἢ *codd.*) sine opere: ἐν ἀπραγίᾳ **β γ** (*Hüb.*)] ἐν εὐπραγίαις **Y** 612 secundum idem: τὸ αὐτὸ **β γ** (*Hüb.*)] τοῦ αὐτοῦ **Y** 613 euenientium: συμβαινόντων **Y** (*Hüb.*)] συμβιούντων **β γ** (*Robb.*) de: ἐξ **Y**] ὡς *praem.* **β γ** (*Hüb.*) 614 (quosdam] τινα *codd.*) accidit: συμβαίνει **M S**, *cf.* Heph.] -νειν **Y D γ**: -νοι (*Robb. Hüb.*) 616 iterum (**β γ**, *Hüb.*)] *om.* **Y** (617 imperfectionem ... ad] τὸ ... ἀτελὲς μὲν πρὸς *codd.*; *cf. totam sententiam in* **G** *cum interpunctione Robb.*) 620 dimissionibus (**Y** Heph.)] ἐν *praem.* **β γ** (*Hüb.*) 621 eos qui corporaliter: τοὺς σωματικῶς **β γ** (*Hüb.*)] τοὺς σωματικοὺς **Y** Heph.

---

601 illam] -a *Ox* sorte] forte *Ox* 607 dominus] *dub.* *Ox* 609 egritudine] -nem *Ox* 610 et[3]] *an* aut* *scrib.* ? 614 aut] autem *Ox* 618 mediocritatem] medicr- *Ox* 619 obuiantes] obseruantes *Ox* 621 non] ut *Ox* 622 quadrangulariter] *an* -larem *scrib.*?

[18] Et primo quidem dandum tempora secundum unamquamque dimissionem ei qui in ipso dimisssorio | gradu extiterit uel configuratus fuerit; si autem non ita habuerit, ei qui propinquissimam precessionem accepit usque ad eum qui aspicit sequentem gradum in consequentia, deinde huic usque ad eum qui deinceps, et in aliis similiter, acceptis ad dominationem et [in] planetis terminos optinentibus.

[19] Dandum autem rursum gradibus distantiarum annos secundum dimissionem quidem horoscopi equalia numero temporibus eleuationum in proprio climate, secundum eam autem que a medio celi equalem numerum temporibus celi mediationum, secundum eas autem que a reliquis proportionaliter secundum eam que ad angulos | propinquitatem eleuationum aut descensionum aut celi mediationum, quemadmodum in uite temporibus determinauimus.

[20] Vniuersales quidem igitur annorum dominatores accipiemus secundum dictum modum, annuales autem proicientes multitudinem | annorum ab unoquoque dimissoriorum locorum in consequentia penes signum unum, et signi facti abiectione occurrentis dominum coassumentes. Idem autem et in mensibus faciemus abicientes rursum et horum multitudinem a natiuitatis mense a | locis dominium anni suscipientibus, secundum signum dies 28; similiter autem et in diebus; hos enim a natiuitatis die proiciemus a mensium locis, secundum signum dies duos et tertiam.

[21] Attendendum autem et ad superingressus qui | fiunt ad loca temporum quasi non contingentes et [in] ipsis coassumptis ad temporum effectus, hiis quidem igitur qui Saturni ad loca uniuersalia temporum, hiis autem qui Iouis ad loca annualium, hiis uero qui solis et Martis et Veneris | et Mercurii ad loca mensium, hiis autem qui lune transitibus ad loca diurna. [22] Tamquam uniuersalibus

---

Y β (= DMS) γ (= ABC) (Γ) 627 huic*: τούτῳ **β γ** (*Hüb.*)] τούτων **Y** Heph. 630 horoscopi (**Y γ** Heph., *Hüb.*)] ἀπὸ *praem.* **β** (*Robb.*) 631 equalem numerum: ἰσάριθμον **Y β** Heph.] ἰσάριθμα **Γ γ** (*Hüb.*) 632-3 secundum eam que ... propinquitatem: κατὰ τὸν ... συνεγγισμὸν (*coni. Robb.*)] ἢ κατὰ τὸν ... συνεγγισμῶν **Y**: τῷ ... συνεγγισμῷ **β γ** (*Hüb.*) 635 Vniversales — accipiemus (**β γ**)] *om.* **Y** 636 annorum] ἀπὸ τῆς γενέσεως *praem. codd.* ab: ἀφ' **β γ** (*Hüb.*)] ἐφ' **Y** 637 unum: ἓν **β γ** (*Hüb.*)] *om.* **Y** Procl. (637-8 signi facti abiectione occurrentis dominum] τοῦ συντελειουμένου ζῳδίου τὸν οἰκοδεσπότην *codd.*; *fort.* **G** *sua sponte textum codicum explicare conatus est*) 641 proiciemus: ἐκβαλ(λ)οῦμεν **β γ** Heph. (*Hüb.*)] -λλομένας **Y** 642 duos et tertiam: δύο καὶ Γ' **D** Procl.: δύο τρίτον **M S** (*Hüb.*)] *cf.* **Y** *et* **γ** 643 ad[2] (**β γ**, *Hüb.*)] *om.* **Y** temporum (**Y S** Heph. Procl., *Hüb.*)] καθολικοὺς μάλιστα **D M γ Γ** 644 contingentes: τὰ τυχόντα **β γ** (*Hüb.*)] τὸ τυχῶν **Y** hiis] καὶ μάλιστα *praem. codd.* 645 (igitur] *om. codd.*) 646 loca[1]: τοὺς **Y γ** Heph. (*Hüb.*)] *om.* **β** et[1] (**Y β** Heph., *Hüb.*)] *om.* **γ Γ** Martis: τοῦ Ἄρεως **D γ** Heph. (*Robb.*)] ταῖς τοῦ Ἄρεως **Y M S** (*Hüb.*) 647 Tamquam (Heph., *Hüb.*)] καὶ *praem. codd.* (*Robb.*)

---

625 ei qui] eique *Ox* 627 huic*] hinc *Ox* 633 descensionum] decencionum *Ox* 636 annorum] *an* a natiuitate* *addend.* ? 638 occurrentis] -remus *ante corr.* *Ox* 640 dies] die *Ox* 641 proiciemus a] proximemus ad *Ox* 644 quasi non] *inv.* *Ox* 645 qui[1]] et *Ox* temporum] + effectus (*exp.*) *Ox*

quidem temporum dominatoribus existentibus principalioribus ad perficiendum effectum, particularibus autem cooperantibus aut impedientibus secundum conuenientiam aut disconuenientiam naturarum, | superingressibus autem intensiones et remissiones simptomatum efficientibus.

[23] Qualitatis quidem enim uniuersalis proprium et temporis protensionem dimissionis locus et uniuersalium temporum dominus cum domino terminorum significat, | quia coappropriatur planetarum unusquisque ab ipsa natiuitate locis, quorum a principio sortiti sunt dominium. Vtrum autem bonum uel contrarium erit simptoma, deprehenditur per naturalem et commixtiuam dominatorum temporis proprietatem, benefactiuam | aut malefactiuam, et per eam que a principio ad locum obtentum coaptationem aut disconuenientiam.

[24] In quibus autem temporibus magis significabitur effectus, ostenditur per proprietatem configurationum signorum annualium et mensium ad loca causatiua, et per configurationes | secundum superingressus planetarum et secundum dimissiones solis et lune [et] ad annualia et mensualia signa.

[25] Concorditer se habentes ad disposita loca ab eo quod in natiuitate principio et <secundum> superingressus concorditer ipsis configurati bonorum | sunt operatiui circa subiectam speciem, quemadmodum si contrarientur, malorum. Qui autem inconcorditer et preter partitionem diametrizantes quidem aut tetragonizantes transitibus malorum sunt causatiui, secundum alias autem configurationes non adhuc.

[26] Et si quidem iidem et temporum et superingressuum dominentur planete, excedens et immoderata fit effectus natura, siue in bonum siue in malum tendat; multo autem magis, si non solum quia dominatores temporum sunt dominantur

---

Y β (= DMS) γ (= ABC) (Γ) 649 aut impedientibus (**β γ**, *Hüb.*)] *om.* **Y** 654 ab: ἀπ' **Y** Heph.] ἐπ' **β γ Γ** *Robb. Hüb.*; cf. *adn. insequ.* (coappropriatur ... ab [*sic*]) ipsa natiuitate locis (quorum): αὐτῆς τῆς γενέσεως τοῖς τόποις **G°** **γ Γ** (*Robb.*)] αὐτοῖς τῆς γενέσεως τοῖς τόποις **Y M S**: αὐτοῖς τοῖς τῆς γενέσεως τόποις **D** (*Boer Hüb.*) 655 quorum: ὧν **Y** Heph. (*Robb. Hüb.*)] + ἂν **β γ** (*Boer*) (sortiti sunt dominium] ἔτυχον λαβόντες τὴν οἰκοδεσποτείαν **β γ**, *cf.* **Y**) Vtrum *hic* **Y S** Heph. (*Hüb.*] *alio loco* **D M γ Γ** (659-60 per proprietatem configurationum] διὰ τῶν ... συσχηματισμῶν **β γ**, *cf.* **Y**) 660 ad (**β γ**, *Hüb.*)] *om.* **Y** 662 dimissiones: ἀφέσεις **Y**] φάσεις **β γ** (*Hüb.*) 664 sunt: εἰσι **Y γ Γ** (*Hüb.*)] εἰς τὸ **β** 667 causatiui: παραίτιοι **Y S** Heph. (*Hüb.*)] αἴτιοι **D M γ Γ** 669 quidem: μὲν **Y** Heph., Co le (*Hüb.*)] *om.* **β γ Γ** superingressuum: ἐπεμβάσεων **M S γ** Heph. (*Hüb.*)] ἐπιβάσεων **Y D** 670 tendat: ῥέπῃ **β γ** (*Hüb.*)] ῥέπει **Γ**: *om.* **Y** Heph. 671 multo autem magis: πολὺ δὲ πλέον **M γ Γ** Heph. (πλεῖον *Hüb.*)] καὶ ἐπὶ πολ(λ)ὺ πλέον **Y G°**, *cf.* **S** si non solum: ἐὰν μὴ μόνον **Y** Heph. (*Hüb.*)] *om.* **β γ Γ**

---

648 perficiendum] profic- *Ox* 650 intensio- *sic Ox* (*alibi plerumque* intencio- / intentio-) 652 uniuersalis] naturalis *ante corr. Ox* 653 uniuersalium] naturalium *Ox* 656 simptoma] -mata *ante corr. Ox* commixtiuam] commixta̅ *Ox* 658 coaptationem] -ni *Ox* 662 annualia] animalia *Ox* 664 superingressus] supergressum *Ox* 671 non] uero *Ox*

890 speciei cause, | sed et quia secundum principium natiuitatis sortiti fuerint do-
minium ipsius. Secundum omnia autem simul infortunantur uel eufortunantur,
quando aut locus unus et idem ab omnibus dimissionibus extiterit comprehensus,
895 aut, hiis differentibus existentibus, eadem tempora omnes aut | plures obuiationes
similiter beneficas aut maleficas habuerint.

897 [27] Modus quidem igitur temporum considerationis talis aliquis utique
C2 fiet | secundum modum congruentem naturalibus negotiis; particulares autem
adiectiones qualitatis temporalium effectuum, multi sudoris et difficilis expli-
C5 cationis existentes, relinquendum, hic autem maxime | pertractationem euen-
turorum accipiendum, propter propositum nobis a principio, mathematici facili
coniectura ad modum commixtiuum potentis adaptare efficaciam uniuersalis na-
ture astrorum et particularibus similiter secundum consequentiam.

C10 Consummata iam geneatici sermonis specie | summatim, bene utique habebit
huic tractatui conuenientem imponere finem.

Explicit liber quartus Ptholomei

Y β (= DMS) γ (= ABC) (Γ) (672 secundum principium] κατ' αὐτὴν τὴν ἀρχὴν *codd.*) 674 omnibus] + ἢ τῶν πλείστων *codd.* 677 talis: τοιοῦτος **Y** (*Robb. Hüb.*)] τοσοῦτος **β γ** (*Boer*) 678 fiet: γένοιτο **Y**] εἴη *cett.* (*Hüb.*) 678-85 secundum — finem: *hanc conclusionem* (κατὰ — τέλος) *hab.* **Y Gº Φ** (= «*prima conclusio*» *Robb.*, «*altera conclusio*» *Hüb.*)] *aliam conclusionem* (τὰ δὲ εἴδη — συγκράσεως) *exhibent cett.* (= «*prima*» *Hüb.*, «*altera*» *Robb.*)

Y Gº Φ 678 naturalibus: φυσικαῖς *coni. Robb. Hüb.*] δυσικαῖς **Y Gº (Φ** *non citatur*) negotiis: *fort.* πραγματείαις *leg.* **G**] χρηματείαις *Robb. Hüb.*: χριμαντείας **Y**: χρωματίαις **Gº Φ** 679 adiectiones: ἐπιβολὰς *Robb. Hüb.*, -βολλὰς **Y**] -βουλὰς **Gº Φ** multi sudoris: πολύχους *coni. Hüb.*] -χρουν **Y**: -χροαν **Gº Φ**: -χόως *coni. Robb.* 680 relinquendum (ὑπολειπτέον)] *om. codd.*; *an sua sponte add.* **G** ? hic* autem: ἔνθα δὲ] ἐνθάδε *codd.* pertractationem: τὸ διεξοδικὸν *Hüb.*: τὸ -ωδικὸν **Y**] τῶν διεξοδικῶν **Gº Φ** 681 accipiendum: ὑποληπτέον **Gº Φ** (*Hüb.*)] ὑπολυπτεόν **Y**: ὑπολειπτέον *coni. Robb.* 681-2 facili coniectura: τῇ ... εὐστοχίᾳ (*Robb.*)] τῇ ... εὐστοχίαν **Gº**: τὴν ... ἐυστοχίαν **Y**: <κατὰ> τὴν ... ἐυστοχίαν (*Hüb.*), *cf.* **Φ** 682 potentis*: δυναμένου **Y Gº Φ** (*Robb.*)] δύναται *coni. Hüb.* adaptare: ἐφαρμόζειν **Gº Φ**] -μώζεν **Y**: -μόζεσθαι *coni. Hüb.* 683 (et] ἔτι καὶ *codd.*) particularibus: τοῖς ἐπὶ μέρους **Y**] τῆς ἐπὶ μέρους **Gº Φ** (*Hüb.*) 684 Consummata: διοδευομένου **Gº Φ** (*Robb.*): διωδευμένου *coni. Hüb.*] διὸ δυομένου **Y** (iam] δὲ καὶ *codd.*) summatim: κεφαλαιωδῶς **Gº** (*Hüb.*): -οδῶς **Φ**] -οδὸ **Y** bene: καλῶς (*suppl. Robb. Hüb.*)] *om. codd.* utique habebit: ἂν ἔχοι (*Robb. Hüb.*)] ἀνέχει **Gº Φ**: ἀνέχη **Y** 685 huic: τῇδε τῇ **Gº Φ** (*Hüb.*)] τι δὲ τῇ **Y** (imponere: ἐπιθεῖναι (*Robb. Hüb.*)] -θῆναι **Y Gº Φ**) 686 Explicit liber quartus Ptholomei] τέλος τοῦ Πτωλεμαίου **Y**, *cf. cett.*

---

672 secundum] *an* ipsum* *addend.* ? 674 aut] autem *Ox* 680 hic*] hunc *Ox* 681 mathematici] mathema$\overline{co}$ *Ox* 682 potentis] -tes *Ox* 684 iam] *an* etiam *scrib.* ? 685 imponere] in tempore *ante corr. Ox* 686 Ptholemei (*cf.* 1.2)] Tho- *Ox*

I. INDEX GRAECO-LATINUS

Verba graeca in editione Hübneriana (sive in textu sive in adnotatione) inueniantur.

Numeralia graeca perquam raro recensentur.

† = verbum in editione Hübneriana non attestatum.

Verba latina, quae in textu edito inter uncis < > inclusa sunt, praeter pauca in indice non recensentur.

Numeri latini: «I.», ...«IV.» = Liber I ... Liber IV; «IV C» = Liber IV, Conclusio altera in editione Hübneriana.

Numeri arabici = numeri linearum editionis Hübnerianae.

Verba latina quae unum et idem vocabulum graecum reddunt, plerumque secundum ordinem textus recensemus, interdum tamen consuetudinem Guillelmi respicientes translationes praecipue usitatas contra ordinem textus ante cetera laudamus.

Numeri reclinati («1304-1308(*6*)» indicant quotiens verbum latinum vocabulum graecum reddens in lineis quibus numeri adscripti sunt, invenitur.

* = cf. adnotationem latinam vel adnotationem greco-latinam nostrae editioni subiectam.

[h] = cf. adnotationem editioni Hübnerianae subiectam.

II. INDEX LATINO-GRAECUS

Quibus locis vel quotiens singula verba inveniantur, quidve signis *, † significetur in primo indice vel in adnotatione ei praemissa dictum est.

a, ab: ἀπό, ἐξ, παρά *cum gen.*, *cum dat.*, ὑπό *cum gen.*

abalienatio ἀπαλλοτρίωσις

abalieno ἀπαλλοτριόω

abicio ἀποστρέφομαι, ἐκβάλλω

abiectio ἀποβολή

ablatio ἀφαίρεσις

ablatiuus ἀφαιρετικός

ablator ἀφαιρέτης

aborsus ἐκτρωσμός

abrumpo ἀποκόπτω

abscessio ἀποχώρησις

abscido: -scisus ἀπόκοπος

abscisio ἀποκοπή, ἀποτομή

absens ἀπών

absoluo: -lute ἀπολελυμένως

abstineo (a) ἀπέχω

ac: — si ὡς ἄν

Acaya, Acaia, Achaya Ἀχαΐα

accedo προσγίνομαι, προσέρχομαι, συμπρόσειμι: — (ad) ἐπαναφέρομαι: -ens συμπροσγενόμενος: -ente προσιόντος†: -ens (sursum) (*sc.* locus) ἐπαναφορά: ad hec -entes (?) πάροδοι

acceptabilis εὐαπόδεικτος

acceptatio λῆψις

acceptio διάληψις, λῆψις

accepto ἀσπάζομαι

accessus ἐπαναφορά, ἐπέμβασις, ἔφοδος, πάροδος, περίοδος

accido ἐγγίνομαι, ἐκπίπτω, ἐπισυμβαίνω, συμβαίνω, συμπίπτω: -ens σύμπτωμα: -entia (*sic*) τὸ σύμπτωμα

accipio διαλαμβάνω, ἐπιλαμβάνω, καταλαμβάνω, λαμβάνω, μεταλαμβάνω, παραλαμβάνω, προσλαμβάνω: accepturus ὁ μέλλων λαμβάνειν: cum acceperimus εἰληφότες: -iendum παραληπτέον, ὑποληπτέον: acceptus ἐπαφρόδιτος

acclamatio περιβοησία

accusabilis: facile — εὐδιάβλητος

accusatio κατηγορία

accusator κατήγορος

acquiesco *cf.* preces

acquiro περιποιέω, πορίζω

acquisitibilis: facile — εὐπόριστος

acquisitio πορισμός

acquisitiuus (adqui-) ἐμπορικός, περιποιητικός

acronictus ἀκρόνυκτος

actio πρᾶξις: que pertinent ad -iones πρακτικά (εἴδη): multe -ionis πολυπράγμων

actitor δράστης

actiuus *cf.* uirtus

actor: malorum — κακοπράγμων

actus ἐνέργεια

aculeus κέντρον

acumen τὸ ὀξύ

acutus ὀξύς: *cf.* intellectus, ira, sentio

ad εἰς, ἐπί (*cum acc.*), κατά (*cum acc.*), μέχρι, παρά (*cum acc.*), περί (*cum acc.*), πρός (*cum acc., cum dat.*: ad hec πρὸς τούτοις): quantum — περί (*cum acc.*), πρός (*cum acc.*), τόν (τῷ) *ante nomen* (*acc. graec.*): quem ad (τόν *ante nomen*): usque — πρός (*cum acc.*), μέχρι

adaptator ἐντρεχής

adapto ἐφαρμόζω

addextratio δορυφόρησις

addextratus: non — ἀδορυφόρητος

amorosus ἐρωτικός
ampliatio νομή
amplior πλείων: *cf.* ineuidentia, plenitudo
amplius ἔτι
ana: — duo ἀνά δύο
Andromeda Ἀνδρομέδα
angularis ἐπίκεντρος: magis — ἐπικεντρότερος: —, -iter (existens) κεκεντρωμένος: *cf.* oppositus
angulatio κέντρωσις
angulus γωνία, κέντρον, κέντρωσις: in -o ἐπίκεντρος: in -o (fuerit existit) κεκεντρωμένος (ᾖ): in -is existentes κεντρωθέντες: loca -orum οἱ τῶν κέντρων: rectus — ὀρθός: *cf.* celum
anima ψυχή: quantum ad -as ταῖς ψυχαῖς: -e (*gen. sing.*) ψυχικός (*e.g.* ψυχικὴ κίνησις anime motus): -e opera τὰ ψυχικά: dolose -e δουλόψυχος: fulgentis -e λαμπρόψυχος
animal ζῴδιον, ζῷον: per proprietatem -lis ζῳδιακῶς: -lium fictor ζῳοπλάστης: *cf.* medius
animalis ψυχικός
animatus ἔμψυχος
animosus θυμικός
animus: boni -i εὔψυχος: parui -i μικρόψυχος: *cf.* femininus , muliebris
annalis: non — ἄχρονος
annotabiliter ἐπίσημως
annualis ἐνιαυσιαῖος, ἐνιαύσιος: -les ἐτήσιοι
annuatim ἐνιαυσίως
annus ἐνιαυτός, ἔτος, χρόνος (anni uite χρόνοι ζωῆς): -i (*sing. gen.*) τὰ ἐνιαύσια: -i X et IX τὰ τῆς ἐννεακαιδεκαετηρίδος ἔτη: — bonus εὐετηρία: -orum dominator χρονοκράτωρ: tempora -i ὧραι, ἐτήσιοι ὧραι: uiginti -i εἰκοσαετηρίς: *cf.* octo
ante πρό *cum gen.*: *cf.* fio
antecedo πρωτεύω
anteferor προαναφέρομαι
antequam πρίν
anterior ἐμπρόσθιος
antesursumferor προαναφέρομαι
Anthares Ἀντάρης
antiquus ἀρχαῖος, παλαιός: -i παλαιότεροι: -orum amator φιλάρχαιος
anxietas σκυλμός
Apollo Ἀπόλλων
apoplexia ἀποπληξία
apostema ἀπόστημα
apoteca ἀπόθεσις
apparentia (*nom. sing.*) φάσις
appareo ἐπιπολάζω, φαίνομαι: non -ens ἀφανής
apparitio ἐπιφάνεια, φάσις
appellatio προσηγορία, φάσις†
appello προσαγορεύω
appetitiuus ἐπιθυμητής, ὀρεκτικός
appetitus ἐπιθυμία: nociuus — λύσσα
applicabilis: facile — εὐεπίβολος
appono παρατίθημι, προστίθημι
appositio πρόσθεσις
appositiuus προσθετικός
apprehensio κατάληψις
appropinquo ἐγγίζω, συνεγγίζω
aptitudo εὐθηνία (εὐφυΐα ?)
apto ἀπονέμω, συνοικειόω: bene aptatus εὐφυής
aptus πρόσφορος: qui est — ὁ οἰκεῖος: bene — εὔρυθμος, εὐφυής: — factus ἐφηρμοσμένος
apud κατά *cum acc.*, παρά *cum dat.*, πρός *cum dat.*
Apulia Ἀπουλία
aqua ὕδωρ: retractatio -arum ἄμπωτις: *cf.* defectus, inundatio
Aquarius Ὑδροχόος
aquaticum ὑδρολόγιον
aquatilis ἔνυδρος
Aquila Ἀετός
aquilonaris *cf.* borealibicus
aquosus κάθυγρος, ὑδατώδης

Arabia Ἀραβία: — felix Ἀραβία (ἡ) εὐδαίμων
arbor: -orum τὰ δενδρικά
Arcturus Ἀρκτοῦρος
arcus τόξον
ardor καῦμα
area ἅλως
argentum: -ti laborator ἀργυροκόπος
arguitor ἐλεγκτικός
Ariane Ἀριανή
Aries Κριός
arma: amator -orum, amatiuus -orum φίλοπλος: amatrices -orum φίλοπλοι: -orum factor ὁπλουργός: -is gesticulator ὁπλορχηστής
Armenia Ἀρμενία
armonizatio: bona — τὸ εὐάρμοστον
aroma ἄρωμα: -atum uenditor ἀρωματοπώλης
arpocraticus ἁρποκρατικός
ars ἀγωγή, ἔφοδος, μέθοδος, τέχνη: -tis amator, -tium amator φιλότεχνος
arteria ἀρτηρία
artificialis: -ior εὐμήχανος
artificiosus: in malum — κακότεχνος
ascendo παροίχομαι: -ens ὡροσκοπία, ὡροσκόπιος, ὡροσκοποῦν, ὡροσκοπήσας: -ens τὸ τοῦ πόλου ἔξαρμα: ratio -entis λόγος ὡροσκοπικός
ascensio ἀναφορά
ascensionalis ἀναφορικός
Asia: Magna — Μεγάλη Ἀσία: — Minor, Parua — Μικρὰ Ἀσία
Asinus Ὄνος
aspectus ἐπιθεώρησις, συνεπιθεώρησις: in maxime remoto -tu (existens) διαμηκίζων: *cf.* diametralis, impedio, longitudo, quadratus, trinus, sextilis
asper δασύς, τραχύς
aspicio ἀποβλέπω, ἀφοράω, βλέπω, ἐπιθεωρέω: *cf.* diametralis, quadrangulariter
assecutibilis: -i modo παρακολουθητικῶς
assequentia ἐπακολούθησις
assequor ἀκολουθέω, ἐξακολουθέω, ἔχομαι, κατακολουθέω, παρακολουθέω: -quens ἀκολούθως (*vel* -†θος)
assertiuus διαβεβαιωτικός
assigno ἀποδείκνυμι, ἀποδίδωμι
assigno ἀπονέμω
assimilo ἀφομοιόω, ἐξομοιόω
Assiria Ἀσσυρία
assisto πάρειμι, προσγίνομαι
assuesco ἐθίζω: qui -etus est (circa) ὁ ἐνεθιζόμενος (*c. dat.*)
assumo ἐπιλαμβάνω, λαμβάνω, παραλαμβάνω, προσλαμβάνω: -endum παραληπτέον, συμπαραληπτέον
asto πρόσειμι: -ans καθεστώς
astrolabium ἀστρολάβος
astrologia: amator -e φιλαστρόλογος
astrologicus ἀστρολογικός
astrologus ἀστρολόγος
astronomia ἀστρονομία
astrum (*plur. ut neutrum sed etiam ut masculinum = planetae*) ἀστήρ
astutus κακοῦργος, πανοῦργος
athleta ἀθλητής
attamen (at tamen) μέντοι
attempto πειράομαι: magna mala -ans cum formidine θρασύδειλος
attendo διαλαμβάνω, ἔχομαι, κατακολουθέω, προσανέχω, προσέχω: -dendum (ad) προσεκτέον: *cf.* reuelatio
attentio ἐπίστασις, τὸ προσεκτικόν
attestatio ἐπιμαρτύρησις
attestor ἐπιμαρτυρέω
attingo: quod difficile -itur τὸ δυσέφικτον: *cf.* celum
attraho νοσφίζω
attribuo ἀπονέμω, προσνέμω: -endum ἀποδοτέον
auarus πλεονέκτης

auctor δράστης, συγγραφεύς
audacia: proprietas -ie τὸ θρασύ
audax θρασύς, τολμηρός: non — ἄτολμος
audeo: -ens magna mala cum formidine θρασύδειλος
audio: -iendum est ἀκουστέον
auditus ἀκοή
auertibilis: facile — εὐανάτρεπτος
aufero (a, ab) ἀφαιρέω, ὑφαιρέω
augeo αὐξάνω, αὔξω
augmentatio αὔξησις, συναύξησις: paulatiua — αὐξομείωσις
augmentatiuus προσθετικός
augmentator: — operis ἀνυ(σ)τικός
augmento αὔξω, συναυξάνω
augmentum αὔξησις
auguratiuus οἰωνοσκοπικός
augurus οἰωνιστής
auis οἰωνός, ὄρνις (ὄρνεον)
Auis Ὄρνις
aurifex χρυσοχόος
Auriga Ἡνίοχος
aurum: multi -i πολύχρυσος
auster νότος: inter -trum et orientem πρὸς νοταπηλιώτην
austeritas τὸ αὐστηρόν
austerus αὐστηρός: -a operans αὐστηρόπρακτος
australis νότιος: -ior (a) νοτιώτερος (*c.gen.*): -issimus νοτιώτατος
australitas τὸ νότιον
austrolibicus νοτολιβυκός
aut ἤ, καί: — (... aut) ἤτοι (... ἤ)
autem γάρ, δέ, δή, καί, μέντοι: — (et) τε (καὶ): *cf.* si, siue
autumpnalis μετοπωρινός: -e (?) μετόπωρον
autumpnus μετόπωρον
auxiliatiuus ἐπίκουρος
auxiliatus βοηθοῦμενος
auxilium βοήθεια, βοήθημα, ἐπικουρία: per -iis subleuatum (esse) βοηθεῖσθαι
Azania Ἀζανία

Babilonia Βαβυλωνία
Bactriane (*acc.* -em) Βακτριανή
balbus τραυλός
balneator ἐκδοχεύς (?)
barba: pauce -e ὑπόψιλος (-οι)
Barsania Βαρσανία
Basilicus Βασιλίσκος
bellicosus πολεμικός: -sissimus πολεμικώτατος
bellicus πολεμικός
bellum πόλεμος
bene εὖ, καλῶς: *cf.* apto, aptus, cessibilis, compono, coniecturabilis, coniecturatiuus, condextrabilis, communicatiuus, compositus, consilior, consummatiuus, conueniens, conuersor, deductibilis, dego, electiuus, formatus, habeo, habundo, industrius, ingressiuus, instituo, moderatio, mobilis, pecuniosus, procliuis, proportionatus, rithmizatus, sapiens, sententians, stabilis, status, temperatus
benefactiuus εὐεργετικός, εὐποιητικός
beneficiatiuus εὐεργετικός
beneficientia ἀγαθοποιία
beneficium εὐεργεσία, εὐποιΐα
beneficus ἀγαθοποιός, ἀγαθοποιῶν, ἀγαθοποιεῖν δυνάμενος, εὐεργετικός
beneuolentia (beni-) εὔνοια: indifferens — εὐήθεια: excessiuus in — εὐήθης: *cf.* sinastria
beneuolus: in excessu — εὐήθης
benignitas φιλανθρωπία, τὸ φιλάνθρωπον
benignus φιλάνθρωπος
Bestia Θηρίον
bestia θηρίον: cum -iis pugno θηριομαχέω: -iarum alitor θηριοτρόφος
bestialis θηριώδης: -ior θηριωδέστερος
bestialitas τὸ θηριῶδες
bicorpor (*abl.*: -pore -poribus) δίσωμος

Bithinia Βιθύνια
blasfemia βλασφημία
bonitas: — (et magna copia) δαψίλεια (καὶ πολυπλήθεια): *cf.* aer
bonus ἀγαθός, καλός: -um εὐετηρίαι: -orum amator φιλόχρηστος†: melior βελτίων: optimus βέλτιστος, ἀγαθός (*post* / *ante* καλός): bonus (melior) *cf.* aer, animus, annus, armonizatio, color, coloratio, conuersatio, consistentia, decentia, fama, habeo, habitudo, inexpertus, magnitudo, moderatio, motus, nauigatio, nuptie, oculus, operatio, os, placentia, proportio, rithmizatio, sensus, sententia, status, temperantia, ualitudo, usus
Bootem Βοώτην
boreaf(p)elioticus (+ id est aquilonaris simul et orientalis, + id est simul borealis et subsolanus) βορραπηλιωτικός
borealibicus (+ id est simul aquilonaris et occidentalis) βορρολιβυκός
borealis βόρειος, πρὸς ἄρκτους: -ior βορειότερος: -issimus βορειότατος
borealitas τὸ βόρειον
boreas βορέας: inter boream et occidentem πρὸς βορρόλιβα: inter (ad) boream et orientem πρὸς βορραπηλιώτην
bos βοῦς
breuis: modo -i ἐπιτετμημένως: -iter διὰ βραχέων
Britannia Βρεττανία
brucus βροῦχος
brutus ἄλογος

cado ἀποκλίνω, ἐκπίπτω, πίπτω: — sub (*c. abl.*) ὑποπίπτω (*cum dat.*)
calamitas συμφορά, συνοχή
Caldaicus Χαλδαϊκός
Caldea Χαλδαία, Χαλδαϊκή
Caldeus Χαλδαῖος: -orum Χαλδαϊκός
calefactiuus θερμαντικός
caleo θερμαίνω
caliditas θερμασία, τὸ θερμόν, θερμότης
calidus θερμός, θερμότερος, πυρεκτικός: -dior θερμός (?), θερμότερος: -dissimus θερμότατος
caliginosus ἀχλυώδης, ὁμιχλώδης
caligo ζόφος
calor τὸ θερμόν, θερμότης
caluus: mediocriter — μεσοφάλακρος
calx: -cem coquens χωνευτής
campester χερσαῖος
campestritas τὸ πεδινόν
canalis: -es habens αὐλίσκος
Cancer Καρκίνος
cancer: lupus aut — κονδύλωμα
Canis Κύων
canis κύων
canna φάρυγξ
cantatiuus ᾠδικός
capillatura: extense -e, tense -e τετανότριχες: medie -e μεσόθριξ: tense -e μεσόθριξ
capillus θρίξ: medius in -is μεσόθριξ: nigrorum -orum μελανότριξ: xantorum -orum ξανθότριξ
capio χωρέω
capitaneus προεστώς
capitularis κεφαλαιώδης: -iter κατὰ τὸ κεφαλαιῶδες
capitulum κεφάλαιον: — primum α′: hec sunt -a in ... libro τάδε ἔνεστιν ἐν τῷ ... βιβλίῳ
Cappadocia Καππαδοκία
Capricornus Αἰγόκερως
caprinus: — oculos αἰγωπός
captiuatio αἰχμαλωσία
captiuitas αἰχμαλωσία
captura λαβή
caput κεφαλή: crispi -itis οὐλοκέφαλος: *cf.* summitas, uertex
carbonarius καύστης

Carchedonia, -am Καρχηδονία, -ίαν
Carchidon: -donem Καρχηδόνα
cardiacus καρδιακός
careo: -entia pluuie ἀνομβρία
carissimus *cf.* deus
carmen: *cf.* sepulcrum
carnifagus κρεωφάγος
caro σάρξ: frustrum carnis σάρκωμα
carpentarius τέκτων
caruana στόλος
Caspiria Κασπειρία
Cassiepeia Κασσιέπεια
castigatiuus κολαστικός
casualiter συμπτωματικῶς, διὰ τὸ τύχον
casus πτωματισμός, σύμπτωσις, τύχη
cauda οὐρά
caueo φυλάσσομαι
causa αἰτία, αἴτιος -ον -οι: rationem -e non habens ἀναιτιολόγητος
causatiuus αἰτιατικός (αἰτιακός[h]), παραίτιος
causidicus δικολόγος
causo αἴτιον ποιέω: -ans αἴτιος
cautela φυλακή
cecus τυφλός
celer ταχύς: -rius τάχιον
celeritas (*sic*) ταχεῖα
celestis οὐράνιος
Celtica Κελτική
celum οὐρανός: commediatio -i συμμεσουράνησις: mediatio -i μεσουράνησις, συμμεσουράνησις: medium -i μεσημβρινός, μεσουράνημα, μεσουράνησις: angulus medii -i μεσουράνημα: in medio -i esse, — medio μεσουρανέω, συμμεσουρανέω: attingens ad medium -i μεσουρανοῦν: quod est in medio -i τὸ μεσουρανοῦν: fuerint in medio -i μεσουρανῶσιν: oppositum medii -i ἀντιμεσουρανοῦν, ἀντιμεσουράνημα, ἀντιμεσουράνησις: in opposito medii celi fuerint ἀντιμεσουρανῶσιν: in opposito solis in medio celi ἀντιμεσουρανοῦντος: oppositum medii -i sub terra τὸ ὑπὸ γῆν μεσουρανοῦν
censor ἐμπρηστής
Centaurus Κένταυρος
centrum κέντρον
Cepheus Κηφεύς
cera *cf.* plasmator
cerebrum ἐγκέφαλος
certitudinaliter ἀκριβῶς: — examino, — nosco ἀκριβόω
certitudo βεβαιότης
cessibilis: bene -e εὐπροχώρητον
Cete (*sic*) Κῆτος (*acc.*)
ceterus: de cetero λοιπόν
Chele, Khele Χηλαί
chirurgicus χειρουργός
cibo τρέφω: non -atus ἄτροφος
cibus: sine -o ἄτροφος
Ciclades, Kyklades Κυκλάδες
Cilicia Κιλικία
cinctorium περίζωμα
Ciprus Κύπρος
ciragra χειράγρα
circa ἐπί *cum acc.*, κατά *cum acc.*, παρά *cum dat.*, περί *cum acc.*, πρός *cum acc.*: quero — ζητέω *cum gen.*: — terram περίγειος: *cf.* assuesco, descriptio
circonstantia (circum-) περίστασις
circularis κυκλικός
circulus κύκλος: -li (*plur.*) κύκλος
circumcingo περιζώννυμαι
circumdo: -data περιεχόμενα
circumduco περιάγω
circumspectus περιθεατικός†(?)
circumuolutus περίκυκλος†
cito ταχύ: -ius τάχιον: *cf.* disco, transmutabilis
ciuilis πολιτικός
ciuis *cf.* oditor

ciuitas πόλις: principalis — μητρόπολις: *cf.* constructio
clamor περιβοησία
clamosus κραυγαστής
claritas λαμπήνη
clarus λαμπρός: -ior λαμπρότερος
claudicatio χώλωσις
clima κλίμα
coaccipio παραλαμβάνω: -ior et facio aliquid συλλαμβάνομαι τι: -ior (cum) συλλαμβάνομαι (*cum dat.*)
coadiuuo συνεφοδιάζω
coaduoco συμφιλοκαλέω
coapproprio συνοικειόω
coaptatio οἰκείωσις, συναρμογή, συνοικείωσις, ἐναλλαγή (ἐναρμοργή†?)
coapto ἐφαρμόζω, οἰκειόω, συναρμόζω, συνεπισχύω, συνοικειόω: -or συνεικάζομαι: -tatum συναρμοργή
coassumo παραλαμβάνω, προσπαραλαμβάνω, συλλαμβάνω, συμπαραλαμβάνω, συμπεριλαμβάνω
cocus μάγειρος
coeo (cum) συνέρχομαι *cum dat.*: parum -euntes νωχελεῖς
coetaneus ὁμῆλιξ
coexisto (-ens) πρόσειμι, συμπάρειμι, σύνειμι (-ών)
cognatio συγγένεια
cognatus (connatus) συγγενής, συγγενικός: *cf.* pater
cognitio γνῶσις
cognosco γινώσκω, διαγινώσκω, ἐπιγινώσκω: per signa — σημειόομαι
cogor ἀναγκάζομαι
cohereo (*cum dat.*) ἔχω (*cum gen.*)
coimpeditus (*sic*) συγκακωθείς
coincido συμβάλλω, συνεμπίπτω
coitiuus ὀχευτικός
coitus ὀχεία, συνέλευσις, συνουσία
Colchica Κολχική: *cf.* Kolchica
colera χολή
colica: -am patiens κωλικός
collatio: sine -ione cum aliis ἀσυμπερίφορος
collectio ἐπισύνθεσις, συγκομιδή, συναγωγή: fructuum — καρποφορηθέντων συγκομιδή
collectiuus (?: -tus) συγκιρνάμενος
colligatio συνεπιπλοκή
colligo συλλαμβάνω, συνάγω, σχηματίζω: qui colligitur ἐπισυναγόμενος: *cf.* tempus
collocutio: -ionis euitatio ὑποσιώπησις
collum τράχηλος
colo σέβω
color *cf.* albus
color χρόα, χροιά, χρῶμα: boni -ris εὔχρους: medii -ris μεσόχρους: mellei -is μελίχρους: nigri -ris μελάγχρους, μελανόχρους, proprii -oris ἰδιόχροος
coloratio: bona — τὸ εὔχρουν
coma κόμη
combustio σύγκαυσις
cometa κομήτης
Commagena Κομμαγηνή
commassatio (*sic*) ψυλλία («vox inaudita»)
commediatio: *cf.* celum
commensuratio συμμετρία: τὸ συμμετρώτερον
commensuratus σύμμετρος
commisceo *cf.* commixtus
commiscitiuus συγκρατικός
commixtio κρᾶσις, μίξις, σύγκρασις, σύγκριμα, σύγκρισις: — seminum σύγκριμα
commixtiuus συγκρατικός: — modus τὸ συγκρατικὸν εἶδος
commixtus συγκιρνάμενος: *cf.* albedo, figuratio
commodum τὸ ὠφέλιμον: — reportans ὀνησιφόρον ἔχων: magis — τὸ ὠφελιμώτερον

commotionalis σεισμώδης
commotiuus κινητικός, συγκινητικός: bene — εὐσυνάλλακτος
communicatiuus κοινωνικός
communico (cum) κοινωνέω (*c. genit.*): -cauerint sibi ἀμφιλελαχότες ὦσιν: -ans ἐπίκοινος
communio κοινωνία
communis ἐπίκοινος, καθόλου, κοινός: -ior δημοτικώτερος: -iter κοινότερον, κοινῶς: multum — πολύκοινος: *cf.* dico, uita
communitas τὸ κοινόν: amator -atis φιλόκοινος†
commutatio συναλλαγή
commutatiuus: magis commutatiuus συναλλακτικώτερος
commutatus (cum) συντρεπόμενος
compactio σπονδή, σύστασις
compactiuus πυκνωτικός
comparatio σύγκρισις: per -ionem ad πρὸς *cum acc.*
comparo παραβάλλω, συγκρίνω, συμβάλλομαι, συνεπικρίνω
compassio συμπάθεια: -ionem efficio συμπαθέω
compassiuus ἐμπαθής
compatior συμπαθέω
compilatio σύνταξις: *cf.* medicinalis
compingo *cf.* nauis
complacens ἐπαφρόδιτος: -entior θυμήρης
complacentia εὐαρέστησις
complementum συμπλήρωσις
completus (cum) συμπληρούμενος
complexio κρᾶσις, σύγκρασις, σύγκρισις: propria —, proprietas -ionis ἰδιοσυγκρισία (-κρασ-)
complicator *cf.* serta
compositor *cf.* farmacum
compositus συγκείμενος: bene— εὐάρμοστος, εὐσχήμων
comprehendo καταλαμβάνω, κατανοέω: -di potest καταλήψεων δυνατός: ea que possunt -di τὰ ἐνδεχομένην ἔχοντα κατάληψιν
comprehensibilis: male — δυσκατάληπτος
comprehensio κατάληψις
comptus καλλωπιστής
compulsor διαναγκάζομαι
concausa συναίτιον
concedo: -cessus ἀπαρτιζόμενος
concipio: facile -iens εὐσύλληπτος
conclamatio περιβοησία
concludo συμπεραίνω
concomitor ἀκολουθέω
concordo: non -ans ἀσύμφωνος
concors: — sum συμφωνέω, σύμφωνος: -corditer συμφώνως
concretio σύγκρισις: penes fetus (*sic*) -ionem κατ' αὐτὴν τὴν σύγκρισιν
concretus συγκραθείς
concubitus συνουσία
concupiscentia ἐπιθυμία
concupiscitiuus ἐπιθυμητικός
concupiscitor ἐπιθυμητής
concurro παροδεύω, συνέρχομαι, συντρέχω
concursus ἐπιπόρευσις
concussio ἐκτίναξις, προσδρομή
condempnatio κατάκρισις
condextrabilis: bene — εὐσυνδεξίαστος
condicio περίστασις
condiminuo συμμειόω
condispono οἰκοδεσποτέω (συνοικο- ?), συνδιατίθημι, συνοικοδεσποτέω
condispositor συνοικοδεσπότης, συνοικοδεσποτῶν
conditor: sacrorum — ἱεροποιός: *cf.* prodigium
condomino συνοικοδεσποτέω
conduco ἐπισυνάγω
confederatio σύστασις
confero συμβάλλομαι, συμφέρω, συμφορέω, συνάπτω, συνδιαλαμβάνω, συνεφοράω: magis -ens συμφορώτερον

configuratio διατύπωσις, συναρμογή, συσχηματισμός, σχῆμα, σχηματισμός, σχηματογραφία, (?) πάροδος

configuratus συνάπτων

configuro συνοικειόω, συσχηματίζω, σχηματίζω: -or συμπροσγίνομαι

confirmo βεβαιόω

confiteor ὁμολογέω

conformis οἰκεῖος, οἰκειούμενος, ὁμοειδής, ὁμόφυλος, σύμφυλος, συμφώνος: -iter ἀκολούθως, οἰκείως

conformitas τὸ οἰκεῖον

confractio σύνθραυσις

confractus σαθρός

confricatrix τριβάς

confricatus παρατριβείς

congenitus συγγεννώμενος

congermino συνακμάζω

congrego προσεπισυγκρίνω, συγκρίνω, συνάγω

congruus ἐγχώριος, οἰκεῖος: — sibi οἰκεῖος, οἰκεῖος αὐτοῦ: -e οἰκείως

coniectatio στοχασμός

coniecto στοχάζομαι, συνεικάζω

coniectura στοχασμός: facilis — εὐστοχία: -am fieri καταστοχάζεσθαι

coniecturabilis: -liter ἐστοχασμένως: bene -liter εὐστόχως

coniecturatiuus: bene — εὐείκαστος, bene — εὔστοχος

coniecturo καταστοχάζομαι, στοχάζομαι: -andus καταστοχαστέος

coniugatio συζυγία

coniugium συναρμοργή

coniunctio συναφή, σύνοδος: ad -ionem festinans συνοδεύων: *cf.* coniunctionalis

coniunctionalis συνοδικός, σύνοδος: coniunctio — (?) συνοδικός

coniungo συνάπτω: -gi (facit) συνιόντας (ποιεῖ): -iunctus συνάπτων: de -iunctis περὶ συνεπιπλοκῶν: *cf.* amicitia

coniurator ἐφορκιστής (ἐξορκιστής)

connectio συμπλοκή

connecto συζεύγνυμι, συνάπτω

connexio ἐπίζευξις

connubium γάμος

connubo συμβάλλομαι

conor ἐπιχειρέω: -ans μεταχειρίζόμενος

conrestitutio συναποκατάστασις

consanguineus *cf.* mater

conscriptor συγγραφεύς

consentio συγγινώσκω

consequens ἀκόλουθος, (ὁ) ἑξῆς, (ὁ) ἐφεξῆς: -ter ἀκολούθως, ἑξῆς, ἐφεξῆς

consequentia ἀκολουθία, τὸ ἀκόλουθον, τὸ ἑξῆς: — (comparatio ?) παραβολή: ratio -tie τὸ ἑξῆς τῆς ἀκολουθίας

consequor (-cuntur) ἀκολουθέω, ἕπομαι, παρακολουθέω, συμπίπτω

conseruo διασῴζω, διαφυλάττω, συντηρέω, φυλάσσομαι

consideratio διόπτευσις, ἐπίσκεψις

considero διασκέπτομαι, ἐπισκέπτομαι, ἐπισκοπέω, θεωρέω, σκεπτόμαι, σκοπέω, συνθεωρέω, συνοράω: -atus, qui -atur (ὁ) νοούμενος: -andum σκεπτέον

consilio: bene -iatus καλοσύμβολος

consimilis ὁμοιότροπος, παραπλήσιος: -iter ὁμοίως, παραπλησίως, ὡσαύτως

consimilitudo τὸ παρόμοιον

consistentia κατάστασις, κατάστημα, σύστασις, σύστημα: *cf.* aer

consolabilis: facile — εὐπαρηγόρητος

consolatio τὸ νουθετικόν, τὸ παραμυθητικόν

consolor παρηγορέω

consonantia συμφωνία

consono συμφωνέω

consonus συμφωνῶν, σύμφωνος: maxime — συμφωνότατος: -ne συμφώνως

conspector συνορατικός

conspeculor συνεπισκέπτομαι

conspicio συνοράω

constituo καθίστημι, κατασκευάζω, συνίστημι: -stitutus σταθείς
constitutio σύστασις
consto συνίστημι: -ans καθεστώς
constrictio ἀγχόνη, συνέχεια (συναχίαι†?), συνοχή
constructio: — ciuitatis κτίσις
consuesco: -euit εἴωθε
consuetudinalis ἔθιμος
consuetudo ἐθισμός, ἔθος
consuetus συνήθης
consummatio: non ad -ionem pertingens ἀποτευκτικός
consummatiuus ἐπιτευκτικός: bene — εὐεπίτευκτος† *vel* εὐεπιτευκτικός†
consummator ἐπιτευκτικός
consummo διοδεύω
consumptio ἀνάποσις, σύντηξις
consumptiuus ἀναπωτικός
consumptor: rerum — διοικητικός
contabefactiuus συντηκτικός
contabescentia σύντηξις
contactus συναφή
contemperatus κεκραμένος, συγκιρνάμενος
contempno καταγινώσκω, καταφρονέω
contemptiuus καταφρονητικός
contemptor καταφρονητικός
contentiosus ἐριστικός
contestificor προσμαρτυρέω, συμμαρτυρέω
contineo περιέχω: -tinens ἐγκρατής, (τὸ) περίεχον: -tentus ἐμπεριεχόμενος: *cf.* uita
contingo: -it ἐνδέχεται, ἔνεστι (ἔνι): ut -it ὡς ἔτυχεν: -ens (ὁ) τυχών: *cf.* euentus
continuitas συνέχεια
continuus συνεχής: -o παρ' αὐτά: -e συνεχῶς
contionator δημηγορικός
contra πρὸς *cum acc.*: *cf.* lex
contraago ἀντιπαθέω, ἀντιπράττω
contraascendens ἀνθωροσκοπῶν
contractus συνεσπασμένος: -ior βραχύτερος
contraho: — cum ζεύγνυμι (*c. dat.*), συμβάλλω (*c. dat.*)
contrapassio ἀντιπάθεια: -ionem infero ἀντιπαθέω
contrapatior ἀντιπαθέω
contrariatio ἐναντίωσις
contrarietas: ἀναλογία (ἀντιλογία ?)
contrarior ἐναντιόομαι
contrarius (-um) ἀντικείμενος, ἐναντίος (τοὐναντίον): -ia agenta τὰ ἀντιπαθοῦντα: in -ium ago ἀντιπράττω: in -ium regimen ἀποτροπιασμός: -e ἐναντίως, τοὐναντίον: *cf.* econtrario
conueniens οἰκεῖος, οἰκειούμενος, προσήκων, συνοικειούμενος: bene — εὐσύνθετος: non — ἀνοίκειος: -tius οἰκειότερον: -ienter οἰκείως
conuenientia τὸ οἰκεῖον, οἰκειότης, οἰκείωσις, συναρμογή, συνοικείωσις
conuenio οἰκειόομαι, συμπίπτω, συνέρχομαι: (quod) -it (τὸ) οἰκεῖον, προσήκει
conuentio συναρμογή, συνέλευσις, συνουσία
conuersatio ἀναστροφή, διαγωγή, δίαιτα: bone -tionis εὐάγωγος: simul — συναναστροφή: in -ione ταῖς διαίταις
conuersor: bene -ans εὐανάστροφος
conuerto (ad) ἐπιστρέφω, συντρέπω
conuictus συμβίωσις
conuolutio συστροφή
coobuiatio συνάντησις
cooperatio τὸ συμπρακτικόν
cooperatiuus συνεργητικός
cooperor ἐπισυνεργέω, συμβάλλομαι, συνεργέω
coordinatio σύνταξις
coorior συνανατέλλω
copia φορά: bona — δαψίλεια
copiosus: -ior ἱκανώτερος

copulatio συναφή
copulo (*ad, cum*) συνάπτω, συζεύγνυμι: -or ἐπισυνάπτω: -atus οἰκειούμενος
coquo: calcem -ens χωνευτής
cor καρδία: cordis refugatio (*sic*) φυγαδεία
corda: -arum tortor χορδοστρόφος
cornu κέρας
Corona: Australis — Νότιος Στέφανος:— Borealis Βόρειος Στέφανος
corona: -arum gestator στεφανηφόρος
corporalis σωματικός, σωματώδης: -libus σωματικῶς (?): -liter σωματικῶς: -liter agere τὸ αὐτουργόν
corpus σῶμα: -oris σωματικός: -oris amatrix φιλοσώματος: -oris formatiua tempora τῆς σωματοποιήσεως χρόνοι: duorum -orum δίσωμος: -orum locator σωματέμπορος: que pertinent ad – σωματικά (?): *cf.* oditor
corrodo διαφθείρω
corrosio ἀνάβρωσις, τὸ διεφθάρθαι: *cf.* depastura
corrumpo φθείρω: -or διαφθείρω
corruptio διαφθορά, φθορά: — (corruptiuum ?) τὸ φθοροποιόν
corruptiuus φθαρτικός, φθοροποιός
corruptor διαφθορεύς
coruscatio ἀστραπή , -αί
Coruus Κόραξ
cotidianus καθ' ἡμέραν
coutor συγχράομαι
Crater Κρατήρ
credo καταξιοπιστεύομαι, πιστεύω
Creta Κρήτη
criminabilis: facile — εὐδιάβλητος
criminatio διαβολή
criminator διαβολικός, διάβολος
crines τρίχες
crispus οὐλόθριξ, οὖλος, *cf.* caput
cristallosus κρυσταλλώδης
cronica τὰ χρονικά
crus μηρός: figura equalium -rum ἰσοσκελία
crustula ἐσχάρα
crux σταυρός
cultor, cultus: *cf.* deus
cum μετά (*cum gen.*), ὅταν, πρός *cum dat.*, *σύν*, *pro dat.*: — ... tum τε καὶ: *cf.* contraho, respectus, sum
cura *cf.* domus
cura ἐπιμέλεια, θεραπεία: *cf.* domus
curabilis ἰάσιμος
curator θεραπευτής
curo θεραπεύω
curro: uacue -entia κενοδρομοῦντα
cursus δρόμος, ἐπιπόρευσις
cuspis (sagitte) ἀκίς: is qui fuerit magis in -pide ὁ ἐπικεντρότερος
Cyrenaica, Cirenaica Κυρηναϊκή

dampnosus ἐπιζήμιος
datio δόσις: amator -onis φιλόδωρος
dator δοτήρ, ἐπιδοτήρ
de ἀπὸ, εἰς, ἐκ (ἐξ), ἐπί (*cum gen.*), περί (*cum gen.*), ὑπέρ (*cum gen.*)
dearticulo διαρθρόω
debeo ὀφείλω: -et ὀφείλων εἶναι: qui debeat μέλλων
debilis ἀδυναμῶν, ἀσθενής, ἀσθενικός: -ior ἀσθενέστερος: -issimus ἀδρανέστατος: -iter ἠρέμα
debilitas ἀδρανία, ἀσθένεια
debitus κατειλεγμένος
decanto: *cf.* sepulcrum
decapitatus ἀποκεφαλιζόμενος
decennium δεκαετία
decens εὐάγωγος, εὔμορφος, εὐσχήμων: -tius τὸ εὐσχημονέστερον
decentia τὸ εὔσχημον: bona — τὸ εὐπρεπές: reuerenda — σεμνοπρέπεια: *cf.* muliebris
deceptibilis σφαλλόμενος

deceptio ἀπάτη
decidentia: minor — τὸ ἀπταιστότερον
decido ἀποκλίνω, καταντάω: -ens κατάφορος: facile -ens εὐκατάφορος† (-φρονητός)
decipio ἐξαπατάω
decisio ἀποτομή, τμῆμα, τομή
declaro δηλόω
declinatio ἔγκλισις, παρακμή
decliuis: <quia> — est κατὰ τὴν ἔγκλισιν
decrepitus πάνυ γέρων
dedo: deis -ditus ἐνθουσιαστής
deductibilis: bene — εὐδιάγωγος
deductio ἀναγωγή, διαγωγή
defectus ἀπορία, ἔλλειψις, ἔνδεια, παράσημος: — aquarum λειψυδρία: — natorum ἀτεκνία
defero: -latus φερόμενος
deficio διαπίπτω, ἐλλείπω, πταίω, οὐ διευθυνέω: — (a) λείπομαι (*cum gen.*): -iens ἀποκρουστικός
defluxus ἀπόρροια
defundo: -ens περιβαλλόμενος
dego: bene -ens εὐδιάγωγος
deiectio καθαίρεσις
deiectus ταπεινός
deinceps ἑξῆς, ἐφεξῆς: id quod — τὸ ἑξῆς
deinde εἶτα, ἔπειτα
delectabilis ἡδύς: — uite ἡδύβιος
delectatio ἡδονή: amor -ionis φιληδονία
delecto *cf.* addisco
Delfinus, Delfis Δελφίς
delicatus τρυφερός, τρυφητής: -e diete τρυφεροδίαιτος
deliciosus: -e uite ἡδύβιος
delinquo καταλείπω
demaceratus κάτισχνος
demento: -tans ἐκστατικός
Demeter Δημήτηρ
demissio παράθεσις, ταπεινότης
democraticus δημοκρατικός
demon δαίμων: malus — (uel infortunium) κακοδαιμονοῦν: -onum inuocator γόης: -onum obsequium δεισιδαιμονία: -onibus obsequiosus, -onibus seruiens δεισιδαίμων
demoniacus δαιμονιόπληκτος
demonizatio δαιμονιοπληξία
demonstratiue ἀποδεικτικῶς
demonstro δείκνυμι
denominatio ὀνομασία
denudatio ἀπογύμνωσις, γυμνητεία
deorsum *cf.* fero
depascens νομή
depastura: — uel corrosio νομή
dependeo ἀρτάομαι: -ens τετανός (*e. g.* τετανὸν τὰς τρίχας)
depono *cf.* feminitas
deportator ὑπεξερχόμενος
depredatio λῃστεία
deprehendo διαλαμβάνω, καταλαμβάνω (-ομαι), καταλείπω, κατανοέω: -di ἐπισημαίνειν: ea que intus -untur τὰ ἐναπειλημμένα
depulsiuus ἀποκρουστικός
derelinquo καταλείπω
descendo καταφέρομαι
descensio καταφορά
descensus καταφορά
describo γράφω, ὑπομνηματίζω
descriptio παράθεσις: tabularis — ἀναγραφή: graduum — (circa) μοιρογραφία (*cum genit.*)
desertus ἔρημος
deses βλακώδης, ῥᾴθυμος
desidia βλακεία
despectatiuus *cf.* formido
despectibilis: facile — εὐκαταφρόνητος
despector καταφρονητικός
despicabilis: non — ἀκαταφρόνητος
despicio καταγινώσκω
desponsor γαμέομαι

destruo: -endum ἀναιρετέον

desuper ἄνωθεν

deterior: — proportio τὸ ἀρρυθμότερον

determino ἀφορίζω, διορίζω

detrimentum φθίσις

detrunco: membris -atus μελ[ε]οκοπούμενος

deuastatio νομή

deuoratio νομή

deuoro δαπανάω

deus θεός: -i cultor θεοσεβής: -i cultus θεοσέβεια: maior -i cultus τὸ θεοσεβέστερον: -i illustratio θεοφορία: -o carissimus φιλοθεώτατος: -o supplices θεοπρόσπλοκοι: sine -o ἄθεος: -i, -orum amator φιλόθεος: -is deditus ἐνθουσιαστής: *cf.* timidus

dexter δεξιός

diameter (*acc.* -trum -tros) διάμετρος, -ον: ea que secundum –trum αἱ καταφοραί

diametralis διάμετρος, -ον: — (est) διαμετρῶν: — oppositio διαμέτρησις: in -li aspectu διαμετρήσας, in -li situ (sum) διαμηκίζω: -liter κατὰ διάμετρον: -liter aspectu διαμετρούμενος: -liter aspicio διαμηκίζων τυγχάνω: -liter aspiciens διαμηκίσας: -liter oppositus (*sc.* locus) διάμετρος

diametrisis: -es διαμετρήσεις

diametrizatio: -ones διαμετρήσεις

diametrizo διαμετρέω, διαμηκίζω

dichotomia διχότομος

dico ἀποφαίνω, εἶπον, εἴρω, λέγω, φάσκω, φημί: dixi προεῖπον: -ctus (is qui -citur) λεγόμενος, εἰρημένος, ῥηθείς: uerum — ἀληθεύω: communiter -ctus κοινῶς φερόμενος: *cf.* figuraliter, omnis, totus

dies (die, de die, in die) ἡμέρα (-ας): -ei ἡμερήσιος: in singulis -bus καθ' ἑκάστην ἡμέραν

dieta δίαιτα: regimen -e διαιτητικὴ ἀγωγή: simplicis -e ἁπλοδίαιτος: -e splendide ἁβροδίαιτος: *cf.* delicatus

differentia διαφορά

differo διαφέρω: -ens διάφορος, τὸ διάφορον: -enter διαφόρως

difficilis, -e *cf.* adibilis, attingo, effugibilis, explicatio, mobilis, nauigatio, preces, redarguibilis

difficultas τὸ δύσκολον, τὸ δυσχερές: *cf.* euitatio, nauigatio, precognitio, usus

difficulter *cf.* regibilis

diffusio διάχυσις

digero πεπαίνω

dignatiuus ἀξιωματικός

dignifico καταξιόω

dignitas ἀξία, ἀξίωμα, κατάστασις, ψῆφος: propria — ἰδιοπροσωπία: in propria -tate ἰδιοπροσωπῶν: rationes -atis (?) ψῆφους οἰκοδεσποτείας

dignitatiuus ἀξιωματικός

dignus ἄξιος: — fide ἀξιοπιστός: traditio -a fide ἀξιοπιστία τῆς παραδόσεως: -um haberi ἠξιῶσθαι

diiudicatiuus διακριτικός

diiudico διακρίνω: bene -catum εὖ διειλημμένως

diligens ἀκριβής: -ter ἀκριβῶς, ἐπιστημονικῶς

diligentia: omnimoda — pertracto ἀκριβόω κατὰ τὴν διέξοδον

diligo: -endus est ἀγαπᾶν προσήκει

diluuium κατακλυσμός

dimidium (τὸ) ἥμισυ

diminutio μείωσις

dimissio ἄφεσις

dimissor ἀφέτης, ἀφέτις, ὁ ἀφιείς

dimissorius ἀφετικός

dimitto ἀφίημι

dinumero καταριθμέω

directibilis: -ius κατορθωτικώτερον

directiuus κατορθωτικός

dirigibilis: id quod peius -e τὸ δυσκατορθώτερον
dirigo ἐφοδιάζω
discerno διαγινώσκω, διακρίνομαι: -endum διακριτέον
disciplina: amator -e φιλομαθής
disco: cito -ens ταχυμαθής
discoloratio χρωματοκρασία
disconuenientia ἀντιπάθεια
discors φίλεχθρος
discursus διαδρομή
discussor διοικητικός
disgrego διακρίνω
dispensatio οἰκονομία
dispensator οἰκονόμος
dispersio: multe -ionis πολύσπορος
displicentia: -am habens δυσάρεστος
dispono (-ponor) διατίθημι (-κειμαι), οἰκοδεσποτέω
dispositus κείμενος
dispositio διάθεσις, οἰκείωσις, περίστασις
disrumpo: -ruptus ἐκρηγνύμενος (περιρρηγνύμενος ?)
dissentericus δυσεντερικός
dissero διέρχομαι
dissidens ἄστατος
dissimilis ἀνοίκειος, ἀνόμοιος: in nullo — ἀπαράλλακτος
dissimilitudo τὸ ἀνόμοιον
dissipatio καθαίρεσις
dissipator διασυρτικός, διάσυρτος
dissoluo διαλύω, παραλύω
dissolutio διάλυσις
distantia ἀποχή, διάστασις, διάστημα
distinctio διάκρισις, διάληψις
distinguo διαλαμβάνω, διακρίνω, διαλαμβάνω, διορίζω, χωρίζω: -uendum διαληπτέον
disto (a, ab, per) ἀπέχω, ἀφίστημι (-σταμαι), διίστημι: *cf.* equalis
distribuo ἀπομερίζω, προσνέμω
ditissimus πλουσιώτατος
diuersitas ἐναλλοίωσις: -tates (continentis) ἐπισημασίαι (τοῦ περιέχοντος)
diuersus διάφορος: -a simulans ἀλλοπρόσαλλος
dives *cf.* ditissimus
diuido διαιρέω
diuinatio μαντεία: pro -one purificans φοιβαστικός
diuinatiuus μαντικός
diuinator μάντις: in mortuis — νεκρόμαντις
diuino: in ollis -ans λεκανόμαντις
diuinus θεῖος
diuisio διαίρεσις, τμῆμα
diuitie πλοῦτος
diurnalis ἡμερήσιος: penes -ia τῶν ἡμερησίων
diurnus ἡμερινός, ἡμερήσιος
diutinus διὰ μακροῦ
diuturnus πολυχρόνιος
diuulgabilis: -ius διαβοητότερον: facile -ius εὐδιαβοητότερον
diuulgatio δημοσίωσις, ἐξαγορία
diuulgatus: multum — πολυθρύλλητος
do δίδωμι, παρέχω: dandum δοτέον: responsa dantes ἀποφθεγγόμενοι
docilis: per se — αὐτοδίδακτος
doctor διδάσκαλος
doctrina διδασκαλία
dolor ἀλγηδών: partus — ἐκτρωσμός: -rum implicator δολοπλόκος: *cf.* matrix
dolosus δόλιος: *cf.* anima
domesticus ἥμερος, οἰκεῖος: amatiuus -orum, amator -orum φιλοίκειος: -e οἰκείως: *cf.* oditor
domina δέσποινα
dominabilitas: maior — τὸ αὐθεντικώτερον
dominatio δεσποτεία, ἐπικράτησις, κυρία, οἰκοδεσποτεία, συνοικοδεσποτεία: -ionis (-tiuis ?) κυρίων
dominatiuus οἰκοδεσποτικός

dominator ἐπικρατήτωρ, κυριεύσας, οἰκοδεσποτήσας, οἰκοδεσπότης: annorum —, temporum — χρονοκράτωρ

dominatorius αὐθεντικός

dominium ἐπικράτησις, κυρία, οἰκοδεσποτεία: optinet — luminarium κατέχει τὰ φῶτα οἰκοδεσποτήσας

domino (-or) ἐπικρατέω κρατέω, κυριεύω, οἰκοδεσποτέω, οἰκοδεσποτίζω

dominus δεσπότης, κύριος, οἰκοδεσπότης

domus οἶκος: -us curam habentes οἰκουροί

donatiuus χαριστικός

donec ἕως

donum δωρέα

dorsum νῶτος

Dracon Δράκων

dubitatio ἀπορία

dubito ἀπορέω, διστάζω

dubius ἀμφίβολος

ducatus ἡγεμονία

duco *cf.* sponsa

ductiuus: *cf.* exercitus

dulcis: — moris χρηστοήθης

duo δύο, δισσοί: singuli —, ana — ἀνὰ δύο: — (et dimidium) β′ (ἥμισυ): *cf.* pars, corpus

duodecim δώδεκα: — signa δωδεκατημόρια

duodecimus: -a pars δωδεκατημόριον

duodenarius: -arie partes δωδεκατημόρια

duodeni: -e partes δωδεκατημόρια

duodennium δωδεκαετία

duplex διπλοῦς, δισσός: *cf.* forma

durabilis: -ior πολυχρόνιος

durabilitas: minoris -atis πρόσκαιρος: modica — ὀλιγοχρονιότης: multa — πολυχρονιότης

duratio παράτασις

duritia ἀντιπάθεια

duro παραμένω, παρατείνω: ad tempus -antes διακρατοῦσαι χρόνον

durus ἀπόκροτος, νηλεής, σκληρός

ebriosus μέθυσος

eclipsis ἔκλειψις

eclipticus ἐκλειπτικός

econtra: *cf.* induco

econtrario ἀνάπαλιν, ἄντικρυς, τὸ ἐναντίον

econuerso ἀνάπαλιν

edentulus νωδός

edificium θεμέλιος, οἰκοδόμημα

eductio ἀναγωγή

effectiuus ἀποτελεστικός, ποιητικός: -a uirtus τὸ ποιητικόν

effectus ἀποτέλεσμα, τὸ ἀποτελούμενον, ἐνέργεια, ἐνέργημα, περίστασις, προτέλεσις, προτέλεσμα: — (*gen.*) κατὰ τὴν ἐπισημασίαν

efficacia τὸ ποιητικόν

efficaciter ἐνεργῶς

efficax ἐπιτευκτικός: -catior δραστικώτερος, -cacissimus δραστικώτερος, δραστικώτατος

efficientia τὸ ποιητικόν

efficio ἀπεργάζομαι, ἀποτελέω, διατίθημι, ἐμποιέω, ποιέω (-έομαι): -ior γίνομαι: quod -itur τὸ ἀποτελούμενον: *cf.* compassio

effugibilis: difficile — δυσφύλακτος

effusio: — maris ἄμπωτις

efimerinus ἀφημερινός (ἐφημερινος)

Egeum pelagum Ἀιγαῖον πέλαγος

Egiptiacus Αἰγυπτιακός

Egiptius Αἰγύπτιος: -iorum Αἰγυπτιακός

Egiptus Αἴγυπτος: -i inferioris regio Αἰγύπτου κάτω χώρα

ego: *cf.* mihi, nos, noster

egritudinalis νοσηματικός, νοσηματώδης, νοσοποιός, νοσώδης

egritudo νόσημα, νόσος: -dines uentris κοιλιακά: -dines splenis σπληνικά: colerica — χολικός

elaboratio ἐξεργασία

elaboro ἐπεξεργάζομαι

electio προαίρεσις

electiuus: bene — εὐπροαίρετος

elefantia ἐλεφαντίασις

elementum στοιχεῖον

elephanticus ἐλεφαντιῶν

eleuatio ἀναφορά, ὕψος: — super horizontem ἀναφορά: a tumultibus — τὸ ἀνακεχωρηκός: tempora -ionum ἀναφορικοί χρόνοι

eleuor ἀναφέρομαι: qui -atur (ὁ) ἐπαναφερόμενος

Ellas Ἑλλάς

elocutio ἐξαγορεία

elongatus μακρός

elucido ἐκφαίνω

embrio: -ionum excisio ἐμβρυοτομία

eminens ἐπιφανής: -entius operationis τὸ ἀνυτικώτερον

emiolius ἡμιόλιος

emissio ἄφεσις

emitriticus ἡμιτριταϊκός

emitto ἀποπέμπω

emoroide αἱμορροΐδες

enim γάρ, δέ

epar ἧπαρ

epaticus ἡπατικός

epilempsia (epilentia) ἐπιληψία, ἐπίληψις

epilenticus ἐπιληπτικός

equalis ἴσος, ἰσόρροπος: in -li tempore ἰσοχρόνως: figura -lium crurum ἰσοσκελία: -liter disto τὸ ἴσον ἀφίστημι: -liter possum ἰσοδυναμέω: *cf.* numerus

equalitas τὸ ἴσον, ἰσότης

equester ἱππικός

equicrurius: -ium ἰσοσκελία: figura -ia ἰσοσκελία

equilaterus ἰσόπλευρος

equilibratus ἰσόσταθμος

equinoctialis ἰσημερινός, μεσημβρινός

equinoctium ἰσημερία, τὸ ἰσημερινόν

equipotens ἰσοδυναμούμενος

equipotentia ἰσοδυναμία

Equus Ἵππος

equus ἵππος

equus ἴσος: ex -quo ἐν ἴσῳ

Eracles Ἡρακλῆς

erarius χαλκεύς

erectus ἀνορθούμενος

ermafroditus ἑρμαφρόδιτος

erratici πλανώμενοι

erraticus: -i πλανώμενοι: non — ἀπλανής

erro: -antes πλανώμενοι: non -ans ἀπλανής

error ἀπορία: sine -re ἄπταιστος

eruca κάμπη

erudibilis: facile — εὐπαίδευτος

eruditio παιδεία

erysipela (-pila) ἐρυσίπελας

esca: ineptitudo ad -as ἀνάβρωσις

estas θέρος

estimo οἴομαι

estiualis θερινός

estiuus θερινός

estuo: -abo καυσόομαι

estuosus καυματώδης, καυσώδης: -issimus καυστικώτατος

estus ἀλέα, καῦμα, καύσων

et καί, δέ (*in* μὲν ... δέ , μὲν ... δὲ ... δέ), δὲ καί, ἤ, ἢ καί, τε, τε καί

etas ἡλικία: perfecte -tis ἀκμαῖος

etiam καί, δέ (*post* καί = et), τε: aut —, uel — ἤ

euaporatiuus ἀναφορικός

euaporo ἀναθυμιάομαι

euenio ἀποβαίνω, γίνομαι, ἐκβαίνω, ἐμπίπτω, ἐκπίπτω, ἐπέρχομαι, προσγίνομαι, συμβαίνω, συμπίπτω: id quod -it τὸ περιγενόμενον: qui -it συνιστάμενος: euenturus (est) ἀποβησόμενος, ἐσόμενος: -ientia συμπτώματα: *cf.* symptoma

euentus ἀποτέλεσμα, τὸ ἀποτελούμενον, ἔκβασις, τὸ ἐπισυμβαῖνον, σύμπτωμα, ἰδίωμα (σύμπτωμα†?): — qui contingit τὸ ἀποτελούμενον: que pertinent ad — (?) συμβατικά

euexia εὐεξία

eufortunia ἐπιτυχία (†εὐτυχία ?)

eufortunor εὐτυχέω

euidens ἐναργές, προφανής: -tior, multum — ἐπιφανέστερος: -tius ἐκφανέστερον: -tissime ἐναργέστατον: -ter ἐναργῶς

euidentia τὸ ἐναργές: maior — τὸ προφανέστερον

euitatio τὸ φυλακτικόν: difficultas -ionis τὸ δυσφύλακτον: *cf.* collocutio

euitatiuus φυλακτικός

eunuchus εὐνοῦχος

Europa Εὐρώπη

ex ἐκ (ἐξ), διά *cum. gen.* (*e. g.* ex toto διὰ πάντος), κατά *cum acc.*, (?) ὑπό *cum gen.*: — equo ἐν ἴσῳ

exagonalis: -le τὸ ἑξάγωνον

exagonus ἑξάγωνος

exalatio ἀναθυμίασις

exalatus: — a calido ἀναπνούμενος† ὑπὸ θερμοῦ

exaltatio ὕψωμα

exalto ὑψόω

examinatus: -e κατὰ τὸ ἀκριβές

examino: certitudinaliter — ἀκριβόω

exarator: historiarum -res (?) λαοξόοι

excedo ὑπερβάλλω: -ens πλεομνάζων

excello: -ens ὑπερέχων: -entior (*sic*) ὁλοσχερέστερος: *cf.* ago

excelsus: -a μετέωρα

excepto πλήν

excessiuus ἐκστατικός: in beneuolentia — εὐήθης

excessus ἀμετρία, ἐπίτασις, ὑπερβολή, ὑπεροχή: in -u beneuolus εὐήθης: — (+ *gen.*) ὁ ἄγαν (+ *subst.*)

excisio: embrionum — ἐμβρυοτομία

excito ἐγείρω

excorians λεπίς

excrescentia ἀνάβασις, ὑπερμετρία

exemplar ἀντίγραφον, παραδειγματισμός

exemplaris παραδειγματικός

exemplum παράδειγμα: -i gratia λόγου ἕνεκεν: non productus in — ἀπαραδειγμάτιστος: *cf.* productibilis

exercitium γυμνάσιον: militaria -ia στρατιωτικαὶ χρεῖαι

exercitus: -tus ductiuus στρατοπεδαρχικός

exhibeo διαδίδωμι, παρέχω, -ομαι

exilis *cf.* uox

existimo (-or) ἡγέομαι, ὑπολαμβάνω, ὑπονοέω: -andum ὑπονοητέον

existo παρατυγχάνω, συνίσταμαι, τυγχάνω, ὑπάρχω: -titerit εἶεν, τυγχάνω ὤν: -istens ὤν, γινόμενος, γενόμενος, σταθείς, συνών: simul -istens συνών: *cf.* explicatio, gloria, plenilunia

exitus ἔξοδος

expansus: -orum pilorum ἁπλόθριξ

expensor δάπανος

expensus: -as δάπανας†

expers ἄμοιρος

expertus ἔμπειρος

explicatio παραδειγματισμός: difficilis -ionis existens δυσερμηνεύτως ἔχων

expono, -or, -ositus ἐφοδεύω, προεκτίθημι, ἐκτίθημι, -μαι, ἐκκεῖμαι, -μενος

expositio ἔκθεσις

expressus αὐθέκαστος

exprimo τυπόω

exprobator μωμητικός

exquero, exquiro ἐξετάζω

ex<s>ecror ὑπερεχθαίρω

exsiccatio ἀναξήρανσις, ξηρασία, ξηρότης: maior — τὸ ξηραντικώτερον

exsiccatiuus ξηραντικός

exsicco ξηραίνω

exsto συνίσταμαι

exsurrectio ἀναστασία
extasis ἔκστασις: -im faciens ἐξοιστικός
extendo: quantum -di potest τὴν ποσότητα τῆς παρατάσεως
extensus: *cf.* capillatura
exterior ὁ ἐκτός, ὁ ἔξω
exterminatio ἀπώλεια
extermino ἀφανίζω
extra παρά *cum acc.*: id quod — τὸ ἐκτός
extraneus: amator -eorum φιλόξενος: in loco -eo ἐπὶ τῆς ξένης
extremitas (τὸ) ἄκρον, ἀκρότης: -tate noctis ἀκρόνυκτος
extremus ἔσχατος: -a τὰ ἄκρα: in -o noctis ἀκρόνυκτος
extrinsecus ὁ ἔξωθεν
exuberantia (aquarum) ἐπομβρία: delicate -tie δαψιλεῖς ἐπομβρίαι

f *cf. etiam ph*
fabricatiuus τεκτονικός
facies πρόσωπον: proprias habere — ἰδιοπροσωπεῖν
facilis: de -i ῥᾳδίως: *cf.* coniectura, motus: -e *cf.* accusabilis, acquisitibilis, adibilis, applicabilis, auertibilis, concipio, consolabilis, criminabilis, decido, despectibilis, diuulgabilis, erudibilis, fatigabilis, fragilis, incidens, ingero, intelligibilis, mobilis, mutabilis, nauigatio, occasionabilis, peccabilis, permutabilis, permutabilitas, procliuis, productibilis, profectibilis, promptus, proportionatus, punibilis, sensibilis, sompnio, transponibilis, uerecundo, uestitus, uisibilis
facilitas τὸ εὐεπίβολον
facio ποιέω, -έομαι, ἀποτελέω, ἐμποιέω, ἐπιτελέω, παρασκευάζω, παρέχω: — ad συμβάλλομαι πρὸς: si factum fuerit συμβεβηκότος: factus (?) συντελειούμενος: *cf.* coaccipior, extasis, inimicitia, mora, sanguis, speculatio, transitum
factiuus (sum) ποιητικός: *cf.* natum, terremotus
factor: *cf.* arma, mirabilis, organum
fallax παράβολος: -acior ἐπισφαλέστερος
fallibilis ἐπισφαλής
fallo διαμαρτάνομαι: -or σφάλλομαι
falsarius πλαστογράφος
falsator: litterarum — πλαστογράφος
falsum ψεῦδος
falsus σφαλερός
fama: bonam -am habeo εὐφημέομαι: male -e κακόφημος
fames λιμός
familiariter γνησίως
fanaticus ἐγκάτοχος
farmacia φαρμακεία: per -ias διὰ τῶν φαρμάκων
farmacum φάρμακον: -corum compositor φαρμακοποιός
fatigabilis: facile — εὐπρόσκοπος
fatum εἱμαρμένη
Fazania Φαζανία
febrilis πυρεκτικός: rigor — ῥιγοπύρετος
febris πυρετός: — cum rigore ῥιγοπύρετος
fecunditas τὸ γόνιμον
fecundus γόνιμος: -issimus γονιμώτατος
fedus: sine -ere ἀδεξίαστος
fel χολή
felicitas δαψίλεια, εὐδαιμονία, τὸ εὔφορον
felix εὐδαιμονίζων, εὐδαίμων
femella θήλεια: -arum procreatio θηλυγονία
femina (ἡ) θήλεια: -e procreatio θηλυγονία: *cf.* iuuenis, senior
femininus θηλυκός, θῆλυς, τεθηλυσμένος: -ius τὸ θηλυκώτερον: magis — θηλυμορφώτερος: — fio, sum θηλύνομαι: — animo θηλύψυχος: *cf.* humilis
feminitas τὸ θηλυκόν, τὸ τεθηλυσμένον: natura que deposuerit -tatem τὸ ἀθήλυντον τῆς φύσεως
feminizatus τεθηλυσμένος

Fenices Φοίνικες
Fenicia (*Fenix) Φοινίκη
feralis ἀνήμερος: -ia θηριώδη
fere σχεδόν
fero ἐπιφέρω, φέρω: — ad ἐπιφέρω *cum dat.*: deorsum -or καταφέρομαι: *cf.* testimonium
ferocitas τὸ δυσήμερον
ferrum σίδηρος
fertilitas εὐφορία
festino *cf.* coniunctio, plenilunia
fetus *cf.* concretio
fictio: -ione sapiens προσποιησίσοφος
fictor *cf.* animal
ficulnea συκῆ
fidelis πιστός
fidens παρρησιαστής, παρρησιαστικός
fides (*sing.*) πίστεις: *cf.* dignus
fiducia παρρησία
fiducialitas: maior — τὸ παρρησιαστικώτερον
figura μορφή, μόρφωσις, σχῆμα, σχηματισμός: similis -e ὁμοιοσχήμων: *cf.* equalis, equicruria, crus
figuraliter τυπωδῶς: — dico ὑποτυπόω: — predico , —premitto προτυπόω
figuratio διατύπωσις, μόρφωμα, μόρφωσις, συσχηματισμός, σχῆμα: commixte -ionis συγκατεσχηματισμένος: similis -ionis sum ὁμοιοσχημονέω: *cf.* passio
figuro σχηματίζω: ante ... — προτυπόω (*c. gen.*)
filia θυγάτηρ
filius τέκνον, υἱός: multitudo -iorum πολυτεκνίαι: *cf.* procreatio
fingo: -ens προσποιούμενος: ficta peccata ἀνεπίπλαστα ἁμαρτήματα
finis τέλος
fio γίνομαι, ἐγγίνομαι: qui fit γινόμενος, ἐπιγινόμενος: non fiunt ἀγένητα (*sc.* ἐστίν): factus γινόμενος, γενόμενος, ἐπιγινόμενος: prius factus, ante factus προγινόμενος
firmitas: maior — τὸ βεβαιότερον
firmus βέβαιος: -ior στερεώτερος: non — ἀβέβαιος
fisiologus φυσιολόγος
fissor: lignorum — ξυλοσχίστης
fistula σῦριγξ: -arum amator φιλόμουσος
fixura ταριχεία
fixus ἀπλανής, στερεός
flauus ξανθός
flegma φλέγμα: multi -atis πολυφλέγματος (-τώδης)
flexiones (lune) κάμπια (καμπίαι)
flo πνέω
floretus ἄνθινος
flos ἄνθος
Flumen Ποταμός
flumen ποταμός
fluuialis ποτάμιος
fluuius ποταμός
fluxus ῥύσις, χύσις: sanguinis — αἱμορραγία
follis: sine –e ἄτρωγλος
fons πηγή
fontalis πηγαῖος
foramen: sine -ine ἄτρητος
fore ἔσεσθαι
forma μορφή, μόρφωμα, μόρφωσις: alterius -e ἑτερόμορφος: duplicis -e δίμορφος: humane -e ἀνθρωποειδής, ἀνθρωπόμορφος: unius -e μονοειδής
formabile μόρφωμα
formatio μόρφωσις: σχηματισμός
formatiuus διαμορφωτικός: *cf.* corpus
formatus: bene — εὐσχήμων
formido δειλία: maior — τὸ εὐλαβέστερον: cum -dine despectatiuus δειλοκαταφρόνητος: *cf.* audeo
formidolosus δειλός
formior: multo autem — existens πολυτροπώτατος δ' ὤν: *cf.* multiformis
formo διαμορφόω

formosus εὔμορφος, εὐσχήμων: -ius εὐμορφώτερον
fornicarium: quod — τὸ πορνικόν
forte ἴσως: si — ... εἰ τύχοιμεν, εἰ τύχοι
fortifico ἰσχυροποιέω
fortis ἰσχυρός, σθεναρός: -ior ἰσχυρότερος: -issimus ἰσχυρότατος: in usu -tior χρηματιστικώτερος
fortitudo ἰσχύς
fortuna τύχη: *cf.* sors
fortunor εὐτυχέω
fragilis: quod facile -e τὸ εὔθρυπτον
frater: -tres ἀδελφοί: qui sunt eiusdem matris — et soror δυὸ ὁμομητρίοι ἀδελφοί: rari fiunt -res σπαναδελφοῦσιν: raritas -rum σπαναδελφία
fraternus: in amicitia -a φιλάδελφος
frequentator *cf.* sacer
frequentius συνεχέστερον
fricator τριβάς
Frigia Φρυγία
frigidatio ψῦξις
frigidatus καταψυχόμενος
frigiditas ψυχρότης
frigidus ψυχρός, κατεψυγμένος, ὑπόψυχρος: -issimus ψυχρότατος: — uenter ψυχροκοίλιος
frigor κρύος, ψῦχος, τὸ ψυχρόν
frons μέτωπον
fructificatio καρποφορία, πληροφόρησις
fructus καρπός, καρποφορηθείς: *cf.* collectio
fruitio ἀπόλαυσις
fruitiuus ἀπολαυστικός
fruitor ἀπολαυστικός
frustrum *cf.* caro
fugio φεύγω
fulgens λαμπρός: -entior λαμπρότερος: -entis anime λαμπρόψυχος
fulgidus λαμπρός
fulmen κεραυνός
fundamentum ἕδρα, θεμέλιος
fundatio κτίσις (-εις): tempus -onis χρόνοι κτίσεων
fur κλέπτης
fusio ῥύσις
fusor *cf.* metallum, plumbus
futuricius μελλητικός
futurus μέλλων: (qui) futurus (est) (ὁ) ἐσόμενος

Gades Herculis πορθμός Ἡρακλείου
Galata Γαλάτης
Galatia Γαλατία
Gallia Γαλλία
Garamantica Γαραμαντική
garrio φλυαρέω
gaudeo χαίρω: non -ens (*sic*) προσκορής
gaudium χαρά
Gedrusia Γεδρουσία
gemellus διδυμόγονος
Gemini Δίδυμοι
geneaticus γενεθλιαλογικός: -ce rationes γενεθλιαλογίαι: — sermo γενεθλιαλογία, γενεθλιαλόγος: -ci sermonis γενεθλιαλογικός: -ci sermones seu species γενεθλιαλογικοί τόποι: -ci sermonis species γενεθλιαλογικός τόπος
generalis: -ior γενικώτερος
generatio γένεσις
genero: qui -atur ὁ γενόμενος
generosus γενναῖος
genitalia τὰ μόρια
genitalis γεννητικός
genitura γονή
genitus γεννώμενος, γινόμενος: duo -ti δίδυμα
gens ἔθνος: -tes ἐθνικοί: sue -tis ἐμφύλιος
genu: In genibus Ἐν γόνασιν
genus γένος: eiusdem -eris ὁμογενής: id quod -eris τὸ γενικόν
geometra (*vel* -tres): -as γεωμέτρας

Germania Γερμανία
Germanus Γερμανός
germen βλαστός
gestator: *cf.* corona, sacer
gesticulatio: amator -onis φιλόστροφος†
gesticulator: armis — ὁπλορχηστής
Getulia Γαιτουλία
gibbositas κύρτωσις
gigno *cf.* genitus
giratio περιστροφή
gladium μάχαιρα
glaucedo ἀπογλαύκωσις, γλαύκωσις: *cf.* oculus
gloria δόξα: amator (-trix) -e φιλόδοξος: in — existo ἐνδόξως ἔχω: super — ἐπίδοξον (ἐπὶ δόξαν ?)
gloriatio οἴησις
gloriator καυχηματίας
gloriosus ἔνδοξος, ἐπίδοξος
gnomon γνώμων
Gorgonum (*acc.*) Γοργόνιον (*acc.*)
gracilis ἰσχνός: -lius ἰσχνότερον
gracilitas: maior — τὸ ἰσχνώτερον
gradalis (gradualis ?) μοιρικός
gradior παραπορεύομαι
gradualiter μοιρικῶς
gradus μοῖρα: per -um μοιρικός: -uum descriptio μοιρογραφία: -uum situs μοιροθεσία: per -um ydemptitates θίξεις (θήξεις): — occumbens τὸ δῦνον: *cf.* horoscopus, terminus
grandinosus χαλαζώδης
grando χάλαζα
gratia ἕνεκεν, χάριν: *cf.* ago, uerbum
gratiosus ἐπιχάριτος (ἐπίχαρις): -ius ἐπιχαριτώτερον
grauis φορτικός: -issimus ἐπαχθής: *cf.* ira
grauitas: — aeris δυσαερία
Grecia Ἑλλάς
grossus παχύς: in -o καθ' ὅλα μέρη, κατὰ τὸ ὁλοσχερέστερον, ὁλοσχερῶς: magis — , -ior ὁλοσχερέστερος

gubernatio οἰκοδεσποτεία
gubernatiua κυβερνητική
guberno βασιλεύω
gustus γεῦσις

ho- vide etiam o-
habeo ἔχω, ἴσχω, λαμβάνω: me — (ad) ἔχω, ἐπέχω: me — quantum ad ἔχω *cum gen.*: -et (*c. acc.*) ἔτυχε (*c. gen.*): -uerint τύχωσιν ἔχοντες: bene se -ens εὐεκτικός: — (situm, distantia) ποιέω (διάστασιν): *cf.* displicentia, facies, fama, mixtura
habilitas: maior — τὸ εὐφυέστερον
habitabilis οἰκουμένη
habitatio οἰκουμένη, οἴκησις
habito κατοικέω: -ans νεμόμενος
habitudo σχέσις, φάσις (σχέσις?): bona — εὐεξία: bone -dinis εὐέκτης: melior —, melioris -dinis τὸ εὐεκτικώτερον: -dine inuoluor ἐμπεριέχομαι
habitus ἕξις: uniuersalis — τὸ ὁλοσχερές
habundantia ἀφθονία, δαψίλεια, εὐθηνία, εὐφορία, πλεονασμός, πλεονέκτημα, ὑπερβολή
habundo: -ans πλεονάζων: -antius δαψιλεστέρως: bene — εὐπορέω
hac (*adv.*) πῇ
hactenus ἐπάνωθεν (ἔτι ἄνωθεν ?)
halo ἅλως
hebes ἀμβλύς
hebetudo: maior — τὸ ἀμβλύτερον
herbarius βοτανικός
hereditas κληρονομία
heredito κληρονομέω
heremus: amator -i, -i sectator φιλέρημος
hic (*pron.*) οὗτος, (?) ἐκεῖνος, (?) τοιοῦτος: — (qui) ὁ, ὅδε: huius κατ' αὐτόν: hoc est τουτέστι: in hoc solo τοῦτο ... μόνον
hic (*adv.*) ἐνθάδε, ἐνταῦθα
hictericus ἰκτερικός
hilaris ἱλαρός
hinc ἔνθεν, ἐντεῦθεν

illicentia: magis proprie -tie αὐθαδέστερος
illicitus: -e occisor ἀθεμιτοφάγος
Illiria Ἰλλυρία
illuminatio φωτισμός
illumino: id quod non -atur τὸ ἀφώτιστον
illustratio: dei — θεοφορία
imber ὄμβρος
imitator μιμητής, μιμητικός
immansiuus ἐπίμονος
immansuetus ἀνήμερος
immaturus ἀδρανής
immeabilis: -ia δυσοδία
immisericors ἀνελεήμων
immissio ἐμβολή
immitto ἐμπεριποιέω
immoderatio ἀμετρία
immoderatus ἄκρατος
immolator θύτης
immolo καθιερόω
immunditia τὸ ἀκάθαρτον
immundus ἀκάθαρτος: -ior ἀκαθαρτότερος
immutabilis ἄτρεπτος
immutabilitas ἀκινησία
immutatio ἐναλλοίωσις, μετατροπή, παραλλαγή, τροπή
immuto συντρέπω: -or τρέπομαι
impedimentum ἐμπόδιον, ἐποχή, λώβησις: sine -to ἀνεμπόδιστος
impedio ἀποσυνεργέω, βλάπτω, ἐμποδίζω, παραποδίζω: -ditus (*sic*) κεκακωμένος: aspectu -iens κατοπτεύων
imperfectio τὸ ἀτελές
imperfectus ἀτελείωτος, ἀτελής
imperforatus ἄτρητος
impermixtus: magis — ἀκρατότερος
impero προστάσσω
impetigo λειχήν
impetuosus καταφερής, ὁρμητής
impetus ὁρμή (-αί)
impius ἀσεβής
impletio πλήρωσις
implicatio ἐπιπλοκή
implicator: *cf.* dolor
impono ἐντίθημι, ἐπιτίθημι
impositorius ἐπιθετικός
impossibilis ἀδύνατος
impostor ἐπιθέτης
impotentia ἀδυναμία
impressio τὸ τεταγμένον
imprimo ἐμποιέω, ἐνδίδωμι, περιποιέω
imprudens ἄφρων
impudicia (-dicitia ?) ἀσέλγεια
impudicus ἀσελγής: magis — ἀσελγέστερος: -a agens ἀσελγαίνων
in (*cum acc.*) εἰς, ἐπί *cum acc.*, πρός *cum acc.* — (*cum abl.*) ἀπό, ἐν, ἐπί *cum gen.*, *cum dat.*, κατά *cum gen.*, *cum abl.*, ὑπό *cum gen.*, *cum acc.*
inanimatus ἄψυχος
inanimosus ἄθυμος
inarticulatio τὸ ἀδιάρθρωτον
incensio ἐκπύρωσις, πρηστήρ
incertus ἐπισφαλής: -tissimus ἐπισφαλέστατος
incestuosus ἄναγνος
inchoatio καταρχή
inchoo ἄρχομαι, κατάρχομαι
incido ἐκπίπτω, ἐμπίπτω, ἐπιβάλλω, ἐπισυμπίπτω, περιπίπτω, περιτυγχάνω, συμπίπτω, ὑποπίπτω: facile -ens εὐέμπτωτος
incipio ἄρχομαι
incisio τομή
inclamatio ἔγκλημα
incolligatus ἀσύνδετος
incommensuratio τὸ ἀσύμμετρον
incommensuratus ἀσύμμετρος
incompactio ἀπηξία
incompatiens ἀσυμπαθής
incompositio τὸ ἄμορφον
incompositus ἀσύνθετος

ingloriatio ἀδοξία
ingloriosus ἀνεπίφαντος
inglorius ἄδοξος: -e ἀδόξως
ingratus βάσκανος
ingressiuus: bene — εὐεπίβολος (-βουλος)
ingressus ἐπέμβασις
ingrossatus πεπαχυμένος
ingulo σφάζω
inhereo *cf.* uoluptas
inhertia νωθρία
inicio *cf.* manus
iniector (-trix): — oculorum ῥιψόφθαλμος
inimicabiliter φιλέχθρως
inimicitia ἔχθρα: -iam faciens ἐχθροποιούμενος
inimicus ἐχθρός
inimpedibilis ἀνεμπόδιστος, ἀπαραπόδιστος
initium (anni) νεομηνία (τοῦ ἔτους)
iniuria ὕβρις: -iam sustinens εὐαδίκητος
iniuriatiuus ὑβριστής
iniuriator ὑβριστής
iniuriosus ἀπελάτης, ὑβριστής
iniustitia: -iam patior ἀδικέομαι
iniustus ἄδικος
innuptus ἄγαμος
inordinatus ἄτακτος
inquietudo ἀκαταστασία
inquinatus ὑβριζόμενος
inquiro ἐπιζητέω: qui -itur ζητούμενος
inquisitiuus ζητητικός
inquisitor ζητητικός
insania μανία
insaniosus μανιώδης
insanus μανιώδης
inseductibilis ἀνεξαπάτητος
insensatus ἀνόητος
insensibilis: -bilius ἀσημότερον
insidiatiuus ἐπιβουλευτικός
insidiator ἐνεδρευτής, ἐπιβουλευτικός
insidie ἐπιβουλαί
insidiositas τὸ ἐπιβουλευτικόν, ἐπιβουλευτικός: -diose ἐπιβουλευτικῶς
insidiosus ἐπιβουλευτικός: -e ἐπιβουλευτικῶς
insignis ἐπιφανής: -ior ἐπιφανέστερος: minus -ne ἀσεμνότερον: non — ἀνεπίφαντος
insinuatio ἐπιφάνεια, τύπος
insinuo δείκνυμι, δηλόω, διασημαίνω, ἐμφαίνω, ἐμφανίζω, ἐπισημαίνω, ὑποβάλλω, ὑπομνηματίζω, ὑποφαίνω: significationem — ἐπισημαίνω: -antes se δοκοῦντες
insisto *cf.* opus
insocialis ὠμός
insolubilis ἄλυτος
inspicio ἐπιθεωρέω
instabilis ἄστατος
instabilitas τὸ ἐπισφαλέστερον, ἀκαταστασία
instituo ἵστημι, καθίστημι, κατασκευάζω, συνίστημι: -or σύγκειμαι: bene -tutus εὐσχήμων
instructio παιδαγωγία
instruo: quando -unt κατὰ τὰ παρατάξεις
insubditus ἀνυπότακτος
insubicibilis ἀνυπότακτος
insubiectus ἀνυπότακτος
insula νῆσος
insum ἔνειμι, ἐνυπάρχω, ὑπάρχω: cum insit ὑπάρχοντος: *cf.* capitulum
insuper ἔτι
insuperatus ἀκαθυπερτέρητος
insurrectio ἐπανάστασις
integer: — uel illesus ἀφελής
intellectiuus διανοητικός, συνετός
intellectualis νοερός
intellectus ἔννοια, σύνεσις: -u acutus νουνεχής: *cf.* muliebris
intelligibilis: facile -e est εὐκατανόητον
intelligo ἐπινοέω: -ens συνετός

intemperantia ἀκρασία, ἀμετρία
intendo σκεπτόμαι, χράομαι σκοπῷ: -ens sapientie φρενήρης
intentio (-tencio, -tensio) ἐπιβολή, ἐπίσκεψις, ἐπίτασις, προαίρεσις, σκοπός
intentum τὸ ὑποκείμενον
inter μεταξύ *cum gen.*, πρὸς *cum acc.* (*cum nominibus ventorum*): *cf.* intermedius
intercessio ἔντευξις
interfector: uirorum — ἀνδροφόνος
interitus ἀπώλεια
intermedius (ὁ) μεταξύ: — inter μεταξύ *cum gen.*
intermitto διαπίπτω
interpolate ἐπιληπτικῶς
interpres ἑρμενεύς
interpretatio ἑρμηνεία
interpretatiuus ἑρμηνευτικός: *cf.* sompnus
interrogo λέγω (?)
intestinus ἐμφύλιος
intransponibilis ἀμετάθετος, ἀμετάπτωτος
introductio ὑφήγησις
introductorius εἰσαγωγικός
intueor: is qui iam -tuitus est ὁ διασκεπτόμενος
intus *cf.* deprehendo
inuasio ἐπίθεσις
inuasor: theatrorum — θεατροκόπος
inuectio ἐπήρεια
inuenio εὑρίσκω
inuentio εὕρεσις
inuentiuus εὑρετικός
inuerecundus ἀναιδής
inuicem (ab, ad, secus) ἀλλήλων -λοις -λους
inuidia φθόνος†
inuidiosus φθονερός
inuiolatus ἀκαθαίρετος
inuisibilis ἀθεώρητος
inundatio ἀνάβασις, ὄμβρος: -iones τὰ ῥεύματα: — aquarum ὑδάτων ὀμβρίων ἐπιφορά: -ionibus (morientes) ὑποβρύχιοι
inuocator: — demonum γόης
inuoluo περικυλίω: habitudine -or ἐμπεριέχομαι
inutilis ἄχρηστος
inutilitas τὸ ἄχρηστον
ioculator θυμελικός
ipse αὐτός, οὗτος: — (qui), — uidelicet ὁ: ipsius ἑαυτοῦ, κατ' αὐτόν
ira χόλος: acute -e ὀξύθυμος: grauis -e δύσμηνις
iracundus θυμικός, ὀργίλος
irrationalis ἄλογος
irregularis ἀνώμαλος
irregularitas (-tates) ἀνωμαλία
irrigator ἀντλητής
is (qui) αὐτός, ὁ: eo quod ἐπειδή: id est τουτέστι: *dat. abl. plur.* hiis
Isis Ἶσις
Issicus Ἰσσικός
iste αὐτός, ὅδε, οὗτος: *cf.* modo
ita οὕτως: tamquam — οὕτως: — ut ὡς, ὥστε
Italia Ἰταλία
itaque δή
iterum πάλιν
Iudea Ἰουδαία
Iudex *cf.* sompnus
iudicatiuus κριτικός
iudicator κριτικός: *cf.* sompnus
iudicialia ἀποτελεσματικά
iudicium διάκρισις, προτέλεσις: sine -o ἄκριτος
iudico κρίνω, νομίζω
Iupiter (Iouis) Ζεύς (Διός): Iouis proles διόσκουροι
iustitia δικαιοσύνη: amator -ie φιλοδίκαιος
iustus δίκαιος: -ius τὸ δικαιότερον: -e δικαίως
iuuencula *cf.* nubo

iuuenilis νεανικός, νέος
iuuenis νέος: iunior νεώτερος: iunior masculus νεώτερος: iunior femina νεωτέρα: *cf.* nubo

katarche (id est optentio principii) καταρχή
kathaclismus κατακλυσμός
Keltogalatia Κελτογαλατία
Kharitarum Χαρίτων
Kolchica Κολχίς
Kyklades *cf.* Ciclades

labor πόνος: amatrix -orum, amatiuus -orum φιλόπονος: *cf.* recusatrix
laborator: argenti — ἀργυροκόπος
laboriositas τὸ ἐπίπονον
laboriosus ἐπίμοχθος, ἐπίπονος, ἐργατικός, πονικός, φιλόπονος
laboro: -are τὸ ἐπίπονον: singulariter -ans μονόπονος
lamentatio θρῆνος
lamentator θρηνῳδός
lamentum θρῆνος
lamento ταλανίζω
Lampaurus Λαμπαύρας
lana ἔριον
languorosus ἐπίνοσος
lapidarius λιθουργός
lapis λίθος: -idum sculptor λιθοξόος
largitiuus χαριστικός
latenter κρύβδην, λάθρᾳ, λεληθότως
lateo λανθάνω: -eat nos λάθωμεν αὐτούς
latio φορά
latitudo πλάτος: maxime -ines ἐπικάμπια
latomus λατόμος
latreuticus λατρευτικός
latro ληστής: -ones ληστήρια
latrocinium κλοπή: -a ληστήρια
latrocinor: -ans ληστρικός
laus ἔπαινος
ledo βλάπτω, κακόω: sensum -ens φρενοβλαβής
legislatiuus νομοθετικός
legitimus νόμιμος
lentigo φακός
Leo Λέων
lepra λέπρα
Lepus Λαγωός
lesiuus βλαπτικός
lesura σίνος: -am patiens ἐπισινής, σεσινωμένος: membrorum -am patiens, membrorum -am sustinens ἐπισινής: -arum illatiuus σινωτικός: -is infectus ἐπισινής: sine — ἀσινής
letatrix: *cf.* quies
letor: non -ans ἀνεύφραντος
leuis κοῦφος
lex νόμος: -gis scriba νομογράφος: contra -em παράνομος: -ge statuo νομοθετέω: sue -gis αὐτόνομος
liber αὐθάδης, αὐθεντικός, ἐλεύθερος
liber βιβλίον, σύνταξις: primus — τὸ α' βιβλίον
liberalis ἐλευθέριος, ἐλεύθερος: -ior ἐλευθερώτερος
liberalitas ἐλευθεριότης
libertas: amator -atis, amatiuus -atis φιλελεύθερος: amor -atis τὸ φιλελεύθερον
Libia Λιβύη
libonotus (liuonotus) λιβόνοτος
Libra Χηλαί
Libra Ζυγός
licet: -ebit ulterius προχωρήσει
Licia Λυκία
Lidia, Lydia Λυδία
ligamen σύνδεσμος
lignum *cf.* fissor
linea γραμμή
lingua γλῶττα
linum λίνον

liquet: -ebit φανείη
Lira Λύρα
lis δίκη
litigium περιβοησία: amator -ii φίλερις
littera γράμμα: amator -arum φιλογράμματος: *cf.* falsator
localis τοπικός: -iter τοπικῶς
locator *cf.* corpus
locus τόπος, ὅρος, χώρα, (?) σχηματισμός: peregrinus — ἀπόκλιμα: in multis -is πολλαχῇ: *cf.* extraneus, luminaria, sol
locusta ἀκρίς
logicus λογικός
logista λογιστής
longeuitas πολυχρονιότης
longeuus πολυχρόνιος
longinquus: prospicio ex -quo διαμηκίζω
longitudo μῆκος: per -dinem disto διαμηκίζω: -dinis aspectu impedior διαμηκίζομαι
longus μακρός: -e precedens nos ἔτι ἄνωθεν
loquor: idem — ταυτολογέω: uix -ens μογγιλάλος
lucratiuus ποριστικός
lucror: turpiter -ans αἰσχροκερδής
lucrosus ἐπικερδής
lucrum ἐπικέρδ(ε)ια, κέρδος, τὸ πορίζειν: sine -o ἀπερίκτητος
lucta παλαίστρα
luctus πένθος
lumbi (*sic*) ὀσφύς
lumen φῶς: -ina τὰ φῶτα
luminaria (*genit. plur.* -ium) φῶτα: locus -ium φωσφορία: loca — οἱ φωσφοροῦντες
luna σελήνη
lunaris σεληνιακός
lupus: — aut cancer (*sic*) κονδύλωμα
lusiuus παιγνιώδης
lux φῶς
luxuria ἐπαφροδισία
luxuriosus λάγνος

Macedonia Μακεδονία
magicus μαγευτικός, μαγικός
magis μᾶλλον, πλεῖον: quanto — ὅσῳ μάλιστα: maxime μάλιστα: quam maxime ὅτι μάλιστα: *cf.* angularis, confero, consono, commodum, commutatiuus, cuspis, grossus, illicentia, imperfectus, latitudo, mediocris, muliebris, naturalis, obseruatiuus, orientalis, perfectus, periodalis, quies, sensibilis, studiosus, sollicitus, uaco
magistralis διδασκαλικός
magistratus (*sic*) καθυπερτερούμενος
magnanimitas μεγαλοψυχία
magnanimus μεγαλόψυχος
magnetes (magnes) μαγνῆτις: ἡ λίθος ἡ μαγνῆτις
magnificus μεγαλοπρεπής: -centius τὸ μεγαλοπρεπέστερον
magnitudo μέγεθος, μεγαλειότης: bone -inis εὐμεγέθης: in -ine τοῖς μεγέθεσιν
magnus μακρός, μέγας: -um facio μεγαλοποιέω: maior μείζων: maius τὸ μᾶλλον: maximus μέγιστος: *cf.* Asia, attempto, audeo, deus, dominabilitas, euidentia, exsiccatio, consono, fiducialitas, firmitas, formido, gracilitas, habilitas, latidudo, nocumentum, oculus, profectus, quam, sapientia, sapio, tarditas, totalitas
magus γόης: -i μάγοι
maioritas *cf.* totus
maledictio λοιδορία
maledicus λοίδορος
malefactiuus κακωτικός
malefactor κακωτής
malefico κακοποιέω
maleficus κακοῦν, κακοποιός, κακωτικός
malignatio κάκωσις

malignitas κάκωσις, τὸ πονηρόν: amator -atis φιλόμοχθος: *cf.* stultus

malignizo: -atus κεκακωμένος: -atum esse κεκακῶσθαι

maligno κακόω: -ans κακοποιοῦν

malignus (-um) κακοῦν, κακοποιός, κακοῦργος, κακωτικός, πονηρός, ὑποπόνηρος: — factus κακωθείς: profunde — βαθυπόνηρος: amator -orum φιλοπόνηρος: *cf.* oditor

maliloquus κακολόγος

malitia κάκωσις: sine — ἄκακος

malus (-um) κακός, φαῦλος: ualde — πάγκακος: -um κακουχία, κάκωσις: -e κακῶς: patiens -orum κακοπαθής: non toleratiuus -orum ἀνεξίκακος: *cf.* (malus, -e, peior) actor, arteficiosus, demon, fama, memor, modus, muliebris, odio, os, sententia, stultitia, sufferentia, zelator

mamilla μαστός

maneo μένω†

mania μανία

manifestatio ἐμφανισμός, τὸ ἐμφανιστικόν

manifesto ἀναδεικνύω, ἐμφανίζω: -or καταφαίνομαι

manifestus καταφανής, πρόδηλος: in -o ἀναφανδόν: -issimus ἐκφανέστατος

mansio ἐπιμονή

mansorius ἔμμονος

mansuetudo τὸ πρᾷον

mansuetus πρᾷος, προσηνής: -a χειροήθη

manus χείρ: sibi ipsis manus inicientes αὐτόχειρες ἑαυτῶν γινόμενοι

marcesco ὑπομαραίνομαι

mare θάλασσα, -αι: effusio -ris ἄμπωτις

marinus θαλάσσιος: *cf.* uas

Marmarica Μαρμαρική

Mars Ἄρης

marturinus αἴλουρος

masculinizo ἀρρενόομαι, ἀρρενοποιέω

masculinus ἀρρενικός (ἀρσεν-), ἄρρην, ἠρρενωμένος: -um τὸ ἄρρεν: — fio (sum) ἀρρενόομαι

masculus ἄρρην: -i (-orum) procreatio ἀρρενογονία: -ius ἐπανδρότερον: *cf.* iuuenis, senior

mater μήτηρ: -ris μητρικός: -ri τῷ μητρικῷ: consanguineus ex parte -ris μητρυιά: eiusdem -tris ὁμομήτριος: *cf.* frater

materia ὕλη

materialis ὑλικός: -ior σωματωδέστερος: -ius ὑλικώτερον

maternalis μητρικός

maternus μητρικός

mathemata μαθήματα

mathematicus μαθηματικός: -ca (*sing.*) τὸ μαθηματικόν: -ca (*neutr. plur.*) μαθήματα

Matiane (*acc.*:-nem) Ματιανή

matrix μήτρα: -icis (deuastatio) ὑστερική (νομή): *cf.* histericus

matutinalis ἑῷος

matutinus (ad) ἑῷος (*cum dat.*)

Mauritania Μαυριτανία

meatus πόρος

mechanicus μηχανικός

Media Μηδία

mediatio *cf.* celum

medicamen: per -ina ταῖς ἰατρείαις

medicinalis ἰατρικός: — compilatio ἰατρομαθηματικὴ σύνταξις: amator -ium φιλίατρος

medicus ἰατρός

medietas μεσότης, τὸ ἥμισυ, τὸ καθ' ἥμισυ μέρος

medio *cf.* celum

mediocris μέτριος: magis — μετριώτερος: *cf.* pinguedo

mediocritas (-tates) μετριότης, συμμετρία

mediocriter *cf.* caluus

meditatio μελέτη

meditatiuus διανοητικός

medius μέσος, ἥμισυς: — inter μεταξύ *cum gen.*: -o ἀνὰ μέσον, μεταξύ *cum gen.*: transiens per -ia animalia ὁ διὰ μέσων: *cf.* capillus, capillatura, celum, color, oculus
Medius: -a Ethiopia μέση Αἰθιοπία
melancolia μελαγχολία
melancolicus μελαγχολικός
melior *cf.* bonus
melleus *cf.* color
melodia μελῳδία
membrum μόριον: *cf.* detrunco
memor: -res malorum, -res mali μνησίκακοι
memoratiuus μνημονικός
mendax ψεύστης
mens διάνοια
mensis μήν (μείς): qui — ὁ μηνιαῖος: -ium (locorum, signorum) τῶν μηνιαίων (τόπων, ζῳδίων): quattuor -es τετράμηνον: *cf.* tot
mensualis μηνιαῖος
mensura μέτρον
mensuratiuus μετρητικός
mensurator *cf.* palma
mentior διαψεύδομαι, καταψεύδομαι
Meotidus: -a palus Μαιῶτις λίμνη
mercator ἔμπορος
Mercurius Ἑρμῆς
meretrix ἑταίρα
meridianus μεσημβρινός
meridies μεσημβρία: *cf.* sol
meridionalis μεσημβρινός
merito εἰκότως
Mesopotamia Μεσοποταμία
Metagonites (-tidis) Μεταγωνῖτις
metallicus μεταλλικός
metallum: -orum fusor μεταλλευτής
meteora μετέωρα
metheorologicus μετεωρολογικός
methodicus μεθοδευτής (-τικός)
mihi: — uidetur μοι δόκει: *cf.* nos
miles στρατιώτης
militaris στρατηγικός: -ia στρατεῖαι
militarius στρατιωτικός
militia στρατεία, στρατηλασία
ministratiuus ὑπηρετικός
minor (minus) ἧττον, νεανικώτερος: -imus ὀλιγοστός: *cf.* animosus, decidentia, durabilitas, insignis
minuo (cum) συμμειόω: -or μειόομαι
mirabilis ἐξαίρετος: -ium factrix θαυματοποιός
miror θαυμάζω
misceo: mixtus κεκραμένος, μεμιγμένος
miseratiuus ἐλεητικός
misericordia τὸ ἐλεητικόν
misericors ἐλεήμων
misterialis μυστηριακός
misterium μυστήριον
misticus μυστικός
Mithram (Solis) Μίθραν (Ἥλιον: †-ίου ?)
mixtura μῖξις: -am habentes ex nigro et uiridi μελάγχλωροι
mobilis: difficile — δυσκίνητος: facile —, bene — εὐκίνητος: facilius -e, facile -ius εὐκινητότερον
moderatio: bone -onis εὐφρόσυνος
moderatiuus σύμμετρος
moderatus σύμμετρος: -ior μετριώτερος
modicus μικρός: *cf.* durabilitas
modus τρόπος, τύπος: commixtiuus — τὸ συγκρατικὸν εἶδος: eodem -o (cum), tamquam eodem -o ὡσαύτως (*cum dat.*): illo -o ἐκείνως: isto -o ... illo οὕτως ... οὕτως: per omnem -um πάντα τρόπον: mali -i κακότροπος: multorum -orum πολύτροποι: similis -i ὁμοιότυπος: singularis -i ἰδιότροπος: subacti in -um mulierum γυναικῶν ἔργα διατιθέμενοι: *cf.* assecutibilis, breuis, neutrum, omnis, solus

molestatio ἐπισημασία, ὄχλησις
molestator ὀχληρός
molestia ὄλεθρος
molesto ἐνοχλέω
mollis μαλακός
monachus: amator -orum φιλομόναχος
monstrum τέρας: *cf.* species
monstruosus τερατώδης
montanus: per -ana ὀρεινῆς ῥάχεως
mora: -am facio παροδεύω
moralis ἠθικός
morbus νόσος
morior ἀποθνήσκω, τελευτάω: mortuus ἀποθνήσκω, κατοιχόμενος, νεκρός: *cf.* diuinator
moror: — in (*cum abl.*) ἐγχρονίζω: *cf.* ibi
morphea ἀλφός
mors θάνατος (-οι): -tis θανατικός: per uiolentiam — βιαιοθανασία
morsa δακετά
mortalis θανατικός
mos ἔθος, ἦθος, τρόπος: dulcis moris χρηστοήθης, φιλόκαλος
motiuus κινητικός
motus κίνησις, πάροδος: boni — εὐκίνητος: facilioris — εὐκινητότερον: uelox — ὀξυκινησία
moueor κινέομαι, τρέπομαι: subtus motus ὑποκεκινημένος
mox εὐθύς
muliebris γυναικεῖος, γυναικώδης: — animo γυναικόθυμος: — intellectus γυναικονοήμων: — decentia magis τὸ γυναικοπρεπωδέστερον: in malum — κακογύναιος
mulier γυνή: -erum (dona, insidie) γυναικεῖαι (δωρέαι, ἐπιβουλαί): a -ibus personis ὑπο θηλυκῶν προσώπων: multarum -ierum πολυγύναιος: *cf.* modus, oditor
muliercula: -e τὰ γυναικεῖα
multiformis πολύμορφος: -iter πολυμερῶς: -ius τὸ πολυτροπώτερον: multo autem formior existens πολυτροπώτατος δ' ὤν
multiplex πολύχους: -plicior πολυμερής
multipliciter πολλαχῶς
multiplico πολλαπλασιάζω, πολυπλασιάζω: in 17 — ἑπτασκαιδεκάκις ποιέω
multitudo πλῆθος: *cf.* filius
multotiens πολλαχῇ
multus -um (-o) περισσός, πολύς -ύ, τὰ πολλὰ (*acc.*): in -is τὰ πολλὰ (*acc.*): ad -um ἐπιπολύ: — ualde παμπληθής: *cf.* actio, agilis, aurum, communis, dispersio, diuulgatus, durabilitas, euidens, flegma, formior, historia, historiosus, ingeniosus, locus, modus, mulier, negotiosus, negotium, notatus, pars, possessio, probrosus, sperma, sudor, tardo, tempus, uir
mundialiter κοσμικῶς
mundus κόσμος, τὸ κοσμικόν: id quod est -di τὸ κοσμικόν: rector -di κοσμοκράτωρ
munitor ὑπονομευτής
murus: -orum suffossor τοιχωρύχος
musa μοῦσα: amatrix -arum φιλόμουσος
musica: amator -e φιλόμουσος
musicum: amator -orum φιλόμουσος
mutabilis: facile — εὐμετάβολος
mutor τρέπομαι

narratio ὑφήγησις
nascor φύομαι: plures -scentes πλειστογόνοι: qui -itur, natus, qui natus (ὁ) γεννώμενος: natus sum πέφυκα: natum τέκνον, τεχθέν: defectus natorum ἀτεκνία: natorum amatrix φιλότεκνος: natorum priuatio ἀτεκνία: natorum factiuus τεκνοποιός: natos procreare τεκνοῦν: *cf.* oditor, paucitas, terra
Nassamonitica Νασαμωνιτική

Nassamonitis: -nitem Νασαμωνῖτιν

natatilis νηκτός

nates *sic* (Leonis) πλόκαμος (τοῦ Λέοντος): nates, posteriora (*sc.* Leonis, *duplex lectio*) τὰ ὀπίσθια (*sc.* μοῖρα Λέοντος)

natiuitas γένεσις: -tatis γενεθλιακός

natura φύσις: per -am φύσει, τὴν φύσιν: secundum -am τὴν φύσιν: inferior — ἡ ἐφεξῆς φύσις: — que deposuerit feminitatem τὸ ἀθήλυντον τῆς φύσεως

naturalis φυσικός: magis -e φυσικώτερον: -iter φυσικῶς: magis -iter φυσικώτερον

naualis ναυπηγικός

nauigatio: bona —, facilis — εὐπλοία: difficilis —, -ionis difficultates δυσπλοία

nauigia (*sic*) ναυάγια, ναυκληρία

Nauis Ἀργώ

nauis: -es compingens ναυπηγός

nauta ναυτιλλόμενος

nautici ναυτικοί

nebula νιφετός

nebulosus νεφελοειδής, νιφετώδης

nec μήτε

necessarium ἀνάγκη

necessarius ἀναγκαῖος: — circa καταγινόμενος περί (*duplex lectio, cf.* oportunus)

necesse ἀνάγκη

necessitas ἀνάγκη: de -ate ἐξ ἀνάγκης

nefarius ἀθέμιτος

nefriticus νεφριτικός

negotiatio ἐμπορία, ἐπικουρία, πραγματεία

negotiatiuus ἐμπορικώτερος

negotiator ἔμπορος, πραγματευτικός

negotior ἐμπορεύομαι, περιεργάζομαι

negotiosus ἔμπρακτος: multum — πολυπράγματος

negotium πρᾶγμα, πραγματεία, χρηματεῖα (?): multorum -iorum πολυπράγμων

neomenia (anni) νεομηνία (τοῦ ἔτους)

nequam πονηρός

nequaquam μηδαμῶς, οὐδαμῶς

neque μή, μηδέ, μήτε, οὐδέ, οὔτε

nequitia: sine — ἀπόνηρος

neruus νεῦρον

neuter οὐδέτερος: -trum μηδέτερον: -tro modo μηδετέρως

nichil μηδέν, οὐδέν

niger μέλας: *cf.* capillus, color, mixtura, oculus

Nilus Νεῖλος

nisi εἰ μή, πλήν

nociuus βλαβερός, βλαπτικός, ἐπιβλαβής: -e βλαπτικῶς: *cf.* appetitus

nocturnalis τῆς νυκτὸς

nocumentum βλάβη, σίνος: maius — τὸ βλαβερώτερον

nodus σύνδεσμος

nomen ὄνομα

nominatio ὀνομασία

nomino ὀνομάζω

non οὐ, οὐκ, οὐχ, μή, μηδέ: aut — ... aut — ἢ μήδ' ... ἢ μή: — iam μηκέτι: non ... neque μήτε ... μήτε: *cf.* addextratus, adhuc, adopertus, adulatrix, adumbratus, amator, amicus, annalis, appareo, audax, concordans, conueniens, deus, erraticus, erro, fio, firmus, gaudeo, honoro, iam, illumino, insignis, leto, pertingo, productus, recogito, saluus, suffero, uertibilis, zelotipus

nos ἡμεῖς: lateat — λάθωμεν ἑαυτούς: *cf.* precedens

nosco: certitudinaliter — ἀκριβόω

noster ἡμῶν, ἡμέτερος: -rorum ἡμῶν

notabilis ἐπίσημος: -ior ἐπιφανέστερος

notapelioticus, notafil-, notafel- (+ id est simul australis et subsolanus, + id est australis simul et orientalis) νοταπηλιωτικός

notitia γνῶσις

noto διασημαίνω, ἐπισημαίνω, παρίστημι: -atur προσημαίνεται: multum -atus πολυθρύλλητος

notolibicus (+ id est simul australis et occidentalis) νοτολιβυκός

nouem ἐννέα

nouus νέος: -issimus πύματος

nox (nocte, de nocte) νύξ (νυκτὸς): in extremo -ctis, extremitate -ctis ἀκρόνυκτος: *cf.* errabundus

nubecula νεφέλιον

nubes νέφος: sine -e ἀνέφελος

nubo: iuuenes -ere (facit), iuuenculas -ere (facit) νεογάμους (ποεῖ): pluries -ens, pluries -ere πολύγαμος: semel -ens, semel -ere (*sc.* factus) μονόγαμος: tarde -ere (facit) βραδυγάμους (ποιεῖ)

nubositas νιφετός, συννέφεια

nudus ψιλός

nuitio πρόσνευσις

nullus μηδείς, οὐδείς: -um οὐδέν: *cf.* dissimilis

Numedia Νουμι(η)δία

numero ἀριθμέω: -andum est τὸ ἀριθμεῖν

numerus ἀριθμός: equalis —, equalis -o, equalis -i ἰσάριθμος

numquam μηδέποτε, οὐδέποτε

nuptialis γαμικός

nuptie γάμος: bone — εὐγαμίαι

nusquam μηδαμῇ

nutrio: is qui non nutritur ὁ ἄτροφος: -iens aquosum τὸ τῶν τροφῶν ὑδατῶδες

Oasis Ὄασις

obedio ἀκούω, ὑπακούω

obfusco θολόω

obliquus *cf.* oculus

obliuiosus ἐπιλήσμων

obscurus ἀμαυρός

obsequiosus *cf.* demon

obseruanter παρατηρητικῶς

obseruatio παρατήρησις: per -ionem παρατηρητικῶς

obseruatiuus εὐεπίβολος: — (misteriorum) ἐπιτελεστικός (μυστηρίων): magis — παρατηρητικώτερος: -e παρατηρητικῶς

obseruator ἐπίβολος†

obseruo -or θεωρέω, παρατηρέω, παραφυλάττω, τηρέω: -andum παρατηρητέον, τηρητέον

obsideo παροδεύω

obsistentia ἀντιδικία, ἐναντίωσις

obtempero ὑπακούω

obtenebratio ἐπισκίασμα, ἐπισκότησις, σκοτισμός

obtineo (optin-) καθυπερτερέω, ἐπέχω, ἐπικρατέω

obuiatio ὑπάντησις

obuio ὑπαντάω

occasio ἀφορμή, πρόφασις, τὸ αἴτιον, τὰ -ια (*sic*)

occasionabilis: facile -biliter εὐπροφασίστως

occasionans ἐπαίτιος

occasiuus δυτικός

occasus δύσις, δυσμαί, τὸ δῦνον, ἑσπέρα, κατάδυσις, λίψ: simul — συγκατάδυσις: regio -uum δυσμαί: que ad -us τὰ λιβυκά

occidens δύσις, δυσμαί, τὸ δῦνον: *cf.* boreas

occidentalis ἀποκρουστικός, δυτικός, πρὸς δυσμάς, ἑσπέριος, λιβυκός: — (angulus) (τὸ) δῦνον (κέντρος): -ior δυτικώτερος, λιβυκώτερος: *cf.* borealibicus, uentus

occidentalitas τὸ λιβυκόν

occido δύνω

occisor: illicite — ἀθεμιτοφάγος

occisor φονεύς: turpis — μιαιφόνος

occultatio (τὸ) ἀπόκρυφον, κρύψις

occultatiuus: — sui ἐνδόμυχος

occulto ἀφανίζω, κρύπτω

occultus ἀπόκρυφος, κρυπτός, κρύφιος

occumbo δύνω, καταδύνω: -ens τὸ δῦνον, τὸ δυτικόν: gradus -ens τὸ δυτικόν

occupo ἀσχολέω, καθικνέομαι: -atus διειλημμένος

occurro ὑπαντάω

occurso συναντάω: -ans ὑπαντικός (vel ὑπαντητικός)

occursus ἀπάντησις, συντυχία, ὑπάντησις

octo ὀκτώ: — anni ὀκταετία

oculus ὀφθαλμός, ὄψις: bonorum -orum εὐόφθαλμος: glaucorum -orum γλαυκόφθαλμος: magnorum -orum μεγαλόφθαλμος: mediorum -orum μεσόφθαλμος: nigrorum -orum μελανόφθαλμος: obliquorum -orum λοξόφθαλμος: paruorum -orum μικρόφθαλμος: *cf.* caprinus, iniector

odio: homines -ens μισάνθρωπος: suos -ens μισοΐδιος: -ens malum μισοπόνηρος

oditor: ciuium — μισοπολίτης: corporis — μισοσώματος: domesticorum — μισοίκειος: malignorum — μισοπόνηρος: mulierum — μισογύναιος: — pulcrorum μισόκαλος: natorum — μισότεκνος

odor ὀσμή: prauitas -oris δυσωδία

offensio: sine -ione ἄπταιστος

olla *cf.* diuino

olus λαχανεία

omnimodus παντοδαπής, παντοῖος, τέλειος: *cf.* diligentia

omnino τὸ ὅλον, ὅλως, παντελῶς, πάντως: — penitus πάντῃ πάντως

omnis ἅπας, πᾶς: ex -ibus ἐκ πάντος: -es ὅλοι: per -em modum πάντα τρόπον: modis -ibus πάντως: ut (quasi) in -ibus, ut ad -e (dicatur) ὡς ἐπίπαν: *cf.* aggredior

omogeneus: -eorum ὁμογενῶν

operatio ἐνέργεια, τὸ ποιητικόν, πραγματεία, πρᾶξις: amator -ionis φιλοπράγμων, φιλόπρακτος: bona — εὐπραγία: eminentius -ionis τὸ ἀνυτικώτερον: *cf.* secundarius

operatiuus ἀπεργαστικός, δράστης, ποιητικός, πρακτικός

operator πρακτικός

operio κατακαλύπτω

operor ἀπεργάζομαι, ἐνεργέω, ἐργάζομαι, κατεργάζομαι, πράσσω (-ττ-): in -ando (adherentes) πραγματείας (ἐχόμενοι): *cf.* austerus

operosus δράστης, ἔμπρακτος, περίεργος, πρακτικός: -ior δραστικώτερος: -issimus πρακτικώτατος

opinio δόξα: uane -ionis κενόδοξος

opinor οἴομαι: uane — κενοδοξέω

oportet δεῖ, χρή, προσήκει

oportunitas χρεία

oportunus ἐπιτήδειος, οἰκεῖος, πρόσφορος, χρήσιμος: -um δέον: -ius εὐκαιρότερον: -e δεόντως, προσηκόντως: — circa καταγινόμενος περί (*duplex lectio, cf.* necessarius)

oppositio διάμετρος, διάμετρος στάσις: secundum -ionem κατὰ τὸ διάμετρον: *cf.* diametralis

oppositio ἀντίθεσις

oppositus ἀντικείμενος: angulariter — ἀντιγώνιος: -e ἀντικειμένως

oppositus: ex -o κατὰ διάμετρον: *cf.* diametralis, celum

oppressio θλῖψις

optentio *cf.* katarche

optentiuus (*sic*) συγκρατικός

optimus *cf.* bonus

optineo (obt-) διακατέχω, διαλαμβάνω, ἐπίσχω, ἐπιτυγχάνω, κατακρατέω, κατέχω, κρατέω: *cf.* dominium

opus πραγματεία, πρᾶξις, ἔργον: sine -ere ἀπραγία, ἄπρακτος: amator -eris φιλεργός: augmentator -eris ἀνυ(σ)τικός: proprio -eri insistens ἰδιοπράγμων: turpium -erum [*sc.* factor] αἰσχροποιός: sine -ere ἀπραγία, ἄπρακτος: *cf.* anima, modus

orbatio πήρωσις

Orchenia Ὀρχηνία

Orchenii Ὀρχήνιοι

ordinatio διάταξις
ordinator διατάκτης
ordinate τεταγμένως
ordo διάταξις, εἱρμός, τάξις: secundum -inem τάξει: — unius post unum τάξις παρ' ἕν: — datus est ponendi unum post unum τάξις ἐδόθη παρ' ἕν
organicus ὀργανικός
organum ὄργανον: -orum factor ὀργανοποιός
oriarius *cf.* hor-
oriens ἀνατολή, ἕως (ἠώς): in -te ἀνατέλλων: *cf.* auster, boreas
orientalis ἀπηλιωτικός, ἀνατολικός, ἑῷος: — (angulus) ἀνατέλλον (κέντρον): -es ἀνατολαί: -ior, magis — ἀπηλιωτικώτερος
Orion Ὠρίων
orior ἀνατέλλω: simul — συνανατέλλω
orizon ὁ ὁρίζων
ornabilis *cf.* uestitus
ornamentum κόσμιον
ornatiuus καλλωπιστής
ornatus κόσμος: amator -us φιλόκοσμος
ortus ἀνατολή, ἐπιτολή
os ὀστέον
os στόμα: boni oris εὔστομος: mali oris κακόστομος
osculatiuus φιλητικός
ostendo ἀναδεικνύω, δείκνυμι, ὑποδείκνυμι
ostensio ἀπόδειξις, ἐπίδειξις
ouis πρόβατον
Oxiane (*acc.* -em) Ὠξίανη

pacificus εἰρηνικός
palam δῆλον: in — ἀναφανδόν
Palestina Παλαιστίνη
palma: ad -am mensurator παλαιστής
palus λίμνη: *cf.* Meotidus
Pamphilia Παμφυλία
parabolicus παράβολος
parabolizans παράβολος
parabolizator παράβολος
parakmes id est transacti status παρακμῆς
paralisis παράλυσις
parallelus παράλληλος
paralogizator παραλογιστής
parelia παρήλια
parens γονεύς
pario κυΐσκω: -or τίκτομαι
pars μέρος, μοῖρα, μόριον, κατὰ μέρος (*sic*): -tes singule μοῖρα ἑκάστη: -tes χρόνοι (*sic*): due -tes δίμοιρον: multarum -tium πολυμερής, πολυμερέστερος: in -te altera ἀνὰ μέρος: secundum -tes κατὰ μέρος (*sic*): *cf.* duodecimus, duodeni, duodenarius, quicumque, triginta, truncatus, uterque
Parthia Παρθία
partialis μερικός, κατὰ μέρος
particeps μέτοχος
participatio: sine -ione ἀμέτοχος
participo κοινωνέω, μεταλαμβάνω: quia -ant μεταλαμβάνοντες
particularis ἐν μέρει, κατὰ μέρος: in -i, ante -ia ἐπὶ μέρους: -ia τὰ καθ' ἕκαστα: -iter ἐπὶ μέρους, κατὰ μέρος: -ior, magis — λεπτομερέστερος
partio μερίζω
partitio αἵρεσις
partus ἔκτεξις, ἐκτροπή, τοκετός: in -u τῆς ἐκτροπῆς: -us dolor ἐκτρωσμός
parum: *cf.* coeo, tardo
paruus μικρός: minor, minus ἐλάσσων, ἧττον: minimus ὀλιγοστός: *cf.* animus, decidentia, durabilitas, insignis, oculus, sermo, uerbum
pascualis νομάς
passibilis ἐμπαθής
passio πάθημα, πάθος, χύμα (?): -ionum illatiuus παθητικός: -iones que eueniunt per ... πάθη τὰ ... διὰ

passionabiliter: ualde — ὑπερπαθῶς
passiuus παθητικός: -e παθητικῶς
pastor νομεύς
patefacio ἀποδεικνύω
patens διαφανής
pater πατήρ: -tris πατρικός: cognatus ex parte -tris πατρῷος
paternalis πατρικός
patior πάσχω: *cf.* colica, iniustitia, lesura, matrix
paucitas: — natorum ὀλιγοτεκνία, temporis — ὀλιγοχρονιότης
paucus ὀλίγος: -i temporis ὀλιγοχρόνιος: -ior ἐλάσσων: *cf.* barba, pilus
pauidus: — pro affectu ἐπτοημένος
paulatim κατὰ μικρόν
paulatiuus *cf.* augmentatio
pausatio τὸ ἀναπαυστικόν
pax εἰρήνη
peccabilis: facile -ius εὐπταιστότερον
peccator ἁμαρτωλός
peccatum ἁμάρτημα, ἁμαρτία: -a πταίσματα: ficta -a ἀνεπίπλαστα ἁμαρτήματα
pectus θώραξ, στῆθος: *cf.* pilosus
pecunia χρῆμα: amator (amatrix) -iarum φιλοκτήματος
pecuniosus: bene — εὐχρηματιστος
peior *cf.* dirigibilis, regimen
penes κατά *cum acc.*, παρά *cum acc.*, περί *cum acc.*: *cf.* diurnalis
penitiuus μεταμελητικός
penitudo: amplior — τὸ μεταμελητικώτερον
penitus παντάπασιν, κατὰ τὸ παντελές, παντελῶς, πάνυ, τέλεον: *cf.* omnino
per διά *cum. gen.*, *cum acc.*, ἐπί *cum gen.*, κατά *cum acc.*, *pro acc. graec.* (*cf.* tot), *pro dat.* (*cf.* medicamen)
perceptio διάληψις
percipio ἀντιλαμβάνομαι, ὑποβάλλω
percunctatiuus περιοδευτικός
percunctor: -ari uolo πολυπραγμονεύω
percuntator (*sic*) ὁ συμβατικός
percurro ἐπέρχομαι: -ere ἐπιδραμεῖν
percussio πληγή
percussor πλήκτης
percutere: promptus — (*sic*) ἐμπάτακτος (εὐπάτ- ?)
perditio ἀπώλεια
perduco *cf.* perfectus
perduro διαμένω
peregre ἐπὶ ξένης
peregrinatio ἀποδημία, ξενιτεία, ὁδοιπορία: in -ione ἐπὶ τῆς ξένης
peregrinus: — locus ἀπόκλιμα
peremptio ἀναίρεσις
peremptiuus ἀναιρετικός
peremptor ἀναιρέτης
peremptorius ἀναιρετικός
pereo ἀπόλλυμαι, φθίνω
perfectio τελείωσις
perfectiuus ἀποτελεστικός, συντελεστικός
perfectus τέλειος: -ior αὐτοτελής, τελειότερος, magis — τελειότερος: -e τέλεον: ad -um perduco τελεσφορέω: -i status συνακμάζων
perficio ἀποτελέω, διατελέω: ad -iendum πρὸς τελείωσιν: non -iens ἀνεπίτευκτος
perflo πνέω
periclitor παρακινδυνεύομαι
periculosus ἐπικίνδυνος
periculum κίνδυνος: sine -o ἀκίνδυνος: *cf.* proicio, proiectio, proiector
perimetrus περίμετρος
perimo ἀναιρέω
periodalis: magis — περιοδικώτερος: -ius τὸ περιοδικώτερον
peri(y)odus περίοδος
peripneumonia περιπνευμονία
periurus ἐπίορκος
permanentia ἐπιμονή

permaneo διαμένω, διατελέω: ad -endum ἐπὶ διαμονῇ
permixtio: sine -ione κατὰ τὸ ἀμιγές
permutabilis: facile — εὐαπάλλακτος
permutabilitas: facile — τὸ εὐμετάβολον
permutatio ἐναλλοίωσις, μεταβολή
permutatiuus μεταβολικός
permuto ἐναλλάσσω, -or μεταβάλλω
perpetuus ἀΐδιος
perscrutabiliter διερευνητικῶς
perscrutatiuus διερευνήτικος
persecutiuus ἐκβιβαστικός
perseuero διαμένω, διατείνω, διατελέω: -ans ἀποτελούμενος
Perseus (-sei *genit.*) Περσεύς
Persia Περσίς
Persis: -side τῷ Πέρσῃ: -sidem Περσίδα
persona πρόσωπον: *cf.* mulier
perspicax συνετός
perstitiuus στασιώδης
persuadeo πιθανολογέω
pertineo *cf.* actio, consummatio, corpus, unusquisque
pertingo διικνέομαι, φθάνω: non ad consummationem -ens ἀποτευκτικός
pertractatio τὸ διεξοδικόν
pertracto διαλαμβάνω, ὑπομνηματίζω: omnimoda diligentia — ἀκριβόω κατὰ τὴν διέξοδον: -atus ἐφωδευμένος
pertranseo διέρχομαι, συνοδεύω
pertransitio διέλευσις
perturbabilis τεταραγμένος
peruagor παροδεύω
peruenio ἥκω, μεταβαίνω, παραγίνομαι, φέρομαι (ἠνέχθην)
peruersor νοθευτής
peruerto συντρέπω
pes πούς
pessimus χείριστος
pestilens: -entia τὰ λοιμικά
pestilentia λοιμός
pestilentialis λοιμικός
philosophia φιλοσοφία: -ie insistens, immiscens se -ie ἐμφιλόσοφος
philosophus φιλόσοφος
p(h)tises φθίσεις
pictor ζωγράφος
piger ὀκνηρός: *cf.* uoluptas
pigritor ἀποκνέω: -ans ὀκνηρός
pilosus: -i pectoris δασύστερνος
pilus expansorum -orum ἁπλόθριξ: paucorum -orum ὑπόθριξ†
pinguedo: mediocris -inis εὐεκτικός
piratus πειρατής
Pisces Ἰχθύες: Australis piscis Νότιος ἰχθύς
piscifagus ἰχθυοφάγος
piscis ἰχθύς: -ium (animalia) ἰχθυακά (ζῴδια)
pithias (?) πίθιας
pius εὐσεβής
placabilitas τὸ προσηνές
placentia: boni — εὐαρέστησις
plaga αἰκία
planctus: amator -us φιλόθρηνος
planeta ἀστήρ: -e ἀστέρες, πλάνητες, πλανῆται, πλανώμενοι
planta φυτόν
plantatio φυτεία
plasmatio διατύπωσις
plasmator: cere — κηροπλάστης
plasmo διαπλάττω
pleniluniaris πανσεληνιακός
plenilunius πανσέληνος: ad -ium existit πανσεληνιάζων τυγχάνει, ad -ium festinans πανσεληνιάζων
plenus πλήρης
Plias (Pliades) Πλειάς (-δες)
plumbus: -i fusor μολυβδουργός
pluries *cf.* nubo
plus τὸ μᾶλλον, (τὸ) πλεῖον: -res (quam) πλείονες (*cum genit.*), πλεῖστοι: -rimus πλεῖστος: quod -rimum τὸ πλεῖον:

per -rimum, secundum -rimum (τὸ) πλεῖστον: secundum -rima τὰ πολλὰ (*acc.*): ut ad -rimum, ut in -ribus, ut in -rimis ὡς ἐπὶ τὸ πολύ: ut in -ribus ὡς ἐπίπαν: *cf.* nascor

pluuia ὑετός: carentia -ie ἀνομβρία

pluuiosus ὀμβρώδης

pneumonicus πνευμονικός

podagra ποδάγρα

poema ποίημα

poeticus ποιητικός

politicus πολιτικός

politor λαοξόος, (?) ὑπουργός

polus πόλος

pono τίθημι, ὑφίστημι: -or τάσσομαι: qui -itur κατατεταγμένος: -situs κείμενος: *cf.* ordo

Pontus Πόντος

popularis ὀχλικός

populus: amator -i φίλοχλος: *cf.* turbator

porcus σῦς

portionator παρασχίστης

positio θέσις

possessio κτῆμα, κτῆσις, πολυκτημοσύνη: -ionis amator φιλοκτήμων: multarum -ionum πολυκτήμονες

possessorius κτητικός

possibilis δυνάμενος, δυνατός:— est καταληπτός: quod -e (est) τὸ δυνατόν: -e est ἔνεστι (ἔνι): οἷόν τε (*c. inf.*)

possibilitas τὸ δυνατόν

possum (qui potest, potens) δύναμαι (-άμενος), ἐγχωρέω, ἰσχύω: qui potest ὁ δυνατός: potens δυναμικός: potentior δυνατώτερος: *cf.* comprehendo, equalis, extendo

post ἀπό, διά *cum. gen.*, ἐπί *cum dat.*, μετά (*cum acc.*)

postea εἶτα

posterior δεύτερος, ὀπίσθιος, ὕστερος: *cf.* nates

postpono: quia -untur διὰ τὸ πλεονεκτεῖσθαι

potabilis πότιμος

potentatus δυναστεία

potentia δύναμις

prauitas *cf.* odor

preacceptio πρόληψις

preaccipio προδιαλαμβάνω, προκαταλαμβάνω, προκαταλέγω

Precanis Προκύων

precaueor: -endi προφυλάσσεσθαι

precedo αὐθεντέω, ἡγέομαι, παροίχομαι, προγίνομαι, προέχω†: -ens προηγούμενος: -entia τὰ ἔμπροσθεν: longe -ens nos ἔτι ἄνωθεν

preceps καταφερής, πρόρρησις

preceptiuus ἐπιτακτικός

preceptum πρόσταγμα

preces: difficile -ibus acquiescens δυσέντευκτος

precessio προήγησις: -iones id est retrogradationes (!) προηγήσεις

precessor ἡγεμονικός, προαγωγός

preciosus τιμώμενος, τιμητικός

precipitatio ἀποκρημνισμός

precipitium κρημνισμός

precipito κατακρημνίζω

precipuus ἐξαίρετος, κυριώτατος: -e ἐξαιρέτως

preclarus πρόδηλος

precognitio πρόγνωσις: difficultas -ionis τὸ δυσθεώρητον

precognosco προγινώσκω

preconsidero προεπισκέπτομαι

predatiuus: -a ληστήρια (ληστρικά)

predator: sacrorum — ἱερόσυλος

predico προλέγω: -ans ἐξαγορευτής: -entes εἰπόντες: -dictus εἰρημένος, προειρημένος, προκείμενος: ut -antur προλέγεσθαι: -endi modus τῶν προρρησεων τρόπος: *cf.* figuraliter

predictio πρόρρησις

predistinguo προδιαλαμβάνω
prediuino προθεσπίζω
predominatio ἐπικράτησις
preeleuatus: — super προαναφερόμενος
preexisto προϋφέστηκα
preexpono προεκτίθημι
prefectura ἐπιτροπεία
prefectus προιστάμενος
prefero προκρίνω
preinductio (*-ta?) ὑφήγησις
preinfrigido προκαταψύχω
preinstituo προτάσσω
preiudico προκρίνω: -andum προκριτέον
prelinguosus πρόγλωσσος
preminentia προαγωγή
premitto προτάσσω, προτίθημι, ὑποτίθημι, πρόκειμαι: *cf.* figuraliter
premium: -ii amator φίλαθλος
prenoto: -atur προσημαίνει
prenumero προκαταλέγω
prenuntiatio πρόρρησις
preoccido: id quod -it τὸ προδῦνον
preordino (ante) προτάσσω
preortum προανατεῖλαν
preostendo προδηλόω
preostensus προαποδεδειγμένος
preparator διασκευαστής
preparo κατασκευάζω, προπαρασκευάζω
prepertracto (pretracto) προδιαλαμβάνω
prescientia πρόγνωσις
presens ἐνεστώς, (τὸ) παρόν
Presepe Φάτνη
preseruatio προφυλακή
preseruatorium φυλακτήριον
preses ἡγεμονεύων, ἡγεμονικός, ἡγεμών
presideo ἄρχω, προίστημι
presignificatio ἐπισημασία
presignificatiuus προδηλωτικός
presignifico προσημαίνω
presulatus ἡγεμονία, -us (*plur.*) ἀξιωματικά
presum *cf.* uir
presuppono προϋποτίθεμαι
preter μετά (*cum gen.*), παρά (*cum acc.*)
pretereo παρίημι
pretergredior παραβαίνω
pretermitto ἀποπέμπω, παραλείπω
pretiosus πολυτελής
pretrado προπαραδίδωμι
preualeo καθυπερτερέω: — super κατακρατέω *cum gen.*
Preuindemiator Προτρυγητήρ
preuisus (*sic*) γνωσθείς
preuolens προπετής
primitiuus προκαταρκτικός
primoalitus πρωτότροφος
primogenitus πρωτοτόκος
primoritas τὰ πρωτεῖα: -atem habeo πρωτεύω
primus -um πρῶτος -ον: -i (*sc.* loci ?) τὰ πρωτεῖα: -o πρῶτον, πρώτα: *cf.* capitulum, liber
princeps ἄρχων
principalis αὐθεντικός, ἡγεμονικός, κύριος: -ior κυριώτερος: -ius τὸ αὐθεντικώτερον: -issimus κυριώτατος, ἀναγκαιότατος: -iter προηγουμένως: *cf.* ciuitas
principatiuus ἀρχικός, ἡγεμονικός
principator ἀρχικός
principium ἀρχή, καταρχή: a principio ἐν ταῖς καταρχαῖς
principo ἄρχω
prior πρότερος, προγενέστερος, προηγούμενος: -ius πρότερον, ἔμπροσθεν, πρῶτον: -iora τὰ πρωτεῖα, τὰ ἔμπροσθεν: *cf.* fio
prioritas πρωτεῖον
priuatio: natorum — ἀτεκνία
priuatiuus στερητικός
pro ἀντί, ἄχρι
probabilis πιθανός
probrosus ἐπίμωμος, ἐπίψογος, ἐπονείδιστος: -ius ἐπονειδιστότερον:

multum — πάναισχρος, πολύψογος: ualde — πάμψογος
processus ἔφοδος, πάροδος, πρόσοδος
procliuis πρόσφορος, καταφερής, κατάφορος: bene —, facile — εὐκατάφορος
procreatio: filiorum — παιδοποιία: masculi (-orum) — ἀρρενογονία: *cf.* femina, femella
procreo: -creor ἀποκυΐσκομαι, ἀποτίκτομαι, κυΐσκομαι: quod -creatum est τὸ γεννώμενον, τὸ γεννώμενον ἐκτέχθεν, τὸ συγγεγενημένον: -creata τὰ βρέφη: *cf.* nascor (natum)
procuratiuus ἐπιτροπικός
procuro περιποιέω, πορίζομαι
prodeo ἔρχομαι
prodigalitas ἀσωτία
prodigiosus αἰνιγματώδης
prodigium: -iorum conditor τερατουργός
prodigus ἄσωτος
proditor προδότης
produco προάγω: qui -ducitur γινόμενος: non -ductus in exemplum ἀπαραδειγμάτιστος
productibilis: facile — in exemplum εὐπαραδειγμάτιστος
productiuus προαγωγικός
profectibilis: facile -bilius εὐπροκοπώτερον
profectus ὄνησις, προκοπή, ὠφέλεια, τὸ ὠφέλιμον: maior — τὸ ὠφελιμώτερον: sine -tu ἀπρόκοπος
profero πληροφορέω
proficio (?) παραπέμπομαι
proficuus ἐπιβλητικός, ὠφέλιμος: *cf.* uita
profluentia φλεγμονή
profundus κοιλούμενος: -i sensus βαθύφρων: -e malignus βαθυπόνηρος: *cf.* Syria
prohemium προοίμιον
prohibeo κωλύω
proicio διεκβάλλω, ἐκβάλλω: pericula -iendo ῥιψοκινδύνως
proiectio κατάθεσις: — periculi ῥιψοκινδυνία: radiorum — ἀκτινοβολία
proiector: — periculi ῥιψοκίνδυνος
proles: sine -e ἄτοκος: *cf.* Iupiter
prolixus μακρός
promoueo: -motus προβιβαζόμενος
promptus πρόχειρος: — percutere ἐμπάτακτος (εὐπάτ- ?): -e ἑτοίμως
promptus: facile in -tu εὐπροφάσιστος
pronosticatio γνῶσις, πρόγνωσις, τὸ προγνωστικόν
pronostico, -or προγινώσκω
pronosticus προγνωστικός
pronuntiatio ἐπαγγελία, ἐπάγγελμα, πρόφασις
pronuntiator: -ores ἀποφθεγγόμενοι
pronuntio: -iandum ἀποφαντέον
pronus καταφερής
prope ἐγγύς, σύνεγγυς: proximo ἐγγύτερῳ: proxime ἔγγιστα
prophetia ἐξαγορεία
propinquitas ἐγγύτης, συνεγγισμός
propinquus ἐγγίζων, ἐγγύς, σύνεγγυς: -ior ἐγγύτερος: -issimus (ὁ) ἔγγιστα, οἰκειότατος: -issime ἔγγιστα: *cf.* terra
propono προτίθεμαι: -positus προκείμενος
proportio τὸ ἀνάλογον: deterior — τὸ ἀρρυθμότερον: melior — τὸ εὐρυθμότερον
proportionaliter ἀνάλογον, ἀναλόγως: idem — ἡ αὐτὴ ἀναλογία
proportionatus ὑπόρρυθμος: bene — εὔρυθμος
propositum πρόθεσις
proprietas τὸ ἴδιον, ἰδιοσυγκρισία (-κρασ-), ἰδιότης, ἰδιοτροπία, τὸ ἰδιότροπον, ἰδίωμα, τὸ οἰκεῖον, οἰκειότης: singulares -tates ἰδιότροποι περίστασεις: *cf.* animal, audacia, complexio

proprius ἴδιος, ἰδιότροπος, κύριος, οἰκεῖος, οἰκείως (?): -ie ἰδίως: magis -ium ἰδικώτερον† (εἰδι-): *cf.* color, complexio, dignitas, facies, illicentia, opus, unusquisque
propter διά *cum. gen., cum. acc.*, παρά *cum acc.*, ἕνεκεν, χάριν: — hoc διὰ τοῦτο: — hoc quia (+ verb. finit.) διὰ τοῦτο (*cum part.*): — quod διό, διόπερ: — quod quidem διόπερ
propterea: — quod, — quia (+ *verb. fin.*) διὰ τὸ + *inf.*
prosequor ἐφοδικῶς ἀναπληρόω, ἐπεξέρχομαι (ἐπέρχομαι), κατακολουθέω
prosneusis id est adnuitio πρόσνευσις
prospicio *cf.* longinquus
prosum ὠφελέω
protendo ἀπομηκύνω: -or παρατείνω
protensio παράτασις
protestor ὑφηγέομαι
proueho: -uectus προβεβηκώς
prouenio περιγίνομαι
prouidentia πρόνοια
prouisio τὸ προνοητικόν
prudens εὐφρόσυνος, φρόνιμος
Ptholomeus (Tholo-) Πτολομαῖος
publicus δημόσιος, ἐπίσημος
Puella Κόρη
puer τὸ παιδικόν, παῖς
puerilis παιγνιώδης, παιδικός
pueritia βρέφος
pugna μάχη
pugnax μάχιμος: -cissimus μαχιμώτατος
pugno: cum bestiis — θηριομαχέω
pulcritudo: amor -inis φιλοκαλία: amator -inis φιλοκαλλωπιστής
pulcrus *cf.* oditor
pulmo πνεύμων
pulsus κροῦσμα
punctum σημεῖον, στιγμή
punibilis: de facili — εὐτιμώρητος
purificatio αἱμαγμός
purifico: pro diuinatione -ans φοιβαστικός
puritas αἰθρία: amor -atis τὸ φιλοκάθαρον
purus καθάριος (-ρός): -ior ἀκρατότερος
puta οἷον, ὡς: ut — οἷον
puto νομίζω: -or δοκέω
putrefacio διασήπω
putrefactio σῆψις

quadrangularis τετράγωνος: -iter τετραγωνικῶς, κατὰ τετράγωνον: aspiciens -iter τετραγωνίσας: -iter aspicio τετραγωνίζων τυγχάνω
quadratura τεταρτημόριον, τὸ τετράγωνον
quadratus τετράγωνος: in -o aspectu τετράγωνος: in -o aspectu (existens) τετραγωνίζων: que (*sc.* configurationes) ex -o οἵ τε (*sc.* συσχηματισμοί) τετράγωνοι
quadriennium τετραετηρίς
quadrilaterus τετράπλευρος
quadrupes τετράπους
qualis οἷος, ὁποῖος, ποῖος, τίς
qualiscumque ὁποῖος ἄν
qualitas τὸ ποῖον (ποιόν), ποιότης
qualiter πῶς, πως
qualitercumque ὅπως ἄν, ὁπωσδήποτε
quam ἤ, ἤπερ: — maxime ὅτι μάλιστα: *add. post comparat.* (*e.g.* plures)
quamdiu ἐφ᾽ ὅσον
quamuis καίτοι, κἄν
quando ἐπάν, ὅταν, ὅτε
quandocumque ὁτεδήποτε
quandoque ποτε
quantitas ποσότης
quantumcumque ὅσον ἄν
quantum ὅσον, τὸ ποσόν, τὴν ποσότητα: -orum in utroque ποταπὸν ἐφ᾽ ἑκατέρου: — ex se (*sic*) ἑαθέν: in — ὡς: — ad *cf.* ad: *cf.* extendo
quare ὥστε

quarta (*sc.* pars) (τὸ) τεταρτημόριον: prima — τὸ πρῶτον τῶν τεταρτημορίων

quartanarius τεταρταϊκός

quartus τέταρτος

quasi ὡς, ὡσεί, ὥσπερ, ὥσπερ ἂν εἰ, ὥσπερ εἰ: *cf.* similis

quatinus ὅπως

quatuor τέσσαρες: *cf.* mensis

que: -que δέ, τε, τε καὶ: -que … -que τε … καί

quemadmodum καθάπερ, ὡς, ὥσπερ

querela: -as tollerans ἐπίμεμπτος

querelo ἐπιμέμφομαι, μέμφομαι

quero ἐπιζητέω: — circa ζητέω *cum gen.*

querulosus μεμψίμοιρος

qui *cf.* is

qui ὅς, ὅσπερ, ὅστις, ποῖος, τίς, τις: (is, ille) — (est), — scilicet ὁ: — quidem ὅσπερ: ad quod οὗ ὅσπερ

quia ἐπεί, ἐπειδήπερ, διότι, ὅτι: — + *verb. finit.* διά *cum part.*, διὰ τὸ *cum inf.*, παρὰ τὸ *cum inf.*

quicumque ἕκαστος, οἱοσδήποτε, ὁποιοσδήποτε (ὁποιοσοῦν), ὁπότερος, ὅς, ὃς ἄν, ὅσος, ὁστισοῦν: quodcumque τὸ τύχον: ad quamcumque partem καθ' ὧν ἄν:

quidam τις: — (*plur.*) ἔνιοι

quidem τε, δέ, μέν, μέντοι: *cf.* qui

quies νώθεια: magis -iete letatrix ῥᾴθυμος

quiete ἠρέμα

quilibet ἕκαστος

quin immo οὐ μὴν ἀλλὰ

quindecim πεντεκαίδεκα

quinque πέντε

quintus πέμπτος

quique ὅσοι

quis *cf.* qui

quod (?) ἵνα

quod διότι, ὅτι, ὡς: *cf.* propter

quomodo πῶς, τίνα τρόπον

quoniam ἐπεί, ἐπειδή, ἐπειδήπερ: — quidem ἐπειδή, ἐπειδήπερ: — itaque ἐπειδήπερ

quot ὅσοι, πόσοι

quotcumque ὅσοι ἄν

quotus πόστος

radius ἀκτίς, αὐγή, βολή: -iorum proiectio ἀκτινοβολία

rapax ἅρπαξ

rapina ἁρπαγή

rapio: raptus (<deo> raptus ?) θεόληπτος

raptor ἅρπαξ

raptus ἁρπαγή

raritas σπάνις, τὸ σπάνιον: *cf.* frater

rarus: -o σπανίως: *cf.* frater

ratio λόγος, τρόπος (*sic*): — ascendentis λόγος ὡροσκοπικός: — consequentie τὸ ἑξῆς τῆς ἀκολουθίας: amator -ionis φιλόλογος: -ionem cause non habens ἀναιτιολόγητος: *cf.* geneaticus, dignitas

ratiocinatio ἐπιλογισμός, λογισμός, λογιστήριον: ratio<ci>natio ἐπιλογισμός

ratiocinatiuus διανοητικός

ratiocinor ἐπιλογίζομαι, σοφίζομαι

rationabilis εὔλογος, λόγιος: -e εὐλόγως

rationalis λογικός

rebellis μάχιμος

recens νεαρός

receptor: — hospitum πανδοκεύς

recipio ἀποδέχομαι, διαλαμβάνω, παραλαμβάνω

recogito: non -ans (*sic*) μὴ προσποιούμενος

rectiuus ἡγεμονικός

rector *cf.* mundus

rectus εὐθύς, ὀρθός: -e ὀρθῶς: *cf.* angulus

recusatrix: — laborum ἀφερέπονος

recuso παραιτέομαι

redarguibilis: difficile — δυσεξέλεγκτος

redditiuus μεταδοτικός

reddo *cf.* testimonium

redeo ἐπιστρέφω
redigo: in summam -or ἐπιμερίζομαι
reditus ἐπάνοδος
reduco ἀνάγω
referor: — ad ἀνήκω *cum dat.*
refertus εὐτραφής
refluo ἀπορρέω
refluxus ἀπόρροια, παλίρροια
refugatio: cordis — (*sic*) φυγαδεία
regalis βασιλικός
reges Ἀνάκτορες
regibilis: difficulter — δυσείκαστος
regimen ἀγωγή, ἀναγωγή: — diete διαιτητική ἀγωγή: id quod peioris -inis τὸ δυσδιαγωγότερον: in contrarium — ἀποτροπιασμός
regio *cf.* occasus
regio ἀπόκλιμα, χώρα, -αι: Egipti inferioris — Αἰγύπτου κάτω χώρα: a, e -ione ἄντικρυς: *cf.* occasus
regno βασιλεύω, ἡγεμονεύω
regressiuus: uix — δυσανακόμιστος
regressus ἐπάνοδος
regula ἔφοδος
regularis ὁμαλός: -iter ὁμαλῶς
regulor διαμορφόομαι
regyratio (*sc.* una reuolutio solaris) ἐνιαυτός
reicio διασύρω
religio θρησκεία: -ione honoro θρησκεύω
relinquo ἀφίημαι, καταλείπω: -endum ὑπολειπτέον†?
reliquum λοιπόν, ὑπόλοιπον
remediatiuus θεραπευτικός
rememoratio ἀνάμνησις
remigator πορθμεύς
reminiscor μιμνήσκω
remissio ἄνεσις
remissus ψυχοῦν: -e ἠρέμα
remotus: a -is ἐκ μακροῦ
renunculi νεφροί
repentinus αἰφνίδιος: -e αἰφνιδίως
repleo *cf.* turbatio
repletio πλήρωσις
repletiuus πληρωτικός
repono ἀνατίθημι
reporto *cf.* commodum
repositio ἀπόθεσις
reprobo ἀποδοκιμάζω
reptilis ἑρπυστικός: -e ἑρπετόν
repudio: -iandum ἀθετητέον
requiro αἰτέω, ἐπιζητέω
res πρᾶγμα: *cf.* consumptor
residuus ἐπίλοιπος, καταλειπόμενος
resolutio: cum -ione ἀναφορικός
respectus: cum -u ad terram μετὰ τῆς γῆς
respicio βλέπω
responsio ἐνθουσιασμός, χρησμός
responsum χρησμός: -a dantes ἀποφθεγγόμενοι
rethor ῥήτωρ
retractio: — aquarum ἄμπωτις
retrogradatio *cf.* precessio
retrogradus ἀναποδιστικός: -gradi (!) προηγούμενοι
reuelatio ἐνθουσιασμός, χρησμός: -ionibus attendens ἐνθουσιαστικός
reuerendus σεμνός: *cf.* decentia
reuerens εὐλαβής, σεμνός: -tior εὐλαβέστερος: -tius τὸ σεμνότερον
reuerentia τὸ σεμνόν, σεμνότης
reuerentialis εὐλαβητικός
reumaticus ῥευματώδης, ῥευματισμός (?)
reumatisma, -mus ῥευματισμός
reuolutio (solaris) περίοδος
reuolutio ἀνακύκλησις: — (solaris) περίοδος
rex βασιλεύς, βασιλεύων
ridiculus γελοῖος
rigidus: — in sententiis ἰσχυρογνώμων
rigor *cf.* febris, febrilis

significo δηλόω, διασημαίνω, ἐπισημαίνω, προσημαίνω, σημαίνω
signo δηλόω
signum δωδεκατημόριον, ζῴδιον, σημεῖον, σκοπός, τμῆμα (*sic*): *cf.* cognosco
siluester ἄγριος
simia (-ius) πίθηκος
similis ὅμοιος, ἐφάμιλλος, παραπλήσιος: -illimus ὁμοιότατος: -iter ὁμοίως: quasi — παρόμοιος: *cf.* figura, figuratio, modus
simplex ἁπλοῦς: -icis diete ἁπλοδίαιτος: -icior ἁπλούστερος: -iciter ἁπλῶς: -iciter existens ἁπλοῦς ὤν
simplicitas τὸ ἁπλοῦν
simul (cum) ἅμα, ὁμοῦ: si — sint συγγενόμενοι: *cf.* conuersatio, occasus, orior, uertibilis
simulator ὑποκριτής, ὑποκριτικός
simulo ὑποκρίνομαι: diuersa -ans, aliud -ans ἀλλοπρόσαλλος
sinastria συναστρία: — uel beneuolentia συναστρία
sine ἄνευ, χωρίς: *cf.* affectio, amor, cibus, collatio, error, fedus, honor, humor, lesura, lucrum, nequitia, nubis, offensio, opus, profectus, prolis, sententia
singularis ἐξαίρετος, μοναχός: -iter ἐξαιρέτως, ἰδικῶς, ἰδίως, κατ' ἰδίαν: *cf.* laboro, modus, proprietas, sententia
singularizo ἰδιάζω
singulus ἴδιος, ἕκαστος, -i ἕκαστος: -i duo ἀνὰ δύο: per -a speculor ἀναθεωρέομαι: *cf.* dies
sinister ἀριστερός, εὐώνυμος
sinus κόλπος
siquidem ἐάνπερ
situs (*subst.*) διάστασις, θέσις, πρόσωπον (*sic*), στάσις: *cf.* diametralis, gradus
situs (*adiect.*) τετραμμένος
siue ἐάν τε: — etiam ἐάν τε:— autem, — uero εἴτε (*in iter.*), — ... — ἤτοι ... ἤ
socior (cum) προσγίνομαι *c. dat.*
socius: -iorum amator φιλέταιρος
sodalitas ἑταιρία
sol ἥλιος: locus -is ὁ ἡλιακός: — in meridie τοῦ ἡλίου μεσουράνησεις
solacium: ea que ad — sunt τὰ ἐκπαιζόμενα†
solaris ἡλιακός
sollers ἀγχίνους, ἀγχίφρων: -ertior ἀγχινούστερος
sollicitudo ἐπιβολή, ἐπικουρία, μέριμνα
sollicitus: -ti (de) πεφροντικότες: magis — ἐπιμελέστερος
solummodo μόνον
soluo λύω
solus -um μόνος -ον: -i κατὰ μόνους: -o modo: μόνου
sompnio: facile -ians εὐόνειρος
sompnus: -iorum iudicator, -iorum iudex ὀνειροκρίτης: -iorum interpretatiuus ὀνειροκριτικός
sophista σοφιστής
sordidus ῥυπαρός
soror ἀδελφή: -roris ἀδελφικός: qui sunt eiusdem matris frater et — δυὸ ὁμομητρίοι ἀδέλφοι
sors (fortune) κλῆρος (τύχης)
sortior ἐπιτυγχάνω, κληρόω, λαγχάνω, τυγχάνω: -iti sunt dominium ἔτυχον λαβόντες τὴν οἰκοδεσποτείαν
spado σπάδων
spasmus σπασμός
spatula: amplarum -arum σπαθηνούς†? (*pro* παθηνούς ?)
species εἶδος: quod -iei τὸ εἰδικόν: unius -iei μονοειδής: monstri — τὸ τοῦ τέρατος ἴδιον: geneatici sermones seu — γενεθλιαλογικοὶ τόποι: geneatici sermonis — γενεθλιαλογικὸς τόπος
speculatio ἐπίσκεψις, θεωρία, σκέψις: -ionem facio σκοπέω: amator -ionis φιλοθέωρος

speculatiuus θεωρητικός: -ue ἐπισκεπτικῶς
speculor ἐπισκέπτομαι, σκοπέω: per singula — ἀναθεωρέομαι: amans -ari φιλοθέωρος
spera σφαῖρα
sperma σπέρμα: multi -atis πολύσπερμος, πολύσπορος
spermaticus σπερματικός
Spica Στάχυς
spina ἄκανθα
spirituosus πνευματώδης
spiritus πνεῦμα
spissus παχύς
splen σπλήν: egritudines -is σπληνικά
splendidus: diete -e ἀβροδιαιτός
spoliatio σύλησις
spondilis: in -ibus ἐν τοῖς σφονδύλοις (σπονδύλοις)
sponsa: -as ducere γάμος
spuitiuus ἀναφορικός
spuo: sanguimen -ere faciens αἱμοπτυϊκός
squinanchia συνάγχη
stabilis: bene — εὐσταθής
stangnifex κασσιτεροποιός
statio στάσις, στηριγμός
stationarius στηρίζων
statuificus ἀνδριαντοποιός
statuo ὑφίστημι: *cf.* lex
status ἀκμαιότης, κατάστασις: bonus — εὐετηρία: *cf.* perfectus
stella ἀστήρ
sterilis ἄγονος, στεῖρος
stipendiarius μισθοφόρος (-φορικός)
sto ἵστημι, στηρίζω: stabam κατατέταγμαι: bene status εὐσταθής
stola στολή
stomachus στόμαχος
strangulatio στραγγαλιά (-λία)
studeo: — circa ἀσκέω *cum acc.*: uirtuti -ens ἀσκητής
studiosus σπουδαῖος: magis — σπουδαιότερος: -ior ζητητικώτερος
studium ἐπίνοια, σπουδή
stultiloquus φλήναφος
stultitia: in -ia malus μωρόκακος
stultus εὐήθης: cum malignitate — μωρόκακος
stuprator διαφθορεύς
sub *cum abl.* ὑπὸ *cum gen., cum acc.*
subamarus ὑπόπικρος
subcalidus ὑπόθερμος
subcaluaster ἀναφαλαντιαῖος
subcorruptor ὑπονοθευτής
subditus ὑποτακτικός
subdolus ὑποπόνηρος
subgaudiosus ὑποχαροποιός
subicio διατίθημι: -ior διατίθεμαι, ὑπόκειμαι: ad hoc quod -iantur πρὸς τὸ διατίθεσθαι: -iectus ὑποκείμενος, ὑποτεταγμένος
subiectio ὑπόβασις, ὑποταγή
subigo διατίθημι: -actus διατιθέμενος
submissio τὸ ὑποτεταγμένον, ὑποβολή
submollis ὑπομάλακος
submotus ὑποκεκινημένος
subrubeus ὑπέρυθρος, ὑπόκιρρος
subrufus ὑπόκιρρος
subsanno: -andum ἀγανακτεῖν
subsiccus ὑπόξηρος
subsistentia ὑπόστασις
subsolanitas τὸ ἀπηλιωτικόν
subsolanus ἀπηλιώτης, *cf.* boreafelioticus
substantia οὐσία: humida — ὑγρασία
subsum ὑποπίπτω
subtiliatus ἀναποθείς
subtilis λεπτός, σπινώδης: -ior λεπτέστερος
subtus *cf.* moueo
subuiridis ὑπόχλωρος
successio διαδοχή
successus ἔφοδος

sudor: multi -oris πολύχους
sufferentia: — mali τὸ κακόπαθον
suffero: -ens ὑπομονητικός: non -ens ἀνύποιστος
sufficiens αὐτάρκης, ἱκανός: — fuerit ἱκανόν
suffocatiuus πνιγώδης
suffocatorius πνιγώδης
suffoco ἀποπνίγω
suffossor: *cf.* murus, sepulcrum
suffusio ὑπόχυσις
Sugdiane (*acc.* -em) Σουγδιανή
sum εἰμί, γίνομαι, τυγχάνω, ὑπάρχω: quod est τὸ τύχον: cum sit γινόμενος: quod sit εἰς τὸ ... εἶναι: sint τυγχάνωσιν ὄντες: fuerit ᾖ, εἴη, τυγχάνῃ: (is) qui est, cum sit, (is qui) fuerit ὤν: ea que fuerunt τὰ ὄντα: fuerint τυγχάνωσιν, ἔτυχον ὄντες: *cf.* fore, futurus
summa τὸ τέλειον: in — καθόλου, ἐν κεφαλαίοις: *cf.* redigor
summatim ἐν κεφαλαίοις, κατὰ τὸ κεφαλαιῶδες, κεφαλαιωδῶς, ἐξ ἐπιδρομῆς
summitas (τὸ) ἄκρον: -tates capitum nostrorum ὁ κατὰ κορυφὴν ἡμῶν τόπος
sumo διαλαμβάνω, λαμβάνω
sumptuosus δάπανος
super (*cum acc.*) ἐπί *cum gen., cum dat., cum acc.*: — (*cum abl.*) ἐπί *cum acc.*: *cf.* aduenio, gloria, preeleuatus
superaccedo ἐπαναφέρομαι: -ens (*sc.* locus) ἐπαναφορά
superafferor ἐπαναφέρομαι
superallatio ἐπαναφορά
superalleuatio (-allatio ?) ἐπαναφορά
superascendens ἐπαναφερόμενος
superaspicio βλέπω (ἐπιβλέπω ?)
superatio καθυπερτέρησις
superbus οἰηματίας, σοβαρός
superemineo καθυπερτερέω
superexaltor καθυπερτερέω
superfero ἐπιφέρω: -or ἐπαναφέρομαι, ἐπιπορεύομαι
superfluus περισσός: -e περιέργως
supergressiuus πλεονεκτικός
supergressor πλεονέκτης
supergressus ἔφοδος
superillatio ἐπαναφορά
superinfero: -or ἐπαναφέρομαι: -illatus ἐπιφερόμενος
superingressus ἐπέμβασις
superiniuriosus ἐφύβριστος
superior ἄνω: -ius ἐπιπάνω: supremus ἀνωτάτω
superioritas καθυπερτέρησις
superlaboriosus ἐπίμοχθος
superlargior: -ientes (*acc.*) ἐπιχαρίτους
superlargitor: -ores (*acc.*) ἐπιχάριτας
supermorior: -ientes ὑπεραποθνήσκοντες
supernociuus ἐπιβλαβής
supero καθυπερτερέω: in -ando καθυπερτερεῖν: -atus κρατούμενος: non -atus ἀκαθυπερτέρητος
superpartiens ἐπιμόριον
superponor ἐπίκειμαι
supersedens ἀντιληπτικός
superstitiosus θρησκευτής
supertripartiens ἐπίτριτος
superuenio ἔπειμι, ἐπέρχομαι, ἐπισκήπτω, ἐπισυμπίπτω, ἐπιφέρομαι, παρατυγχάνω, προσγίνομαι, συμβαίνω
supplex: sacris — ἱεροπρόσπλοκος: *cf.* deus
suppono διατίθημι, ὑποτίθεμαι: -or (-positus) ὑπόκειμαι (-μενος)
supputo: -andum λογιστέον
surdus κωφός
sursum *cf.* accedens
sursumferor ἀναφέρομαι
suscipio ἀναλαμβάνω, ἐκδέχομαι, ἐπέχω, ἐπιδείκνυμι, ἐπιδέχομαι, ἐπιλαμβάνω, ἐπίσχω, λαμβάνω, παραδέχομαι, παραλαμβάνω

suspectus ὕποπτος

suspicio ὑποψία

suspiciosus ὕποπτος

sustineo *cf.* iniuria, lesura

suus ἴδιος, ἑαυτοῦ, αὐτοῦ: suo ταύτης: *cf.* gens, lex, manus, occultatiuus, odio

symptoma, simp- σύμπτωμα, σύμπτωσις: -ata que eueniunt (secundum ...) συμπτώματα τὰ (καθ' ...)

Syria (Siria) Συρία: Profunda — Κοίλη Συρία

Syrus (Sirus) Σύρος

tabernarius οἰνέμπορος

tabularis πινακικός: -iter πινακικῶς

tactus ἁφή

talis ὅμοιος, τοιόσδε, τοιοῦτος: -liter τοιῷδε

tam οὕτως

tamen ἀλλά, γε, δέ (*post* οὐ), μέντοι, ὅμως: — ... non οὐδέ: at — μέντοι

tamquam ὡς, ὥσπερ: — eodem modo ὡσαύτως

tandem τέλεον, ἐπὶ τέλει

tantus τηλικοῦτος, τοσοῦτος

tarditas τὸ βραδύτερον, νωχελία: maior — τὸ νωχελέστερον

tardo: multum — πολυχρονέω: parum -ans ὀλιγοχρόνιος, -ιώτερος

tardus: -ior βραδύτερος: -ius βράδιον, περιοδικώτερον: *cf.* nubo

Taurus Ταῦρος

tediosus ληρώδης

telonearius τελώνης

temperamentum συμμετρία

temperantia εὐκρασία, τὸ εὔκρατον: bona — εὐκρασία

temperatus εὔκρατος, σωφρονικός, σώφρων: bene — εὔκρατος

temperies τὸ εὔκρατον

tempestas κεραυνός, χειμών

temporalis καιρικός, πρόσκαιρος, χρονικός: -es κατὰ καιρούς: -ior καιρικώτερος

temporaneitas: -tates ὧραι

temptatiuus πειρατικός

tempto πειράομαι

tempus καιρός, περίοδος, χρόνος, ὥρα: — fundationis χρόνοι κτίσεων: hyemale — χειμῶνες: -oris, -orum χρονικοί: pro -ore κατὰ καιρόν: per -ora κατὰ χρόνον: per -ra, secundum -ra κατὰ καιρούς: multi -oris πολυχρόνιος: oraria per -ora ὡριαῖοι χρόνοι: -ora anni ὧραι: -ora eleuationum ἀναφορικοί χρόνοι: -ora que colliguntur οἱ σχηματιζόμενοι χρόνοι: -orum dominator χρονοκράτωρ: *cf.* annus, corpus, duro, equalis, paucitas, paucus

tendo ῥέπω: qui -unt ad (οἱ) πρός *cum dat.*

tener ἁπαλός, τρυφητής

tenerosus (?): -ius τὸ τρυφερώτερον

tensus τετανός: *cf.* capillatura

tenuis συντετηγμένος

termino λήγω, ὁρίζω

terminus (-i) ὅριον, πέρας: gradus -orum ὅρια

terra γῆ: -e nascens ἐπίγειος: sub -a ὑπόγειος: circa -am περίγειος: propinquissima -e περιγειοτάτη (*sc.* σελήνη): uicinitas ad -am περιγειότης

terremotus σεισμός: — factiuus σεισμοποιός, σεισμώδης

terrestris ἐπίγειος, χερσαῖος

tertianus τριταϊκός

tertius τρίτος: -ia pars τριτημόριον

testificor μαρτυρέω

testimonium μαρτυρία: — reddo, — fero μαρτυρέω

tetragonalis τετράγωνος

tetragonizo τετραγωνίζω

textor ὑφάντης

theatrum: -orum inuasor θεατροκόπος

Thebais (-idem -ida) Θηβαΐς (-ΐδα)

theorema: -ata μαθήματα
theoria θεωρία
theoricus θεωρητικός
thesaurizatiuus θησαυριστικός
Thracia Θρᾴκη
thronus θρόνος
tibi σοί
timidus δειλός: — ad deos δεισιδαίμων: ad minima — ψοφοδεής
timor φόβος
tinctor βαφεύς
tinctura βαφή
tipus τύπο
tirannicus τυραννικός
toleratiuus ἀνεκτικός, οἰστικός: non — malorum ἀνεξίκακος
tollero *cf.* querela
tonitruosus βροντῶδης, κεραυνώδης
tonitruum βροντή
tormentum: -orum amator φιλοβάσανος
tortor: cordarum — χορδοστρόφος
tot τοσοῦτοι: — mensibus (uiuet) τοὺς ἴσους μῆνας (ἐπιζήσεται): per — horas τὰς ἴσας ὥρας
totalis ὅλος, ὁλοσχερής: -ior ὁλοσχερέστερος: -liter ὅλως
totalitas ὁλοσχέρεια, τὸ ὁλοσχερές: magis in — tate ὁλοσχερέστερον, ὁλοσχερῶς
totus καθολικός, ὅλος, πᾶς: ex -o τέλεον: ut in -um (dicatur), ut in -o, quasi in -um ὡς ἐπίπαν: maioritas -ius ὁλοσχερέστεραι ἐπιτάσεις
trabs δοκίς
tractatus πραγματεία
tracto διεξέρχομαι, ἐφοδεύω
traditio παράδοσις, πραγματεία: — digna fide ἀξιοπιστία τῆς παραδόσεως
trado ἐφοδεύω, παραδίδωμι, περιοδεύω: -endus ὑποδειχθησόμενος: -or διαδίδομαι
traho ἕλκω
traiectator ῥᾳδιουργός
tranquillitas εὐστάθεια
transactus *cf.* parakmes
transeo (a uita) μεθίστημι (τοῦ ζῆν): -iens *cf.* medius
transfiguro μετασχηματίζω
transglutio κατάποσις
transitorius παροδικός
transitus μετάβασις, πάροδος: -um facio παροδεύων τυγχάνω
transmutabilis μετάβλητος: cito — ταχυμετάβολος
transmutatio διαστροφή, μεταβολή, μετατροπή
transponibilis μετάπτωτος: facile — εὐμετάθετος
trapezita τραπεζίτης
tres τρεῖς
triangularis τριγωνικός, τρίγωνος
Triangulum Τρίγωνον
triginta: — partium τριακοντάμοιρον
trigonalis τριγωνικός, τρίγωνος
trigonalitas τὸ τρίγωνον
trinus τρίγωνος: in -o aspectu τριγωνίζων, τρίγωνος
triplicitas τὸ τρίγωνον
tripudiator πηδητής
tristitia λύπη
Trogloditica Τρωγλοδυτική
tropicum τροπή
tropicus τροπικός
truncatus: -arum partium μελ[ε]οκοπούμενος
tuba σάλπιγξ
tumulator ἐνταφιαστής
tumultus: amator -us φιλοθόρυβος: a -ibus eleuatio τὸ ἀνακεχωρηκός
tunc τότε
turba ὄχλος
turbatio θόρυβος: -ione repleri τὸ ταραχῶδες

turbator κυκητής, καναπιστής (κυκητής†?): populi — ὀχλοκόπος

turpis αἰσχρός: -ia agens αἰσχροποιοῦν: -ium operum αἰσχροποιός: -iter lucrans αἰσχροκερδής: *cf.* occisor

turpitudo τὸ αἰσχρόν

tussis: cum -i βηχικός

tussitiuus, tussiuus βηχικός

Tyrrenia, Tirenia Τυρρηνία

uaco: magis -ans κατακορέστερος

uacuus: -e *cf.* curro

ualde ἄγαν, πάνυ, σφόδρα: *cf.* malus, multus, passionabiliter, probrosus

ualeo δύναμαι, ἴσχύω

ualetus (*sic*) ὑπηρέτης

ualidus σφοδρός: -ior σφοδρότερος

ualitudo ἰσχύς: melior — τὸ εὐεκτικώτερον

uanus κενός: *cf.* opinio, opinor

uaporatiuus ἔνικμος

uaporosus ἔνικμος

uarietas παραλλαγή, τὸ ποικίλον

uarius ποικίλος

uas: — marinum στόλος

ubi ὅπου

ubique πανταχῇ

uegetatiuus: maxime — θρεπτικώτατος: -issimus θρεπτικώτατος

uel ἤ, ἤτοι, καί: — etiam καί

uellus πόκος

uelocitas τάχος

uelox: *cf.* motus

uelut(i) ἅτε, οἷον, καθάπερ, ὥσπερ, ὡς

uena φλέψ

uenditor: aromatum — ἀρωματοπώλης: ueneni — φαρμακοπώλης: uestimentorum — ἱματιοπώλης: unguenti — μυροπώλης

ueneficus ἰώδης, φαρμακός

uenenator φαρμακευτής

uenenosus ἰοβόλος, φαρμακός

uenenum: -i uendator φαρμακοπώλης

uenerabilis σεβάσμιος, σεμνός

uenerandus σεμνός

uenereus ἀφροδισιακός, ἐπαφρόδιτος: -ea (τὰ) ἀφροδίσια

ueneritas ἐπαφροδισία

uenio ἔρχομαι: ea que -iunt in alimentum τὰ εἰς τροφὴν

uenter κοιλία: frigidus — ψυχροκοίλιος: -tris egritudines κοιλιακά

uentosus ἀνεμώδης, πνευματικός: — factus πνευματούμενος

uentus ἄνεμος, πνεῦμα: occidentalis — λίψ

Venus (Veneris) Ἀφροδίτη (ὁ τῆς -ης)

uer ἔαρ

ueraciter ἀληθῶς

uerbum: -a τὸ κατὰ λέξιν: parui affectus — μικρολογία: uerbi gratia λόγου ἕνεκεν, λόγου χάριν, uerbi gratia (dicere) φέρε (εἰπεῖν)

uerecundia αἰδώς, τὸ αἰδημονικόν: sine — sunt ἀναισχυντοῦσιν

uerecundor: facile -ans εὐέντρεπτος

uerecundus αἰδήμων

uergo ἀποκλίνω

uerisimilatiuus εἰκαστικός

ueritas ἀλήθεια, -αι: amor -atis φιλαλήθης

uernalis ἐαρινός

uero γοῦν, δέ (*post* autem), γοῦν, μέντοι, τε,

uersatiuus (?) τρεπτικός

uersio τροπή

uersus πρός *cum acc., cum dat.*

uertex κορυφή: — capitum nostrorum κορυφὴ ἡμῶν

uertibilis: simul — συντρεπόμενος: non — ἄτρεπτος

uerto, -or τρέπω, -ομαι

uerumtamen πλήν

uerus ἀληθής: -e ἀληθῶς: -um dico ἀληθεύω

uesica κύστις

uespertinus ἑσπέριος
uestimentum: -orum uenditor ἱματιοπώλης
uestitus: facile -u ornabilis εὐπερίστολος: in -u περὶ τὰς ἐσθῆτας
uetulus πρεσβύτερος
uexillum: -a ἐφαπτίδες (*vel* πτέρυγες) (Τοξότης)
uia ἔφοδος
uicinitas ἐγγύτης: *cf.* terra
uicinus σύνεγγυς
uicissim ἐναλλάξ
uicissitudino: -ans (loca) ἐνηλλαχώς (τοὺς τόπους)
uicissitudo ἐναλλαγή, παραλλαγή
uictoria νίκη: -e amator φιλόνεικος
uidelicet δηλονότι, ὁ (*pro repet. in parenthese*): — (... et) τε (... καὶ)
uideo θεωρέω, ὁράω: -or δοκέω, καταφαίνομαι, φαίνομαι: -etur ἔοικε: quod amiciter -etur (*sic*) τὸ φιλάνθρωπον
uigil νήπτης
uiginti: — anni εἰκοσαετηρίς: — quinque εἴκοσι πέντε
uilis ἀνεπίφαντος, ἄτιμος
uinea ἄμπελος
uinum οἶνος
uiolentia βία: per -iam mors βιαιοθανασία
uiolentus βίαιος
uir ἀνήρ: multi -i πολύανδροι: -orum amatrix φίλανδρος: -orum interfector ἀνδροφόνος: presunt -is ἔπανδροι (sc. εἰσιν)
uirga ῥάβδος
Virgo Παρθένος
uirgo παρθένος
uiridis ὑπόχλωρος, ὕπωχρος, χλωρός: *cf.* mixtura, oculus
uirilis ἀνδρώδης, ἔπανδρος, ὕπανδρος: non – ἄνανδρος
uirilitas: in -tate ἔπανδρος
uirtuosus δυναμικός
uirtus δύναμις, ἰσχύς, διάδοσις (*sic*): actiua — τὸ δραστικόν: secundum -tutem εἰς δύναμιν: *cf.* effectiuus, studeo
uisibilis: facile -e est εὐσύνοπτον
uisus ὅρασις, ὄψις
uita βίος (-οι), τὸ ζῆν, ζωή: -e proficuus βιωφελής: deliciose -e, delectabilis -e ἡδύβιος: communis -e κοινόβιος: paucis pro — contentus ἐφημερόβιος: *cf.* annus
uitalis βιώσιμος
uitari μεταπεσεῖν
uitor φυλάσσομαι
uituperosus ἐπονείδιστος
uiuo ἐπιζάω, ζάω: -ens, -us ζῶν
uix μόλις: *cf.* loquor, regressiuus
ulcerosus ἑλκώδης
ulcus ἕλκος, ἕλκωσις
ulterius *cf.* licet
ultimus ἔσχατος, τελευταῖος, ὕστατος
ultra παρά *cum acc.*, πέρα
ultramotus παρακεκινημένος
umquam: non ... — οὐ... ποτε
unde ἔνθεν, ὅθεν
undecumque ὁθενδήποτε, πανταχόθεν
unguentarius μυρεψός
unguentum μύρον: *cf.* uenditor
ungula: -e χηλαί
unicus μοναχός
uniformis μονοειδής
uniuersalis καθολικός, (ὁ) καθόλου: -ior καθολικώτερος: -iter καθόλου, ὅλως: — habitudo (*redd. ad sensum* ?) ἡ καθόλου πρόσνευσις: *cf.* habitus
unus εἷς: — post unum παρ' ἕν: *cf.* forma, modus, ordo, species
unusquisque ἕκαστος, εἷς ἕκαστος: propriis unicuique ἑκάστοις οἰκείως (-οις ?): ea que pertinent ad unumquodque τὰ καθ' ἕκαστον: *cf.* semper
uociferatio περιβοησία
uoco καλέω, προσαγορεύω

uolatilis πτηνός
uolo βούλομαι, θέλω, προαιρέω: *cf.* percunctor
uoluptas: -tate piger, -tati inherens ῥᾴθυμος
uoluptuosus: -e ἀπολαυστικῶς
uox: exilis -cis σπινός, σπινώδης
uro καίω
ursa ἄρκτος (-οι)
Vrsa maior Ἄρκτος μεγάλη: Vrsa minor Ἄρκτος μικρά
usque μέχρι: — ad ἕως, μέχρι, πρός (*cum acc.*)
ustio τὸ καυστικόν
usus χρεία, χρῆσις: boni -us εὔχρηστος: -us difficultas τὸ δύσχρηστον: in -u fortior χρηματιστικώτερος: in -u fortior fuerit συγκεχρηματικώς
ut ἵνα, οἷον, ὅπως, ὡς, ὥσπερ: — (+ *coniunct.*) ἕνεκεν τοῦ (+ *inf.*): ita — ὅπως, ὥστε (+ *coniunct.*): — puta ὡς: — *cum verb. finit. pro infinit.*: *cf.* predico *e. a.*
uterque ἑκάτερος: ex utraque parte ἐφ' ἑκάτερα, ἑκατέρωθεν
uterus ἀποκύησις, γαστήρ
utilis εὔχρηστος, χρήσιμος, χρηστικός
utilitas χρεία, τὸ χρήσιμον
utique ἄν, δή, δή ... ἄν: — et τε ... καί
utor καταχράομαι, συγχράομαι, χράομαι: -ens μεταχειριζόμενος: -endum προσχρηστέον
utrimque ἑκατέρωθεν
utrum πότερον
uulgabilior περιβοητότερος
uulnus τραῦμα
uxor γυνή
xantus *cf.* capillus

Y *cf.* etiam i
Yas (Yadis Yade Yadem) Ὑάς (Ὑάδος, Ὑάδα)
yconomicus οἰκονομικός
ydemptitas: per gradum -tates αἱ μοιρικαὶ θίξεις (θήξεις)
ydoneitas ἐπιτηδειότης, ἐπιτήδευσις
ydoneus ἐπιτήδειος
ydria κάλπις
ydropisis ὑδρωπικός (νόσος)
Ydrus Ὕδρος
ypocrita ὑποκριτικός
Yrcania, Hircania Ὑρκανία
yris (hyrys) ἶρις
ystoria, hist- ἱστορία
Ytalicus Ἰταλός

zelatiuus ζηλωτικός
zelator ζηλωτής: mali — κακόζηλος
zelotipus ἐπίζηλος, ζηλότυπος: non — ἄζηλος
zephirus ζέφυρος
zodiacus ζῳδιακός, ζῳδιακός κύκλος, (ὁ) διὰ μέσων τῶν ζῳδίων, ὁ διὰ μέσων, ὁ διὰ μέσων κύκλος

III. SUPPLEMENTUM LATINUM

Multis locis verba latina a Guillelmo ad meliorem intellectum textus Ptolemaei, maxime difficilis et ardui, addita sunt. Exempla plurium harum additionum secundum lineas nostrae editionis praebemus. Primo (**A**) afferuntur verba a Guillelmo iterata quae in textu graeco implicite (per articulum vel aliter) indicata sunt, secundo (**B**), quamquam non plene separabiliter, verba quae Guillelmus suo arbitrio addidit, tertio (**C**) verba *scilicet, uidelicet,* quarto (**D**) explicationes verborum graecorum.

A. Verba a Guillelmo iterata (exceptis verbis ut *a(b), ad, circa, de, ex, in, per, propter, secundum*)

I.19 signis — 47 criminationes — 52 mensuram — 57 naturam — 106 naturam — 127 pronosticationem —148 circa rationes — 208 oportet — 357 figurationes — 365 magis — 393 uirtute — 401 uirtuti — 402 uirtuti — 424 stellarum — 451 stellarum — 481 stelle — 565 equinoctio— 593 principium — 633 quia — 668 existentem — 776(2) dominis — 845 dominus — 846 domini — 874 terminorum — 884(3) domini — 886 dominationem — 896 graduum — 898 gradus — 957 (2) ex proprietate — 960 ex eo quod sunt

II.82 zodiacum — 110 trigonalitas (*pro* τὸ *sc.* τρίγωνον) — 113 trigonalitas — 116-7 illa uero trigonalitas que est (*pro* τὸ δὲ) — 119 trigonalitas — 139 Quarte — 148 stella — 164 non ut — 210 coitus — 386 ascendens — 453 locum — 454 locum — 462 dominus — 530 commixtionem — 633(2) loca — 634 locis — 705-6 (coaptationem ...) conuenientiam (*sic*) — 851 sunt significatiue

III.75 illum uidelicet modum (τὸν) — 113 gradum — 120 gradum — 123 gradui ... graduum — 127 gradum — 142 cum sermone (τῷ) — 159-160 capitulo (τῷ) ... locum (τὴν) ... capitulo ... locum — 166 naturas — 213 si — 265 fuerint — 281 loco — 309 dispositionem[2] — 315 a natura — 317 ab habitudine — 334 angulum — 337 in generatione — 411 uiuet — 426 ex perceptione — 427 ex perceptione — 435 gradibus — 473 gradus — 476 locis — 487 perimunt — 571 locum — 607 eleuationibus — 623 partem — 624 tempora — 655 naturam — 689 oculos — 745 qualecumque ... insinuant — 907 nature

IV.100 a proprietate — 296 locum[2] — 354 loca — 356 loca — 389 intentionem — 515 loco — 516 locis — 600 dimissionem — 609 tempus — 646 loca[2] — 647 loca — 653 domino — 661 per configurationes

B. Verba quae Guillelmus suo intellectu addidit, exceptis verbis *est* (*e.g.* I.5 6 31 94 173 175 198 268 541 542 573 ...), *et* (*e.g.* I.213 862 1251 II.482 III.158 928 sqq.), *quidem* (*e.g.* I.853 III.412), *autem* (*e.g.* III.1138)

I.14 (*et* 523*) **orizontis** angulorum — 62 transmutationes **que fiunt** per tempora anni — 66 **locum qui est** secundum uerticem — 69 inundationes **suas** — 107 et sufficiens **fuerit** — 135 **accidit** esse — 139-41 conrestitutione ... **non** concessa, aut non totaliter aut non secundum tempus sensibile ... — 176 symptomata **que eueniunt** secundum — 208-9 Sed **oportet** tamquam quidem **tempore** motus celestium ... a seculo perseuerante —

363* distantie due **scilicet dies et nox** — 426 **ipsius** ale — 521 que **tendunt** ad dissolutionem — 530-1 igneitatem **que fit** a sole — 581-82 ordo ... datus est ipsis **ponendo** unum post unum — 603-4 qualitas que ex talibus **occurrit** figurationibus ... oportuna **esse** uideatur — 605 De signis duodecim que **ad inuicem** configurantur — 618 *et* 621+2 **per comparationem** ad — 663-5 Quoniam enim ... **ideo** — 696-8 Quoniam enim .... **ideo** — 733 propter domum **Martis, scilicet** Scorpionem — 835 **In hac tabula** (*sc.* «termini secundum Egiptios») **tractatur de dispositione terminorum planetarum seu de diuisione** (*cf. adn. gr.-lat.*) — 871 Iam uero incidimus nos **cuidam** exemplari antiquo — 873-4 numerus collectorum **terminorum** — 876 propter corrosionem **libri** — 890 quia in ordine **quantum ad alia signa** postponuntur — 960 ex eo quod sunt ablatiui **a proprio motu** (*pro* ἀφαιρετικούς)

II.142 ut ex hiis **consequitur** — 172 unde amor libertatis **conuenit** ipsis — 222 qui ibi **morantur** — 224 **proprietas** habitantium regionem **est** incompositio — 317-9 Relique autem huius quarte partes ... **site** ad boream — 379 Regiones **omnes** sunt 72 — 385 quemadmodum in **hominis** generatione — 437-9 Particularibus autem remissionibus ... **similiter consideratis** (*cf.* Robb.: «we deduce from») — 441 configurata **fuerint** — 460 **Si autem ambo easdem habent rationes** (*cf. adn. graeco-lat.*) — 483-4 ea **que ueniunt** in alimentum hominum — 499 **significationes** insinuant — 532 **et hoc** — 603 **Per comparationem** autem ad continens— 607-8 Et corruptionis que ... **euenit** per hec — 621 regiones et ciuitates quibus simptoma **futurum** significatur

III.3 Prohemium **de ordine sequentis intentionis ad precedentem** — 51 secundum **tempus illud** (κατ' αὐτὸν) — 65-6 Quamuis .... **tamen** — 94 penes **fetus** concretionem — 103-4 propter **mutationes** <positionum> **et propter** ... transmutationes — 151-2 exponentes **ita** — 174 eleuationes **super orizontem** — 204-5 Mars adiungatur soli **uel lune de die** Saturnus autem lune **de nocte** (*sed cf. adn. gr.-lat.*) — 230 passiones **que eueniunt** per molestationem — 271 non iam **erit** causa — 299 ea **que pertinent** ad unumquodque — 432-3 gradibus **qui dicuntur** boni demonis — 435 gradibus **acceptis** — 461-2 Si autem ... **tunc** luminarium — 516 ex omnibus considerare **oportet** — 518-9 **accipitur <autem>** unumquodque temporum ualere — 617-8 quoniam ... 70 et 58 **partium** excessus — 763 quales species **future sunt** — 775 matutini quidem ad solem **existentes** — 844-5 si autem et ipsi **benefici** — 1112-3 quomodo se habent ad inuicem et ad angulos et ad **alias** stellarum

IV.20-1 sicut possessoria quidem corporis conuenientiis, **ita** dignitatiua hiis que anime — 25 solo **modo** — 149 proprietates secundum speciem coaccipiuntur **et faciunt** aliquid — 435 Obseruata **principalitate** (παρατηρουμένης) — 463 et ab erisipelis et **aliis** molestiis — 562-3 secundus et secundam sortitus **speram** Mercurius — 637-8 et signi facti **abiectione occurrentis** (*sc. anni*) dominum coassumentes, (*cf. Robb.*: «and taking the ruler of the last sign») — 659-60 per **proprietatem** configurationum

C. Exempla (lib. I tantum) verborum *scilicet* et *uidelicet* (quandoque pro τε)

I.51-2 possibilitatis uidelicet et utilitatis — 57 ignis scilicet et aeris — 59 terram scilicet et aquam — 300 solis scilicet et lune — 337 calidum uidelicet et humidum — 338 siccum scilicet et frigidum — 339 Iouem scilicet et Venerem — 348-9 masculinum scilicet et femininum — 364 que uidelicet diei masculina facta — 506 uere scilicet et estate et — 558-9 illud scilicet primum — 563-4 primum uidelicet a uernali equinoctio — 564 scilicet Aries — 565 scilicet Chelarum — 570 scilicet Taurus, Leo, Scorpius ... — 618 medietate scilicet et tertia — 621 emiolii scilicet et supertripartientis — 626 scilicet ex omnibus femininis — 651-2 diametralis scilicet et trigonalis et — 676-7 Aquarius scilicet et Capricornus — 684 Aries scilicet et Scorpius — 692-3 Gemini scilicet et Virgo — 698 equinoctiali scilicet et duobus tropicis — 718 scilicet Capricornus — 732 scilicet Marti — 733 scilicet Scorpionem — 751 Ariete uidelicet — 753 Tauro uidelicet — 754 Scorpius scilicet — 794-5 ascensionale ratione, scilicet in omni climate — 795 quod scilicet tempora — 846 Saturnus scilicet et Mercurius — 847 trigoni domimus, scilicet Mars

(D) Explicationes verborum graecorum

II.111 borealibica (id est simul aquilonaris et occidentalis)
II.114 notafiliotica (id est simul australis et subsolana)
II.117 boreafiliotica (id est simul borealis et subsolana)
II.120 notolibica (id est simul australis et occidentalis)
II.177 notafiliotice (id est australis simul et orientalis)
II.246 boreafiliotice (id est aquilonari simul et orientali)
II.443 precessiones (id est retrogradationes)
III.60 katarchen (id est optentionem principii)
III.819 histericas (id est matricis dolores patientes)
IV.393 mali demonis (uel infortunii)
IV.581 parakmes (id est transacti status)
IV.46 per prosneusim (id est adnuitionem)

DE WULF-MANSION CENTRE
ANCIENT AND MEDIEVAL PHILOSOPHY

Ancient and Medieval Philosophy - Series 1

Ancient and Medieval Philosophy - Series 1 presents state-of-the-art research on the history of Western thought from its origins to the beginnings of modernity. The series' focus lies primarily on the continuity and thematic transitions between ancient, medieval and renaissance thought, but also includes volumes on specific topics related to any of those periods. 'Philosophy' is understood broadly to include metaphysics, ethics, natural philosophy, epistemology and philosophy of mind, and philosophical theology. Since the series is particularly devoted to making available Latin and Greek source texts, we welcome critical text editions as well as contributions that are based on philological scrutiny of those texts, without losing sight of their philosophical importance.

Ancient and Medieval Philosophy - Series 2
Henrici de Gandavo Opera Omnia

Henry of Ghent, who taught initially at the Faculty of Arts in Paris before 1276 and subsequently at the Faculty of Theology in Paris from 1276 until his death in 1293, was an original and extremely important thinker. A major influence on John Duns Scotus, Henry's ideas had a direct impact on the scholastic discussion into the early modern period. His two major works, namely his - fifteen *Quodlibeta* and his *Quaestiones Ordinariae (Summa)*, along with works of his that have never before been printed, are being published in modern critical edition in the *Ancient and Medieval Philosophy - Series 2, Henrici de Gandavo Opera omnia.*

Ancient and Medieval Philosophy - Series 3
Francisci de Marchia Opera Philosophica et Theologica

Francis of Marchia (born ca. 1290) was a highly innovative thinker. Recent work has highlighted facets of his philosophical theology, natural philosophy, political philosophy, philosophy of mind, and metaphysics that show his creativity and deserve more study. A lively discussion of aspects of Marchia's thought can be traced in the second quarter of the fourteenth century, but because Marchia's works are for the most part unprinted, scholars have been hampered in their efforts to reconstruct his thought, let alone trace its influence. This new series of critical editions of Francis of Marchia's most important works will go a long way to improving that situation. Marchia's major works are versions of his *Sentences* commentaries, most of which will be published as part of *Ancient and Medieval Philosophy - Series 3*.

Visit www.lup.be for a list of published publications in the series.

These series are available on standing order. To sign up for a subscription or for more information, please contact info@lup.be.

Leuven University Press
Minderbroedersstraat 4
3000 Leuven
Belgium
www.lup.be